AF356000

Mental Health Promotion and Illness Prevention in Vulnerable Populations

Mental Health Promotion and Illness Prevention in Vulnerable Populations

Editors

Carlos Laranjeira
Ana Querido

Basel • Beijing • Wuhan • Barcelona • Belgrade • Novi Sad • Cluj • Manchester

Editors

Carlos Laranjeira
ciTechCare - Center for
Innovative Care and Health
Technology
Polytechnic University
of Leiria
Leiria
Portugal

Ana Querido
ciTechCare - Center for
Innovative Care and Health
Technology
Polytechnic University
of Leiria
Leiria
Portugal

Editorial Office
MDPI
St. Alban-Anlage 66
4052 Basel, Switzerland

This is a reprint of articles from the Special Issue published online in the open access journal *Healthcare* (ISSN 2227-9032) (available at: www.mdpi.com/journal/healthcare/special_issues/mental_health_promotion).

For citation purposes, cite each article independently as indicated on the article page online and as indicated below:

Lastname, A.A.; Lastname, B.B. Article Title. *Journal Name* **Year**, *Volume Number*, Page Range.

ISBN 978-3-7258-0622-5 (Hbk)
ISBN 978-3-7258-0621-8 (PDF)
doi.org/10.3390/books978-3-7258-0621-8

Contents

About the Editors

Carlos Laranjeira

Associate professor in the School of Health Sciences at the Polytechnic University of Leiria, Portugal.

Principal Investigator in ciTechCare—Center for Innovative Care and Health Technology, Polytechnic University of Leiria, Portugal.

Ana Querido

Associate professor in the School of Health Sciences at the Polytechnic University of Leiria, Portugal.

Investigator in ciTechCare—Center for Innovative Care and Health Technology, Polytechnic University of Leiria, Portugal.

Editorial

Mental Health Promotion and Illness Prevention in Vulnerable Populations

Carlos Laranjeira [1,2,3,*] and Ana Querido [1,2,4]

1   Centre for Innovative Care and Health Technology (ciTechCare), Polytechnic University of Leiria, Campus 5, Rua das Olhalvas, 2414-016 Leiria, Portugal; ana.querido@ipleiria.pt
2   School of Health Sciences, Polytechnic University of Leiria, Campus 2—Morro do Lena, Alto do Vieiro—Apart. 4137, 2411-901 Leiria, Portugal
3   Comprehensive Health Research Centre (CHRC), University of Évora, 7000-801 Évora, Portugal
4   Center for Health Technology and Services Research (CINTESIS), NursID, University of Porto, 4200-450 Porto, Portugal
*   Correspondence: carlos.laranjeira@ipleiria.pt

Citation: Laranjeira, C.; Querido, A. Mental Health Promotion and Illness Prevention in Vulnerable Populations. *Healthcare* **2024**, *12*, 554. https://doi.org/10.3390/healthcare12050554

Received: 10 February 2024
Accepted: 27 February 2024
Published: 28 February 2024

Several cases of social and health inequity have occurred in recent centuries. Despite major innovations and significant changes in the general population's quality of life, the most vulnerable populations continued to survive [1]. Vulnerability is a general concept that, in the context of health, means susceptibility to developing a certain health problem. Thus, vulnerable populations are those at greater risk of developing physical, psychological, or social health problems due to their marginalized socio-cultural status, their limited access to economic resources, or due to individual characteristics such as gender or age [2]. Vulnerability, as a human being's ontological condition [3,4], can be aggravated by experiencing a transition, exposing the client to potential danger, a problematic or excessively long recovery, or an inadequate or delayed adaptation process [5,6].

Vulnerable groups are those who are socially, psychologically, and/or materially more susceptible to social exclusion due to reasons of health, sexual orientation, religion, culture, ethnicity, gender, physical or mental disability, among others. Thus, they are more likely to develop health problems and needs than the rest of the population. In this vein, vulnerability is also a dynamic and multidimensional process, with interactions between personal characteristics and social and environmental conditions. Many socially explicable risk factors and co-morbidities—such as low socioeconomic status or substandard living conditions—are present in vulnerable groups. Rather than being attributable to personal traits, vulnerability is the product of social processes that may lead to greater exposure to different types of risk, higher sensitivity to negative consequences, and a lower ability to react by coping or adapting [7,8]. Therefore, to tackle health disparities, public health interventions should focus on the underlying social and historical factors that perpetuate poor health in marginalized communities. Frohlich and Potvin [9] offer a broad overview of these interventions, emphasizing that they should be intersectoral and participatory [2,10]. However, further investigation is required to determine the specific requirements for promoting fair outcomes for vulnerable populations.

Ensuring the prevention and promotion of mental health is crucial, particularly to mitigate the increasing prevalence of mental disorders. Although health promotion and disease prevention are well-recognized principles in the field of public health, their effective implementation to promote and prevent mental health is sometimes challenging [11]. Thankfully, there has been significant advancement in recent decades, including substantial study on the subject. Given these advancements, academics, care providers, governments, and policymakers are increasingly focused on using preventative techniques to enhance the availability, accessibility, and effectiveness of these services for the population in a less stigmatizing, culturally congruent, and accessible manner.

This Special Issue covers a diverse array of topics and comprises 23 papers that garnered a total of 29,911 views worldwide before drafting this editorial. The studies encompass different populations, including people with disabilities, women with breast cancer, survivors of humidifier disinfectant damage, patients with hematologic malignancies and candidates for allogeneic hematopoietic stem cell transplantation, individuals with depression, suicidal ideation, and other mental health conditions. Additionally, it includes studies on populations facing psychopathological symptoms, such as general populations, older adults, women facing high-risk pregnancies, postpartum women, prison staff, and other vulnerable people in the community. Moreover, this Special Issue covers carers, including adult children of parents with mental illness, adult-child caregivers, parents, caregivers, and teachers of adolescents in secondary school, caregivers of persons with mental disorders, and health care providers. It also encompasses studies of various student populations, including first-year medical students, college students, and university students.

This Special Issue of papers is a unique contribution to the literature on mental health promotion and illness prevention in vulnerable populations. This broad research topic also educates professionals in the fields of healthcare, education, and social services on the significance and advantages of their activities in enhancing mental health within a population. It encourages them to serve as facilitators, mediators, and advocates for mental health across different sectors. These studies from different cultural backgrounds further illuminate and enhance our understanding of how to assess and recognize populations susceptible to mental and behavioral disorders, and how to foster mental well-being and socio-emotional health across all stages of life (such as children, adolescents, adults, elderly individuals, and other vulnerable groups), providing valuable insights for both scholars and practitioners.

Funding: This work was supported by FCT—Fundação para a Ciência e a Tecnologia, I.P. (UIDB/05704/2020 and UIDP/05704/2020) and by the Scientific Employment Stimulus—Institutional Call—[https://doi.org/10.54499/CEECINST/00051/2018/CP1566/CT0012, accessed on 30 December 2023].

Conflicts of Interest: The authors declare no conflicts of interest. The funders had no role in the design of the study; in the collection, analyses, or interpretation of data; in the writing of the manuscript; or in the decision to publish the results.

References

1. Rod, M.H.; Rod, N.H.; Russo, F.; Klinker, C.D.; Reis, R.; Stronks, K. Promoting the health of vulnerable populations: Three steps towards a systems-based re-orientation of public health intervention research. *Health Place* **2023**, *80*, 102984. [CrossRef] [PubMed]
2. Kuran, C.; Morsut, C.; Kruke, B.; Krüger, M.; Segnestam, L.; Orru, K.; Nævestad, T.; Airola, M.; Keränen, J.; Gabel, F.; et al. Vulnerability and vulnerable groups from an intersectionality perspective. *Int. J. Disaster Risk Reduct.* **2020**, *50*, 101826. [CrossRef]
3. Herring, J. *Vulnerable Adults and the Law*; Oxford University Press: Oxford, UK, 2016.
4. Longo, M.; Lorubbio, V. Vulnerability. From the Paradigmatic Subject to a New Paradigm of the Human Condition? An Introduction. *Int. J. Semiot. Law* **2023**, *36*, 1359–1369. [CrossRef] [PubMed]
5. Lindmark, U.; Bülow, P.H.; Mårtensson, J.; Rönning, H.; A.D.U.L.T. Research Group. The use of the concept of transition in different disciplines within health and social welfare: An integrative literature review. *Nurs. Open* **2019**, *6*, 664–675. [CrossRef] [PubMed]
6. Laranjeira, C.; Coelho, C.; Ferreira, C.; Pereira, M.; Ribeiro, M.I.; Cordeiro, L.; Querido, A. How Do Professionals Regard Vulnerable People in a Portuguese Community Setting? A Qualitative Content Analysis. *Soc. Sci.* **2023**, *12*, 499. [CrossRef]
7. McCartney, G.; Popham, F.; McMaster, R.; Cumbers, A. Defining health and health inequalities. *Public Health* **2019**, *172*, 22–30. [CrossRef] [PubMed]
8. Cappelli, F. Investigating the origins of differentiated vulnerabilities to climate change through the lenses of the Capability Approach. *Econ. Polit.* **2023**, *40*, 1051–1074. [CrossRef]
9. Frohlich, K.L.; Potvin, L. Transcending the known in public health practice: The inequality paradox: The population approach and vulnerable populations. *Am. J. Public Health* **2008**, *98*, 216–221. [CrossRef]

10. Siller, H.; Aydin, N. Using an Intersectional Lens on Vulnerability and Resilience in Minority and/or Marginalized Groups During the COVID-19 Pandemic: A Narrative Review. *Front. Psychol.* **2022**, *13*, 894103. [CrossRef] [PubMed]
11. Singh, V.; Kumar, A.; Gupta, S. Mental Health Prevention and Promotion-A Narrative Review. *Front. Psychiatry* **2022**, *13*, 898009. [CrossRef]

Article

Quality of Life and Incidence of Clinical Signs and Symptoms among Caregivers of Persons with Mental Disorders: A Cross-Sectional Study

Vasiliki Oikonomou [1], Evgenia Gkintoni [2,*], Constantinos Halkiopoulos [3] and Evangelos C. Karademas [1,4]

1   School of Social Sciences, Hellenic Open University, 26335 Patras, Greece;
    v_oikonomou@windowslive.com (V.O.); karademas@uoc.gr (E.C.K.)
2   Department of Psychiatry, University General Hospital of Patras, 26504 Patras, Greece
3   Department of Management Science and Technology, University of Patras, 26334 Patras, Greece;
    halkion@upatras.gr
4   Department of Psychology, University of Crete, 74100 Rethymnon, Greece
*   Correspondence: evigintoni@upatras.gr

Abstract: Background: Caring for individuals with mental disorders poses significant challenges for caregivers, often leading to compromised quality of life and mental health issues such as stress, anxiety, and depression. This study aims to assess the extent of these challenges among caregivers in Greece, identifying which demographic factors influence their well-being. Method: A total of 157 caregivers were surveyed using the SF-12 Health Survey for quality-of-life assessment and the DASS-21 questionnaire for evaluating stress, anxiety, and depression symptoms. *t*-tests, Kruskal–Wallis tests, Pearson's correlation coefficients, and regression analyses were applied to understand the associations between demographics, quality of life, and mental health outcomes. Results: The study found that caregivers, especially women and younger individuals, faced high levels of mental health challenges. Marital status, educational level, and employment status also significantly influenced caregivers' well-being. Depression was the most significant factor negatively correlating with the mental component of quality of life. The magnitude of the burden experienced by caregivers highlighted the urgency for targeted social and financial support, as well as strategic treatment programs that consider caregiver well-being. Conclusions: Caregivers of individuals with mental disorders endure significant stress, anxiety, and depression, influencing their quality of life. Demographic factors such as age, gender, marital status, education, and employment status have notable impacts. Findings emphasize the need for society-wide recognition of caregivers' roles and the creation of comprehensive support and intervention programs to alleviate their burden, particularly in the context of the COVID-19 pandemic.

Keywords: clinical symptoms; caregivers; mental disorders; quality of life; stress; anxiety; depression; social support

Citation: Oikonomou, V.; Gkintoni, E.; Halkiopoulos, C.; Karademas, E.C. Quality of Life and Incidence of Clinical Signs and Symptoms among Caregivers of Persons with Mental Disorders: A Cross-Sectional Study. *Healthcare* **2024**, *12*, 269. https://doi.org/10.3390/healthcare12020269

Academic Editors: Carlos Laranjeira and Ana Querido

Received: 13 December 2023
Revised: 15 January 2024
Accepted: 16 January 2024
Published: 20 January 2024

1. Introduction

Mental disorder has extensive repercussions for both patients and their caretakers. The burden of caregiving has been evaluated in the literature using studies whose criteria vary. However, there is widespread agreement that caregivers are heavily burdened by their demanding daily lives and are frequently unsupported in their caring role. Emotional burden and burnout can negatively impact the mental health and quality of life of mental disorder caregivers and worsen the clinical course of their patients.

The majority of previous research has focused on caregivers of individuals with long-term physical illnesses, resulting in a significant lack of knowledge regarding the experiences of those who care for individuals with mental disorders. Comparable research can shed light on the distinct difficulties encountered by these caretakers. It is essential to examine the correlation between providing care for individuals with mental health disorders and the prevalence of depressive and anxiety symptoms in caregivers.

For example, according to research [1], long-term care caregivers may significantly reduce the likelihood of increased hospitalizations for people with mental disorders. Through this research, it is essential to understand how caregivers of people with mental disorders experience anxiety and stress due to the long-term care they provide, their quality of life, and their interactions with the healthcare system [2].

Undoubtedly, the care provided by caregivers to individuals may also impact their own mental well-being. Additional research can highlight the direct impact of caregiving on mental well-being, underscoring the need for mental health services and interventions specifically targeted towards caregivers. The findings obtained from research carried out on caregivers could be pivotal in influencing healthcare policies and practices designed to enhance the mental well-being of this specific subgroup [3]. They have the ability to guide the progress of healthcare models that are more comprehensive, placing patients at the forefront and taking into account the welfare of caregivers. By recognizing the difficulties faced by these caregivers, such studies can help to empower them. It can enhance awareness regarding their needs and facilitate the development of resources and networks that provide the necessary assistance [4].

Examining the well-being of caregivers of individuals with mental disorders and the prevalence of clinical signs and symptoms is essential because it dramatically affects the caregivers' quality of life and highlights the necessity for comprehensive support initiatives. The caregivers of individuals with mental disorders frequently encounter substantial levels of stress, anxiety, and depression, which can significantly impact their overall well-being [4]. It is essential to comprehend the influence of caregiving on the well-being and psychological state of caregivers of individuals with mental disorders. This understanding is vital for creating specific interventions and support systems catering to their needs and difficulties.

Additionally, studies conducted in Greece have assessed caregivers' burden on people with chronic diseases; however, the caregiver experience of people with mental disorders has received scant attention [5]. Therefore, it is anticipated that this article will significantly contribute to recognizing the difficulties caregivers of people with mental disorders face and will encourage the development of educational programs to support and empower them. Furthermore, this study seeks to investigate the relationship between the quality of life of mental disorder caregivers and the occurrence of depressive and anxiety symptoms [6].

In recent years, caregivers' perceptions of health and quality of life have been the subject of extensive research. This study will concentrate on the quality of life of mental illness caregivers so that this issue can be better documented for this particular group of caregivers. Because in recent years, a significant portion of the literature has focused on the caregivers of people with chronic diseases, the collection of information about the quality of life and the incidence of depressive and anxiety symptoms of caregivers of people with a mental disorder will provide meaningful data for this specific group of caregivers.

The purpose of this work is to investigate the quality of life of caregivers of people with a mental disorder, as well as its correlation with the occurrence of depressive and anxiety symptoms. For the implementation and design of this research protocol, the following research questions were formulated:

(RQ1) In which domain (physical/physical or mental) is there a more significant burden?
(RQ2) Which demographic factors are significantly associated with the level of quality of life of caregivers of people with a mental disorder?
(RQ3) Is there a significant correlation between the levels of depressive and anxiety symptoms and the level of quality of life of caregivers of people with a mental disorder?

2. Literature Review

2.1. Quality of Life of Caregivers

Quality of life is essential for modern societies and economies, but it remains a profoundly anthropocentric and complex concept because it attempts to link human needs and activities. Everyone's interpretation of the term quality of life is highly subjective. From a scientific and social perspective, quality of life comprises multiple variables and factors contributing to a quality whole. Quality of life is a blueprint for physical health, psychological parameters, and social interaction [3,4]. The components of physical health are mobility, independence, the capacity to work, and the optimal operation of vital organs. It manifests itself in family life, romantic relationships, and other social manifestations in the social aspect of quality of life. This term became more prevalent in the 1980s as an indicator of social research aimed at eliminating inequalities and ensuring the equitable distribution of resources.

According to the research, physical and mental stress among caregivers contributes to a decline in their quality of life. It has been established that the reactions and consequences of caregiving make the role of caregiver challenging and, consequently, a source of anxiety [5]. Schizophrenia caregivers report a lower quality of life than non-caregivers and caregivers of people with other illnesses [6]. The objective and subjective burden of informal caregiving is substantial, resulting in diminished quality of life [7]. Reports that 38.2% of mental disorder caregivers experience a severe burden in their caregiving role. Caregivers of individuals with a mental disorder spend an average of 22 or more hours per week providing care. Some researchers [8] found a significant correlation between caregiver burden and quality of life.

According to another study [9], caregiver age, gender, education level, job loss due to caregiving demands, income, relationship with the patient, frequency of caregiving, and duration of patient illness are significant correlates of caregiving burden. In addition, the severity of the patient's symptomatology, its duration, the number of needs, the extent of the disability, and the diminished social interests compound the burden for mental disorder caregivers [10,11].

Low burden of care and social and professional support have been identified as positive predictors of the quality of life of caregivers of individuals with mental disorders [8]. Caregivers believe their quality of life would improve if they had more time for themselves and maintained a greater distance from the individual they are tending to. In addition, the caregiver's independence is essential to pursue interests and activities interrupted by caregiving [12]. Special consideration has been given to the mental health of caregivers and the difficulties they may experience, such as anxiety, depression, and distress, since their caregiving responsibilities present several obstacles [13].

2.2. Anxiety, Depression, and Burden in Caregivers

Family members with a close relationship with the patient provide more extended care, devoting most of their time to caregiving, a significant source of stress and burden. In comparison to caregivers who did not receive any educational support, burden, and depression symptoms are significantly lower among those who have received an educational intervention for standard care [14–16]. Within three months, caregivers who received training experienced a significant reduction in depressive symptoms, from 36% to 17%. In contrast, the prevalence of depressive symptoms among untrained caregivers increased from 22% to 50% over the same time frame [17].

It is true that some caregivers' feelings of isolation and helplessness can hinder their ability to provide quality care, limit their goals and life activities, diminish their quality of life, and contribute to mental health issues such as anxiety and depression [18]. The experience of providing care is multifaceted and intricate. Caregivers play a crucial role in assisting individuals with severe mental illness. Supporting caregivers by reducing their burden and enhancing their mental health enables them to continue providing care for their loved ones despite the difficulties of caregiving. Educating caregivers can alleviate their responsibilities and improve their quality of life.

Informal care provides daily health care to dependent individuals by family members, friends, neighbors, or anyone else in the immediate social network. However, they typically do not receive monetary compensation for their assistance [19]. Typically, informal caregivers are members of a close-knit family; however, it is sometimes observed that this responsibility is fragmented and shared by multiple individuals. In order to meet this new challenge, families who primarily provide informal care develop coping mechanisms for their increased stress levels. These include passive coping, reframe, spiritual, social, and health service support [20].

In recent years, caregiver burden and the difficulties of providing care have been the subject of growing research. There are numerous ways to present caregiver burden. However, the terms objective and subjective burden are the most prevalent. Objective burden represents periods of absence from caregiving that are observable and verifiable, whereas subjective burden represents personal feelings of burden [21].

Numerous studies have demonstrated that caregivers experience elevated levels of anxiety and depression. For instance, caregivers of people with schizophrenia are likelier than those with other chronic illnesses to experience sleep difficulties, insomnia, pain, and anxiety [5]. The prevalence of depression in long-term caregivers is 15–32%, 1–10% higher than in the general population. Due to a lack of understanding of the disease and the non-use or absence of social support systems, the incidence of depressive symptoms among family members increases immediately following a disease diagnosis [22,23]. In research conducted in 2018 [17] in Japan, the rate of depression among long-term dementia caregivers before attending a training program was 36%, while in the United Kingdom, it was 29.4% [24].

Moreover, it has been observed that caregivers of individuals with dementia often encounter feelings of anxiety, as emphasized in a comprehensive analysis [25]. The study revealed a positive correlation between caregiver burden and poorer caregiver physical health with heightened levels of anxiety. These factors were additionally correlated with an elevated susceptibility to depression. Furthermore, another study [26] discovered that caregivers of children with asthma who had a low family socioeconomic status, low-income family functioning, and a heightened tendency to experience shame exhibited elevated levels of anxiety and depression.

Numerous academic studies [25–29] have examined the correlation between mental disorders, specifically anxiety and depression, and the consequential influence on the caregivers' quality of life. In general, the available evidence indicates that mental disorders, including anxiety and depression, have a detrimental effect on the quality of life experienced by caregivers. Various factors, including caregiver burden, physical health, and family functioning, can potentially contribute to the onset and progression of mental disorders. Acknowledging and resolving caregivers' mental health requirements are imperative for enhancing their overall welfare and the standard of care they deliver. In a specific investigation [28], a correlation was discovered between the emotional distress experienced by spousal caregivers, such as symptoms of depression and anxiety, and the diminished quality of life observed in the patients they care for.

3. Materials and Methods

3.1. Participants

The sample of participants in the present study includes:

- All caregivers of people with a mental disorder who visit the psychiatric department;
- The outpatient departments of the psychiatric department;
- The psychiatric emergency department of a General Hospital in Greece.

The study's sample comprises all caregivers of individuals with a mental disorder who receive therapy at the psychiatric department, outpatient clinics, and psychiatric emergency department of the General Hospital of the Peloponnese Region in Greece. The characteristics of the sample are described in more detail in the findings/results of this research. The inclusion criteria for the study were the following: (a) consent to participate in the research and (b) being caregivers (close relatives) of people with a mental disorder.

This study was carried out at a General Hospital from October to December 2022. The health unit's selection criteria consisted of the everyday duties of the hospital's psychiatric clinic. The study included the outpatient clinics of the psychiatric clinic and the emergency department dedicated to addressing psychiatric manifestations.

3.2. Research Hypotheses

The research hypotheses formulated are that: (RH1) there is a difference in the burden of caregivers of people with a mental disorder in the physical/physical sector and the mental sector; (RH2) demographic factors have an effect on the occurrence of depressive and anxiety symptoms in the caregivers of people with a mental disorder; (RH3) there is a significant correlation between the levels of depressive and anxiety symptoms and the level of quality of life of the caregivers of people with a mental disorder. For this research, a quantitative methodology was implemented. The quantitative method permits an objective evaluation of a phenomenon (in this case, the quality of life and depressive and anxiety symptoms). In addition, the time horizon of the study is a crucial research parameter. There are two primary approaches: cross-sectional and longitudinal studies. It was decided to conduct a cross-sectional study due to time constraints and the need to assess the quality of life and the occurrence of depressive and anxiety symptoms in a specific time frame for a specific group of caregivers (caregivers of individuals with a mental disorder). As indicated in another study [30], note that the most significant advantage of these studies is that they are generally fast and reliable; this methodology was utilized in the present study.

Furthermore, the most effective approach to establish connections between variables (specifically, the relationship between quality of life and symptoms of depression or anxiety) is to carry out cross-sectional studies. Given these circumstances, conducting such research is essential for the ongoing investigation.

3.3. Diagnostic Assessments

As a research instrument, a questionnaire with closed-ended questions was utilized and administered by the primary researcher in an in-person interview. The questionnaire comprises three sections in total. The first section of the survey documented the caregivers' demographic information (gender, age, level of education, marital status, income, and occupation).

In the second section, the patient's quality of life was evaluated utilizing the SF-12 Health Survey [31,32], and in the third section, the patient's anxiety and depression levels were evaluated utilizing the DASS-21 [33,34].

3.4. Description of Psychometric Scale SF12

The SF-12 questionnaire was developed as a short version of the original SF-36 for large-scale studies, and its reliability and validity have been established [31,32]. In Greek, some researchers [35] standardized the SF-36 and SF-12 instruments. The SF-12 consists of 12 questions in total. The 12 queries assess the patient's physical (PCS) and mental health (MCS). Higher values indicate a higher health-related quality of life, while the mean value is 50 with a standard deviation of 10. Based on the PCS and MCS scales, the SF-12 provides a brief assessment of the patient's health-related quality of life, for which it is widely acknowledged scientifically. Together, mental and physical health assess subscales of physical functioning, vitality, social functioning, general health, physical pain, mental health, and limitations in social functioning due to emotional problems [35–38].

3.5. Description of Psychometric Scale DASS-21

The DASS-21 was developed as a short form of the DASS-42 and has been reported to have slightly improved psychometric properties compared to the entire DASS [39]. The DASS-21 is a 21-item self-report scale that measures levels of depression, stress, and anxiety in the population. Each seven (7)-item scale has four (4) response options ranging from zero (0) (does not apply to me at all) to three (3) (applies to me a lot or most of the time). Its maximum score is 42. The DASS-21 scale score has excellent internal consistency, and interpretive scores have good construct validity [40–43].

Standardization of the DASS-21 involves the assessment of its psychometric properties, including reliability (consistency of results over time) and validity (accuracy in measuring what it is supposed to measure). The Cronbach's alpha coefficients, a measure of internal consistency, are typically high for all three subscales, indicating that the items within each subscale reliably measure the same underlying construct. Also, the DASS-21 has been validated against other established measures of depression and anxiety, showing good convergent and discriminant validity [44]. It is effective in differentiating between the symptoms of depression, anxiety, and stress, which often overlap in other measures.

3.6. Analysis

Data analysis was performed using SPSS v25 software. The prevalence of anxiety, depression, and caregivers' quality of life will be described using descriptive indicators (frequency and percentage or mean value and standard deviation). Using the *t*-test of independence and *t*-test for equality of two means, as well as logistic regression, we investigated the relationship between anxious and depressive symptoms and the risk of low quality of life among caregivers of persons with mental disorders. Using the *t*-test of independence and multiple regression models, the effect of demographic and clinical characteristics of patients on their quality of life and risk of developing depressive and anxiety symptoms was investigated. Lastly, Pearson's correlation coefficient and multiple regression models were used to examine the relationship between the quality of life and the level of anxiety and depression.

4. Results

4.1. Descriptive Data and Preliminary Analyses

The survey was conducted between October and December 2022 and was answered by caregivers (N = 157) of people with a mental disorder. The following results (Table 1) have been extracted using the SPSS statistical package.

Table 1. Demographics.

Demographics	Categories	N	f%
Sex	Male	48	30.6
	Female	108	68.8
	Other	1	0.6
Age	Below 25 years old	5	3.2
	26–35 years old	22	14.0
	36–45 years old	40	25.5
	46–55 years old	53	33.8
	Above 55 years old	37	23.6
Nationality	Greek	155	98.7
	European Union	2	1.3
Family Status	Single	40	25.5
	Married—Cohabitation Agreement	94	59.9
	Divorced—Termination of Cohabitation Agreement	15	9.6
	Widowed	8	5.1
Educational Level	Illiterate	1	0.6
	Primary Graduate	28	17.8
	Secondary Graduate	40	25.5
	Bachelor	59	37.6
	Master	27	17.2
	Ph.D.	2	1.3

Table 1. *Cont.*

Demographics	Categories	N	f%
Job Occupation	Unemployed	7	4.5
	Household	20	12.7
	Self-employed	10	6.5
	Private Employee	63	40.1
	State Employee	39	24.8
	Retired	18	11.5
Income	EUR 0–11,000	41	26.1
	EUR 11,001–13,000	36	22.9
	EUR 13,001–23,000	51	32.5
	EUR > 23,001	15	9.6
	I do not wish to answer	14	8.9
Years of Caregiving	0–10 years	116	73.9
	11–20 years	19	12.1
	21–30 years	17	10.8
	30 years and above	5	3.2

Table 1 presents the demographics of the sample. In particular, of the N = 157 people, the majority (68.8%, N = 108) were women, and 30.6% (N = 48) were men. Regarding their educational level, most (37.6%) are university graduates. Regarding their work, 40.1% are employed in the private sector, 24.8% are civil servants, 12.7% are domestic workers, 11.5% are retired, 6.4% are self-employed, and, finally, 4.5% are unemployed.

4.2. Caregiver Quality of Life Analysis

In this section, an attempt will be made to analyze the quality of life of caregivers. Initially, the reliability of the questionnaire was checked through Cronbach's alpha coefficient using the SPSS statistical program. For the Physical Health subscale (PCS), Cronbach's alpha index was found to be 0.825, while for the Mental Health subscale (MCS), the index equals 0.824. Since values above 0.7 are considered acceptable, it is judged that the analysis shows reliability. Table 2 presents the basic descriptive statistics of the individual questions of the SF-12 questionnaire. In particular, Table 2 presents the mean, standard deviation, median, and frequency of the respondents' answers, which is confirmed by the corresponding graphs. Higher mean values also indicate a higher level of quality of life.

Table 2. Descriptive statistics of SF-12 questionnaire questions.

Factors	Mean	SD	Median	Frequency				
				1	2	3	4	5
Moderate intensity activities (PF)	2.57	0.653	3.00	8.9	25.5	65.6	-	-
Climbing a few flights of stairs (PF)	2.54	0.615	3.00	6.4	33.8	59.9	-	-
Complete fewer tasks (RP)	3.76	1.123	4.00	3.8	10.8	22.3	31.8	31.2
Restricted work (RP)	3.78	1.184	4.00	5.1	12.7	14.0	35.0	33.1
Pain effect (BP)	4.16	0.930	4.00	0.6	8.3	7.6	41.4	42.0
General health status (GH)	3.64	0.735	4.00	-	8.3	26.8	58.0	7.0
Activity (VT)	3.22	1.04	3.00	2.5	26.8	26.8	33.8	10.2
Social activities (SF)	3.64	1.138	4.00	3.8	14.0	23.6	31.2	27.4
Complete fewer tasks (RE)	3.50	1.269	4.00	7.00	19.7	16.6	29.9	26.8
Careless operations (RE)	3.61	1.279	4.00	7.6	16.6	12.7	33.1	29.9
Tranquility (MH)	3.16	1.04	3.00	6.4	21.0	29.9	35.7	7.0
Bad mood/melancholy (MH)	3.01	1.174	3.00	11.5	24.2	26.1	28.7	9.6

The following section of the analysis analyzes the PCS and MCS subscales to determine the domain with the more significant burden. The descriptive statistics of caregiver responses to the PCS and MCS subscales for the physical and mental health of the SF-12 are presented in Table 3. Physical health has a mean of 48.68 and a standard deviation of 8.41, whereas mental health has a mean of 35.37 and a standard deviation of 10.73.

Table 3. Descriptive statistics of PCS and MCS subscales.

Descriptive Statistics	PCS	MCS
Mean	48.68	35.37
Standard Deviation	8.41	10.73
Median	46.99	35.97
Min.	27.59	13.01
Max.	65.94	56.10

Nonetheless, the most crucial aspect of the analysis is to identify the demographic factors that may influence the two domains of the quality of life of mental disorder caregivers. The subsequent step of the analysis investigates whether the sample's demographic characteristics influence the mean quality-of-life scores. At this juncture, the following inspections will be conducted:

Regarding gender, the independent sample t-test's parametric control will be computed because the variable is dichotomous and there are more than 30 observations in each category. The null hypothesis is that the mean values are equal, and the alternative hypothesis is that they differ. This control does not include the observation regarding the individual who selected "Other", as a single observation cannot account for any difference between the means.

Regarding the demographic factors of age, income, marital status, educational level, and employment, the non-parametric Kruskal–Wallis test is used because there are more than 2 categories and fewer than 30 observations per category. The null hypothesis is that the variables in question are independent, while the alternative is that they are dependent. Despite the initial preference for the ANOVA test, it was discovered that some subcategories of the variables contained a small number of observations. Therefore, the Kruskal–Wallis test, which is more powerful in detecting differences in this data pattern, was favored. The significance level in all tests was set at 5%. Therefore, the original hypothesis is accepted when the p-value (p) $\geq$ 0.05 and rejected when the p-value (p) < 0.05. For the case of gender, a statistically significant difference according to gender is found only in terms of the mental component (MCS) (Table 4).

Table 4. Independent sample t-test controls for Gender.

	T	Sig.
PCS	−1.154	0.251
MCS	−3.740	0.000

The next analysis stage examines the possible differentiation based on the Kruskal–Wallis test that presents the relevant results, from which the following emerges (Table 5):

- There is a statistically significant difference by age for the PCS component;
- There is a statistically significant difference in both factors of the SF-12 scale depending on the respondents' marital status, educational level, and employment;
- There is a statistically significant difference in income for the PCS component.

4.2.1. Assessment of Anxiety and Depression Level

Table 6 presents the descriptive statistics for each questionnaire question and the basic descriptive statistics of the individual questions of the DASS-21 questionnaire. Also, Table 6 presents the mean, the standard deviation, the median, and the frequency of the responses given by the respondents, which is also confirmed by the corresponding graphs.

Table 5. Kruskal–Wallis test values.

Demographic Factors	Scales	t	Sig.
Age	PCS	21.77	0.000
	MCS	8.17	0.086
Marital Status	PCS	9.55	0.023
	MCS	7.86	0.049
Educational Level	PCS	14.70	0.012
	MCS	15.89	0.007
Employment	PCS	32.66	0.000
	MCS	37.17	0.000
Income	PCS	14.17	0.007
	MCS	1.69	0.792

Table 6. Descriptive statistics of DASS-21 questionnaire questions.

Items	Mean	Standard Deviation	Median	Frequency Response			
				0	1	2	3
I couldn't calm myself down.	0.78	1.00	0.762	40.8	41.4	16.6	1.3
My mouth felt dry.	0.34	0.00	0.607	72.0	22.3	5.1	0.6
I could not experience a positive feeling.	0.65	1.00	0.741	49.7	36.9	12.1	1.3
I was having trouble breathing.	0.45	0.00	0.692	66.2	23.6	9.6	0.6
I found it difficult to take the initiative to do some things.	0.81	1.00	0.863	43.9	35.7	15.9	4.5
I had a tendency to overreact to the situations I was faced with.	1.23	1.00	0.854	21.0	41.4	31.2	6.4
I felt shaky.	0.39	0.00	0.695	71.3	20.4	6.4	1.9
I often felt nervous.	1.16	1.00	0.836	19.1	54.8	17.2	8.9
I worried about situations where I might panic and look foolish to others.	0.72	1.00	0.838	49.7	31.8	15.3	3.2
I felt like I had nothing to look forward to.	0.75	1.00	0.831	47.1	34.4	15.3	3.2
I found myself feeling annoyed.	1.22	1.00	0.821	17.2	51.0	24.2	7.6
It was hard for me to relax.	1.20	1.00	0.851	23.6	37.6	34.4	4.5
I felt depressed and disappointed.	0.99	1.00	0.820	29.3	46.5	19.7	4.5
I couldn't stand anything that kept me from continuing what I was doing.	0.67	1.00	0.683	43.3	48.4	6.4	1.9
I felt very close to panic.	0.46	0.00	0.772	67.5	21.7	7.6	3.2
Nothing could make me feel excited.	0.65	0.00	0.808	51.6	36.3	7.6	4.5
I felt like I wasn't worth much as a person.	0.58	0.00	0.878	63.1	21.0	10.8	5.1
I felt that I was quite irritable.	1.17	1.00	0.90	23.6	45.2	21.7	9.6
I could feel my heart beating without any previous physical exercise.	0.76	1.00	0.804	41.4	47.1	5.7	5.7
I felt scared for no reason.	0.57	0.00	0.745	54.8	37.6	3.8	3.8
I felt that life had no meaning.	0.25	0.00	0.598	80.9	14.6	2.5	1.9

It should be mentioned that higher mean values also indicate a higher level of stress/anxiety/depression. As can be seen, the symptoms that have a more significant effect on these variables are difficulty in relaxing, the tendency of the respondents to overreact to the situations they faced, and feelings of discomfort, irritability, nervousness, and frustration, which are also confirmed by the relatively higher averages compared to the rest of the questions.

The analysis continues with the descriptive statistics of the three subscales of the questionnaire, as presented in Table 7. First, regarding the stress scale, it is observed that the respondents show mild levels of stress with a mean value equal to 14.87 and a standard deviation of 9.08 units. Next, examining the anxiety scale, it appears that the mean value of anxiety equals 7.38 with a standard deviation of 7.36 points. This value again indicates a mild level of stress for the respondents. Finally, checking the depression scale, it is observed that normal levels of depression characterize the respondents with a mean value of 9.36 and a standard deviation of 8.46 units. Therefore, it could be said that there does not seem to be a very high level of anxiety and depression in the sample.

Table 7. Descriptive statistics of stress, anxiety and depression subscales.

	Stress	Anxiety	Depression
Mean	14.87	7.38	9.36
Standard deviation	9.08	7.36	8.46
Median	14.00	6.00	6.00
Min.	0.00	0.00	0.00
Max.	40.00	38.00	42.00

It is then examined whether there is a statistically significant relationship between demographic factors and the values of the individual scales of the DASS-21 questionnaire. Table 8 presents the *t*-test for the case of gender. In this case, a statistically significant difference according to gender is found only in terms of the mental component (MCS).

Table 8. Independent sample *t*-test controls for Gender (MCS).

MCS	*t*	Sig.
Stress	0.66	0.512
Anxiety	5.51	0.000
Depression	2.80	0.006

The next analysis stage examines the possible differentiation based on the Kruskal–Wallis test. Table 9 presents the relevant results, from which the following emerges:

- There is a statistically significant difference in age regarding anxiety and depression parameters;
- There is a statistically significant difference in terms of all three parameters of the DASS-21 scale depending on marital status and educational level;
- There is a statistically significant difference depending on employment regarding the stress parameter;
- There is no statistically significant difference with any parameter according to income.

Table 9. Kruskal–Wallis test values demographics statistical analysis.

Demographic Factors	MCS	*t*	Sig.
Age	Stress	7.25	0.123
	Anxiety	13.44	0.009
	Depression	13.95	0.007
Marital Status	Stress	15.37	0.002
	Anxiety	26.54	0.000
	Depression	21.82	0.000
Educational Level	Stress	19.57	0.002
	Anxiety	26.57	0.000
	Depression	21.71	0.001
Job Occupation	Stress	23.48	0.000
	Anxiety	10.92	0.053
	Depression	6.89	0.229
Income	Stress	5.30	0.258
	Anxiety	7.54	0.110
	Depression	2.66	0.617

4.2.2. Correlation of Depressive and Anxiety Symptoms of Caregivers' Quality of Life

The last stage of the analysis examines the correlation found between the two subscales of the SF-12 questionnaire and the three categories of the DASS-21 questionnaire. At the first level, the Pearson's correlation coefficient is examined. Table 10 presents the correlation coefficient between the subscales of the SF-12 questionnaire and the variables resulting from the analysis of the DASS-21 questionnaire. The analysis showed the following:

- Regarding the physical factor (PCS), there was a negative correlation with all three categories of the DASS-21 questionnaire, which is statistically significant at the 1% level for the stress and depression variables and the 5% level for the anxiety variable. However, the correlation is weak since the correlation coefficients are small.
- Regarding the mental factor (MCS), there was a statistically significant negative correlation with all three categories at the 1% significance level.
- Finally, it was found that the three variables of the DASS-21 scale are positively correlated, which shows that the increase or decrease in one of the three factors can lead to a corresponding increase or decrease in another factor.

Table 10. Pearson's correlation coefficient.

	PCS	**MCS**	**Stress**	**Anxiety**
MCS	0.056			
Stress	−0.212 **	−0.596 **		
Anxiety	−0.200 *	−0.457 **	0.697 **	
Depression	−0.224 **	−0.580 **	0.759 **	0.758 **

* Statistically significant association at the 5% level; ** statistically significant association at the 1% level.

5. Discussion

In this study, the quality of life of mental disorder caregivers and their correlation with depressive and anxious symptoms were examined. Additionally, the research questions (RQs) that were posed in the study are presented and indicated in this section. One hundred fifty-seven caregivers of individuals with mental disorders who visited the mental health department of a General Hospital participated in the study. Regarding the participant profile, 68.8% were female, 30.6% were male, and 0.6% identified as gender neutral. According to research [45], women are twice as likely as males to fulfill this informal obligation for their family members. Other studies have found the same, attributing the responsibility of care primarily to women, daughters, and spouses, either because of their deep-seated belief that this role is theirs or because of the social environment.

Based on similar research, it has been observed that caregivers of individuals with mental disorders may encounter notable impacts on their quality of life [46]. These effects may manifest in diminished quality of life resulting from heightened psychological strain, compromised physical well-being, and socioeconomic difficulties. Numerous scholarly investigations have been conducted to assess caregiving's influence on caregivers' overall well-being, yielding consistent empirical evidence in support of this assertion. A study conducted amidst the COVID-19 pandemic revealed a decline in care providers' physical and mental well-being. Another study [47] also emphasized the significant impact of social isolation on various dimensions of caregivers' quality of life, such as heightened levels of depression and anxiety. Other researchers conducted a study [48] that examined the relationship between the mental and physical health of advanced cancer patients and their family caregivers, revealing a potential interdependence between the two parties.

According to the SF-12 questionnaire, the quality of life of caregivers of individuals with mental disorders was analyzed in the two main categories of physical/somatic health and mental health, along with their subcategories. For example, energy, serenity, and irritability/depression have an intermediate impact on the respondents' quality of life. (RQ1) In contrast, a substantial proportion of respondents assert that their life quality is unaffected by factors corresponding to their firm belief regarding the general state of health.

Furthermore, a research investigation in England yielded compelling evidence indicating that the provision of care significantly impacts the mental well-being of those who assume the caregiving role [49]. The caregivers of individuals with mental illness experience a significant burden, which can have detrimental effects on both their quality of life and the well-being of those they care for [50]. In summary, a multitude of studies have yielded empirical evidence that substantiates the assertion that the well-being of caregivers for individuals with mental disorders may be substantially impacted. Caregivers may encounter heightened psychological distress, physical health concerns, and socioeconomic difficulties. The reciprocal relationship between patients' and caregivers' mental and physical health has been observed in diverse contexts.

According to the findings of the present research, participants reported more concerns with mental health compared to physical health. Caregivers of people with mental disorders appear to have poorer mental health than caretakers of people with chronic illnesses and non-carers [49,51] (RQ1).

Numerous studies have been conducted to investigate the implications of caregiving on the mental well-being of individuals, yielding consistent and compelling evidence supporting this assertion. Significant mental health burden was observed among caregivers of individuals diagnosed with mental disorders, such as schizophrenia and anorexia nervosa. The act of providing care was found to be linked with increased levels of depression, anxiety, and stress in caregivers [52,53]. Moreover, the ongoing COVID-19 pandemic has further intensified the mental health challenges faced by individuals who provide care to others. Research has indicated that individuals responsible for the care of those with mental disorders encountered mental health challenges amidst the pandemic [54,55]. The ongoing global pandemic has resulted in a notable escalation of stress, anxiety, and depression among individuals who provide care, thus underscoring the imperative for supplementary assistance and available resources. Caregivers may encounter heightened levels of stress, anxiety, and emotional burdens concerning the ongoing provision of care and support. The responsibility of providing care can potentially contribute to the development of psychiatric disorders and have adverse effects on the overall well-being of individuals fulfilling the caregiving role.

According to the findings, the caregivers of persons with mental disorders who participated in this study have moderate stress and anxiety levels, with average values of 14.87 and 7.38 (RQ3). In terms of depression, the respondents have normal levels. In contrast, the findings of a study [56] indicated that the highest proportion of caregivers of individuals with chronic diseases exhibited depressive symptoms due to high psychological strain. In terms of demographic characteristics, female caregivers exhibit higher levels of tension, anxiety, and depression symptoms than male caregivers and younger caregivers than older ones. Moreover, those who have lost their partner, i.e., are widowed and are engaged in housework, experience more significant stress, anxiety, and depression than those who are married and employed, who experience less clinical signs. (RQ2).

Numerous academic studies have explored the correlation between caregiving responsibilities and the subsequent impact on the mental well-being of caregivers, thereby furnishing substantiating evidence in favor of this assertion. Previous studies have indicated that individuals who provide care for those with mental disorders, such as schizophrenia, bipolar affective disorder, and substance use disorder, tend to exhibit elevated levels of depressive and anxiety symptoms [57–60]. The combination of the responsibility of providing care and the societal disapproval linked to specific mental disorders can lead to heightened depressive symptoms among individuals who assume the role of caregivers [61]. According to a study [62], caregivers of individuals with dementia were observed to exhibit psychopathological symptoms, such as depression and anxiety. There is a consistent association between the intensity of the care recipient's symptoms and elevated levels of caregiver strain and mental health symptoms, as indicated by multiple studies [63–65]. Several research studies conducted in Sri Lanka and the Netherlands have revealed that individuals who provide care for individuals with mental disorders tend to experience elevated levels of depressive symptoms and heightened caregiver strain [66,67]. In a similar vein, research conducted in Brazil and Spain has revealed that caregivers of individuals with mental disorders demonstrated elevated levels of emotion expressed and experienced significant levels of stress [68,69].

In an attempt to determine whether the demographic characteristics of the participants influence the two domains of physical and mental health of caregivers of people with mental disorders, it was discovered that gender influences both domains of physical and mental health, with men reporting higher quality of life. (RQ2) The physical, mental, and general health status of informal female caregivers has been documented by a study [70] and is consistent with the results of the current study. Informal care is detrimental to the health of caregivers, who are predominantly women; 27.2% of female informal caregivers report health issues. In addition, the same researchers [70] have documented a correlation between a more significant perceived burden and poorer overall health in women, notably when social support is lacking.

Regarding age, it was discovered that participants of all ages experience superior physical health to mental health. Those with Greek citizenship also enjoy a superior quality of life than those without Greek citizenship. Regardless of the family circumstance, the level of physical health is consistently higher, indicating a higher quality of life.

Concerning the educational level of the caregivers of individuals with mental disorders, the graduates of university and college have a higher quality of life regarding the physical component. In comparison, secondary education graduates have a higher quality of life regarding their mental component. (RQ2) Researchers [71] argued that caregivers with a higher level of education are associated with a more significant burden, attributing it to the increased demands and expectations they may place on themselves and others. In contrast, in another study, researchers [72] and colleagues argue that the more education a caregiver has, the better equipped they are to handle the challenges of caregiving. In addition, a high level of education is associated with greater sociability, improved working conditions, higher pay, and increased family income, all of which contribute to a sense of security (RQ2).

Regarding employment, those engaged in domestic work and earning up to EUR 11,000 reported the highest quality of living in the physical domain. At the same time, retired caregivers enjoy a higher mental quality of life. In contrast, the perceived quality of life of informal caregivers of older citizens increased as their income increased, according to survey results. The DASS-21 self-administered questionnaire was used to evaluate the level of tension, anxiety, and depression. It was discovered that the factors that have a significant impact on the symptoms of stress, anxiety, and depression for caregivers of people with mental disorders are the inability to relax, the tendency to overreact to the situations they face, the feeling of discomfort, irritability, nervousness, and frustration. (RQ2).

The association between demographic factors of caregivers and mental health outcomes has also been documented in research examining caregivers of individuals with severe mental illness. Other researchers [73] discovered a correlation between the female gender and younger age of caregivers and an increased likelihood of experiencing mental distress. Moreover, the burden experienced by caregivers can be influenced by various factors, including the nature of the caregiver–patient relationship and cultural and ethnic variables [74]. The examination of the influence of demographic variables on the mental well-being of caregivers has also been explored within the framework of the COVID-19 pandemic. According to a study conducted by researchers [75], it was observed that female caregivers and older children experienced a greater degree of adverse effects on their mental well-being amidst the pandemic. The influence of socioeconomic factors, specifically lower socioeconomic status, has been recognized as a contributing element to the increased vulnerability of caregiver mental health [76].

The correlation between the two subscales of the SF-12 questionnaire (physical health and mental health) and the three categories of the DASS-21 questionnaire (stress, anxiety, and depression) was also examined. Regarding physical and mental health, it was discovered that there is a negative correlation with all three DASS-21 categories. (RQ3) This correlation is also supported by researchers [56], who discovered that caregivers with depressive symptoms report substantially lower levels of both the physical and mental aspects of quality of life. Caregivers of persons with a mental disorder report low levels of mental health by recording symptoms of anxiety, stress, and depression in percentages ranging from 30 to 40 percent, particularly in the first years following the diagnosis of the disorder and in early psychoses [51]. For caregivers of individuals with schizophrenia, this symptomatology occurs at a rate of 72–83% for stress [77].

Consistent with the recent findings of researchers [78], the coefficients for the variables stress, anxiety, and depression were found to have a negative sign, indicating that their eventual increase causes a decrease in the quality of life of caregivers of people with mental disorders. (RQ3) Several researchers have extensively examined the burden of informal caregivers due to the manifestation of anxiety symptoms, confirming the present study's findings. Caregivers must provide care for an adult relative with a mental disorder, but they need assistance to do so adequately. Providing informal caregivers of individuals with mental disorders with support services can improve the caregivers' physical and mental health and the patients' through better management.

The impact of caregivers on the quality of life of individuals with mental disorders is substantial. Numerous studies have examined the correlation between caregiving and the prevalence of clinical manifestations and symptoms among individuals providing care for individuals with mental disorders. Researchers [79] conducted a comprehensive review and meta-analysis to investigate the extent of care responsibilities experienced by caregivers of Iranian individuals with chronic illnesses. The research revealed that individuals caring for patients with mental disorders, such as Alzheimer's, encountered a more significant caregiving burden. The responsibility of providing care was correlated with a heightened susceptibility to mental disorders among individuals fulfilling the role of caregivers [79]. In England, a study was conducted by researchers [80] to examine the prevalence of mental and physical illness among caregivers. According to the same researchers [80], the research revealed that individuals in the role of caregivers exhibited elevated levels of psychiatric symptomatology compared to those not in a caregiving role. According to the same research [80], there is a correlation between the extent of caregiving and negative impacts on mental health, as well as an increase in psychiatric symptoms. The research conducted in [81] investigated the firsthand accounts of caregivers regarding individuals with severe mental disorders residing in rural areas of Ghana. According to this research [81], caregivers encountered a range of difficulties, such as managing the indications and manifestations of mental disorders, shouldering emotional burdens, facing instances of violence, adapting to changing roles, confronting societal stigma, and navigating disrupted family dynamics [82]. The researchers [83] conducted a study to examine the caregiving responsibilities faced by families of individuals diagnosed with schizophrenia. According to previous research [83], caregivers encountered a noteworthy burden, resulting in detrimental effects on their mental well-being. Also, the same researchers [84] proposed the implementation of family interventions and psychosocial support as potential strategies to tackle the challenges mentioned above effectively. The burden experienced by caregivers of individuals diagnosed with bipolar disorder was examined in a study conducted by researchers [85]. According to this research [85], a high caregiver burden was revealed, significantly impacting the well-being of patients and caregivers. The researchers [85] underscored the significance of acknowledging and addressing the burden experienced by caregivers within clinical and psychosocial interventions for individuals with bipolar disorder [86].

In their study [52], researchers investigated the extent of familial burden experienced by individuals providing care for patients diagnosed with schizophrenia. The research conducted by the same researchers [52] revealed a substantial occurrence of caregiver burden, which correlated with various socio-demographic factors. Researchers in this study [52] emphasized the necessity of implementing comprehensive interventions to alleviate the burden experienced by caregivers. In a qualitative investigation [87] in Saudi Arabia, caregivers expressed encountering a range of difficulties, encompassing the management of indicators and manifestations of mental disorders, emotional strain, instances of violence, alterations in familial roles, societal stigma, and disruptions in family dynamics [87]. The challenges mentioned above have played a role in exacerbating the difficulties individuals face in the role of caregivers [87]. In general, these studies underscore caregiving's substantial influence on the overall well-being of individuals who care for individuals with mental disorders. Caregivers frequently encounter heightened psychiatric symptomatology, burden of care, and a range of challenges that have the potential to impact their mental well-being. The importance of addressing the caregiver burden and implementing suitable support and interventions cannot be overstated in terms of enhancing the welfare of caregivers and the standard of care delivered to individuals with mental disorders.

Numerous research studies have been conducted to explore this association, yielding consistently congruent findings. Other researchers [88] conducted a meta-analysis wherein they examined the relationship between levels of depressive and anxiety symptoms in caregivers and their quality of life. The findings of the study indicated that caregivers with elevated levels of depressive and anxiety symptoms experienced a diminished quality of life. In another study [89], it was discovered that caregivers who experienced elevated levels of anxiety and depression exhibited a diminished health-related quality of life. The study [27] employed a cross-sectional design to examine the relationship between anxiety and depression levels in patient–caregiver dyads. The study's findings revealed a noteworthy correlation between these psychological symptoms, further linked to a diminished quality of life. The association between symptoms of depression and anxiety and the overall quality of life has been documented in specific populations of caregivers as well. An investigation [90] revealed that caregivers of individuals diagnosed with schizophrenia who exhibited elevated levels of depressive symptoms also reported diminished family functioning and a reduced quality of life. The examination of the relationship between symptoms of depression and anxiety and the overall quality of life has also been explored within the framework of particular conditions. Another study [91] revealed that caregivers of individuals with mental illness who exhibited elevated levels of depressive symptoms experienced diminished vitality and poorer overall health, both of which are integral aspects of quality of life. In a separate investigation [92], it was discovered that the association between neuropsychiatric symptoms in individuals with dementia and caregiver mental health was mediated by caregiver burden and affiliate stigma. Furthermore, the impact on quality of life was subsequently influenced by the caregiver's mental health. Caregivers who experience higher levels of these symptoms tend to have a diminished quality of life [93–95].

The present Investigation endeavored to communicate with other researchers on a related topic. In this context, the possible absence of similar research on caregivers of individuals with a mental disorder prevented the comparison of the results with other findings of a similar nature in order to identify any similarities or differences in the sample's attitudes towards the subject of the study. It has also been a deficiency that comparable surveys of caregivers for chronically ill patients have utilized various research methods. In addition, the sample size is limited to a single institution in the country (General Institution), preventing the generalization of the results. Indirectly indicating the quality of life of caregivers of individuals with mental disorders, the results of this study play a crucial role in supporting development policies. All these findings highlight the importance of addressing the mental health needs of caregivers and providing appropriate support to improve their overall well-being and quality of life.

Notwithstanding its limitations, such as restricted sample size and confinement to a single institutional setting, which may impede its generalizability, this study provides valuable perspectives on comprehending the burden experienced by caregivers. The statement effectively communicates the importance of prioritizing the well-being of caregivers to enhance patient management. Moreover, this study promotes the need for ongoing research in this crucial domain, intending to implement focused approaches that directly influence the well-being of caregivers.

An additional limitation of the current research project was the absence of a control group integration. By conducting separate analyses of subgroups, one can gain significant insights into discrete demographic cohorts. Transparency is essential for the precise interpretation of the findings.

Caregivers of individuals with mental disorders are significantly burdened, and problems arise in all aspects of their lives. Multiple studies have documented the high burden and its association with low quality of life, and the present study confirms this association. This burden is anticipated to increase if we consider the factors that have considerably impacted families in recent years, such as the COVID-19 pandemic and the ongoing economic crisis in our country. It is regarded as necessary to provide social and financial support and develop strategic treatment programs to manage patients better and improve their overall health [96–98].

Future work stemming from this research should aim to address the underlying causes of the stress, anxiety, and depression identified among caregivers, potentially by implementing and evaluating intervention programs focused on support and stress reduction. This could include more in-depth investigations examining how different types of support (psychological, financial, community based, etc.) affect caregiver well-being [99–103]. Moreover, a longitudinal perspective could provide insight into how caregiving affects mental health over time and if and how this impact changes as the patient's condition evolves [104–109]. It is important to emphasize the need for a specific national strategic plan to be established for providing assistance to caregivers in Greece. In recent years, there has been a strong emphasis on the use of psychoeducation by primary healthcare facilities, both public and private mental health centers, for the most part.

Additionally, the research could be broadened to look at populations in various geographic and cultural contexts to account for differing healthcare systems, social support structures, and caregiving norms, which could influence caregivers' quality of life and mental health [110–114]. There is also the opportunity to use mixed methods approaches by incorporating qualitative research, which can provide a richer, more nuanced understanding of the caregivers' experiences [115].

6. Conclusions

In conclusion, this research paper underscores the profound and complex burden that caregiving for individuals with mental disorders imposes on caregivers. This study's analysis, drawn from 157 caregivers, reveals a discernible deterioration in the quality of life for these caregivers, punctuated by prominent symptoms of stress, anxiety, and depression. It highlights the intrinsic link between the mental health of the caregiver and demographic factors such as gender, age, educational level, and employment. Notably, the study showcases that female caregivers are particularly vulnerable to the psychological distress associated with caregiving.

The findings of this research articulate the urgent need for targeted support interventions. The correlation between poorer quality of life and increased levels of stress, anxiety, and depression suggests that strategic treatment programs, along with comprehensive social and financial support systems, are essential. These interventions are vital not only for ameliorating the mental health of caregivers but also for ensuring the sustained, effective care of individuals with mental disorders.

Additionally, the exacerbation of challenges posed by the COVID-19 pandemic has only intensified the necessity for such support. Future research is imperative to optimize and tailor these support systems to the unique needs of mental disorder caregivers. Through continued investigation and the implementation of robust support measures, we can hope to improve the lives of both caregivers and their loved ones.

Therefore, while caregivers undertake their crucial roles with resilience and determination, the responsibility falls to healthcare systems, policymakers, and the broader community to recognize and act upon the silent trials they endure. We should ensure that these unsung heroes receive the comprehensive assistance they need to maintain the well-being of those they care for and their own health and quality of life.

Author Contributions: Conceptualization, V.O. and E.C.K.; methodology, V.O. and E.C.K.; formal analysis, V.O., E.G., C.H. and E.C.K.; investigation, V.O.; resources, V.O. and E.G.; writing—original draft preparation, V.O., E.G. and C.H.; writing—review and editing, E.G., C.H. and E.C.K.; visualization, E.G., C.H. and E.C.K.; supervision, E.G., C.H. and E.C.K.; project administration, E.C.K. All authors have read and agreed to the published version of the manuscript.

Funding: This research received no external funding.

Institutional Review Board Statement: This study was approved on 30 September 2022 by the Institutional Review Board (IRB) at Scientific Council of the General Hospital (approval No. 22349).

Informed Consent Statement: Informed consent was obtained from all subjects involved in this study.

Data Availability Statement: Data available on request due to privacy/ethical restrictions.

Conflicts of Interest: The authors declare no conflicts of interest.

References

1. Konetzka, R.T.; Spector, W.; Limcangco, M.R. Reducing Hospitalizations from Long-Term Care Settings. *Med. Care Res. Rev.* **2008**, *65*, 40–66. [CrossRef] [PubMed]
2. Gkintoni, E.; Ortiz, P.S. Neuropsychology of Generalized Anxiety Disorder in Clinical Setting: A Systematic Evaluation. *Healthcare* **2023**, *11*, 2446. [CrossRef] [PubMed]
3. Gkinton, E.; Telonis, G.; Halkiopoulos, C.; Boutsinas, B. Quality of Life and Health Tourism: A Conceptual Roadmap of Enhancing Cognition and Well-Being. In *Tourism, Travel, and Hospitality in a Smart and Sustainable World, Proceedings of the IACuDiT 2022, Syros, Greece, 1–3 September 2023*; Katsoni, V., Ed.; Springer Proceedings in Business and Economics; Springer: Cham, Switzerland, 2023. [CrossRef]
4. Jeyagurunathan, A.; Sagayadevan, V.; Abdin, E.; Zhang, Y.; Chang, S.; Shafie, S.; Rahman, R.F.A.; Vaingankar, J.A.; Chong, S.A.; Subramaniam, M. Psychological status and quality of life among primary caregivers of individuals with mental illness: A hospital-based study. *Health Qual. Life Outcomes* **2017**, *15*, 106. [CrossRef] [PubMed]
5. Nooraeen, S.; Bazargan-Hejazi, S.; Naserbakht, M.; Vahidi, C.; Shojaerad, F.; Mousavi, S.S.; Malakouti, S.K. Impact of COVID-19 pandemic on relapse of individuals with severe mental illness and their caregiver's burden. *Front. Public Health* **2023**, *11*, 1086905. [CrossRef] [PubMed]
6. Gkintoni, E.; Pallis, E.G.; Bitsios, P.; Giakoumaki, S.G. Neurocognitive performance, psychopathology and social functioning in individuals at high-genetic risk for schizophrenia and psychotic bipolar disorder. *Int. J. Affect. Disord.* **2017**, *208*, 512–520. [CrossRef]
7. Rahmani, F.; Roshangar, F.; Gholizadeh, L.; Asghari, E. Caregiver burden and the associated factors in the family caregivers of patients with schizophrenia. *Nurs. Open* **2022**, *9*, 1995–2002. [CrossRef]
8. Ribé, J.M.; Salamero, M.; Pérez-Testor, C.; Mercadal, J.; Aguilera, C.; Cleris, M. Quality of life in family caregivers of schizophrenia patients in Spain: Caregiver characteristics, caregiving burden, family functioning, and social and professional support. *Int. J. Psychiatry Clin. Pract.* **2018**, *22*, 25–33. [CrossRef]
9. Peng, M.-M.; Zhang, T.-M.; Liu, K.-Z.; Gong, K.; Huang, C.-H.; Dai, G.-Z.; Hu, S.-H.; Lin, F.-R.; Chan, S.K.W.; Ng, S.; et al. Perception of social support and psychotic symptoms among persons with schizophrenia: A strategy to lessen caregiver burden. *Int. J. Soc. Psychiatry* **2019**, *65*, 548–557. [CrossRef]
10. Grandón, P.; Jenaro, C.; Lemos, S. Primary caregivers of schizophrenia outpatients: Burden and predictor variables. *Psychiatry Res.* **2008**, *158*, 335–343. [CrossRef]
11. Aranda-Reneo, I.; Oliva-Moreno, J.; Vilaplana-Prieto, C.; Hidalgo-Vega, A.; Gonzalez-Dominguez, A. Informal care of patients with schizophrenia. *J. Ment. Health Policy Econ.* **2013**, *16*, 99–108.
12. Vellone, E.; Piras, G.; Venturini, G.; Alvaro, R.; Cohen, M.Z. The Experience of Quality of Life for Caregivers of People with Alzheimer's Disease Living in Sardinia, Italy. *J. Transcult. Nurs.* **2012**, *23*, 46–55. [CrossRef] [PubMed]
13. Chang, S.; Zhang, Y.; Jeyagurunathan, A.; Lau, Y.W.; Sagayadevan, V.; Chong, S.A.; Subramaniam, M. Providing care to relatives with mental illness: Reactions and distress among primary informal caregivers. *BMC Psychiatry* **2016**, *16*, 80. [CrossRef] [PubMed]
14. Martín-Carrasco, M.; Martín, M.F.; Valero, C.P.; Millán, P.R.; García, C.I.; Montalbán, S.R.; Vázquez, A.L.; Piris, S.P.; Vilanova, M.B. Effectiveness of a psychoeducational intervention program in the reduction of caregiver burden in Alzheimer's disease patients' caregivers. *Int. J. Geriatr. Psychiatry* **2009**, *24*, 489–499. [CrossRef] [PubMed]
15. Gaugler, J.E.; Roth, D.L.; Haley, W.E.; Mittelman, M.S. Can counseling and support reduce burden and depressive symptoms in caregivers of people with Alzheimer's disease during the transition to institutionalization? Results from the New York University caregiver intervention study. *J. Am. Geriatr. Soc.* **2008**, *56*, 421–428. [CrossRef] [PubMed]
16. Mittelman, M.S.; Roth, D.L.; Clay, O.J.; Haley, W.E. Preserving health of Alzheimer caregivers: Impact of a spouse caregiver intervention. *Am. J. Geriatr. Psychiatry* **2007**, *15*, 780–789. [CrossRef]

17. Terayama, H.; Sakurai, H.; Namioka, N.; Jaime, R.; Otakeguchi, K.; Fukasawa, R.; Sato, T.; Hirao, K.; Kanetaka, H.; Shimizu, S.; et al. Caregivers' education decreases depression symptoms and burden in caregivers of patients with dementia. *Psychogeriatrics* **2018**, *18*, 327–333. [CrossRef] [PubMed]

18. Bowman, S.; Alvarez-Jimenez, M.; Wade, D.; Howie, L.; McGorry, P. The positive and negative experiences of caregiving for siblings of young people with first episode psychosis. *Front. Psychol.* **2017**, *8*, 730. [CrossRef]

19. Masanet, E.; La Parra, D. Relationship between the Number of Hours of informal care and the Mental Health Status of caregivers. *Rev. Esp. Salud Publica* **2011**, *85*, 257–266. [CrossRef]

20. Golforto, L.A. *Meeting the Challenge of Disability on Chronic Illness: A Family Guide*; Brookes Publishing: Baltimore, MD, USA, 1986.

21. Kuipers, E.; Onwumere, J.; Bebbington, P. Cognitive model of caregiving in psychosis. *Br. J. Psychiatry* **2010**, *196*, 259–265. [CrossRef]

22. Chesla, C.A. Nursing science and chronic illness: Articulating suffering and possibility in family life. *J. Adv. Nurs.* **2005**, *11*, 371–387. [CrossRef]

23. Cuijpers, P. Depressive disorders in caregivers of dementia patients: A systematic review. *Aging Ment. Health* **2005**, *9*, 325–330. [CrossRef] [PubMed]

24. Coope, B.; Ballard, C.; Saad, K.; Patel, A.; Bentham, P.; Bannister, C.; Graham, C.; Wilcock, G. The prevalence of depression in the carers of dementia sufferers. *Int. J. Geriatr. Psychiatry* **1995**, *10*, 237–242. [CrossRef]

25. Cooper, C.; Balamurali, T.B.S.; Livingston, G. A systematic review of the prevalence and covariates of anxiety in caregivers of people with dementia. *Int. Psychogeriatr.* **2006**, *19*, 175. [CrossRef] [PubMed]

26. Zhou, T.; Yi, C.; Zhang, X.; Wang, Y. Factors impacting the mental health of the caregivers of children with asthma in China: Effects of family socioeconomic status, symptoms control, proneness to shame, and family functioning. *Fam. Process* **2014**, *53*, 717–730. [CrossRef] [PubMed]

27. Li, Q.; Lin, Y.; Xu, Y.; Zhou, H. The impact of depression and anxiety on quality of life in Chinese cancer patient-family caregiver dyads, a cross-sectional study. *Health Qual. Life Outcomes* **2018**, *16*, 320. [CrossRef] [PubMed]

28. Chung, M.L.; Moser, D.K.; Lennie, T.A.; Rayens, M.K. The effects of depressive symptoms and anxiety on quality of life in patients with heart failure and their spouses: Testing dyadic dynamics using actor–partner interdependence model. *J. Psychosom. Res.* **2009**, *67*, 29–35. [CrossRef]

29. Locatelli, G.M.; Rebora, P.; Occhino, G.; Ausili, D.; Riegel, B.P.; Cammarano, A.; Uchmanowicz, I.P.; Alvaro, R.M.; Vellone, E.P.; Zeffiro, V. The impact of an intervention to improve caregiver contribution to heart failure self-care on caregiver anxiety, depression, quality of life, and sleep. *J. Cardiovasc. Nurs.* **2023**, *38*, 361–369. [CrossRef]

30. Olsen, J.; Kaare, C.; Murray, J.; Ekbom, A. The cross-sectional study. In *An Introduction to Epidemiology for Health Professionals*; Springer Series on Epidemiology and Public Health; Springer: New York, NY, USA, 2010; Volume 1, p. 79.

31. Ware, J.E.; Kosinski, M.; Keller, S.D. A 12-Item Short-Form Health Survey: Construction of scales and preliminary tests of reliability and validity. *Med. Care* **1996**, *34*, 220–233. [CrossRef]

32. Ware, J.E.; Kosinski, M.; Keller, S.D. *How to Score the SF-12 Physical and Mental Health Summary Scales*, 2nd ed.; The Health Institute, New England Medical Center: Boston, MA, USA, 1995.

33. Lovibond, S.H.; Lovibond, P.F. *Manual for the Depression Anxiety Stress Scales*, 2nd ed.; Sydney Psychology Foundation: Sydney, Australia, 1995.

34. Osman, A.; Wong, J.L.; Bagge, C.L.; Freedenthal, S.; Gutierrez, P.M.; Lozano, G. The Depression Anxiety Stress Scales—21 (DASS-21): Further examination of dimensions, scale reliability, and correlates. *J. Clin. Psychol.* **2012**, *68*, 1322–1338. [CrossRef]

35. Pappa, E.; Kontodimopoulos, N.; Niakas, D. Validating and norming of the Greek SF-36 health survey. *Qual. Life Res.* **2005**, *14*, 1433–1438. [CrossRef]

36. Kontodimopoulos, N.; Pappa, E.; Niakas, D.; Tountas, Y. Validity of SF-12 summary scores in a Greek general population. *Health Qual. Life Outcomes* **2007**, *5*, 55. [CrossRef] [PubMed]

37. Melville, M.R.; Lari, M.A.; Brown, N.; Young, T.; Gray, D. Quality of life assessment using the short form 12 questionnaire is as reliable and sensitive as the short form 36 in distinguishing symptom severity in myocardial infarction survivors. *Heart* **2003**, *89*, 1445–1446. [CrossRef] [PubMed]

38. Bindawas, S.M.; Vennu, V.; Al Snih, S. Differences in health-related quality of life among subjects with frequent bilateral or unilateral knee pain: Data from the Osteoarthritis Initiative study. *J. Orthop. Sports Phys. Ther.* **2015**, *45*, 128–136. [CrossRef]

39. Steenstrup, T.; Pedersen, O.B.; Hjelmborg, J.; Skytthe, A.; Kyvik, K.O. Heritability of health-related quality of life: SF-12 summary scores in a population-based nationwide twin cohort. *Twin Res. Hum. Genet.* **2013**, *16*, 670–678. [CrossRef] [PubMed]

40. Antony, M.M.; Bieling, P.J.; Cox, B.J.; Enns, M.W.; Swinson, R.P. Psychometric properties of the 42-item and 21-item versions of the Depression Anxiety Stress Scales in clinical groups and a community sample. *Psychol. Assess.* **1998**, *10*, 176–181. [CrossRef]

41. Musa, R.; Ramli, R.; Abdullah, K.; Sarkarsi, R. Concurrent validity of the depression and anxiety components in the Bahasa Malaysia version of the Depression Anxiety and Stress scales (DASS). *ASEAN J. Psychiatry* **2011**, *12*, 66–70.

42. Shea, T.L.; Tennant, A.; Pallant, J.F. Rasch model analysis of the Depression, Anxiety and Stress Scales (DASS). *BMC Psychiatry* **2009**, *9*, 21–31. [CrossRef]

43. Sinclair, S.J.; Siefert, C.J.; Slavin-Mulford, J.M.; Stein, M.B.; Renna, M.; Blais, M.A. Psychometric evaluation and normative data for the depression, anxiety, and stress scales-21 (DASS-21) in a nonclinical sample of U.S. adults. *Eval. Health Prof.* **2012**, *35*, 259–279. [CrossRef]

44. Henry, J.; Crawford, J. The Short-Form Version of the Depression Anxiety Stress Scales (DASS-21): Construct Validity and Normative Data in a Large Non-Clinical Sample. *Br. J. Clin. Psychol.* **2005**, *44*, 227–239. [CrossRef]
45. Donelan, K.; Falik, M.; DesRoches, C.M. Caregiving: Challenges and Implications for Women's Health. *Women's Health Issues* **2001**, *11*, 185–200. [CrossRef]
46. Clipp, E.; George, K. Dementia and cancer: A comparison of spouse caregivers. *Gerontologist* **1993**, *33*, 534–541. [CrossRef] [PubMed]
47. Bertuzzi, V.; Semonella, M.; Bruno, D.; Manna, C.; Edbrook-Childs, J.; Giusti, E.M.; Castelnuovo, G.; Pietrabissa, G. Psychological support interventions for healthcare providers and informal caregivers during the COVID-19 pandemic: A systematic review of the literature. *Int. J. Environ. Res. Public Health* **2021**, *18*, 6939. [CrossRef] [PubMed]
48. Kershaw, T.; Ellis, K.R.; Yoon, H.; Schafenacker, A.; Katapodi, M.C.; Northouse, L.L. The interdependence of advanced cancer patients' and their family caregivers' mental health, physical health, and self-efficacy over time. *Ann. Behav. Med.* **2015**, *49*, 901–911. [CrossRef]
49. Smith, L.M.; Onwumere, J.; Craig, T.; Kuipers, E. Caregiver correlates of patient-initiated violence in early psychosis. *Psychiatry Res.* **2018**, *270*, 412–417. [CrossRef] [PubMed]
50. Walke, S.C.; Chandrasekaran, V.; Mayya, S.S. Caregiver burden among caregivers of mentally ill individuals and their coping mechanisms. *J. Neurosci. Rural Pract.* **2018**, *9*, 180–185. [CrossRef] [PubMed]
51. Onwumere, J.; Glover, N.; Whittaker, S.; Rahim, S.; Man, L.C.; James, G.; Khan, S.; Afsharzadegan, R.; Seneviratne, S.; Harvey, R.; et al. Modifying illness beliefs in recent onset psychosis carers: Evaluating the impact of a cognitively focused brief group intervention in a routine service. *Early Interv. Psychiatry* **2018**, *12*, 1144–1150. [CrossRef]
52. Lasebikan, V.O.; Ayinde, O.O. Family burden in caregivers of schizophrenia patients: Prevalence and socio-demographic correlates. *Indian J. Psychol. Med.* **2013**, *35*, 60–66. [CrossRef]
53. Ohara, C.; Komaki, G.; Yamagata, Z.; Hotta, M.; Kamo, T.; Ando, T. Factors associated with caregiving burden and mental health conditions in caregivers of patients with anorexia nervosa in Japan. *BioPsychoSocial Med.* **2016**, *10*, 21. [CrossRef]
54. Wang, C. Mental health and social support of caregivers of children and adolescents with ASD and other developmental disorders during COVID-19 pandemic. *J. Affect. Disord. Rep.* **2021**, *6*, 100242. [CrossRef]
55. Ferizi, M.M.; Aali, S.; Bigdeli, I.; Talab, F.R.; Tavalaei, A.M. COVID-19 pandemic and the mental health of caregivers of the elderly with chronic diseases. *Iran. Rehabil. J.* **2022**, *20*, 225–236. [CrossRef]
56. Cheng, W.-L.; Chang, C.-C.; Griffiths, M.D.; Yen, C.-F.; Liu, J.-H.; Su, J.-A.; Lin, C.-Y.; Pakpour, A.H. Quality of life and care burden among family caregivers of people with severe mental illness: Mediating effects of self-esteem and psychological distress. *BMC Psychiatry* **2022**, *22*, 672. [CrossRef] [PubMed]
57. Lee, G.K.; Krizova, K.; Shivers, C.M. Needs, strain, coping, and mental health among caregivers of individuals with autism spectrum disorder: A moderated mediation analysis. *Autism* **2019**, *23*, 1936–1947. [CrossRef] [PubMed]
58. Timko, C.; Lor, M.C.; Rossi, F.; Peake, A.; Cucciare, M.A. Caregivers of people with substance use or mental health disorders in the US. *Subst. Abus.* **2022**, *43*, 1268–1276. [CrossRef] [PubMed]
59. Ali, F.; Baig, W.; Saad, M.; Tabassum, S.; Aslam, M.S.; Zahra, F.; Choudhary, A.K. Depression in the primary caregivers of patients with substance use disorder. *Pak. J. Med. Health Sci.* **2022**, *16*, 913–915. [CrossRef]
60. Maina, G.; Ogenchuk, M.; Phaneuf, T.; Kwame, A. "I can't live like that": The experience of caregiver stress of caring for a relative with substance use disorder. *Subst. Abus. Treat. Prev. Policy* **2021**, *16*, 11. [CrossRef] [PubMed]
61. Tyo, M.B.P.; McCurry, M.K.P.; Horowitz, J.A.P.; Elliott, K.D. Predictors of burden and resilience in family caregivers of individuals with opioid use disorder. *J. Addict. Nurs.* **2023**, *34*, E8–E20. [CrossRef] [PubMed]
62. Vázquez, F.; Otero, P.; Blanco, V.; López, L.; Torres, Á. Psychopathological symptoms in caregivers of demented and nondemented patients. In *Caregiving and Home Care*; Mollaoglu, M., Ed.; IntechOpen: London, UK, 2018. [CrossRef]
63. Brannan, A.M.; Brennan, E.M.; Sellmaier, C.; Rosenzweig, J.M. Employed parents of children receiving mental health services: Caregiver strain and work–life integration. *Fam. Soc. J. Contemp. Soc. Serv.* **2018**, *99*, 29–44. [CrossRef]
64. Litzelman, K.; Skinner, H.G.; Gangnon, R.E.; Nieto, F.J.; Malecki, K.; Witt, W.P. The relationship among caregiving characteristics, caregiver strain, and health-related quality of life: Evidence from the survey of the health of Wisconsin. *Qual. Life Res.* **2014**, *24*, 1397–1406. [CrossRef] [PubMed]
65. Accurso, E.C.; Garland, A.F.; Haine-Schlagel, R.; Brookman-Frazee, L.; Baker-Ericzén, M.J. Factors contributing to reduced caregiver strain in a publicly funded child mental health system. *J. Emot. Behav. Disord.* **2014**, *23*, 131–143. [CrossRef]
66. Rodrigo, C.; Fernando, T.; Rajapakse, S.; De Silva, V.; Hanwella, R. Caregiver strain and symptoms of depression among principal caregivers of patients with schizophrenia and bipolar affective disorder in Sri Lanka. *Int. J. Ment. Health Syst.* **2013**, *7*, 2. [CrossRef]
67. Tuithof, M.; Have, M.T.; van Dorsselaer, S.; de Graaf, R. Emotional disorders among informal caregivers in the general population: Target groups for prevention. *BMC Psychiatry* **2015**, *15*, 23. [CrossRef] [PubMed]
68. Araujo, A.d.S.; Pedroso, T.G. A relação entre emoção expressa e variáveis sociodemográficas, estresse precoce e sintomas de estresse em cuidadores informais de pessoas com transtornos mentais. *Cad. Bras. Ter. Ocup.* **2019**, *27*, 743–754. [CrossRef]
69. Di Lorito, C.; Bosco, A.; Godfrey, M.; Dunlop, M.; Lock, J.; Pollock, K.; Harwood, R.H.; van der Wardt, V. Mixed-methods study on caregiver strain, quality of life, and perceived health. *J. Alzheimer's Dis.* **2021**, *80*, 799–811. [CrossRef] [PubMed]
70. Díaz, M.; Estévez, A.; Momeñe, J.; Ozerinjauregi, N. Social support in the relationship between perceived informal caregiver burden and general health of female caregivers. *Ansiedad Estres* **2019**, *25*, 20–27. [CrossRef]

71. Teixeira, L.A.; Borges, M.d.C.; de Abreu, D.P.H.; Ribeiro, K.B.; Shimano, S.G.N.; Martins, L.J.P. Caregivers of older adults in palliative care: Level of burden and depressive symptoms. *Fisioter. Mov.* **2022**, *35*. [CrossRef]

72. Casarez, R.L.; Barlow, E.; Iyengar, S.M.; Soares, J.C.; Meyer, T.D. Understanding the role of m-health to improve well-being in spouses of patients with bipolar disorder. *J. Affect. Disord.* **2019**, *250*, 391–396. [CrossRef]

73. Sintayehu, M.; Mulat, H.; Yohannis, Z.; Adera, T.; Fekade, M. Prevalence of mental distress and associated factors among caregivers of patients with severe mental illness in the outpatient unit of Amanuel hospital, Addis Ababa, Ethiopia, 2013: Cross-sectional study. *J. Mol. Psychiatry* **2015**, *3*, 9. [CrossRef]

74. Ayalew, M.; Workicho, A.; Tesfaye, E.; Hailesilasie, H.; Abera, M. Burden among caregivers of people with mental illness at jimma university medical center, southwest ethiopia: A cross-sectional study. *Ann. Gen. Psychiatry* **2019**, *18*, 10. [CrossRef]

75. Price, A.M.; Measey, M.-A.; Hoq, M.; Rhodes, A.; Goldfeld, S. Child and caregiver mental health during 12 months of the COVID-19 pandemic in Australia: Findings from national repeated cross-sectional surveys. *BMJ Paediatr. Open* **2022**, *6*, e001390. [CrossRef]

76. Ribeiro, W.S.; Romeo, R.; King, D.; Owens, S.; Gronholm, P.C.; Fisher, H.L.; Laurens, K.R.; Evans-Lacko, S. Influence of stigma, sociodemographic and clinical characteristics on mental health-related service use and associated costs among young people in the United Kingdom. *Eur. Child Adolesc. Psychiatry* **2022**, *32*, 1363–1373. [CrossRef]

77. Sharma, R.; Sharma, S.C.; Pradhan, S.N. Assessing caregiver burden in caregivers of patients with schizophrenia and bipolar affective disorder in Kathmandu Medical College. *J. Nepal Health Res. Counc.* **2018**, *15*, 258–263. [CrossRef] [PubMed]

78. Lewis, C.; Zammit, S.; Jones, I.; Bisson, J.I. Prevalence and correlates of self-stigma in post-traumatic stress disorder (PTSD). *Eur. J. Psychotraumatology* **2022**, *13*, 2087967. [CrossRef] [PubMed]

79. Rezaei, H.; Niksima, S.H.; Gheshlagh, R.G. Burden of care in caregivers of Iranian patients with chronic disorders: A systematic review and meta-analysis. *Health Qual. Life Outcomes* **2020**, *18*, 261. [CrossRef] [PubMed]

80. Smith, L.; Onwumere, J.; Craig, T.; McManus, S.; Bebbington, P.; Kuipers, E. Mental and physical illness in caregivers: Results from an English national survey sample. *Br. J. Psychiatry* **2014**, *205*, 197–203. [CrossRef] [PubMed]

81. Ae-Ngibise, K.A.; Doku, V.C.K.; Asante, K.P.; Owusu-Agyei, S. The experience of caregivers of people living with serious mental disorders: A study from rural Ghana. *Glob. Health Action* **2015**, *8*, 26957. [CrossRef] [PubMed]

82. Karademas, E.C. A new perspective on dyadic regulation in chronic illness: The dyadic regulation connectivity model. *Health Psychol. Rev.* **2021**, *16*, 1–21. [CrossRef] [PubMed]

83. Caqueo-Urízar, A.; Gutiérrez-Maldonado, J. Burden of care in families of patients with schizophrenia. *Qual. Life Res.* **2006**, *15*, 719–724. [CrossRef]

84. Caqueo-Urízar, A.; Gutiérrez-Maldonado, J.; Miranda-Castillo, C. Quality of life in caregivers of patients with schizophrenia: A literature review. *Health Qual. Life Outcomes* **2009**, *7*, 84. [CrossRef]

85. Perlick, D.A.; Rosenheck, R.A.; Miklowitz, D.J.; Chessick, C.; Wolff, N.; Kaczynski, R.; Ostacher, M.; Patel, J.; Desai, R.; the STEP-BD Family Experience Collaborative Study Group. Prevalence and correlates of burden among caregivers of patients with bipolar disorder enrolled in the systematic treatment enhancement program for bipolar disorder. *Bipolar Disord.* **2007**, *9*, 262–273. [CrossRef]

86. Gkintoni, E. Clinical neuropsychological characteristics of bipolar disorder, with a focus on cognitive and linguistic pattern: A conceptual analysis. *F1000Research* **2023**, *12*, 1235. [CrossRef]

87. Sharif, L.; Basri, S.; Alsahafi, F.; Altaylouni, M.; Albugumi, S.; Banakhar, M.; Mahsoon, A.; Alasmee, N.; Wright, R.J. An exploration of family caregiver experiences of burden and coping while caring for people with mental disorders in Saudi Arabia—A qualitative study. *Int. J. Environ. Res. Public Health* **2020**, *17*, 6405. [CrossRef] [PubMed]

88. Del-Pino-Casado, R.; Frías-Osuna, A.; Palomino-Moral, P.A.; Ruzafa-Martínez, M.; Ramos-Morcillo, A.J. Social support and subjective burden in caregivers of adults and older adults: A meta-analysis. *PLoS ONE* **2018**, *13*, e0189874. [CrossRef] [PubMed]

89. van Beusekom, I.; Bakhshi-Raiez, F.; de Keizer, N.F.; Dongelmans, D.A.; van der Schaaf, M. Reported burden on informal caregivers of ICU survivors: A literature review. *Crit. Care* **2016**, *20*, 16. [CrossRef]

90. Hsiao, C.; Tsai, Y. Factors of caregiver burden and family functioning among Taiwanese family caregivers living with schizophrenia. *J. Clin. Nurs.* **2014**, *24*, 1546–1556. [CrossRef] [PubMed]

91. Kim, D. Relationships between Caregiving Stress, Depression, and Self-Esteem in Family Caregivers of Adults with a Disability. *Occup. Ther. Int.* **2017**, *2017*, 1686143. [CrossRef]

92. Chen, Y.-J.; Su, J.-A.; Chen, J.-S.; Liu, C.-H.; Griffiths, M.D.; Tsai, H.-C.; Chang, C.-C.; Lin, C.-Y. Examining the association between neuropsychiatric symptoms among people with dementia and caregiver mental health: Are caregiver burden and affiliate stigma mediators? *BMC Geriatr.* **2023**, *23*, 27. [CrossRef]

93. Yfantopoulos, J.; Sarris, M. Health related quality of life: Measurement methodology. *Arch. Hell. Med.* **2001**, *18*, 218–229.

94. Andrews, G.; Sanderson, K.; Slade, T.; Issakidis, C. Why does the burden of disease persist? Relating the burden of anxiety and depression to effectiveness of treatment. *Bull. World Health Organ.* **2000**, *78*, 446–454. [CrossRef]

95. Gaugler, J.E.; Kane, R.L.; Kane, R.A.; Clay, T.; Newcomer, R.C. The effects of duration of caregiving on institutionalization. *Gerontologist* **2005**, *45*, 78–89. [CrossRef]

96. Awad, A.G.; Voruganti, L.N. The burden of schizophrenia on caregivers: A review. *Pharmacoeconomics* **2008**, *26*, 149–162. [CrossRef]

97. Karambelas, G.J.; Filia, K.; Byrne, L.K.; Allott, K.A.; Jayasinghe, A.; Cotton, S.M. A systematic review comparing caregiver burden and psychological functioning in caregivers of individuals with schizophrenia spectrum disorders and bipolar disorders. *BMC Psychiatry* **2022**, *22*, 422–446. [CrossRef]

98. Ochoa, S.; Vilaplana, M.; Haro, J.M.; Villalta-Gil, V.; Martinez, F.; Negredo, M.C.; Casacuberta, P.; Paniego, E.; Usall, J.; Dolz, M.; et al. Do needs, symptoms or disability of outpatients with schizophrenia influence family burden? *Soc. Psychiatry Psychiatr. Epidemiol.* **2008**, *43*, 612–618. [CrossRef] [PubMed]

99. Zhang, Y.; Subramaniam, M.; Lee, S.P.; Abdin, E.; Sagayadevan, V.; Jeyagurunathan, A.; Chang, S.; Shafie, S.B.; Rahman, R.F.A.; Vaingankar, J.A.; et al. Affiliate stigma and its association with quality of life among caregivers of relatives with mental illness in Singapore. *Psychiatry Res.* **2018**, *265*, 55–61. [CrossRef] [PubMed]

100. Le Fevre, M.; Matheny, J.; Kolt, G.S. Eustress, distress and interpretation in occupational stress. *J. Manag. Psychol.* **2003**, *18*, 726–744. [CrossRef]

101. Polit, D.F.; Beck, C.T. *Essentials of Nursing Research: Appraising Evidence for Nursing Practice*, 7th ed.; Wolters Kluwer Health/Lippincott Williams & Wilkins: Philadelphia, PA, USA, 2010.

102. Savithri, J.; Ragothaman, C.B. Stress management—A study on soft competencies affecting stress. *ZENITH Int. J. Bus. Econ. Manag. Res.* **2016**, *6*, 24–30.

103. Stavrianou, A.; Kafkia, T.; Mantoudi, A.; Minasidou, E.; Konstantinidou, A.; SapountziKrepia, D.; Dimitriadou, A. Informal caregivers in Greek Hospitals: A unique phenomenon of a health system in financial crisis. *Mater. Sosio-Medica* **2018**, *30*, 147–152. [CrossRef]

104. Mitsonis, C.; Voussoura, E.; Dimopoulos, N.; Psarra, V.; Kararizou, E.; Latzouraki, E.; Zervas, I.; Katsanou, M.-N. Factors associated with caregiver psychological distress in chronic schizophrenia. *Soc. Psychiatry Psychiatr. Epidemiol.* **2012**, *47*, 331–337. [CrossRef]

105. Allen, S.M.; Lima, I.C.; Goldscheider, F.K.; Roy, J. Primary caregiver characteristics and Transitions in community-based care. *J. Gerontol. Ser. B Psychol. Soc. Sci.* **2012**, *67*, 362–371. [CrossRef]

106. Boyer, L.; Caqueo-Urizar, A.; Richieri, R.; Lancon, C.; Gutierrez-Maldonado, J.; Auquier, P. Quality of life among caregivers of patients with schizophrenia: Across-cultural comparison of Chilean and French families. *BMC Fam. Pract.* **2012**, *13*, 42. [CrossRef]

107. Clibbens, N.; Berzins, K.; Baker, J. Caregivers' experience of service transitions in adult mental health: An integrative qualitative synthesis. *Health Soc. Care Community* **2019**, *27*, 535–548. [CrossRef]

108. Onwumere, J.; Sirykaite, S.; Schulz, J.; Man, E.; James, G.; Afsharzadegan, R.; Khan, S.; Harvey, R.; Souray, J.; Raune, D. Understanding the experience of "burnout" in first-episode psychosis carers. *Compr. Psychiatry* **2018**, *83*, 19–24. [CrossRef]

109. Stanley, S.; Balakrishnan, S.; Ilangovan, S. Psychological distress, perceived burden and quality of life in caregivers of persons with schizophrenia. *J. Ment. Health* **2017**, *26*, 134–141. [CrossRef] [PubMed]

110. Broeze van Groenou, M.I.; de Boer, A. Providing informal care in a changing society. *Eur. J. Ageing* **2016**, *13*, 271–279. [CrossRef] [PubMed]

111. World Health Organization. *Annual Report 2019: Health for All*; World Health Organization Office in Lebanon: Beirut, Lebanon, 2020.

112. World Health Organization. *The World Health Report 2001: Mental Health, New Understanding, New Hope*; World Health Organization: Geneva, Switzerland, 2001.

113. World Health Organization. Chapter V: Mental and behavioral disorders. In *International Statistical Classification of Diseases and Related Health Problems 10th Revision (ICD-10)—WHO Version for 2016*; World Health Organization: Geneva, Switzerland, 2019.

114. American Psychiatric Association. *Diagnostic and Statistical Manual of Mental Disorders*, 5th ed.; American Psychiatric Association: Washington, DC, USA, 2013.

115. Goodhead, A.; McDonald, J. *Informal Caregivers Literature Review. A Report Prepared for the National Health Committee Health Services Research Centre: Victoria University of Wellington*; New Zealand Ministry of Health: Wellington, New Zealand, 2007.

 *healthcare*

Article

Novel Telehealth Adaptations for Evidence-Based Outpatient Suicide Treatment: Feasibility and Effectiveness of the Crisis Care Program

J. Conor O'Neill *, Erin T. O'Callaghan, Scott Sullivan and Mirène Winsberg

Brightside Health, 2471a Peralta Street, Oakland, CA 94607, USA; mimi.winsberg@brightside.com (M.W.)
* Correspondence: conor.oneill@brightside.com

Abstract: Background: Suicide rates in the United States have escalated dramatically over the past 20 years and remain a leading cause of death. Access to evidenced-based care is limited, and telehealth is well-positioned to offer novel care solutions. The Crisis Care program is a suicide-specific treatment program delivered within a national outpatient telehealth setting using a digitally adapted version of the Collaborative Assessment and Management of Suicidality (CAMS) as the framework of care. This study investigates the feasibility and preliminary effectiveness of Crisis Care as scalable suicide-specific treatment model. **Methods:** Patient engagement, symptom reduction, and care outcomes were examined among a cohort of patients (n = 130) over 16 weeks. The feasibility of implementation was assessed through patient engagement. Clinical outcomes were measured with PHQ-9, GAD-7, and the CAMS SSF-4 rating scales. **Results:** Over 85% of enrolled patients were approved for Crisis Care at intake, and 83% went on to complete at least four sessions (the minimum required to graduate). All patient subgroups experienced declines in depressive symptoms, anxiety symptoms, suicidal ideation frequency, and suicide-specific risk factors. **Conclusions:** Results support the feasibility and preliminary effectiveness of Crisis Care as a suicide-specific care solution that can be delivered within a stepped-care model in an outpatient telehealth setting.

Keywords: suicide intervention; telehealth; crisis intervention; outpatient; evidence-based treatment; CAMS

Citation: O'Neill, J.C.; O'Callaghan, E.T.; Sullivan, S.; Winsberg, M. Novel Telehealth Adaptations for Evidence-Based Outpatient Suicide Treatment: Feasibility and Effectiveness of the Crisis Care Program. *Healthcare* **2023**, *11*, 3158. https://doi.org/10.3390/healthcare11243158

Academic Editors: Carlos Laranjeira and Ana Querido

Received: 4 November 2023
Revised: 7 December 2023
Accepted: 9 December 2023
Published: 13 December 2023

1. Introduction

Suicide is a public health crisis and is the second leading cause of death among adults aged 18–45 in the United States [1]. Despite ongoing national suicide prevention and intervention initiatives, estimates suggest approximately 50,000 people (14.3 per 100,000) in the United States died by suicide in 2022, representing the highest national rate of suicide in decades [2]. Trends in suicide within the United States are alarming, with a 37% increase in suicide rates from 2000–2018, followed by a slight decline (−5%) from 2019–2020 [3]. Unfortunately, suicide rates have risen significantly in 2021 and 2022 post-COVID-19 pandemic (CDC, 2023), consistent with evidence suggesting that suicide rates may decline during disasters but increase in the period that follows [4]. Given the substantial societal and individual impacts of the COVID-19 pandemic within the past three years, feasible and effective treatments for suicide risk are needed. However, social perceptions of suicide can further exacerbate risk with affected individuals contributing to increased isolation and limited engagement with societal structures that can help improve access to essential care [5]. Shifting to an ecological understanding of suicide risk to emphasize comprehensive prevention methods within health systems and existing social structures is critical.

Further exacerbating this crisis in the United States, the proportion of individuals who experience suicidal thinking is 200 times higher than those who die by suicide and serves as a primary treatment target for healthcare systems. In 2021, 1.7 million adults made a suicide attempt, 3.5 million adults made a plan to attempt suicide, and 12.3 million adults reported serious suicidal ideation [2]. Contemporary suicide theories and empirical evidence suggest ideation-to-action pathways; that is, individuals who think about suicide are more likely to engage in suicidal behavior, with suicide attempt history serving as one of the most salient risk factors for death by suicide [6–9]. With over 12 million adults contemplating suicide, treatments that can effectively screen, detect, assess, and manage suicidal thinking are paramount to interrupt the ideation-to-action pathway and reduce suicide deaths.

With such significant needs across a spectrum of suicide risk, stepped care models that offer treatment options proportionate to risk in settings with fewer restrictions are indicated [10]. Still, many individuals are evaluated through emergency departments when presenting with suicide risk [11]. While such settings serve a vital role in stabilizing the imminent risk of self-harm, hospitals are generally not designed to provide ongoing treatment of suicide risk [10]. Rather, hospitals are tasked with connecting patients to outpatient treatment that can effectively address suicidality [10]. This transition in care is critical and has been well established as a particularly vulnerable and high-risk time period [12,13]. Data show that in the month following discharge from inpatient hospitalization, suicide death rates are 200 times higher than those of the general public, and a heightened risk of suicidal behavior and death can persist for up to three months [14,15]. Yet, patients are often lost to follow-up during this period. Further, estimates suggest approximately one-third of patients fail to connect to an outpatient appointment within 30 days of discharge [16].

Access and connection to care is limited for individuals at risk for suicide, with a recent meta-analysis suggesting that fewer than 26% of individuals who died by suicide had contact with inpatient or outpatient mental health services within the year prior to their death [17,18]. Barriers to accessing mental healthcare services disproportionately impact certain populations, such as those in rural areas and minority populations [19,20]. To mitigate such barriers and enhance access within a stepped-care model, digital interventions and telehealth treatments have been identified as valuable care options [19,21,22]. Increasing care options serves as an important component of comprehensive suicide prevention and may reduce potential deaths [23]. By increasing access to care for those who truly need it, health systems may provide lifesaving care while reducing access gaps for vulnerable populations and deliver evidence-based care in a timely manner [24,25].

Telehealth treatment provides several benefits that can support access to care, including flexibility in scheduling, time savings and convenience, and geographical flexibility [26]. Evidence suggests that patients are more likely to attend their appointments virtually [27,28] with equivalent effectiveness in psychotherapy outcomes compared with in-person treatment in both outpatient and intensive outpatient levels of care [28,29]. However, there remains a dearth of telehealth suicide treatment options with notable hesitancy among clinicians in providing such specialty care in a telehealth setting [30].

Additionally, there are relatively few evidence-based approaches for suicide specific treatment (Dialectical Behavioral Therapy, Cognitive Therapy for Suicide Prevention, and the Collaborative Assessment and Management of Suicidality), none of which were developed specifically for telehealth delivery [31–35]. Despite these challenges, evidence for the effective treatment of suicide risk through telehealth settings is emerging.

The Collaborative Assessment and Management of Suicidality (CAMS) has demonstrated robust evidence in the treatment of suicide risk over the past several decades, especially with regard to reducing suicidal ideation to promote safety [36]. The CAMS framework adopts a collaborative, patient-centered approach that is empathic, therapeutic, and sensitive to unique suicide risk factors within each patient. During each session, CAMS utilizes the Suicide Status Form-4 (SSF-4) to guide a collaborative, therapeutic risk assessment, followed by the development and implementation of a suicide specific, individualized treatment plan that includes stabilization planning (e.g., a safety plan) [34]. CAMS is intended to be short-term, lasting approximately 6–12 sessions, and can be effectively implemented as a stepped-care option to treat suicide risk within outpatient settings [37].

CAMS has been studied in seven randomized controlled trials (RCTs), and in a recent meta-analysis, CAMS was found to have outperformed alternative interventions in significantly reducing suicidal ideation and depression and increasing hope [36,38–44]. Patients are more likely to remain in care and report better satisfaction in care compared with treatment as usual [36].

CAMS has also successfully been adapted for telehealth use across a range of settings and is being evaluated in multiple RCTs as a result of necessary conversions from in-person care during the COVID-19 pandemic [45–48]. However, to date, there have been no published studies demonstrating the feasibility and effectiveness of telehealth adaptations to CAMS in a diverse cohort of outpatient behavioral health patients.

This study aimed to evaluate the feasibility and preliminary effectiveness of the Crisis Care program; a digitally adapted version of the CAMS framework for patients presenting with elevated suicide risk within a national outpatient telehealth treatment setting.

Crisis Care seeks to provide timely access to treatment, increase access to care, and effectively treat patients with intermediate suicide risk who do not require immediate hospitalization, but do require suicide specific specialty care. Situated within a comprehensive tele-mental health treatment framework, Crisis Care offers continuity of care with step-down services and care coordination support for escalations to higher levels of care. The goals of the present study are as follows:

1.   Examine time to treat, patient engagement, retention and response to treatment for patients with intermediate suicide risk within a telehealth outpatient setting.
2.   Investigate preliminary clinical outcomes specific to suicide risk and mental health functioning.
3.   Explore continuity of care outcomes for patients post discharge from Crisis Care.

The findings of this investigation provide valuable contributions to the field and demonstrate a promising telehealth care model to treat those at elevated risk for suicide.

2. Materials and Methods

2.1. Participants

Participants who were enrolled in Crisis Care and completed their initial session during an 8-month period (October 2022–June 2023) were included in the study. Eligible participants included existing patients receiving care at a national telehealth company who were referred and enrolled in Crisis Care, as well as new patients who were enrolled in Crisis Care when initiating treatment with the telehealth company.

Existing patients were enrolled in Crisis Care if they presented with elevated suicide risk, such as the endorsement of intense or persistent suicidal ideation, suicidal ideation with significant risk or precipitating event(s), suicidal thinking with a degree of planning or intent, or recent suicidal behavior (e.g., attempt). New patients were enrolled in Crisis Care if they reported suicidal ideation with a degree of planning endorsed through item 9 of the PHQ-9 or if they reported a suicide attempt within the past 12 months. Exclusionary criteria included active psychosis or mania, active primary substance use disorder diagnosis, active primary eating disorder diagnosis, or individuals who did not present with suicide specific treatment needs. Crisis Care is not a substitute for emergency care, and individuals reporting imminent risk of self-harm were referred to emergency services and not included in the present study.

The sample (n = 130) included adults ages 18+ (m = 31, SD = 10.3) and was diverse in educational attainment and income (see Table 1). The majority of patients were female (58.46%), primarily diagnosed with major depression (66.92%), and included geographical diversity across 25 different states in the US. Approximately 42% (n = 54) of participants were new patients who enrolled in Crisis Care when initially signing up for services with the telehealth company, while 58% of participants (n = 76) were existing patients of the telehealth company who were referred internally to Crisis Care.

Table 1. Demographics.

Characteristic	Study Sample (n = 130)
Age	m = 31.1 (SD = 10.3)
Sex	
Female	58.5%
Male	41.5%
Ethnicity	
White	62.3%
Hispanic/Latino	13.1%
Black	10.8%
Asian	3.1%
Other	9.2%
Not available	1.5%
Education	
<High School	3.9%
High School	46.9%
Associate's Degree	16.2%
Bachelor's Degree	24.6%
Advanced Degree	6.9%
Not available	1.5%
Annual Income	
<30 K	20.0%
30 K–60 K	29.2%
60 K–100 K	13.1%
<100 K	14.6%
Not available	23.1%
Geographic Region	
South	34.6%
Northeast	33.1%
West	24.6%
Midwest	7.7%

2.2. Procedures

The study was approved by the WCG Institutional Review Board. A 16-week cohort was established to monitor patient engagement and response to treatment over time. Consistent with the CAMS model, Crisis Care does not have concrete requirements for length of care, as such decisions are informed by response to treatment and states of patient risk. Still, Crisis Care is intended to be completed within approximately 12 sessions. Therefore, a 16-week time frame was examined for all patients, with the expectation that the majority of patients would complete treatment prior to the conclusion of the 16-week window.

2.3. Patient Status

Patient status was monitored throughout the course of treatment. Patients were enrolled after consenting to terms and scheduling their initial Crisis Care intake session. Patients were either accepted or declined from Crisis Care during the initial intake session. Patients were declined if they did not meet eligibility criteria or elected to not move forward with Crisis Care treatment and were provided with appropriate referrals for further care. Accepted patients scheduled their follow-up sessions and began Crisis Care treatment.

The course of treatment was adherent to the CAMS model. Patients were encouraged to attend weekly Crisis Care appointments for approximately 12 sessions. Successful completion of Crisis Care was based on the CAMS resolution guidelines with specified, clinical criteria associated with safety, stability, and reductions in suicide risk. This included three consecutive 'low overall suicide risk scores', evidenced by self-report ratings on the SSF-4 (see below), the absence of suicidal behavior in the past week, and the management of suicidal thoughts in the past week (if applicable). Clinical discretion was also applied for graduation so that patients may meet graduate eligibility criteria but continue in Crisis Care until the therapist and patient agree on program completion at the appropriate time. Patients were eligible for graduation from Crisis Care in as few as 4 sessions based on these criteria.

2.4. Adaptations to CAMS

The CAMS intervention was digitally adapted with fidelity and implemented within a national telehealth company's care process model. A digital version of each CAMS form was developed with screen-sharing capabilities to support the collaborative treatment experience. Documentation and completed forms were made available for clinicians and patients to readily access. Additional digital mental health tools including clinical check-ins (i.e., PHQ-9, GAD-7), asynchronous messaging, clinical alerts and notifications, and dedicated on-call clinical supervisor support were incorporated throughout the course of treatment to allow for real-time detection, assessment, and response to patient suicide risk and clinical symptoms.

2.5. Crisis Care Treatment Components

The CAMS framework was implemented via video-based telehealth treatment sessions. In CAMS, the Suicide Status Form-4 (SSF-4) is used to guide patients through the completion of a suicide risk self-assessment in partnership with the therapist, while the therapist helps the patient identify key drivers related to their suicide risk that are used to construct an individualized treatment plan. CAMS is flexible in that various treatment techniques can be implemented during treatment to address the corresponding suicide drivers. A stabilization plan is developed in the first session and reviewed continuously throughout treatment. The stabilization plan is comparable to other safety planning interventions and consists of coping techniques, means restriction measures, social contacts, and emergency resources that the patient can utilize to support safety and reduce the risk of acting on suicidal thoughts or feelings. Suicide risk, response to treatment, and stabilization planning are evaluated throughout care in each CAMS session using the SSF-4.

In the present study, all core components of the CAMS intervention using the SSF-4 were implemented in each Crisis Care session (i.e., collaborative risk assessment, suicide-specific treatment planning, stabilization planning). Crisis Care patients also received a psychiatric evaluation by a psychiatric provider. Patients were prompted to complete weekly asynchronous clinical check-ins (i.e., PHQ-9, GAD-7), attend weekly live video-based Crisis Care treatment sessions (60-min intake, 45-min follow-up sessions), and had access to their clinicians through asynchronous messaging. Patients were instructed to use messaging as a supportive resource to communicate with their clinicians and not as a form of text therapy or for emergencies. All Crisis Care therapists were trained in CAMS. Ongoing consultation and support from CAMS-trained clinical leadership with expertise in suicide risk assessment and intervention was available to Crisis Care clinicians.

2.6. Measures

Clinical outcomes were measured using three self-report instruments. The PHQ-9 is a 9-item, 4-point Likert scale self-report measure of depression symptom severity within the past two weeks. The PHQ-9 includes a specific item related to the frequency of suicidal thoughts (item 9). If suicidal ideation is endorsed, a follow-up item that screens for suicide attempt planning is administered. In the PHQ-9, item 9 was used as one data point for assessing suicidal ideation frequency in the past two weeks, and an additional assessment of suicide risk was measured using components of the CAMS intervention outlined below. Higher scores on the PHQ-9 indicate increased symptom severity and frequency of symptoms. The PHQ-9 has strong reliability and validity, with 88% sensitivity and 88% specificity for major depressive disorder (MDD) [49].

The GAD-7 is a 7-item, four-point Likert scale self-report measure of Generalized Anxiety Disorder (GAD) symptoms within the past two weeks. Higher scores reflect increased levels of anxiety. The GAD-7 has strong psychometric properties, with 89% sensitivity and 82% specificity for GAD [50].

The SSF-4 from the CAMS intervention is a 6-item, 5-point Likert scale self-report assessment measure that includes five domains of suicide risk grounded in suicide theory (self-hatred, agitation, hopelessness, psychological pain, distress) and an overall suicide risk score. Higher scores suggest higher levels of related symptoms and increased risk. The SSF-4 demonstrates strong convergent and criterion–predictive validity and moderate test–retest reliability, which is expected given the nature of suicide risk [51–53].

Patients were prompted to complete the PHQ-9 and GAD-7 at baseline and throughout the course of treatment at weekly intervals. Patients completed the SSF-4 at the start of each Crisis Care session in accordance with the CAMS intervention model.

2.7. Data Analysis

The feasibility of implementing Crisis Care across a diverse cohort of patients was evaluated using descriptive methods. The average time to treat was calculated by observing the length of time from enrollment to the program to completion of the first Crisis Care session.

Of the patients who were accepted to Crisis Care after the initial session, patient engagement and response to care were evaluated to identify three distinct subgroups: (1) those who graduated, (2) those who exited the program without graduating, and (3) those who remained in Crisis Care at the conclusion of the 16 weeks.

Specific clinical outcomes were examined across each of the three subgroups, including changes in PHQ-9 overall scores, PHQ-9 item 9 (suicidal ideation frequency), GAD-7 scores, and the six domains of the SSF-4 at baseline and last-completed rating in order to calculate the average change in respective scores over time.

The average time to graduation was calculated by examining the frequency distributions of the number of sessions completed at the point of graduation and the average length of time in weeks to complete the sessions. Reasons for exiting the program prior to meeting graduation criteria were qualitatively identified through medical record review, and continuity of care post graduation was assessed by observing attendance to an outpatient appointment within 30 days following step-down from Crisis Care to ongoing treatment within the telehealth company.

3. Results

A total of 168 patients consented to terms and enrolled in Crisis Care, 130 went on to complete their initial session, and 112 (86%) patients were approved to continue with Crisis Care treatment after the initial session. Of those approved, 93 (83%) patients completed at least four Crisis Care sessions. The average wait time for treatment from enrollment to attending the first session was 3.9 days (SD = 3.2) (Figure 1).

3.1. Graduated

Of those approved for Crisis Care (n = 112), approximately 40% (n = 45) of patients graduated within a 16-week timeframe, and of those who attended at least 4 sessions (n = 93), 48% successfully graduated from the program in an average of 7.4 sessions. Thirty-one percent (n = 14) of patients graduated by the third follow-up session, which is the minimum number of sessions required to reach clinical graduation criteria. An additional 49% (n = 13) of patients graduated between four and eight sessions (m = 6.73) in an average of 8 weeks (SD = 3), and 20% (n = 9) graduated in nine or more sessions (m = 14.11) within 12 weeks on average (SD = 2). Taken together, over 50% (n = 23) of patients graduated by the fourth follow-up session (m = 4.07) within 6 weeks on average (SD = 3).

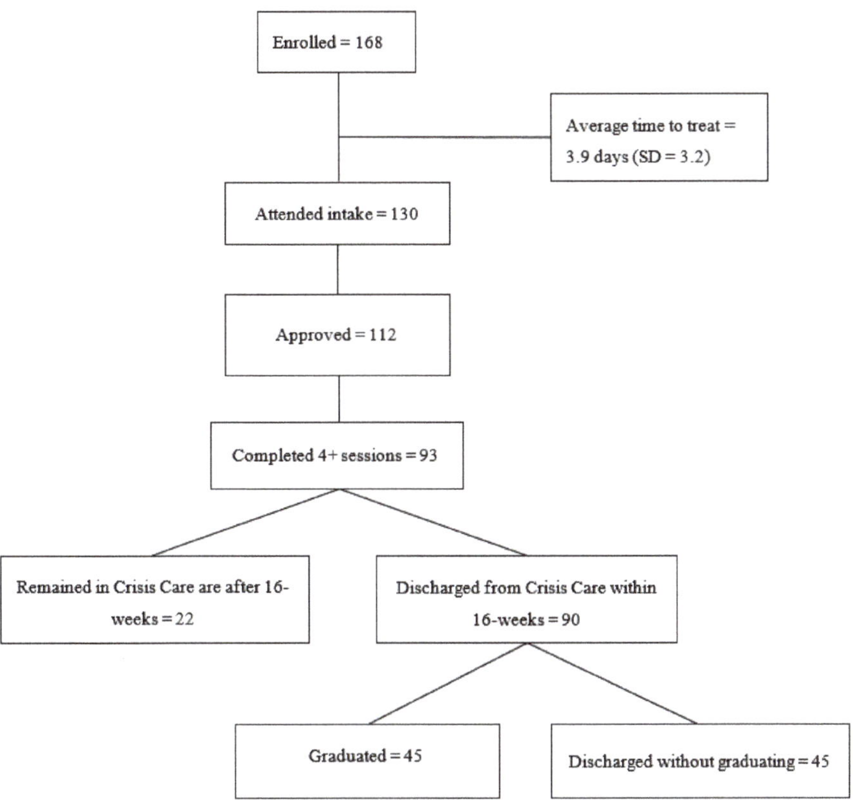

Figure 1. Crisis Care Patient Pathways.

Regarding clinical outcomes for all graduates (n = 45), the PHQ-9 total scores on average reduced by 46.5% from the baseline (m = 19.4, SD = 4.86) to the last score follow-up (m = 10.38, SD = 6.76) (Figure 2, Table 2). For suicidal ideation frequency, as measured by the PHQ-9 item 9, average scores reduced from 1.62 (SD = 0.96) at baseline to 0.47 (SD = 0.87), marking a 71.2% reduction in frequency ratings of suicidal thinking over the past two weeks (Figure 3). Average GAD-7 ratings reduced by 42.2% from the baseline (m = 15.31, SD = 4.49) to the last score (m = 8.84, SD = 6.18; Figure 4).

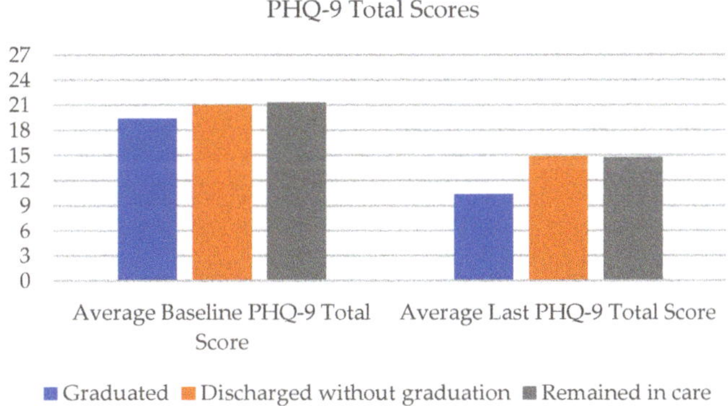

Figure 2. Average PHQ-9 total scores from baseline to last rating.

Table 2. Engagement and Clinical Outcomes.

Approved for Crisis Care (n = 112)	Graduated	Discharged without Graduation	Remained in Crisis Care
Patient sample	45 (40.2%)	22 (19.6%)	45 (40.2%)
Average # of sessions	7.4	4.3	8.6
Graduation criteria met	45 (100%)	12 (54.5%)	31 (68.9%)
Clinical Outcomes	**M (SD)**		
Baseline PHQ Total	19.4 (4.9)	21 (3.9)	21.29 (4.1)
Baseline PHQ-9, Item 9 (SI)	1.62 (0.96)	1.82 (1.14)	2 (1.02)
Baseline GAD Total	15.31 (4.5)	15.41 (4.9)	16.6 (3.7)
Last PHQ Total	10.38 (6.8)	14.86 (7.9)	14.73 (7.4)
Last PHQ-9, Item 9 (SI)	0.47 (0.87)	0.91 (1.15)	1.02 (1.16)
Last GAD Total	8.84 (6.2)	12.91 (6.7)	11.82 (6.3)
PHQ-9 Total % Diff Baseline vs. Last	−46.51%	−29.22%	−30.79%
PHQ-9, Item 9 % Diff Baseline vs. Last	−71.23%	−50.00%	−48.89%
GAD-7 Total % Diff Baseline vs. Last	−42.24%	−16.22%	−28.78%
SSF-4	**M (SD)**		
Baseline Agitation	3.16 (1.33)	3.25 (1.48)	3.05 (1.4)
Baseline Hopelessness	3.4 (1.25	3.65 (1.09)	3.51 (1.26)
Baseline Psychological Pain	3.13 (1.12)	3.67 (0.91)	3.62 (1.13)
Baseline Self Hate	3.24 (1.37)	3.8 (1.2)	3.47 (1.39)
Baseline Stress	3.96 (1.17)	4 (0.97)	3.73 (1.23)
Baseline Overall Risk of Suicide	1.78 (0.82)	2.1 (1.12)	1.84 (0.81)
Last rating—Agitation	2.24 (1.11)	2.82 (1.5)	2.71 (1.38)
Last rating—Hopelessness	1.76 (1.05)	2.73 (1.32)	2.76 (1.46)
Last rating—Psychological Pain	1.82 (0.96)	2.77 (1.34)	2.98 (1.37)
Last rating—Self-Hate	1.87 (1.04)	2.86 (1.55)	2.71 (1.42)
Last rating—Stress	2.56 (1.25)	3.23 (1.27)	3.24 (1.3)
Last rating—Overall Risk of Suicide	1.04 (1.01)	1.5 (1.01)	1.44 (0.66)
Agitation % Diff Baseline vs. Last	−28.87%	−17.02%	−10.85%
Hopelessness % Diff Baseline vs. Last	−48.37%	−43.14%	−20.55%
Psychological Pain % Diff Baseline vs. Last	−41.84%	−38.18%	−18.47%
Self-Hate % Diff Baseline vs. Last	−42.47%	−42.00%	−20.69%
Stress % Diff Baseline vs. Last	−35.39%	−31.48%	−13.46%
Overall Risk of Suicide % Diff Baseline vs. Last	−41.25%	−37.93%	−21.05%

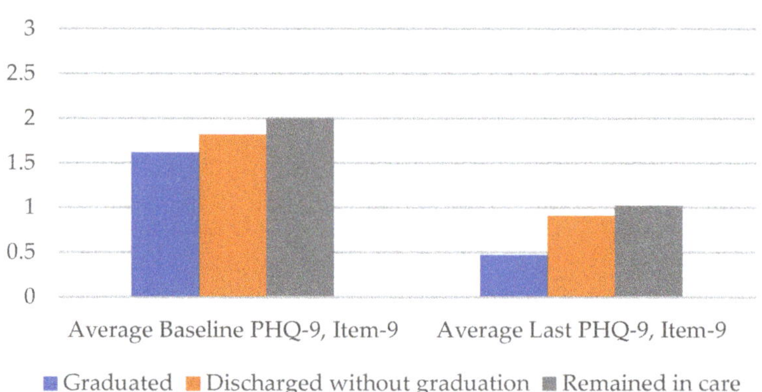

Figure 3. Average PHQ-9, item 9 (suicidal ideation frequency) from baseline to last rating.

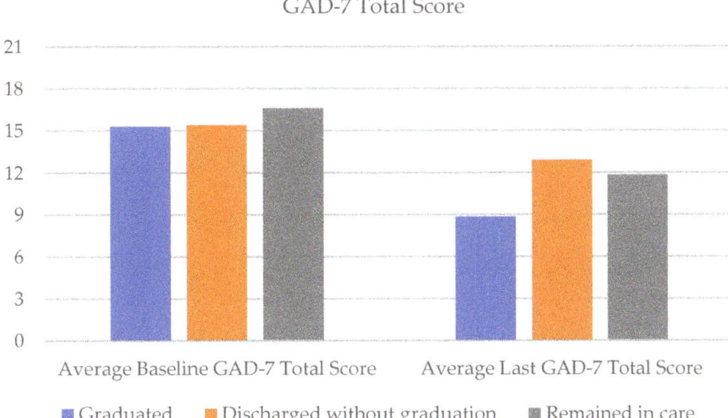

Figure 4. Average GAD-7 total scores from baseline to last rating.

Significant reductions were observed from baseline to program completion on all items of the SSF-4 among graduated patients. On average, self-reported agitation ratings reduced by 28.9%, hopelessness reduced by 48.4%, psychological pain decreased by 41.8%, self-hate declined by 42.5%, and stress decreased by 35.4%. Notably, overall suicide risk was reduced on average by 41.3% per patient self-report (Figure 5). Of the 45 patients that graduated, 100% stepped down to ongoing care within the telehealth company, and 87% (n = 39) attended at least one outpatient appointment within 30 days of completing Crisis Care.

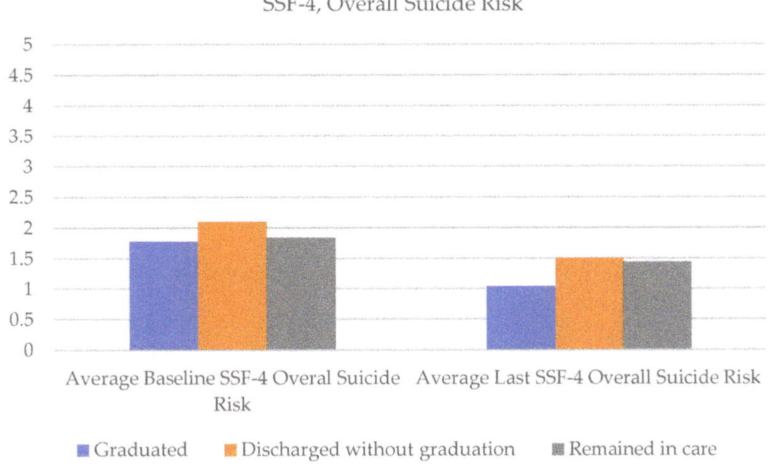

Figure 5. Average SSF-4, overall suicide risk scores from baseline to last rating.

3.2. Discharged without Graduation

Approximately 20% (n = 22) of patients who were approved for Crisis Care (n = 112) exited the program without graduating. The term "exit" refers to patients who left the program without achieving graduation status. Ten (45%) of those patients exited the program within the first three sessions. Reasons for exiting the program were mixed, with the majority of patients disengaging from treatment (60%, n = 13), 9% (n = 2) requiring a higher level of care, and 32% (n = 7) for other reasons (e.g., patient was appropriate for step-down prior to meeting graduation criteria or no longer consented to terms).

Results suggest clinical symptom reduction even for patients who exited without graduating. On average, PHQ-9 total scores reduced by 29% (baseline m = 21.00, SD = 3.89; last score m = 14.86, SD = 7.85; Figure 2), item 9 of the PHQ-9 (suicidal ideation frequency) reduced by 50% (baseline m = 1.82, SD = 1.14; last score m = 0.91, SD = 1.15; Figure 3), and GAD total scores reduced by 16% (baseline m = 15.41, SD = 4.92; last score m = 12.91, SD = 6.73) from baseline to the last completed ratings (Figure 4). With regard to SSF-4 items, reductions were observed on average across all items from baseline to the last completed score. Agitation decreased by 17%, hopelessness by 43%, psychological pain by 38%, self-hate by 42%, stress by 31%, and overall suicide risk by 38% on average compared with baseline ratings (Figure 5).

3.3. Remained in Crisis Care

Forty percent (n = 45) of approved patients (n = 112) remained in Crisis Care after 16 weeks. This subgroup of patients attended an average of 8.58 sessions over the 16-week period. However, nearly 70% (n = 31) of patients met the clinical criteria for graduation despite not having been formally processed for graduation from the program. Further, these patients experienced notable reductions in clinical symptoms. On average, total PHQ-9 scores declined by 31% from baseline (m = 21.29, SD = 4.12) to the last completed score (m = 14.73, SD = 7.35; Figure 2). The frequency of suicidal ideation declined by an average of 49% (baseline m = 2.00, SD = 1.02; last score m = 1.02, SD = 1.16) across the subgroup of patients (Figure 3), and GAD scores reduced by an average of 29% (baseline m = 16.6, SD = 3.74; last score m = 11.82, SD = 6.26; Figure 4).

SSF-4 scores also declined, though to a lesser extent than those who exited the program (due to graduation or otherwise). On average, agitation was reduced by 11%, hopelessness by 21%, psychological pain by 18%, self-hate by 21%, stress by 13%, and overall suicide risk declined by 21% from baseline to the last completed Crisis Care session (Figure 5).

Overall, results demonstrate that clinical symptoms, as measured by PHQ-9, GAD-7, and SSF-4 scores, improved from baseline to the last score in all Crisis Care clinical patient groups.

4. Discussion

This study investigated the feasibility of implementing a novel telehealth solution for the treatment of elevated suicide risk within a cohort of outpatient behavioral health patients using a digitally adapted version of the CAMS intervention. Complementing the CAMS model with digital mental health tools and comprehensive care (e.g., med management, care coordination), Crisis Care was designed to deliver suicide specific treatment within a stepped-care model to quickly, safely, and effectively treat patients with escalated risk for suicide in the least restrictive setting within a scalable and sustainable framework.

The results of the current study are encouraging with respect to the feasibility of implementation, preliminary effectiveness of clinical symptom reduction, and continuity of care. Patients at risk of suicide are at heightened risk when transitioning care [13] and should be seen within one week of referral [10]. Telehealth treatment has been identified as a viable option to create and support a fast connection to care [22]. In the present study, over 75% of enrolled patients attended their initial session, with an average time to treat of 4 days. These results support the use of telehealth treatment within a stepped-care framework as an effective solution to address key access gaps and expedite care connection for vulnerable populations at risk of suicide.

While initial care connection and time-to-treat outcomes were promising, results related to engagement and adherence to treatment were mixed. The vast majority of patients (86%) who were seen for an initial Crisis Care evaluation were deemed appropriate for the program, suggesting accurate referral pathways. Over 80% of patients went on to complete at least four sessions, which is the minimum number needed for graduation, in addition to satisfying clinical criteria and clinician judgment. Less than half the sample graduated within the specified time period, and 80% of graduates completed treatment within eight sessions. This is consistent with the length of treatment completion rates established in the literature for the CAMS intervention [40] and suggests that digital adaptations of CAMS within a telehealth outpatient setting are feasible for a considerable portion of patients. Furthermore, all graduated patients stepped down to ongoing care at the telehealth company and attended at least one tele-behavioral healthcare visit within 30 days of discharge, supporting the advantages of a stepped-care model in providing continuity of care for those at risk of suicide. This is particularly encouraging given the risk of suicide remains particularly high for up to three months after a step-down transition in care [15]. By situating specialty suicide treatment within telehealth care models, health systems may close key gaps in care and support ongoing treatment to detect potential escalations of risk and respond accordingly in a least restrictive setting.

Conversely, there was a decline in treatment engagement among a portion of patients within the first three sessions, and a quarter of patients who enrolled in Crisis Care did not move forward with completing the initial intake. It is unclear specifically why this portion of patients did not continue with treatment, and this warrants further examination. While creating quick access to care is critical, these results also suggest that additional efforts should target initial engagement in care at the point of referral and shortly after treatment commences.

In addition, 40% of patients remained in treatment after 16 weeks, and 20% of patients were discharged before meeting graduation criteria. Upon further examination of these subgroups, it was clear that most patients were discharged due to disengagement, and those who remained in care without graduating attended an average of nine sessions over the 16-week timeframe, which is less than the recommended weekly session cadence. The finding that more than half of the patients who were discharged without formally graduating met the clinical graduation criteria set by CAMS, along with the 70% of those who remained in care after 16 weeks who also met the criteria, suggests that a majority of these patients indeed benefited from care and could explain reduced session attendance. Given that this study did not examine clinical outcomes beyond the 16-week timeframe, it is unknown whether this subgroup of patients ultimately graduated. Other possible explanations for patient disengagement could include patient-related factors such as time commitment challenges, preference for in-person care, limited acceptability of the treatment program, or treatment program related factors, such as difficulties accommodating preferred appointment times. However, reasons for patient disengagement were not overtly observable in the present study. Future research may support the identification of barriers to treatment adherence. Regardless of the reason, care coordination services and referrals were provided to all patients who either did not engage in Crisis Care or disengaged at any point in treatment. Given the portion of patients who experience difficulties adhering to care, it is recommended that care coordination services be included in specialty suicide treatment programs.

Reductions in depressive symptoms, anxiety symptoms, suicidal ideation frequency, and suicide specific risk factors were measured for all patient subgroups (i.e., those who graduated, discharged without graduating and those who remained in care). These findings may be consistent with evidence that even single-session brief interventions to address suicide risk can be effective in reducing clinical symptoms and supporting safety [54].

This study was limited to a preliminary investigation of the feasibility and clinical effectiveness of the Crisis Care program. While efforts were made to track patient status, there was limited visibility as to why some patients discontinued care or were less engaged in treatment. Future studies focused on evaluating reasons for engagement in care will be beneficial. This study did not track clinical outcomes beyond the 16-week mark, and additional research evaluating the long-term results of the CAMS intervention administered within a stepped-care telehealth model should be explored. Finally, the clinical outcomes are preliminary and warrant further investigation using more rigorous study designs and analytic methods, such as a randomized controlled trial to determine the extent to which clinical symptom reduction is attributable solely to intervention components. However, only two patients in the study required referral to a higher level of care, which suggests that telehealth treatment of suicide risk in an outpatient setting is safe, feasible, and shows significant promise in effectiveness. Additional research to further evaluate the clinical effectiveness of suicide-specific telehealth treatment using inferential statistics is recommended to expand on these encouraging preliminary results.

5. Conclusions

This work extends beyond and contributes to the existing literature in a number of ways. To our knowledge, this is the first study to examine the feasibility of treating a cohort of behavioral health patients using a telehealth-adapted version of CAMS within an established mental health services delivery model. Previous works have offered recommendations for telehealth adaptations to the CAMS framework [45–48], and the Crisis Care program has implemented these effectively within an established telehealth care model. In doing so, this study demonstrates feasible application of recommended adaptations to evidence-based treatment of suicide risk within a telehealth setting and offers a potential framework to scale such efforts and increase critical care options for people at risk of suicide.

Author Contributions: Conceptualization, M.W., J.C.O. and E.T.O.; methodology, J.C.O. and S.S.; software, S.S.; validation, S.S. and J.C.O.; formal analysis, S.S.; investigation, S.S.; resources, J.C.O.; data curation, S.S.; writing—original draft preparation, J.C.O.; writing—review and editing, J.C.O., M.W., E.T.O. and S.S.; visualization, J.C.O.; supervision, M.W. and E.T.O.; project administration, J.C.O., E.T.O. and M.W.; funding acquisition, M.W. All authors have read and agreed to the published version of the manuscript.

Funding: The authors declare that this study received funding from Brightside Health. The authors are employees of Brightside Health, but aside from employment status the funder was not involved in the study design, interpretation of data, or the decision to submit to for publication.

Institutional Review Board Statement: This study was conducted according to the guidelines of the Declaration of Helsinki and approved by the Institutional Review Board of WCG, protocol code 1308524 and date of approval: 1 May 2023.

Informed Consent Statement: Informed consent was obtained for all subjects involved in this study.

Data Availability Statement: The datasets analyzed for the current study are available from the corresponding authors upon reasonable request.

Conflicts of Interest: O'Neill, O'Callaghan, Winsberg, and Sullivan all hold stock in Brightside Health, Inc. and are all employees of Brightside Health, Inc. The authors declare that this study received funding from Brightside Health. The authors are employees of Brightside Health, but aside from employment status the funder was not involved in the study design, interpretation of data, or the decision to submit to for publication.

References

1. Centers for Disease Control and Prevention. Web-Based Injury Statistics Query and Reporting System (WISQARS) Fatal Injury Reports. 2020. Available online: https://webappa.cdc.gov/sasweb/ncipc/mortrate.html (accessed on 19 October 2023).
2. Centers for Disease Control and Prevention. Suicide Data and Statistics. 2020. Available online: https://www.cdc.gov/suicide/suicide-data-statistics.html (accessed on 19 October 2023).
3. Centers for Disease Control and Prevention. Suicide Data and Statistics. 2020. Available online: https://www.cdc.gov/injury/wisqars/fatal/trends.html (accessed on 19 October 2023).
4. Kessler, R.; Galea, S.; Gruber, M.; Sampson, N.; Ursano, R.; Wessely, S. Trends in mental illness and suicidality after Hurricane Katrina. *Mol. Psychiatry* **2008**, *13*, 374–384. [CrossRef]
5. Nuttman-Shwartz, O.; Lebel, U.; Avrami, S.; Volk, N. Perceptions of suicide and their impact on policy, discourse and welfare. *Eur. J. Soc. Work* **2010**, *13*, 375–392. [CrossRef]
6. Hubers, A.A.M.; Moaddine, S.; Peersmann, S.H.M.; Stijnen, T.; van Duijn, E.; van der Mast, R.C.; Dekkers, O.M.; Giltay, E.J. Suicidal ideation and subsequent completed suicide in both psychiatric and non-psychiatric populations: A meta-analysis. *Epidemiol. Psychiatr. Sci.* **2018**, *27*, 186–198. [CrossRef] [PubMed]
7. Jobes, D.A.; Joiner, T.E. Reflection on suicidal ideation. *Crisis* **2019**, *40*, 227–230. [CrossRef] [PubMed]
8. Nock, M.K.; Borges, G.; Bromet, E.J.; Cha, C.B.; Kessler, R.C.; Lee, S. Suicide and suicidal behavior. *Epidemiol. Rev.* **2008**, *30*, 133–154. [CrossRef] [PubMed]

9. O'Connor, R.C.; Kirtley, O.J. The integrated motivational-volitional model of suicidal behaviour. *Philos. Trans. R. Soc. Lond. Ser. B Biol. Sci.* **2018**, *373*, 20170268. [CrossRef]

10. National Action Alliance for Suicide Prevention. Best Practices in Care Transitions for Individuals with Suicide Risk: Inpatient Care to Outpatient Care. 2019. Available online: https://www.samhsa.gov/sites/default/files/suicide-risk-practices-in-care-transitions-11192019.pdf (accessed on 19 October 2023).

11. Kuehn, B.M. Rising emergency department visits for suicidal ideation and self-harm. *JAMA* **2020**, *323*, 917. [CrossRef]

12. Chung, D.; Hadzi-Pavlovic, D.; Wang, M.; Swaraj, S.; Olfson, M.; Large, M. Meta-analysis of suicide rates in the first week and the first month after psychiatric hospitalisation. *BMJ Open* **2019**, *9*, e023883. [CrossRef]

13. Forte, A.; Buscajoni, A.; Fiorillo, A.; Pompili, M.; Baldessarini, R.J. Suicidal risk following hospital discharge: A review. *Harv. Rev. Psychiatry* **2019**, *27*, 209–216. [CrossRef]

14. Chung, D.T.; Ryan, C.J.; Hadzi-Pavlovic, D.; Singh, S.P.; Stanton, C.; Large, M.M. Suicide rates after discharge from psychiatric facilities: A systematic review and meta-analysis. *JAMA Psychiatry* **2017**, *74*, 694–702. [CrossRef]

15. Walter, F.; Carr, M.J.; Mok, P.L.H.; Antonsen, S.; Pedersen, C.B.; Appleby, L.; Fazel, S.; Shaw, J.; Webb, R.T. Multiple adverse outcomes following first discharge from inpatient psychiatric care: A national cohort study. *Lancet Psychiatry* **2019**, *6*, 582–589. [CrossRef]

16. National Committee for Quality Assurance. Follow-up after Hospitalization for Mental Illness (FUH). Available online: https://www.ncqa.org/hedis/measures/follow-up-after-hospitalization-for-mental-illness/ (accessed on 19 October 2023).

17. Walby, F.A.; Myhre, M.Ø.; Kildahi, A.T. Contact with mental health services prior to suicide: A systemic review and meta-analysis. *Psychiatr. Serv.* **2018**, *69*, 737–839. [CrossRef]

18. Vahratian, A.; Blumberg, S.J.; Terlizzi, E.P.; Schiller, J.S. Symptoms of anxiety or depressive disorder and use of mental health care among adults during the COVID-19 pandemic—United States, August 2020-February 2021. *MMWR Morb. Mortal. Wkly. Rep.* **2021**, *70*, 490–494. [CrossRef]

19. Tang, S.; Reily, N.M.; Arena, A.F.; Batterham, P.J.; Calear, A.L.; Carter, G.L.; Mackinnon, A.J.; Christensen, H. People who die by suicide without receiving mental health services: A systematic review. *Front. Public Health* **2022**, *9*, 736948. [CrossRef] [PubMed]

20. Goldstein, E.V.; Brenes, F.; Wilson, F.A. Critical gaps in understanding firearm suicide in Hispanic communities: Demographics, mental health, and access to care. *Health Aff. Sch.* **2023**, *1*, qxad016. [CrossRef]

21. Torok, M.; Han, J.; Baker, S.; Werner-Seidler, A.; Wong, I.; Larsen, M.E.; Christensen, H. Suicide prevention using self-guided digital interventions: A systematic review and meta-analysis of randomised controlled trials. *Lancet Digit. Health* **2020**, *2*, e25–e36. [CrossRef] [PubMed]

22. Zerio Suicide. Least Restrictive Care. Available online: https://zerosuicide.edc.org/toolkit/treat/least-restrictive-care (accessed on 19 October 2023).

23. Hung, P.; Busch, S.; Shih, Y.W.; McGregor, A.J.; Wang, S. Changes in community mental health services availability and suicide mortality in the US: A retrospective study. *BMC Psychiatry* **2020**, *20*, 188. [CrossRef] [PubMed]

24. Ahmedani, B.K.; Vannoy, S. National pathways for suicide prevention and health services research. *Am. J. Prev. Med.* **2014**, *47*, 222–228. [CrossRef] [PubMed]

25. Mok, K.; Riley, J.; Rosebrock, H.; Gale, N.; Nicolopoulos, A.; Larsen, M.; Armstrong, S.; Heffernan, C.; Laggis, G.; Torok, M.; et al. *The Lived Experience of Suicide: A Rapid Review*; Black Dog Institute: Sydney, NSW, Australia, 2020. Available online: https://www.suicidepreventionaust.org/wp-content/uploads/2020/11/The-lived-experience-perspective-of-suicide-A-rapid-review.pdf (accessed on 19 October 2023).

26. Siegel, A.; Zuo, Y.; Moghaddamcharkari, N.; McIntyre, R.S.; Rosenblat, J.D. Barriers, benefits and interventions for improving the delivery of telemental health services during the coronavirus disease 2019 pandemic: A systematic review. *Curr. Opin. Psychiatry* **2021**, *34*, 434–443. [CrossRef] [PubMed]

27. Cuthbert, K.; Parsons, E.M.; Smith, L.; Otto, M.W. Acceptability of telehealth CBT during the time of COVID-19: Evidence from patient treatment initiation and attendance records. *J. Behav. Cogn. Ther.* **2022**, *32*, 67–72. [CrossRef]

28. Bulkes, N.Z.; Davis, K.; Kay, B.; Riemann, B.C. Comparing efficacy of telehealth to in-person mental health care in intensive-treatment-seeking adults. *J. Psychiatr. Res.* **2022**, *145*, 347–352. [CrossRef] [PubMed]

29. Fernandez, E.; Woldgabreal, Y.; Day, A.; Pham, T.; Gleich, B.; Aboujaoude, E. Live psychotherapy by video versus in-person: A meta-analysis of efficacy and its relationship to types and targets of treatment. *Clin. Psychol. Psychother.* **2021**, *28*, 1535–1549. [CrossRef] [PubMed]

30. Gilmore, A.K.; Ward-Ciesielski, E.F. Perceived risks and use of psychotherapy via telemedicine for patients at risk for suicide. *J. Telemed. Telecare* **2019**, *25*, 59–63. [CrossRef] [PubMed]

31. Linehan, M.M.; Korslund, K.E.; Harned, M.S.; Gallop, R.J.; Lungu, A.; Neacsiu, A.D.; McDavid, J.; Comtois, K.A.; Murray-Gregory, A.M. Dialectical behavior therapy for high suicide risk in individuals with borderline personality disorder: A randomized clinical trial and component analysis. *JAMA Psychiatry* **2015**, *72*, 475–482. [CrossRef]

32. Brown, G.K.; Ten Have, T.; Henriques, G.R.; Xie, S.X.; Hollander, J.E.; Beck, A.T. Cognitive therapy for the prevention of suicide attempts: A randomized controlled trial. *JAMA* **2005**, *294*, 563–570. [CrossRef]

33. Wenzel, A.; Jager-Hyman, S. Cognitive therapy for suicidal patients: Current status. *Behav. Ther.* **2012**, *35*, 121–130.

34. Jobes, D.A. *Managing Suicidal Risk: A Collaborative Approach*; Guilford Press: New York, NY, USA, 2006.

35. Jobes, D.A. The Collaborative Assessment and Management of Suicidality (CAMS): An evolving evidence-based clinical approach to suicidal risk. *Suicide Life-Threat. Behav.* **2012**, *42*, 640–653. [CrossRef]

36. Swift, J.K.; Trusty, W.T.; Penix, E.A. The effectiveness of the Collaborative Assessment and Management of Suicidality (CAMS) compared to alternative treatment conditions: A meta-analysis. *Suicide Life-Threat. Behav.* **2021**, *51*, 882–896. [CrossRef]

37. Jobes, D.A.; Gregorian, M.J.; Colborn, V.A. A stepped care approach to clinical suicide prevention. *Psychol. Serv.* **2018**, *15*, 243–250. [CrossRef]

38. Comtois, K.A.; Jobes, D.A.; O'Connor, S.S.; Atkins, D.C.; Janis, K.; Chessen, C.E.; Landes, S.J.; Holen, A.; Yuodelis-Flores, C. Collaborative assessment and manage of suicidality (CAMS): Feasibility trial for next day appointment service. *Depress. Anxiety* **2011**, *28*, 963–972. [CrossRef]

39. Andreasson, K.; Krogh, J.; Wenneberg, C.; Jessen, H.K.; Krakauer, K.; Gluud, C.; Thomsen, R.R.; Randers, L.; Nordentoft, M. Effectiveness of dialectical behavior therapy versus collaborative assessment and management of suicidality treatment for reduction of self-harm in adults with borderline personality traits and disorders: A randomized observer-blinded clinical trial. *Depress. Anxiety* **2016**, *33*, 520–530. [CrossRef]

40. Jobes, D.A.; Comtois, K.A.; Gutierrez, P.M.; Brenner, L.A.; Huh, D.; Chalker, S.A.; Ruhe, G.; Kerbrat, A.H.; Atkins, D.C.; Jennings, K.; et al. A randomized controlled trial of the collaborative assessment and management of suicidality versus enhanced care as usual with suicidal soldiers. *Psychiatry* **2017**, *80*, 339–356. [CrossRef]

41. Ryberg, W.; Zahl, P.H.; Diep, L.M.; Landrø, N.I.; Fosse, R. Managing suicidality within specialized care: A randomized controlled trial. *J. Affect. Disord.* **2019**, *249*, 112–120. [CrossRef]

42. Pistorello, J.; Jobes, D.A.; Gallop, R.; Compton, S.N.; Locey, N.S.; Au, J.S.; Noose, S.K.; Walloch, J.C.; Johnson, J.; Young, M.; et al. A randomized controlled trial of the collaborative assessment and management of suicidality (CAMS) versus treatment as usual (TAU) for suicidal college students. *Arch. Suicide Res.* **2021**, *25*, 765–789. [CrossRef]

43. Comtois, K.A.; Hendricks, K.E.; DeCou, C.R.; Chalker, S.A.; Kerbrat, A.H.; Crumlish, J.; Huppert, T.K.; Jobes, D. Reducing short term suicide risk after hospitalization: A randomized controlled trial of the Collaborative Assessment and Management of Suicidality. *J. Affect. Disord.* **2023**, *320*, 656–666. [CrossRef]

44. Santel, M.; Neuner, F.; Berg, M.; Steuwe, C.; Jobes, D.A.; Driessen, M.; Beblo, T. The collaborative assessment and management of suicidality compared to enhanced treatment as usual for inpatients who are suicidal: A randomized controlled trial. *Front. Psychiatry* **2023**, *14*, 1038302. [CrossRef]

45. Waltman, S.H.; Landry, J.M.; Pujol, L.A.; Moore, B.A. Delivering evidence-based practices via telepsychology: Illustrative case series from military treatment facilities. *Prof. Psychol. Res. Pract.* **2020**, *51*, 205–213. [CrossRef]

46. Marshall, E.; York, J.; Magruder, K.; Yeager, D.; Knapp, R.; De Santis, M.L.; Burriss, L.; Mauldin, M.; Sulkowski, S.; Pope, C.; et al. Implementation of online suicide-specific training for VA providers. *Acad. Psychiatry* **2014**, *38*, 566–574. [CrossRef]

47. Tipton, M.V.; Brenner, J.; Crumlish, J.; Moore, M.; Jobes, D.A. Online suicide-focused treatment: The telehealth use of CAMS. In *Advances in Online Therapy: Emergence of a New Paradigm*, 1st ed.; Weinberg, H., Rolnick, A., Leighton, A., Eds.; Routledge: New York, NY, USA, 2022. [CrossRef]

48. Jobes, D.A.; Crumlish, J.A.; Evans, A.D. The COVID-19 pandemic and treating suicidal risk: The telepsychotherapy use of CAMS. *J. Psychother. Integr.* **2020**, *30*, 226–237. [CrossRef]

49. Kroenke, K.; Spitzer, R.L.; Williams, J.B. The PHQ-9: Validity of a brief depression severity measure. *J. Gen. Intern. Med.* **2001**, *16*, 606–613. [CrossRef]

50. Spitzer, R.L.; Kroenke, K.; Williams, J.B.W.; Löwe, B. A brief measure for assessing Generalized Anxiety Disorder: The GAD-7. *Arch. Intern. Med.* **2006**, *166*, 1092–1097. [CrossRef] [PubMed]

51. Jobes, D.A.; Jacoby, A.M.; Cimbolic, P.; Hustead, L.A.T. Assessment and treatment of suicidal clients in a university counseling center. *J. Couns. Psychol.* **1997**, *44*, 368–377. [CrossRef]

52. Conrad, A.K.; Jacoby, A.M.; Jobes, D.A.; Lineberry, T.W.; Shea, C.E.; Arnold Ewing, T.D.; Schmid, P.J.; Ellenbecker, S.M.; Lee, J.L.; Fritsche, K.; et al. A psychometric investigation of the Suicide Status Form II with a psychiatric inpatient sample. *Suicide Life-Threat. Behav.* **2009**, *39*, 307–320. [CrossRef]

53. Brausch, A.; O'Connor, S.S.; Powers, J.T.; McClay, M.M.; Gregory, J.A.; Jobes, D.A. Validating the suicide status form for the collabortive assessment and management of suicidality in a psychiatric adolescent sample. *Suicide Life Threat. Behav.* **2019**, *50*, 263–276. [CrossRef]

54. Doupnik, S.K.; Rudd, B.; Schmutte, T.; Worsley, D.; Bowden, C.F.; McCarthy, E.; Eggan, E.; Bridge, J.A.; Marcus, S.C. Association of suicide prevention interventions with subsequent suicide attempts, linkage to follow-up care, and depression symptoms for acute care settings: A systematic review and meta-analysis. *JAMA Psychiatry* **2020**, *77*, 1021–1030. [CrossRef]

 *healthcare* 

Article

Relationship between Subjective Health, the Engel Coefficient, Employment, Personal Assets, and Quality of Life for Korean People with Disabilities

Kyung-A Sun [1] and Joonho Moon [2,*]

[1] Department of Tourism Management, Gachon University, Seongnam 13120, Republic of Korea; kasun@gachon.ac.kr
[2] Department of Tourism Administration, Kangwon National University, Chuncheon 24341, Republic of Korea
* Correspondence: joonhomoon0412@gmail.com

Abstract: The aim of this research is to examine the effect of subjective health on the quality of life of Korean people with disabilities. The second goal of this study is to examine the effect of the Engel coefficient on quality of life. Additionally, this study is conducted to inspect the effect of employment and personal assets on quality of life. Further, in this work, the moderating effect of personal assets on the association between employment and quality of life for people with a disability is explored. The Panel Survey of Employment for the Disabled served as the source of data. The study period ranges from 2016 to 2018. To test the research hypotheses, this study adopted econometric analyses, namely, ordinary least squares, fixed effect, and random effect models. The results revealed that the quality of life for people with disabilities is positively influenced by subjective health, employment, and personal assets. In contrast, the Engel coefficient exerts a negative impact on quality of life. Plus, the finding indicates that personal assets negatively moderate the relationship between employment and quality of life for people with disabilities. This research is aimed at presenting policy implications for the welfare of people with disabilities.

Keywords: quality of life; subjective health; the Engel coefficient; employment; personal assets; people with disability

Citation: Sun, K.-A.; Moon, J. Relationship between Subjective Health, the Engel Coefficient, Employment, Personal Assets, and Quality of Life for Korean People with Disabilities. *Healthcare* **2023**, *11*, 2994. https://doi.org/10.3390/healthcare11222994

Academic Editors: Carlos Laranjeira, Ana Querido and Feng Gao

Received: 6 October 2023
Revised: 15 November 2023
Accepted: 17 November 2023
Published: 19 November 2023

1. Introduction

The definition of a person with disabilities in Korea is an individual who is limited in social life due to both mental and physical disabilities [1]. According to Statistics Korea [2], the number of disabled Koreans is approximately 2.65 million, and the population size of disabled Koreans has been increasing steadily since 2018. Such growth indicates that the proportion of disabled individuals in Korea has increased. Hence, policy design for people with disabilities is likely to become an even more imperative issue in Korean society because the welfare budget for people with disabilities is constrained. To provide a guideline for policy making aimed at people with disabilities, this research is conducted to inspect the characteristics of people with disabilities. The dependent variable of this work is quality of life. Scholars have defined quality of life as an indicator of individual welfare and happiness [3,4]. Numerous studies have also scrutinized the influential attributes on quality of life [5–8]. Such bountiful works implied that investigating the characteristics of quality of life is valuable. Additionally, quality of life is likely to function as an attribute when determining the characteristics of the disabled.

In this research, the determinants of quality of life were identified as subjective health, the Engel coefficient, employment, and personal assets. Previous studies have uncovered the significant and positive effect of subjective health on quality of life, arguing that healthy conditions are indispensable for improving quality of life [9–11]. This study is thus aimed at ensuring the effect of subjective health on quality of life in the domain of people with disabilities. Next, this study inspects the impact of the Engel coefficient on quality of life. The Engel coefficient serves as evidence of poor life conditions because a higher Engel coefficient indicates an insufficient budget for other areas of life, such as hobbies, recreation, education, and medical services [12–14]. Despite this argument, few studies have demonstrated the effect of the Engel coefficient on quality of life in the case of individuals with disabilities. This research is aimed at minimizing such a research gap by presenting empirical evidence. The third area of this research is employment. The extant literature has shown that employment plays a significant role in ensuring a better life because it satisfies human needs, such as social needs, achievement needs, esteem, and financial gain [15–17]. From this perspective, this study examines the effect of employment on the quality of quality of life of people with disabilities. In addition, this research uses personal assets as a determinant of quality of life because scholars contended that wealth is an essential element in improving individual quality of life [18,19]. Furthermore, this research tests the moderating effect of personal assets on the relationship between employment and quality of life. Labor might exert varied effects on individual life depending on the relevant wealth condition. Specifically, earning from labor is likely to be more desperate for poor individuals than for wealthy individuals because it is directly linked with survival. Namely, the meaning of labor could vary depending on the condition of wealth, and personal assets enable the approximation of individual wealth conditions in this study. The moderating effect of wealth, which clarifies the meaning of labor in the case of Korean people with disabilities, is also scrutinized.

This research sheds light on the literature by elucidating the moderating effect of personal assets on the association between employment and quality of life. This is because prior works exploring the effects of disabilities have rarely assessed such an effect. Moreover, this research is valuable in that it presents empirical evidence on the effects of subjective health, the Engel coefficient, and employment on the quality of life of the disabled. Such efforts might enable confirming the findings in the extant literature. The outcomes of this work are likely to lead policy makers to design a more adequate welfare system for people with disabilities.

2. Review of the Literature and Hypothesis Development

2.1. Quality of Life

Prior studies have addressed the fact that quality of life is an individual evaluation of individual's current life conditions and their level of satisfaction with their living condition [3–5]. Moreover, prior studies addressed that quality of life reflects the overall satisfaction of individual living [20,21]. Numerous works have adopted quality of life as the main attribute when determining individual behavior. Guida and Carpentieri [22] explored the quality of life using Milan city residents. Zhang and Ma [8] inspected the quality of life by employing Chinese participants. Ravens-Sieberer et al. [7] when researching German primary school students, revealed the antecedents of quality of life. Alsubaie et al. [23] and Berdida and Grande [24] examined influential attributes of quality of life for university students. Uysal and Sirgy [25] investigated the characteristics of the life quality of travelers. Additionally, Moon et al. [26] used quality of life as a dependent variable to scrutinize the behavior of Korean elderly individuals. Moreover, many prior studies have empirically explored the characteristics of people with a disability using quality of life as a principal indicator [27–29]. Specifically, Steptoe and Di Gessa [28] inspected the determinants of quality of life by employing people with physical disabilities; Verdugo et al. [29] implemented empirical research to identify the determinants of quality of life by studying people with intellectual disabilities. Given the numerous studies on quality of life, one can see that quality of life is an imperative attribute in various domains.

2.2. Subjective Health and Quality of Life

Subjective health is the degree of mental and physiological healthiness assessed from one's own point of view [9,11,30]. Subjective health is an essential aspect of better living because healthiness is a precondition for improving individuals' lives [31,32]. A vast body of the literature has empirically demonstrated the effect of subjective health on quality of life. For instance, Qazi et al. [33] showed the positive impact of subjective health on quality of life by researching older women. Miyakawa and Hamashima [34], while researching high school students, suggested that subjective health exerted a positive effect on quality of life. Moon et al. [26] also found a positive association between subjective health and quality of life for Korean senior citizens. In addition, Low et al. [35] explored older people in 20 countries, and their results documented the fact that quality of life is positively affected by subjective health. Next, Skevington et al. [36] examined cross-cultural data, and the results indicated that subjective health is a significant antecedent for quality of life. Also, Kim et al. [10] exposed a significant and positive link between the subjective health of students and their quality of life. Plus, Heyne et al. [37] found that the quality of life for patients is significantly influenced by subjective health conditions. From the literature review, it can be inferred that subjective health is likely to become an essential attribute in assessing individual quality of life. Therefore, the following hypothesis is proposed:

Hypothesis 1. *Subjective health positively impacts quality of life.*

2.3. Engel Coefficient and Quality of Life

The Engel coefficient refers to the proportion of food cost to the total cost of living [6,13]. A low Engel coefficient means that individuals possess more surplus money to make life better because food expenditure is an opportunity cost that coexists with other living-related budgets [14,38]. Ma et al. [39] alleged that the Engel coefficient lowers the overall living quality because of the constrained resource conditions. Yang et al. [13] and Xie et al. [12] also argue that a high Engel coefficient reflects significantly worse individual living conditions because a high proportion of food costs causes resource constraints and increases the likelihood of diseases such as obesity, diabetes, high blood pressure, and heart attack. Also, Shi et al. [40] and Qin et al. [41] showed that the Engel coefficient negatively impacted on quality of life by inspecting the Chinese population. Moon et al. [6] revealed that a higher Engel coefficient reflects a negative impact on the quality of life of Korean elderly individuals. Furthermore, regarding the literature review, it is presumed that a higher Engel coefficient is likely to reflect a more negative quality of life. Thus, the following hypothesis is proposed:

Hypothesis 2. *A higher Engel coefficient reflects a poorer quality of life.*

2.4. Employment and Quality of Life

Prior research has documented the fact that employment is the status of working with economic value [42–44]. Working is an instrument for improving quality of life because it provides financial rewards and opportunities to satisfy higher levels of needs, such as social needs, esteem, and self-actualization [15,16,45]. Many studies have shown the effect of employment on quality of life. For example, Carlier et al. [46], when examining Dutch individuals, indicated that quality of life is enhanced by labor. Blalock et al. [47] also documented that employment played an essential function in the improvement of life quality when studying patients with serious diseases. In a similar vein, Kim and Feldman [48] revealed a positive relationship between quality of life and employment by researching retired people. Kober and Eggleton [44] implemented research by employing persons with disability, the findings documented that quality of life is improved by employment. Plus, Beyer et al. [49] disclosed that the quality of life for people with intellectual disabilities was enhanced by employment. Cocks et al. [17] found the significant impact of employment on the living quality of people with disabilities because of various elements: benefits from work, social connection, and job satisfaction. Additionally, previous studies have shown that the employment of people with disabilities is an imperative attribute for the enhancement of their life quality [17,50]. With respect to the literature review, the following hypothesis is proposed:

Hypothesis 3. *Employment positively impacts quality of life.*

2.5. Personal Assets and Quality of Life

Wealth is indispensable for living because wealth allows individuals to consume products and services [18,19,51]. Wealthier people are more likely to consume better-quality services and goods because they have greater financial power [52–54]. Luburić and Fabris [20] contended that wealth is a critical element for better living. Diwan [18] also claimed that wealth plays a significant attribute in a better life because it can cause emotional security in an individual. An et al. [55] uncovered the fact that financial status is positively associated with quality of life. Xiao et al. [56] showed that students' quality of life is positively influenced by financial power. Zafar et al. [19] conversely demonstrated that financial burden exerted a negative impact on quality of life. Regarding the literature review, the following research hypothesis is proposed:

Hypothesis 4. *Personal assets positively impact quality of life.*

2.6. Moderating Effect of Personal Assets on the Relationship between Employment and Quality of Life

The law of diminishing marginal utility posits that individuals gain a higher level of utility in consumption when their resources are scarce [57,58]. Kober and Eggleton [44] demonstrated that the quality of life is influenced in varied manners by the situation of workers by scrutinizing people with disability. Gautié and Schmitt [59] documented that the meaning of labor appeared in different manners depending on the wealth condition because labor is compulsory for people with lower levels of wealth. Also, labor functions to provide the money necessary for living [43,44], and this resource is likely to exert a stronger impact on quality of life in the case of individuals with fewer assets, according to the law of diminishing marginal utility [60,61]. Namely, the impact of labor is likely to vary depending on the relevant wealth condition. Thus, the following hypothesis is proposed:

Hypothesis 5. *Personal assets negatively moderate the relationship between employment and quality of life.*

3. Methods

3.1. Research Model and Data Collection

Figure 1 exhibits the research model. The dependent variable is quality of life. Quality of life is positively impacted by subjective health. The Engel coefficient negatively affects quality of life. Employment is also positively associated with quality of life. Additionally, personal assets exert a positive effect on quality of life, and personal assets further moderate the relationship between labor and quality of life.

The data collection was implemented using the Panel Survey of Employment for the Disabled, which was published by the Employment Development Institute. The Panel Survey of Employment for people with disabilities provides researchers with survey information about people with disabilities. The study period spanned from 2016 to 2018. After 2018, the Employment Development Institute did not continue to report such survey information. The data appeared as panel data which consists of multiple periods and times [62]. Panel data refer to data that references multiple participants and multiple time points. In the case of this study, the data appeared as unbalanced panels because all participant information was not matched throughout the entire study period [62,63]. The initial observation of this research was 12,225. In the data cleaning process,836 observations were eliminated because of no response in the dataset. Thus, the total number of valid observations for data analysis was 11,389.

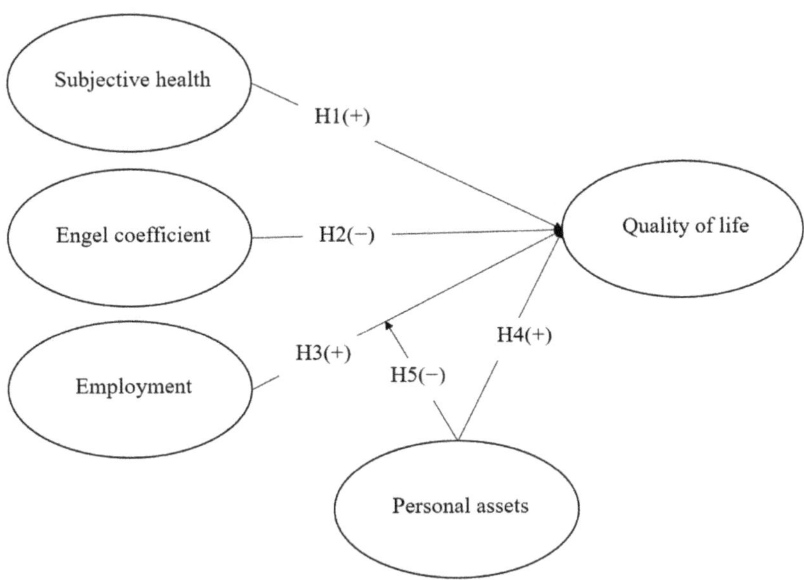

Figure 1. Research model.

3.2. Variable Illustration and Analytic Instrument

Table 1 describes the measurements of the variables. Quality of life (QL) was measured with a five-point scale (from 1 = very dissatisfied to 5 = very satisfied). The question for QL is 'how are you satisfied with your living?' The measurement of subjective health (SH) is a four-point scale (from 1 = very poor to 4 = very good). The question is 'how do you assess your overall health condition?'. The Engel coefficient is measured by monthly food expenditure over monthly total living expenses. Employment (EM) appeared as a binary variable (0 = unemployed, 1 = employed). Personal asset (PA) was measured by a natural log of the amount of wealth possessed by survey participants, and its unit was 10,000 KRW. With regard to the control variables, disability level (DL) (0 = mild, 1 = severe) and gender (GN) (0 = male, 1 = female) were both measured as dummy variables. Age was the physical age (AG) of the survey participant.

Table 1. Variable description (N = 11,389).

Name	Code	Description (Unit)
Quality of life	QL	(1 = very dissatisfied, 5= very satisfied)
Subjective health	SH	(1 = very poor, 4 = very good)
Engel coefficient	EG	Monthly food expense/Monthly total expense
Employment	EM	(0 = unemployed, 1 = employed)
Personal assets	PA	(Personal assets) (10,000 KRW)
Gender	GN	(0 = male, 1 = female)
Age	AG	Physical age of survey participants
Disability level	DL	(0 = mild, 1= severe)

Note: KRW denotes Korean won.

First, descriptive statistical analysis was conducted to compute the basic information of the study variables. A Spearman correlation matrix was chosen to examine the correlation coefficients between variables. To test the hypotheses, three econometric instruments, namely, ordinary least squares, fixed effects model, and random effects model, were applied. Ordinary least squares are a regression model that minimizes the sum of squared residual in the estimation [62,64]. The fixed effect model incorporates multiple time-related variables in the form of dummy variables into the model for the minimization of omitted variable bias in relation to time. Next, the random effect model refers to a regression model that contains unobservable effects in the estimation [63,64]. To test the moderating effect, two variables were multiplied, EM and PA (EM × PA). Then, the EM × PA variable was incorporated into the regression model. Furthermore, median split analysis (personal assets median value = 6000) was used to further scrutinize the direction of the moderating effect. Moreover, this work used three control variables (e.g., gender (GN), age (AG), and disability level (DL)) in the regression model. Plus, this research implemented the Lagrangian multiplier test and Hausman test to select the best model for panel data estimation. The results of both tests were not statistically significant. It suggests that ordinary least squares are the most appropriate econometric instrument for parameter estimation [62,63]. Moreover, this research performed ordinal logit analysis for the robustness check because it is more lenient for the normality assumption [62,63]. The regression equation model is as follows:

$$QL_{it} = \beta_0 + \beta_1\,SH_{it} + \beta_2\,EG_{it} + \beta_3\,EM_{it} + \beta_4\,PA_{it} + \beta_5\,EM \times PA_{it} + \beta_6\,GN_{it} + \beta_7\,AG_{it} + \beta_8\,DL_{it} + \varepsilon_{it} \tag{1}$$

where ε is the residual, i is the ith participants, t is the tth year, QL is quality of life, SH is subjective health, EG is the Engel coefficient, EM is Employment, PA is personal assets, GN is gender, AG is physical age, DL: disability level.

4. Main Findings and Discussion

4.1. Descriptive Statistics and Correlation Matrix

Table 2 illustrates the descriptive statistics. The number of observations is 11,389. The mean values (standard deviation) of QL, SH and EG are 3.27 (SD = 0.69), 2.48 (SD = 0.66), and 0.31 (SD = 0.13), respectively. Fifty percent of the survey participants were employed, as presented in Table 2. The mean and standard deviation of PA are 13,787.44 and 26,001.21, respectively. Table 2 also presents the information for GN (mean = 0.65, SD = 0.47), AG (mean = 43.71, SD = 12.60), and DL (mean = 0.32, SD = 0.46). The mean value of AGE is 43.71, and its standard deviation is 12.60, with minimum and maximum values of 15 and 66, respectively.

Table 3 exhibits the spearman correlation matrix. QL is positively correlated with SH (r = 0.429, $p < 0.05$), EM (r = 0.343, $p < 0.05$), PA (r = 0.348, $p < 0.05$), and GN (r = 0.036, $p < 0.05$). However, QL negatively correlated with EG (r = −0.232, $p < 0.05$), AG (r = −0.122, $p < 0.05$), and DL (r = −0.165, $p < 0.05$). Next, SH is positively correlated with EM (r = 0.367, $p < 0.05$), PA (r = 0.281, $p < 0.05$) and GN (r = 0.096, $p < 0.05$), while SH is negatively correlated with EG (r = −0.228, $p < 0.05$), AG (r = −0.266, $p < 0.05$), and DL (r = −0.140, $p < 0.05$). Additionally, EG is negatively correlated with EM (r = −0.295, $p < 0.05$) and PA (r = −0.355, $p < 0.05$, whereas it is positively correlated with AG (r = 0.102, $p < 0.05$) and DL (r = 0.156, $p < 0.05$).

Table 2. Descriptive statistics (N = 11,389).

Variable	Mean	SD	Minimum	Maximum
QL	3.27	0.69	1	5
SH	2.48	0.66	1	4
EG	0.31	0.13	0	1
EM	0.50	0.50	0	1
PA	13,787.44	26,001.21	0	600,000
GN	0.65	0.47	0	1
AG	43.71	12.60	15	66
DL	0.32	0.46	0	1

Note: SD denotes standard deviation, QL: quality of life, SH: subjective health, EG: Engel coefficient, EM: employment, PA: personal assets, GN: gender, AG: physical age, DL: disability level.

Table 3. Spearman correlation matrix.

Variable	1	2	3	4	5	6	7
1. QL	1						
2. SH	0.429 *	1					
3. EG	−0.232 *	−0.228 *	1				
4. EM	0.343 *	0.367 *	−0.295 *	1			
5. PA	0.348 *	0.281 *	−0.355 *	0.281 *	1		
6. GN	0.036 *	0.096 *	−0.018	0.220 *	0.034 *	1	
7. AG	−0.122 *	−0.266 *	0.102 *	0.017	−0.091 *	−0.067 *	1
8. DL	−0.165 *	−0.140 *	0.156 *	−0.305 *	−0.173 *	−0.041 *	−0.189 *

Note: * $p < 0.05$, QL: quality of life, SH: subjective health, EG: Engel coefficient, EM: employment, PA: personal assets, GN: gender, AG: physical age, DL: disability level.

4.2. Results of Hypotheses Testing

Table 4 shows the results of testing the hypotheses. The dependent variable is QL. All three econometric models are statistically significant regarding F-values and Wald χ^2 ($p < 0.05$). SH exerts a positive effect on QL ($\beta = 0.345$, $p < 0.05$). EG negatively affects QL ($\beta = -0.338$, $p < 0.05$), whereas EM positively impacts QL ($\beta = 0.270$, $p < 0.05$). It also appears that QL is positively influenced by PA ($\beta = 3.47 \times 10^{-6}$, $p < 0.05$). Moreover, the moderating variable EM $\times$ PA ($\beta = -8.14 \times 10^{-7}$, $p < 0.05$) exerts a negative effect on QL. Namely, PA negatively moderates the association between EM and QL. The results are consistent in all three models in terms of significance and direction. Therefore, all the proposed hypotheses are supported.

Table 4. Results of hypotheses testing using linear regression.

Variable	Model 1 (O) β (t-stat)	Model 2 (F) β (t-stat)	Model 3 (R) β (wald)
Intercept	2.502 (61.46) **	2.502 (61.29) **	2.502 (61.46) **
SH	0.345 (35.77) **	0.345 (35.78) **	0.345 (35.77) **
EG	−0.338 (−7.45) **	−0.338 (−7.43) **	−0.338 (−7.45) **
EM	0.270 (18.42) **	0.270 (18.42) **	0.270 (18.42) **
PA	3.47×10^{-6} (10.26) **	3.47×10^{-6} (10.27) **	3.47×10^{-6} (10.26) **
EM × PA	-8.14×10^{-7} (−1.83) *	-8.14×10^{-7} (−1.84) *	-8.14×10^{-7} (−1.83) *
GN	−0.062 (−5.14) **	−0.062 (−5.14) **	−0.062 (−5.14) **
AG	−0.002 (−4.28) **	−0.002 (−4.22) **	−0.002 (−4.28) **
DL	−0.077 (−5.88) **	−0.077 (−5.87) **	−0.077 (−5.88) **
F-value	491.36 *	491.41 *	
Wald χ^2			3930.88 *
R^2	0.2579	0.2579	0.2579

Note: Dependent variable: QL, * $p < 0.1$, ** $p < 0.05$, O is ordinary least square, F is fixed effect, R is random effect, QL: quality of life, SH: subjective health, EG: Engel coefficient, EM: employment, PA: personal assets, GN: gender, AG: physical age, DL: disability level.

Table 5 presents the results of testing the hypotheses using ordinal logit. The dependent variable is QL. All models are statistically significant regarding F-values and Wald χ^2 ($p < 0.05$). SH exerts a positive effect on QL ($\beta = 1.136$, $p < 0.05$). EG negatively affects QL ($\beta = -1.171$, $p < 0.05$), whereas EM positively impacts QL ($\beta = 0.860$, $p < 0.05$). It also appears that QL is positively influenced by PA ($\beta = 1.36 \times 10^{-5}$, $p < 0.05$). The results are consistent in all models in terms of significance and direction.

Table 5. Results of hypotheses testing using ordinal logit.

Variable	Model 4 (OL) β (t-stat)	Model 5 (OF) β (t-stat)
SH	1.136 (32.95) **	1.137 (32.96) **
EG	−1.171 (−7.66) **	−1.171 (−7.63) **
EM	0.860 (16.73) **	0.860 (18.42) **
PA	1.36×10^{-5} (9.31) **	1.37×10^{-5} (9.34) **
EM × PA	-1.59×10^{-6} (−0.86)	-1.61×10^{-6} (−1.87)
GN	−0.194 (−4.73) **	−0.194 (−4.74) **
AG	−0.006 (−4.10) **	−0.006 (−4.02) **
DL	−0.244 (−5.53) **	−0.243 (−5.51) **
LR χ^2	3319.91 *	3321.56 *
Psuedo R^2	0.1408	0.1409

Note: Dependent variable: QL, * $p < 0.1$, ** $p < 0.05$, OL is ordinal logit, OF is ordinal logit with fixed effect, QL: quality of life, SH: subjective health, EG: Engel coefficient, EM: employment, PA: personal assets, GN: gender, AG: physical age, DL: disability level.

Table 6 depicts the additional analysis to inspect the moderating effect of personal assets. The mean values of the high assets group and unemployed and that of the high assets group and employed are 3.26 and 3.61, respectively. Additionally, the mean value of the low assets group and the unemployed group is 2.88 with 0.70 as the standard deviation. Finally, the mean value of the low assets and the employed group is 3.34. Additionally, Figure 2 presents a graphical illustration of the moderating effect of personal assets.

Table 6. Results of moderating effect of personal assets.

Variable	Unemployed Mean (SD)	Employed Mean (SD)
High-assets group	3.26 (0.66)	3.61 (0.55)
Low-assets group	2.88 (0.70)	3.34 (0.61)

Note: Dependent variable: QL.

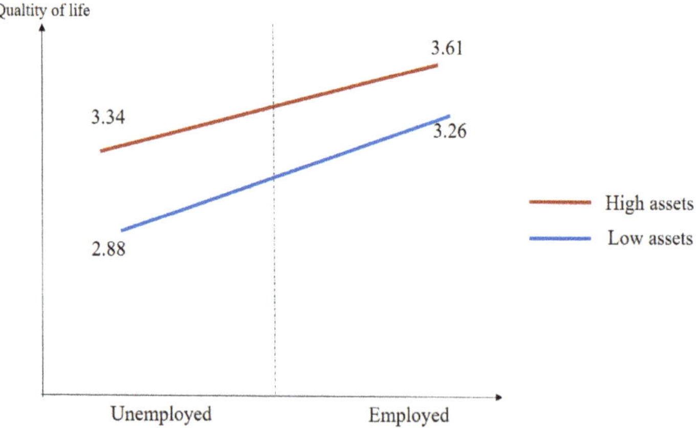

Figure 2. Graphical presentation for the results of moderating effect.

5. Conclusions

The purpose of this study is to examine the determinants of quality of life in people with disabilities in Korea. The results show that subjectively healthier participants are better in their lives. This implies that subjective health becomes a precondition for a better quality of life for people with disabilities in Korea. The findings of this research appeared similar compared with the research implemented by employing an international sample [35] and performing cross-cultural research [36]. It can be inferred that the positive effect of subjective health on quality of life is also crucial in the case of Korean people with disabilities. Moreover, the findings document the fact that the Engel coefficient reflects a negative effect on the quality of life for people with disability in Korea. Specifically, food costs could become a burden for people with disabilities from the perspective of achieving a higher quality of life. The results displayed a similar pattern in the Korean case as compared to the case researched by the Chinese population [40,41]. The results also reveal that employed participants had a better quality of life than unemployed participants. Namely, this suggests that employment plays an essential role in improving the quality of quality of life of people with disabilities. Additionally, the results indicate that wealth is a crucial attribute for a better quality of life for people with disabilities in Korea.

The crux of this work is the moderating effect of personal assets on the relationship between employment and quality of life. With regard to the moderating effect, it was found that the group with fewer assets was more sensitive to labor than the group with greater assets. That is, the gap between employment and unemployment was larger in the case of possessing fewer assets. Such a fact may be the reason for the negative moderating effect of personal assets on the relationship between employment and quality of life. It can also be inferred that labor might become more important in the case of those with disabilities who possess fewer assets for elevating quality of life. In other words, the value of labor could be varied whether it is compulsory or voluntary. The high-level personal assets group is more likely to become a voluntary group, whereas the low-level personal assets group is more likely to be a compulsory group. Given the moderating effect of personal assets, it could be inferred that working might become a necessary means of survival for the compulsory group.

From the results of including the control variables, it can be inferred that quality of life decreased for males, older individuals, and those with severe disabilities. It is aligned with the findings of previous research. In detail, Shim et al. [65] showed that people with severe disability showed worse appraisal for their living because their life is more constrained than people with mild disabilities. The findings of the extant literature also support the findings of this work in that individual quality of life varies by gender and age [26,27,65,66].

This research contributes to the literature by ensuring the moderating effect of personal assets on the association between employment and quality of life. Because labor plays a significant role in obtaining funds, the value of labor is likely to vary depending on individual financial conditions. However, few studies have examined the moderating effect of personal assets on the relationship between labor and quality of life. This study is thus aimed at minimizing this research gap, and this work unveils the significant link among variables. This achievement represents the value of this research by elucidating the association between variables. Scholars have argued that the Engel coefficient reflects an important effect on the quality of life of individuals because a higher Engel coefficient indicates a reduction in other living areas aside from food [15,16,26]. Despite such an argument, prior studies have rarely reported empirical evidence for the relationship between the Engel coefficient and the quality of life of individuals with disabilities. Hence, the next contribution of this research is to demonstrate the effect of the Engel coefficient on the quality of life of people with disabilities and thus to fill this research gap. In addition, this study contributes to the literature by documenting the significant impact of subjective health on quality of life, which supports the results of the extant literature [6,10,33]. Thus, this study externally validates the findings of previous studies in terms of the link between subjective health and quality of life.

This work has policy implications. First, government budgets for people with disabilities might need to be allocated to enhance both mental and physiological health conditions. Such budgets can enable investments in medical care, social work, and counseling care services for people with disabilities. Also, government resources could be allotted to outdoor activities for people with disabilities such as leisure and cultural activities because they can refresh the individual mental condition. Moreover, policy makers may be able to consider investing resources in supporting food expenditures because food costs hinder people with disabilities from spending their money on other areas and thus improving their quality of life. This could be accomplished by direct food aid and issuing food coupons for people with disabilities. Next, policy makers need to more efficiently allocate government money to job creation. In detail, policy makers might be able to reduce the unemployment rate of disabled individuals with disabilities by offering jobs with adequate wage levels. This could be accomplished by investigating the type of disabilities that are prevalent in the population because the potential for employment varies across types of disabilities. Such an action would enable policy makers to find ways to use constrained resources in a more effective manner. Furthermore, policy makers might be able to contemplate the resource for poor people with disabilities more because their labor is compulsory, which is likely to degrade the life quality. However, the criteria for detecting financially distressed people needs to be designed in a more delicate manner. Otherwise, the government budget is likely to be allocated inadequately.

This study has some limitations. First, the measurement of labor was a binary measurement. Future research might be able to contemplate various aspects of labor, such as salary level, job stability, and working conditions. This inclusion might be useful for attaining more concrete implications for policy design. Additionally, this study depended on archival data. Future research should consider directly surveying people with disabilities, which could become an avenue for acquiring more robust estimation results using more advanced statistical instruments. Last, the sample of this work is limited to the case of Korea. Future works might be able to consider cases from other nations; ensuring the link between variables could make the results of this study more generalizable.

Author Contributions: Formal analysis, K.-A.S.; writing—original draft, J.M.; writing—review and editing. All authors have read and agreed to the published version of the manuscript.

Funding: This research received no external funding.

Institutional Review Board Statement: Not applicable.

Informed Consent Statement: Not applicable.

Data Availability Statement: Data are contained within the article.

Conflicts of Interest: The authors declare no conflict of interest.

References

1. Disabilityin. Republic of Korea Disability Definition. 2023. Available online: https://disabilityin.org/country/republic-of-korea/ (accessed on 23 August 2023).
2. Statistics Korea. Nationwide Status of Disabled in Korea. 2023. Available online: https://kosis.kr/statHtml/statHtml.do?orgId=117&tblId=DT_11761_N003 (accessed on 26 August 2023).
3. Hargreaves, S.M.; Raposo, A.; Saraiva, A.; Zandonadi, R.P. Vegetarian diet: An overview through the perspective of quality of life domains. *Int. J. Environ. Res. Public* **2021**, *18*, 4067. [CrossRef]
4. Shek, D.T. COVID-19 and quality of life: Twelve reflections. *App. Res. Qual. Life* **2021**, *16*, 1–11. [CrossRef]
5. Barlow, M.A.; Wrosch, C.; McGrath, J.J. Goal adjustment capacities and quality of life: A meta-analytic review. *J. Pers.* **2020**, *88*, 307–323. [CrossRef] [PubMed]
6. Moon, J.; Hwang, J.; Lee, W.S. Relationship between the Engel coefficient, life satisfaction, and subjective health for senior citizens in Korea: Moderating effect of COVID-19. *Behav. Sci.* **2022**, *13*, 22. [CrossRef]
7. Ravens-Sieberer, U.; Kaman, A.; Erhart, M.; Devine, J.; Schlack, R.; Otto, C. Impact of the COVID-19 pandemic on quality of life and mental health in children and adolescents in Germany. *Eur. Child Adolesc. Psychiatry* **2022**, *31*, 879–889. [CrossRef]
8. Zhang, Y.; Ma, Z.F. Impact of the COVID-19 pandemic on mental health and quality of life among local residents in Liaoning Province, China: A cross-sectional study. *Int. J. Environ. Res. Public Health* **2020**, *17*, 2381. [CrossRef]
9. Elran-Barak, R.; Mozeikov, M. One month into the reinforcement of social distancing due to the COVID-19 outbreak: Subjective health, health behaviors, and loneliness among people with chronic medical conditions. *Int. J. Environ. Res. Public Health* **2020**, *17*, 5403. [CrossRef]
10. Kim, S.R.; Nho, J.H.; Kim, H.Y. Influence of Type D personality on quality of life in university students: The mediating effect of health-promoting behavior and subjective health status. *Psychol. Sch.* **2020**, *57*, 768–782. [CrossRef]

11. Poortinga, W.; Bird, N.; Hallingberg, B.; Phillips, R.; Williams, D. The role of perceived public and private green space in subjective health and wellbeing during and after the first peak of the COVID-19 outbreak. *Landsc. Urban Plan.* **2021**, *211*, 104092. [CrossRef] [PubMed]
12. Xie, Q.; Yi, F.; Tian, X. Disparate changes of living standard in China: Perspective from Engel's coefficient. *China Agric. Econ. Rev.* **2022**. *ahead-of-print.* [CrossRef]
13. Yang, B.; Xu, T.; Shi, L. Analysis on sustainable urban development levels and trends in China's cities. *J. Clean. Prod.* **2017**, *141*, 868–880. [CrossRef]
14. Wang, Z.; Yang, L. Indirect carbon emissions in household consumption: Evidence from the urban and rural area in China. *J. Clean. Prod.* **2014**, *78*, 94–103. [CrossRef]
15. Stum, D.L. Maslow revisited: Building the employee commitment pyramid. *Strat. Leadersh.* **2001**, *29*, 4–9. [CrossRef]
16. Benson, S.G.; Dundis, S.P. Understanding and motivating health care employees: Integrating Maslow's hierarchy of needs, training and technology. *J. Nurs. Mgt.* **2003**, *11*, 315–320. [CrossRef] [PubMed]
17. Cocks, E.; Thoresen, S.H.; Lee, E.A. Pathways to employment and quality of life for apprenticeship and traineeship graduates with disabilities. *Int. J. Disabil. Dev. Educ.* **2015**, *62*, 422–437. [CrossRef]
18. Diwan, R. Relational wealth and the quality of life. *J. Soc. Econ.* **2000**, *29*, 305–340. [CrossRef]
19. Zafar, S.Y.; McNeil, R.B.; Thomas, C.M.; Lathan, C.S.; Ayanian, J.Z.; Provenzale, D. Population-based assessment of cancer survivors' financial burden and quality of life: A prospective cohort study. *J. Oncol. Prac.* **2015**, *11*, 145–150. [CrossRef]
20. Luburić, R.; Fabris, N. Money and the quality of life. *J. Central Bank. Theory Prac.* **2017**, *6*, 17–34. [CrossRef]
21. Pukeliene, V.; Starkauskiene, V. Quality of life: Factors determining its measurement complexity. *Engi. Econ.* **2011**, *22*, 147–156. [CrossRef]
22. Guida, C.; Carpentieri, G. Quality of life in the urban environment and primary health services for the elderly during the COVID-19 pandemic: An application to the city of Milan (Italy). *Cities* **2021**, *110*, 103038. [CrossRef]
23. Alsubaie, M.M.; Stain, H.J.; Webster, L.A.; Wadman, R. The role of sources of social support on depression and quality of life for university students. *Int. J. Adolesc. Youth* **2019**, *24*, 484–496. [CrossRef]
24. Berdida, D.J.; Grande, R.A. Academic stress, COVID-19 anxiety, and quality of life among nursing students: The mediating role of resilience. *Int. Nur. Rev.* **2023**, *70*, 34–42. [CrossRef]
25. Uysal, M.; Sirgy, M.J. Quality-of-life indicators as performance measures. *Ann. Tour. Res.* **2019**, *76*, 291–300. [CrossRef]
26. Moon, J.; Lee, W.S.; Shim, J. Exploring Korean middle-and old-aged citizens' subjective health and quality of life. *Behav. Sci.* **2022**, *12*, 219. [CrossRef]
27. Nota, L.; Ferrari, L.; Soresi, S.; Wehmeyer, M. Self-determination, social abilities and the quality of life of people with intellectual disability. *J. Intellect. Disabil. Res.* **2007**, *51*, 850–865. [CrossRef]
28. Steptoe, A.; Di Gessa, G. Mental health and social interactions of older people with physical disabilities in England during the COVID-19 pandemic: A longitudinal cohort study. *Lancet Public Health* **2021**, *6*, e365–e373. [CrossRef]
29. Verdugo, M.A.; Navas, P.; Gómez, L.E.; Schalock, R.L. The concept of quality of life and its role in enhancing human rights in the field of intellectual disability. *J. Intellect. Disabil. Res.* **2012**, *56*, 1036–1045. [CrossRef]
30. Marques, A.; Demetriou, Y.; Tesler, R.; Gouveia, É.R.; Peralta, M.; Matos, M.G. Healthy lifestyle in children and adolescents and its association with subjective health complaints: Findings from 37 countries and regions from the HBSC Study. *Int. J. Environ. Res. Public Health* **2019**, *16*, 3292. [CrossRef] [PubMed]
31. Assari, S.; Bazargan, M. Educational attainment and subjective health and wellbeing; diminished returns of lesbian, gay, and bisexual individuals. *Behav. Sci.* **2019**, *9*, 90. [CrossRef] [PubMed]
32. Park, S.; Lee, H.J.; Jeon, B.J.; Yoo, E.Y.; Kim, J.B.; Park, J.H. Effects of occupational balance on subjective health, quality of life, and health-related variables in community-dwelling older adults: A structural equation modeling approach. *PLoS ONE* **2021**, *16*, e0246887. [CrossRef] [PubMed]
33. Qazi, S.L.; Koivumaa-Honkanen, H.; Rikkonen, T.; Sund, R.; Kröger, H.; Isanejad, M.; Sirola, J. Physical capacity, subjective health, and life satisfaction in older women: A 10-year follow-up study. *BMC Geriatr.* **2021**, *21*, 1–9. [CrossRef] [PubMed]
34. Miyakawa, M.; Hamashima, Y. Life satisfaction and subjective health status of young carers: A questionnaire survey conducted on Osaka prefectural high school students. *Jap. J. Pub. Health* **2020**, *68*, 157–166.
35. Low, G.; Molzahn, A.E.; Schopflocher, D. Attitudes to aging mediate the relationship between older peoples' subjective health and quality of life in 20 countries. *Health Qual. Life Outcomes* **2013**, *11*, 1–10. [CrossRef] [PubMed]
36. Skevington, S.M.; Emsley, R.; Dehner, S.; Walker, I.; Reynolds, S.E. Does subjective health affect the association between biodiversity and quality of life? Insights from international data. *App. Res. Qual. Life* **2019**, *14*, 1315–1331. [CrossRef]
37. Heyne, S.; Haufe, E.; Beissert, S.; Schmitt, J.; Günther, C. Determinants of depressive symptoms, quality of life, subjective health status and physical limitation in patients with systemic sclerosis. *Acta Derm.* **2023**, *103*. [CrossRef] [PubMed]
38. Yu, X. Engel curve, farmer welfare and food consumption in 40 years of rural China. *China Agri. Econ. Rev.* **2018**, *10*, 65–77. [CrossRef]
39. Ma, L.; Liu, S.; Fang, F.; Che, X.; Chen, M. Evaluation of urban-rural difference and integration based on quality of life. *Sustain. Cities Soc.* **2020**, *54*, 101877. [CrossRef]
40. Shi, X.; Cheong, T.S.; Yu, J.; Liu, X. Quality of life and relative household energy consumption in China. *China World Econ.* **2021**, *29*, 127–147. [CrossRef]

41. Qin, L.; Chen, W.; Sun, L. Impact of energy poverty on household quality of life--based on Chinese household survey panel data. *J. Clean. Prod.* **2022**, *366*, 132943. [CrossRef]

42. O'Neill, J.; Hibbard, M.R.; Broivn, M.; Jaffe, M.; Sliwinski, M.; Vandergoot, D.; Weiss, M. The effect of employment on quality of life and community integration after traumatic brain injury. *J. Head Trauma Rehabil.* **1998**, *13*, 68–79. [CrossRef]

43. Hyde, S.; Satariano, W.A.; Weintraub, J.A. Welfare dental intervention improves employment and quality of life. *J. Dent. Res.* **2006**, *85*, 79–84. [CrossRef] [PubMed]

44. Kober, R.; Eggleton, I.R. The effect of different types of employment on quality of life. *J. Intellect. Disabil. Res.* **2005**, *49*, 756–760. [CrossRef]

45. Bravata, D.M.; Keeffe, E.B. Quality of life and employment after liver transplantation. *Liver Transplant.* **2001**, *7*, S119–S123. [CrossRef] [PubMed]

46. Carlier, B.E.; Schuring, M.; Lötters, F.J.; Bakker, B.; Borgers, N.; Burdorf, A. The influence of re-employment on quality of life and self-rated health, a longitudinal study among unemployed persons in the Netherlands. *BMC Public. Health* **2013**, *13*, 503–507. [CrossRef] [PubMed]

47. Blalock, A.C.; Mcdaniel, J.S.; Farber, E.W. Effect of employment on quality of life and psychological functioning in patients with HIV/AIDS. *Psychosomatics* **2002**, *43*, 400–404. [CrossRef]

48. Kim, S.; Feldman, D.C. Working in retirement: The antecedents of bridge employment and its consequences for quality of life in retirement. *Acad. Manag. J.* **2000**, *43*, 1195–1210. [CrossRef]

49. Beyer, S.; Brown, T.; Akandi, R.; Rapley, M. A comparison of quality of life outcomes for people with intellectual disabilities in supported employment, day services and employment enterprises. *J. Appl. Res. Intellect. Disabil.* **2010**, *23*, 290–295. [CrossRef]

50. Eggleton, I.; Robertson, S.; Ryan, J.; Kober, R. The impact of employment on the quality of life of people with an intellectual disability. *J. Vocat. Rehabil.* **1999**, *13*, 95–107.

51. Diener, E.D.; Diener, C. The wealth of nations revisited: Income and quality of life. *Soc. Indic. Res.* **1995**, *36*, 275–286. [CrossRef]

52. Zagorski, K.; Evans, M.D.; Kelley, J.; Piotrowska, K. Does national income inequality affect individuals' quality of life in Europe? Inequality, happiness, finances, and health. *Soc. Indic. Res.* **2014**, *117*, 1089–1110. [CrossRef]

53. Munson, J.; Scholnick, J. Wealth and well-being in an ancient maya community: A framework for studying the quality of life in past societies. *J. Archaeol. Method Theory* **2021**, *29*, 1–30. [CrossRef]

54. Parmenter, T.R. An Analysis of the Dimensions of Quality of Life for People with Physical Disabilities. In *Quality of Life for Handicapped People*; Routledge: Oxfordshire, UK, 2021.

55. An, H.Y.; Chen, W.; Wang, C.W.; Yang, H.F.; Huang, W.T.; Fan, S.Y. The relationships between physical activity and life satisfaction and happiness among young, middle-aged, and older adults. *Int. J. Environ. Res. Pub. Health* **2020**, *17*, 4817. [CrossRef]

56. Xiao, J.J.; Tang, C.; Shim, S. Acting for happiness: Financial behavior and life satisfaction of college students. *Soc. Indic. Res.* **2009**, *92*, 53–68. [CrossRef]

57. Easterlin, R.A. Diminishing marginal utility of income? Caveat emptor. *Soc. Indic. Res.* **2005**, *70*, 243–255. [CrossRef]

58. Lin, C.C.; Peng, S.S. The role of diminishing marginal utility in the ordinal and cardinal utility theories. *Aust. Econ. Pap.* **2019**, *58*, 233–246. [CrossRef]

59. Gautié, J.; Schmitt, J. *Low-wage Work in the Wealthy World*; Russell Sage Foundation: Manhattan, NY, USA, 2010.

60. Beattie, B.R.; LaFrance, J.T. The law of demand versus diminishing marginal utility. *App. Econ. Pers. Pol.* **2006**, *28*, 263–271.

61. Hu, Z.; Wang, M.; Cheng, Z.; Yang, Z. Impact of marginal and intergenerational effects on carbon emissions from household energy consumption in China. *J. Clean. Prod.* **2020**, *273*, 123022. [CrossRef]

62. Gujarati, D.; Porter, D. *Basic Econometrics*; McGraw-Hill International Edition: New York, NY, USA, 2009.

63. Wooldridge, J. *Introductory Econometrics: A Modern Approach*; Cengage Learning: Boston, MA, USA, 2009.

64. Baltagi, B. *Econometric Analysis of Panel Data*; John Wiley &Sons: Hoboken, NJ, USA, 2008; Volume 1.

65. Shim, H.; Lee, W.S.; Moon, J. The Relationships between Food, Recreation Expense, Subjective Health, and Life Satisfaction: Case of Korean People with Disability. *Sustainability* **2023**, *15*, 9099. [CrossRef]

66. Campos, A.C.V.; Ferreira, E.F.; Vargas, A.; Albala, C. Aging, gender and quality of life (AGEQOL) study: Factors associated with good quality of life in older Brazilian community-dwelling adults. *Health Qua. Life Outcomes* **2014**, *12*, 1–11. [CrossRef]

 *healthcare*

Article

Exploring the Reciprocal Relationship between Depressive Symptoms and Cognitive Function among Chinese Older Adults

Jiehua Lu [1] and Yunchen Ruan [2,*]

1 Department of Sociology, Peking University, No. 5 Yiheyuan Road, Haidian District, Beijing 100871, China
2 School of Humanities and Social Sciences, Fuzhou University, No. 2 Xueyuan Road, Fuzhou 350108, China
* Correspondence: ruanyunchen@fzu.edu.cn

Abstract: (1) Objectives: This study aims to investigate the bidirectional relationship between depressive symptoms and cognitive function among older adults in China, addressing a research gap in the context of developing nations. (2) Methods: A total of 3813 adults aged 60 and older participating in 2013, 2015, and 2018 waves of the China Health and Retirement Longitudinal Study (CHARLS) were included. A fixed-effects model and cross-lagged panel model (CLPM) was utilized. (3) Results: First, the results indicated that a significant negative correlation existed between depressive symptoms and cognitive function in older adults during the study period ($\beta = -0.084, p < 0.001$). Second, after controlling for unobserved confounding factors, the deterioration and improvement of depressive symptoms still significantly affected cognitive function ($\beta = -0.055, p < 0.001$). Third, using the cross-lagged panel model, we observed a reciprocal relationship between depressive symptoms (Dep) and cognitive function (Cog) among Chinese older adults (Dep2013 → Cog2015, $\beta = -0.025$, $p < 0.01$; Dep2015 → Cog2018, $\beta = -0.028, p < 0.001$; Cog2013 → Dep2015, $\beta = -0.079, p < 0.01$; Cog2015 → Dep2018, $\beta = -0.085, p < 0.01$). (4) Discussion: The reciprocal relationship between depressive symptoms and cognitive functioning in older adults emphasizes the need for integrated public health policies and clinical interventions, to develop comprehensive intervention strategies that simultaneously address depressive symptoms and cognitive decline.

Keywords: depressive symptoms; cognitive function; reciprocal relationship; Chinese older adults; cross-lagged panel model

Citation: Lu, J.; Ruan, Y. Exploring the Reciprocal Relationship between Depressive Symptoms and Cognitive Function among Chinese Older Adults. *Healthcare* **2023**, *11*, 2880. https://doi.org/10.3390/healthcare11212880

Academic Editors: Carlos Laranjeira and Ana Querido

Received: 26 September 2023
Revised: 29 October 2023
Accepted: 31 October 2023
Published: 1 November 2023

1. Background

Preserving optimal cognitive functioning among older individuals is paramount for the realization of healthy aging. China, considering its dramatic demographic-aging trends, is facing significant issues concerning the cognitive health of its older adult populace. Recent estimates have indicated that, as of 2018, the number of individuals aged 60 and above afflicted with dementia in China stood at approximately 15.07 million. Within this demographic, approximately 9.83 million individuals were diagnosed with Alzheimer's disease (AD) [1]. Furthermore, there were approximately 38.77 million cases of mild cognitive impairment (MCI) among individuals aged 60 and above, yielding a prevalence rate of approximately 15.5%. Cognitive health issues have a profound impact on the well-being of older adults in their later years. On the one hand, cognitive decline frequently co-occurs with physical health ailments and psychological distress in older adults, thereby elevating the risk of mortality. On the other hand, diminished cognitive function can curtail the societal and familial roles that older adults fulfill, diminishing their ability to navigate the challenges associated with aging. Consequently, it has become imperative to address the challenges posed by cognitive health risks in older adults, delay the decline in cognitive function among older adults in China, and mitigate the proportion of older adults affected by dementia.

1.1. The Impact of Late-Life Depressive Symptoms on Cognitive Function among Older Adults

In recent years, an increasing number of scholars have directed their attention toward examining the ramifications of late-life depressive symptoms on the cognitive function of older adults [2]. Late-life depressive symptoms stand out as a significant contributor to the global disease burden and represents a profound mental health concern that markedly influences quality of life in older adults. It manifests through physical symptoms, a waning interest in and enthusiasm for daily activities, and in severe cases, even suicidal tendencies [3]. Psychological and epidemiological investigations have established a connection between heightened symptoms of late-life depressive symptoms and decline in various cognitive domains, encompassing information-processing capacity, memory function, and executive function, among others [2,4].

Information-processing speed is the initial cognitive domain impacted by late-life depressive symptoms, rendering it a sensitive indicator of cognitive decline. Information-processing capacity pertains to an individual's competency in receiving external stimuli, and replicating, processing, encoding, and retaining information gleaned from the external environment [5,6]. Depression is tightly intertwined with declines in memory function. Multiple studies have posited that older adults manifesting significant depressive symptoms often exhibit diminished activation within the working-memory-network region [7]. Furthermore, the inhibitory impact of depressive symptoms on episodic memory in older adults assumes greater prominence when contrasted with younger cohorts [8]. Late-life depressive symptoms manifest associations with impairments in executive function. Extensive research has underscored the fact that late-life depressive symptoms frequently co-occur with various cognitive deficits, prominently including executive function deficiencies [9]. Among the diverse components comprising executive function, inhibition function has emerged as the element most susceptible to the impact of late-life depressive symptoms [10]. Furthermore, late-life depressive symptoms not only exacerbate the decline of various cognitive functions, but also culminates in more-severe cognitive impairments; for example, late-life depressive symptoms constitute a pivotal risk factor for MCI [11]. Additionally, a profound nexus exists between late-life depressive symptoms and AD, with select studies proposing that late-life depressive symptoms exert a substantial influence on the likelihood of developing AD [12].

Late-life depressive symptoms expedite cognitive-function decline and exhibit a close association with conditions such as MCI and Alzheimer's dementia [13]. Furthermore, certain studies have suggested the potential existence of a bidirectional causal link between late-life depressive symptoms and cognitive function [14–16]. However, to date, there remains insufficient empirical evidence to robustly substantiate this bidirectional causality.

1.2. The Bidirectional Causal Causality between Late-Life Depressive Symptoms and Cognitive Function

Compared to the impact of late-life depressive symptoms on cognitive functioning, how cognitive function may influence late-life depressive symptoms has only been explored to a limited extent [6,17]. Scholars have investigated this relationship using longitudinal data from a survey tracking older adults in the United States [18]. Their findings revealed a noteworthy correlation between cognitive impairments in older adults and subsequent depressive symptoms. Similarly, a previous longitudinal study observed a significant association between cognitive impairments—including memory function—recorded during baseline surveys and the subsequent emergence of heightened depressive symptoms in older adults [19]. Research has indicated that older adults often perceive cognitive decline and impairment as akin to physical decline, viewing it as an irreversible weakening process which, in turn, exacerbates depressive symptoms. In terms of specific cognitive domains, some researchers have delved into the connection between these cognitive functions and late-life depressive symptoms [20]. Their study demonstrated that declines in information-processing speed significantly elevate late-life depressive symptoms in older adults.

In summary, although many scholars have discussed the relationship between late-life depressive symptoms and cognitive function, the existing literature has mostly focused on the correlation between the two [21,22] or the influence of the former on the latter [4], lacking sufficient examination of the bidirectional causal relationship between the level of late-life depressive symptoms and cognitive function in older adults.

Furthermore, the results of the few studies exploring bidirectional causal relationships are inconsistent [16,17], and few studies have attempted to effectively control for the omitted variable bias between the two, leading to difficulties in establishing causal identification. Some studies have found that late-life depressive symptoms can predict cognitive function, while others have come to the opposite conclusion. For instance, a study utilizing data from the population-based Longitudinal Aging Study Amsterdam indicated that depressive symptoms in older adults at baseline could predict subsequent cognitive function, but the reverse was not true [6]. However, a longitudinal study on older Hispanic adults in the United States indicated that a decline in cognitive abilities could predict depressive symptoms among older adults, but the converse was not substantiated [14].

1.3. Theoretical Perspectives and Hypothesis

This study aims to investigate the bidirectional relationship between depressive symptoms and cognitive function among Chinese older adults. However, discussing the bidirectional relationship between depressive symptoms and cognitive function is only part of the work examining the declining health in older adults, which features multi-dimensionality. We hope to preliminarily explore the theoretical framework for constructing a multi-dimensional pattern of declining health in older adults by introducing theoretical perspectives such as healthy lifestyle theory and stress process theory.

From the perspective of a healthy lifestyle, late-life depressive symptoms may affect cognitive function in older adults through their daily lifestyle choices [23]. Healthy lifestyle theory suggests that higher levels of depression hinder older adults from maintaining healthy lifestyles, leading to reduced sleep quality, decreased frequency of physical exercise and social engagement, disruption of adherence to medication schedules, and an increased likelihood of engaging in health-risk behaviors such as alcohol abuse [24]. For example, existing research has found that late-life depressive symptoms can affect sleep quality in older adults [25], and sleep quality is highly correlated with cognitive function [26]. Based on these viewpoints, we propose the following hypothesis:

Hypothesis 1. *In terms of causality, depressive symptoms impact cognitive function in older adults, meaning that the higher the level of depression in older adults, the lower their cognitive function.*

Stress process theory suggests that the process of cognitive decline in older adults is similar to the process of disability and can lead to gradually restricted functional abilities, thus acting as a long-term chronic stressor [27]. A decline in cognitive function not only limits an individual's functional abilities but also weakens their social-support network. The extent to which an older adult's functional abilities are limited directly affects their difficulty in engaging in social interactions and fulfilling social roles. As a long-term stressor, cognitive decline can lead to continuous hardships in maintaining independent living, depleting psychological resources such as sense of control and self-esteem, which further exacerbates late-life depressive symptoms [28]. Based on these viewpoints, we propose the following hypothesis:

Hypothesis 2. *In terms of causality, cognitive function impacts late-life depressive symptoms, meaning that the lower the cognitive function in older adults, the higher the level of depression.*

2. Methods

2.1. Data Sources

This study uses data from three waves of the China Health and Retirement Longitudinal Study (CHARLS), including data from 2013, 2015, and 2018. CHARLS is a large-scale nationally representative household survey led by the National School of Development at Peking University. The survey covers 28 provinces (municipalities, autonomous regions) and 450 villages in China. The CHARLS questionnaire covers a wide range of individual and family information for middle-aged and older adults, including variables representing mental and physical health status, such as depressive symptoms, cognitive function, and self-rated health, as well as demographic variables like gender, age, marital status, and education level. These data provide support for exploring the causal relationship between depressive symptoms and cognitive function in Chinese older adults.

The baseline survey of CHARLS began in 2011, followed by tracking surveys in 2013, 2015, and 2018. The sample sizes for CHARLS 2013, 2015, and 2018 were 18,605, 21,095, and 19,816, respectively. Leveraging the advantages of CHARLS tracking data, this study combined data from 2013, 2015, and 2018 and processed it into panel data. Based on the research objectives, this study retained samples of older adults aged 60 and above, and removed samples with missing values in the variables, resulting in a final sample of 3813 individuals. The demographic and health characteristics of the participants are shown in Table 1.

Table 1. Demographic and health characteristics of participants at baseline (N = 3813).

	N (%)	Mean (SD)
Cognitive function		9.59 (3.73)
Mental intactness		5.73 (2.67)
Memory ability		3.86 (1.76)
Depressive symptoms		7.77 (5.67)
Gender		
Male	2050 (53.76)	
Female	1763 (46.24)	
Age		66.11 (5.18)
Region		
Urban	983 (25.78)	
Rural	2830 (74.22)	
Marital status at baseline		
Currently married	3261 (85.52)	
Others	552 (14.48)	
Educational level at baseline		
No formal education	1797 (47.13)	
Primary school	1059 (27.77)	
Junior middle school and above	957 (25.10)	
Self-reported health at baseline		
Good	853 (22.37)	
Normal or poor	2960 (77.63)	

Note: Data: CHARLS 2013.

2.2. Variable Definitions

The measurement of depressive symptoms in older adults was based on previous research [29,30]. We used the Center for Epidemiologic Studies Depression Scale (CES-D-10), which consists of 10 questions asking respondents about their feelings and behaviors within the past week. Eight questions are concerned with the frequency of various depression-related symptoms, while the other two questions are about the frequency of positive emotions in the past week. For each of the 10 questions, there are four possible answers: rarely or none of the time; some or a little of the time; occasionally or a moderate amount of the time; and most or all of the time. These answers are scored 0 (rarely or none of the time) to 3 (most or all of the time). After reversing the 2 positive-mood items, the scores for all 10 questions are summed. The depression scale score ranges from 0 to 30, with higher scores indicating higher levels of depression.

The dependent variable in this study is the cognitive function of the older adults. Cognitive function in older adults refers to the mental process of acquiring and processing information through senses, memory, reasoning, and decision making. Following previous research [31], we used two indicators—mental intactness and memory ability—to measure cognitive function in older adults. First, in the measurement of mental intactness using CHARLS questionnaire items, we mainly relied on nine questions related to the calculation, orientation, and drawing ability of respondents. For example, questions such as "What is 100 minus 7?", "What day of the week is it today?", and "Draw two overlapping pentagrams" are included. The number of correctly answered questions represents the mental intactness score, which ranges from 0 to 9. Second, in the measurement of memory ability, the CHARLS questionnaire provides a list of 10 common words and asks respondents to recall these words twice, once immediately and again after four minutes. The average of the correct answers to both recall sessions yields the memory ability score, ranging from 0 to 10. Overall, the sum of mental intactness and memory ability scores yields the cognitive function score for older adults, ranging from 0 to 19, with higher scores indicating stronger cognitive function.

2.3. Data Analysis

The data analysis consisted of three main steps:

Step 1. Cross-sectional OLS analysis: In this step, we examined the results in existing research concerning the correlation between late-life depressive symptoms and cognitive function using an ordinary least squares (OLS) regression model. This helped to validate the relationships found in previous studies, which is that there is a significant correlation between depressive symptoms and cognitive function among Chinese older adults.

Step 2. Individual-level fixed-effects model analysis: The primary purpose of this step was to address the omitted variable bias. In the individual-level fixed-effects model, we introduce dummy variables for each individual to control for unobservable individual characteristics, which helps us to handle omitted variable bias in panel data. We used an individual-level fixed-effects model to control for unobserved individual-level factors that may affect both late-life depressive symptoms and cognitive function, such as genetic inheritance and personality traits. These individual-level factors were assumed to be time-invariant and not influenced by other variables in the study. By employing the fixed-effects model, we obtained more-reliable estimates of the causal relationship between late-life depressive symptoms and cognitive function.

Step 3. Cross-lagged panel model analysis: Considering the possibility of bidirectional causality, we used a CLPM to examine the causal relationship between late-life depressive symptoms and cognitive function over time. Cross-lagged panel model (CLPM) is a recognized powerful method for exploring dynamic relationships between variables. CLPM reflects the dynamic effects between variables through cross-lagged paths, that is, it constructs a path of effect (known as autoregressive effect) from the previous level of a variable to its current level, as well as a path of effect (known as cross-lagged effect) to the current level of another variable. The CLPM allowed us to explore the temporal ordering of the variables and determine whether the relationship between late-life depressive symptoms and cognitive function was consistent with a causal direction. This step provided further insight into the causal relationship between the two variables.

By employing these three steps in the analysis, we aimed to gain a comprehensive understanding of the causal relationship between late-life depressive symptoms and cognitive function in Chinese older adults, while addressing potential endogeneity issues.

2.4. Summary

Based on three waves of the China Health and Retirement Longitudinal Study (CHARLS), this study attempts to address potential endogeneity issues using individual-level fixed-effects and cross-lagged panel models (CLPM) to explore the reciprocal relationship between late-life depressive symptoms and cognitive function in older adults.

3. Results

3.1. Cross-Sectional OLS Analysis

This study aimed to investigate the bidirectional causal relationship between late-life depressive symptoms and cognitive function in Chinese older adults. Therefore, we replicated the conclusions of previous research and analyzed the correlation between the two variables as a basis for further examining their causal relationship. The correlation analysis serves as a bridge to eliminate the possibility of inconsistent research results in the subsequent discussion of causal relationships due to data and measurement issues.

Table 2 presents the cross-sectional OLS estimation results regarding the relationships between late-life depressive symptoms and cognitive function, mental intactness, and memory ability in the study year (2018). The results of the correlation analysis indicated a significant negative association between depressive symptoms and cognitive function, and mental intactness and memory ability in older adults.

Table 2. Ordinary least squares (OLS) regression model of cognitive function (N = 3813).

Variables	Cognitive Function (M1)		Mental Intactness (M2)		Memory Ability (M3)	
	Coefficient	SE	Coefficient	SE	Coefficient	SE
Depressive symptoms	−0.084 ***	0.009	−0.055 ***	0.007	−0.029 ***	0.005
Male	0.738 ***	0.108	0.798 ***	0.079	−0.059	0.057
Age	−0.092 ***	0.010	−0.050 ***	0.007	−0.042 ***	0.005
Urban	1.305 ***	0.128	0.750 ***	0.093	0.555 ***	0.067
Married	0.123	0.150	0.079	0.109	0.044	0.078
Education						
Primary school	2.287 ***	0.125	1.668 ***	0.091	0.618 ***	0.065
Junior middle school and above	3.303 ***	0.142	2.183 ***	0.104	1.119 ***	0.074
Self-reported health	0.109	0.125	0.048	0.091	0.061	0.065
R2	0.303		0.279		0.147	
Adj-R2	0.301		0.278		0.145	

Note: Data = CHARLS 2013; R2, R-squared; Adj-R2, Adjusted R-squared; SE, standard error; *** $p < 0.001$.

In Model 1 of Table 2, with cognitive function as the dependent variable, the results showed that, regardless of whether control variables were included, there existed a significant negative association between late-life depressive symptoms and current cognitive function, indicating that an increase in depressive symptoms led to a significant decrease in current cognitive function ($\beta = -0.084$, $p < 0.001$).

The results for Model 2 demonstrated a significant negative correlation between late-life depressive symptoms and current mental intactness, regardless of the inclusion of control variables ($\beta = -0.055$, $p < 0.001$). Similarly, Model 3 supported a negative relationship between late-life depressive symptoms and current memory ability ($\beta = -0.029$, $p < 0.001$).

Based on the above results, we repeated the analysis using cross-sectional data from 2013 and 2015, and the consistent OLS regression results further supported the robustness of these relationships. Although the current correlation analysis could not test the research hypotheses directly, it provided a foundation for the next step of examining Hypotheses 1 and 2, specifically discussing the causal relationship between late-life depressive symptoms and cognitive function.

3.2. Fixed-Effects Model Analysis

The first step towards causal inference in this study was addressing the issue of omitted variable bias, which was achieved using panel data to control for individual-level confounding factors that do not vary over time, regardless of whether these confounding factors are observable. By employing a fixed-effects model, the results in Table 3 reveal that, after controlling for time-invariant confounding factors, an increase in depressive symptoms among older adults led to a significant decline in cognitive function, mental intactness, and memory ability. This finding was consistent with the existing literature, as the significant impact of late-life depressive symptoms on cognitive function has been supported by numerous scholars [6].

Table 3. Fixed-effects models of cognitive function (N = 3813).

Variables	Cognitive Function (M1)		Mental Intactness (M2)		Memory Ability (M3)	
	Coefficient	SE	Coefficient	SE	Coefficient	SE
Depressive symptoms	−0.055 ***	0.009	−0.038 ***	0.007	−0.017 ***	0.005
Age	−0.113 ***	0.018	−0.067 ***	0.014	−0.049 ***	0.011
Married	0.393	0.269	0.162	0.203	0.231	0.161
Self-reported health	0.050	0.111	0.070	0.084	−0.020	0.067

Note: Data: CHARLS 2013 & 2015; SE, standard error; *** $p < 0.001$.

Table 3 presents the fixed-effect estimates for the effects of late-life depressive symptoms on the cognitive function, mental intactness, and memory ability. An increase in depressive symptoms resulted in a significant decline in cognitive function for Chinese older adults, as well as markedly suppressing both mental intactness and memory ability.

Model 1 in Table 3 treats the cognitive function as the dependent variable. The results indicated that, for older adults, an increase in the severity of depressive symptoms was associated with a significant decline in cognitive function ($\beta = -0.055, p < 0.001$).

Similarly, Models 2 and 3 in Table 3 treat mental intactness and memory ability as the dependent variables, respectively. In both cases, the results indicated that, an increase in depressive symptoms among older adults significantly impaired their mental intactness ($\beta = -0.038, p < 0.001$) and memory ability ($\beta = -0.017, p < 0.001$).

3.3. Cross-Lagged Panel Model Analysis

Based on the estimation using the fixed-effects model as a foundation, we employed a CLPM to examine the bidirectional causal relationship between depressive symptoms and the cognitive function among older adults. Based on the statistical results of the CLPM, we observed a significant bidirectional causal relationship between depressive symptoms and the cognitive function. Specifically, current depressive symptoms significantly affected the cognitive function in the next period, while the current cognitive function could also predict depressive symptoms in the subsequent period. This mutual influence is also applicable to the relationship between depressive symptoms and mental intactness, as well as memory ability.

Figure 1 displays the results of the analysis regarding the relationship between depressive symptoms and the cognitive function. Both depressive symptoms and the cognitive function exhibited a strong autoregressive effect (Dep2013 → Dep2015, $\beta = 0.572, p < 0.001$; Dep2015 → Dep2018, $\beta = 0.496, p < 0.001$; Cog2013 → Cog2015, $\beta = 0.480, p < 0.001$; Cog2015 → Cog2018, $\beta = 0.519, p < 0.001$). Furthermore, while it remains unobservable, the co-variation of the two residuals (for instance, ε_1 and ε_2) indicates a co-movement between depressive symptoms and cognitive function, even after accounting for potential confounding variables. The bidirectional causal relationship of interest in this study is represented by two sets of cross-lagged coefficients.

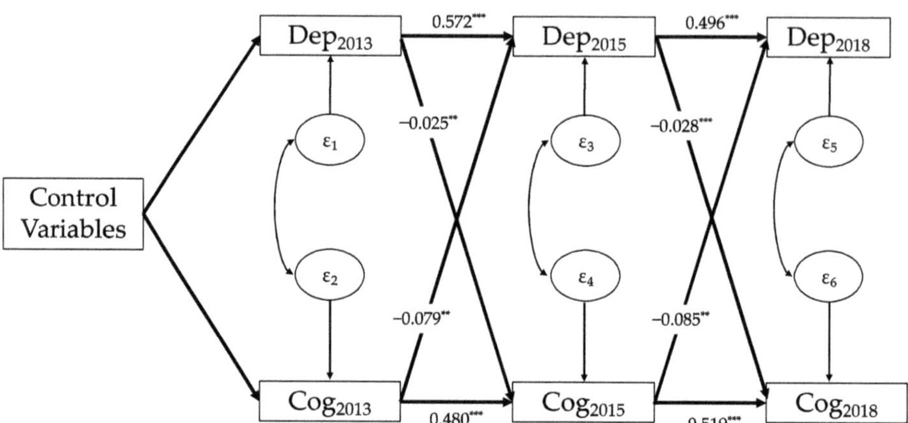

Figure 1. Cross-lagged panel model of depressive symptoms and cognitive function (N = 3813). Note: Data: CHARLS 2013, 2015, and 2018; Dep: depressive symptoms; Cog: cognitive function; subscript number of the variables mean survey wave; control variables: age, gender, region, educational level at baseline, marital status at baseline, self-reported health at baseline; ε_i: residuals; and ** $p < 0.01$, *** $p < 0.001$.

Regarding the influence of depressive symptoms on cognitive function, both coefficients were negative and significant, indicating a significant negative impact of depressive symptoms on the cognitive function in the subsequent period (Dep2013 → Cog2015, $\beta = -0.025, p < 0.01$; Dep2015 → Cog2018, $\beta = -0.028, p < 0.001$). Concerning the effect of the cognitive function on depressive symptoms, both coefficients were also negative and significant, indicating that cognitive function among older adults can well-predict depressive symptoms in the subsequent period (Cog2013 → Dep2015, $\beta = -0.079, p < 0.01$; Cog2015 → Dep2018, $\beta = -0.085, p < 0.01$).

Figure 2 shows that depressive symptoms in 2013 had a significant negative impact on mental intactness in 2015 ($\beta = -0.014, p < 0.05$), and the depressive symptoms in 2015 also had a significant negative impact on mental intactness in 2018 ($\beta = -0.017, p < 0.01$). Additionally, mental intactness in 2013 significantly predicted depressive symptoms in 2015 ($\beta = -0.095, p < 0.01$), and mental intactness in 2015 also significantly predicted depressive symptoms in 2018 ($\beta = -0.08, p < 0.05$).

As can be seen from Figure 3, depressive symptoms in 2013 had a significant negative impact on memory ability in 2015 ($\beta = -0.018, p < 0.001$), while depressive symptoms in 2015 also had a significant negative impact on memory ability in 2018 ($\beta = -0.021, p < 0.001$). Furthermore, memory ability in 2013 significantly predicted depressive symptoms in 2015 ($\beta = -0.105, p < 0.05$), and memory ability in 2015 also significantly predicted depressive symptoms in 2018 ($\beta = -0.156, p < 0.01$).

By combining the results from Figures 1–3, we found that, regardless of the cognitive function, mental intactness, or memory ability, the bidirectional causal patterns between depressive symptoms and the three cognitive indicators among Chinese older adults were very similar; that is, depressive symptoms among older adults had a significant predictive effect on the cognitive function, mental intactness, and memory ability in the subsequent period. Simultaneously, all three cognitive indicators had a significant impact on the level of depressive symptoms in the subsequent period.

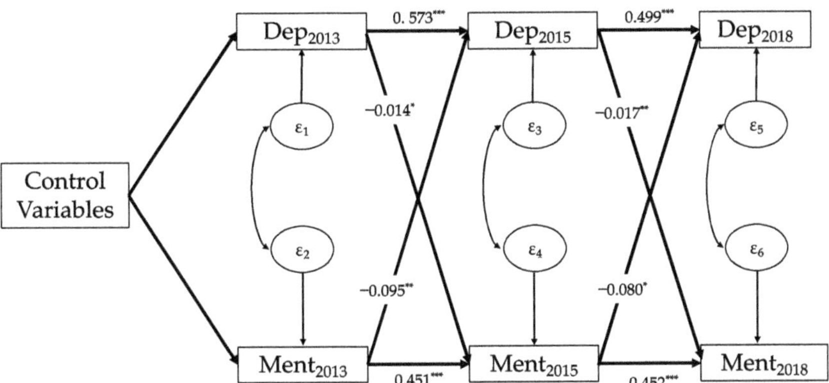

Figure 2. Cross-lagged panel model of depressive symptoms and mental intactness (N = 3813). Note: Data: CHARLS 2013, 2015, and 2018; Dep, depressive symptoms; Ment, mental intactness; subscript number of the variables mean survey wave; control variables: age, gender, region, educational level at baseline, marital status at baseline, self-reported health at baseline; ε_i: residuals; and * $p < 0.05$, ** $p < 0.01$, *** $p < 0.001$.

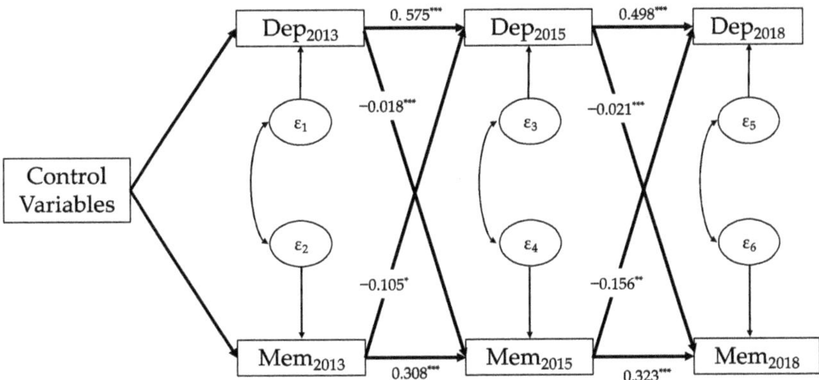

Figure 3. Cross-lagged panel model of depressive symptoms and memory ability (N = 3813). Note: Data: CHARLS 2013, 2015, and 2018; Dep, depressive symptoms; Mem, memory ability; subscript number of the variables mean survey wave; control variables: age, gender, region, educational level at baseline, marital status at baseline, self-reported health at baseline; ε_i: residuals; and * $p < 0.05$, ** $p < 0.01$, *** $p < 0.001$.

4. Discussion

The main aims of the present study were (1) to retest the correlation between late-life depressive symptoms and cognitive function among older adults in existing research; (2) on the basis of addressing omitted variable bias, to examine the impact of changes in depressive symptoms on changes in cognitive function; (3) and to explore the bidirectional relationship between late-life depressive symptoms and cognitive function among older adults in China.

The results showed that (1) a significant negative correlation existed between depressive symptoms and cognitive function among Chinese older adults; (2) after controlling for unobserved confounding factors, the deterioration and improvement of depressive symptoms still significantly affected cognitive function; (3) and a reciprocal relationship existed between depressive symptoms and cognitive function among Chinese older adults.

Hence, the findings suggest that depressive symptoms influence cognitive function among older adults, and vice versa. The Hypotheses 1 and 2 are confirmed.

4.1. Association between Depression and Cognitive Function of Older Adults

The correlation analysis based on cross-sectional data demonstrated the existence of a significant negative correlation between the level of depression and the current cognitive function in older adults. This finding is consistent with previous research, as there is relative unanimity in the academic community regarding the negative relationship between depressive symptoms in Chinese older adults and various dimensions of cognitive function [32].

The results of this study not only support existing research on Chinese older adult populations but are also in line with findings from other countries [33]. For example, using cross-sectional data, some researchers observed a highly significant correlation between depression and declining cognitive function in older adults in India, another developing country [34]. A study based on developed countries suggested that, in the older adult population in the United Kingdom, depressive symptoms were significantly correlated with cognitive decline [35].

4.2. The Fixed Effect of Depressive Symptoms on the Cognitive Function of Older Adults

The results of the fixed-effects model analysis indicated that, after controlling for unobserved confounding factors, an increase in the level of depression in older adults led to a significant decline in cognitive function. At the level of the causal relationship, we found support for Hypothesis 1. Based on the results from Table 3, we believe that Hypothesis 1 is partially supported; meaning that, after controlling for some confounding factors, changes in depression significantly affected changes in the cognitive function of older adults.

The role of depressive symptoms in cognitive function has been extensively discussed by scholars [6]. Using longitudinal data, studies have pointed out the predictive effect of depressive symptoms on cognitive function in older adult women and found a significant impact of depression on various dimensions of cognitive function [36]. Some other scholars also utilized longitudinal data to examine the influence of baseline depression levels on subsequent cognitive function in older adults, and their results also supported a significant effect of depression on cognitive function [37]. The statistical results in Table 3 are consistent with the empirical findings of the aforementioned studies.

4.3. The Reciprocal Relationship between Depressive Symptoms and the Cognitive Function of Older Adults

Using the CLPM, this study revealed a bidirectional causal relationship between the level of depression in Chinese older adults and their cognitive functioning. Specifically, it was found that the level of depression significantly predicted cognitive functioning in the following period, and, conversely, cognitive functioning also had a significant impact on the subsequent level of depression. Based on the results obtained from the CLPM analysis, this study provides support for Hypotheses 1 and 2. This means that, at the causal level, it was evident that depression significantly affected the cognitive function of the older adults and, conversely, the cognitive function had a statistically validated role in influencing depressive symptoms.

To date, only a limited number of studies have explored the bidirectional causal relationship between depressive symptoms in older adults and their cognitive functioning [6,16]. One study examined the temporal associations between depressive symptoms and cognitive decline among Chinese older adults, and found that baseline depressive symptoms influenced the subsequent cognitive function [16]. However, over time, the decline in baseline cognitive function does not predict depressive symptoms. Our study results suggested that not only can depressive symptoms in older adults significantly predict cognitive function, but the reverse relationship also holds true.

4.4. Theoretical Perspectives: Exploration Multi-Dimensional Patterns of Declining Health in Older Adults

In this study, we explored the bidirectional causal relationship between late-life depressive symptoms and cognitive function, and the results are in line with the perspectives from health lifestyle theory, stress process theory, and pathophysiological processes theory. Our research findings first indicated that depressive symptoms in older adults lead to a decline in cognitive function. According to the health lifestyle hypothesis, late-life depressive symptoms significantly restrict healthy lifestyle choices, including reduced physical exercise and diminished sleep quality, ultimately exacerbating cognitive decline in older adults [38,39]. In addition to the health lifestyle hypothesis, the pathophysiological processes hypothesis suggests that long-term depression in older adults can induce toxic reactions related to glutamate or steroids, leading to pathophysiological processes such as atrophy of the frontal lobe and hippocampus that weaken cognitive function [40,41].

Furthermore, cognitive function also has a predictive role in depressive symptoms. Stress process theory posits that the decline in cognitive function in later life serves as a chronic stressor affecting the mental health of older individuals. Such a decline in cognitive function threatens their social role, performance, and engagement, thus weakening their sense of control and self-esteem at the psychological level [28]. This study discussed the bidirectional causal relationship between depressive symptoms and cognitive function in Chinese older adults, particularly highlighting the role of cognitive function in depressive symptoms from the perspective of stress process theory.

By examining the bidirectional relationship between late-life depressive symptoms and cognitive decline, this study provides a potential foundation for the integration of perspectives from health lifestyle theory, stress process theory, and pathophysiological processes theory, and offers a preliminary exploration into the construction of a theoretical framework for examining the multi-dimensional patterns of declining health in older adults [42].

4.5. Contribution, Limitations and Future Directions

Despite the consensus in the academic community on the correlation between depressive symptoms and cognitive function in older adults, there is an urgent need to extend the investigation from correlation analyses to establishing causal relationships between the two variables. Existing research on the bidirectional causal relationship between depressive symptoms and cognitive function is scarce and yields inconsistent conclusions [6,16]. In comparison to the existing literature, this study not only fills gaps in content but also advances the causal identification of the relationship between the two considered variables.

It is important to note that this study has two potential limitations. First, due to data constraints, the cognitive function-related measures discussed in this study only included measures of mental intactness and memory abilities. Future studies should further explore the predictive effects of late-life depressive symptoms on specific cognitive function, such as executive function and processing speed. Second, the analytical strategy of this study was to estimate the causal relationship in two steps, separately addressing omitted variable bias and bidirectional causality, without considering these two issues simultaneously. Therefore, in the research findings, the reciprocal relationship between depressive symptoms and cognitive function among older adults may be affected by omitted variable bias, including the existence of a common mechanism between depression and cognitive function or even the concurrence of multiple co-causes. Future research could address these issues by employing advanced statistical methods, such as maximum likelihood for cross-lagged panel models with fixed effects.

4.6. Implications

The study reveals a bidirectional causal relationship between depressive symptoms and cognitive functioning among Chinese older adults. This discovery has significant implications for public health policy and clinical practice. Firstly, the understanding of this reciprocal relationship deepens our knowledge of the complex interplay between mental health and cognitive function among older adults. It suggests that these two aspects should not be treated independently, but rather, an integrated approach should be adopted. Secondly, this study calls for government intervention in the mental health and cognitive well-being of the elderly population. There is a pressing need to develop comprehensive intervention strategies that simultaneously address depressive symptoms and cognitive decline. These strategies could include mental health promotion, improved access to mental health services, cognitive training programs, and strengthening of social-support networks.

5. Conclusions

Examining the relationship between late-life depressive symptoms and cognitive function in Chinese older adults has significant clinical implications for improving cognitive health in the older adult population in China, thus delaying cognitive decline. In this study, we revealed a significant negative correlation between the severity of late-life depressive symptoms and current cognitive function. After controlling for unobserved confounding factors, an increase in the level of depression was found to lead to a significant decline in cognitive function among Chinese older adults. Furthermore, a bidirectional causal relationship was identified between the severity of late-life depressive symptoms and cognitive function, with the severity of depression significantly predicting subsequent cognitive function, while cognitive function also exerted a significant influence on subsequent depression levels.

Author Contributions: This paper was completed by two authors. Data curation, J.L.; language modification, Y.R.; formal analysis, Y.R.; funding acquisition, J.L.; methodology, Y.R.; and resources, J.L. All authors have read and agreed to the published version of the manuscript.

Funding: This research was supported by the Major Project of Beijing Social Science Fund, (Project No. 20ZDA32).

Institutional Review Board Statement: The CHARLS study was approved by the Biomedical Ethics Committee of Peking University (IRB00001052-11015).

Informed Consent Statement: All subjects gave their informed consent for inclusion before they participated in the CHARLS study.

Data Availability Statement: Data are derived from public domain resources.

Conflicts of Interest: The authors declare no conflict of interest.

References

1. Jia, L.; Du, Y.; Chu, L.; Zhang, Z.; Li, F.; Lyu, D.; Li, Y.; Zhu, M.; Jiao, H.; Song, Y. Prevalence, risk factors, and management of dementia and mild cognitive impairment in adults aged 60 years or older in China: A cross-sectional study. *Lancet Public Health* **2020**, *5*, e661–e671. [CrossRef] [PubMed]
2. Riddle, M.; Potter, G.G.; McQuoid, D.R.; Steffens, D.C.; Beyer, J.L.; Taylor, W.D. Longitudinal cognitive outcomes of clinical phenotypes of late-life depression. *Am. J. Geriatr. Psychiatry* **2017**, *25*, 1123–1134. [CrossRef] [PubMed]
3. Cieri, F.; Esposito, R.; Cera, N.; Pieramico, V.; Tartaro, A.; di Giannantonio, M. Late-life depression: Modifications of brain resting state activity. *J. Geriatr. Psychiatry Neurol.* **2017**, *30*, 140–150. [CrossRef] [PubMed]
4. Wang, S.; Blazer, D.G. Depression and cognition in the elderly. *Annu. Rev. Clin. Psychol.* **2015**, *11*, 331–360. [CrossRef] [PubMed]
5. Simon, H.A. Information processing models of cognition. *Annu. Rev. Psychol.* **1979**, *30*, 363–396. [CrossRef]
6. Van den Kommer, T.N.; Comijs, H.C.; Aartsen, M.J.; Huisman, M.; Deeg, D.J.; Beekman, A.T. Depression and cognition: How do they interrelate in old age? *Am. J. Geriatr. Psychiatry* **2013**, *21*, 398–410. [CrossRef] [PubMed]
7. Dumas, J.A.; Newhouse, P.A. Impaired working memory in geriatric depression: An fMRI study. *Am. J. Geriatr. Psychiatry* **2015**, *23*, 433–436. [CrossRef] [PubMed]
8. Evans, J.; Charness, N.; Dijkstra, K.; Fitzgibbons, J.M.; Yoon, J.-S. Is episodic memory performance more vulnerable to depressive affect in older adulthood? *Aging Neuropsychol. Cogn.* **2019**, *26*, 244–263. [CrossRef] [PubMed]
9. Liao, W.; Zhang, X.; Shu, H.; Wang, Z.; Liu, D.; Zhang, Z. The characteristic of cognitive dysfunction in remitted late life depression and amnestic mild cognitive impairment. *Psychiatry Res.* **2017**, *251*, 168–175. [CrossRef] [PubMed]
10. Rajtar-Zembaty, A.; Sałakowski, A.; Rajtar-Zembaty, J.; Starowicz-Filip, A. Executive dysfunction in late-life depression. *Psychiatr. Pol.* **2017**, *51*, 705–718. [CrossRef]

11. Lara, E.; Koyanagi, A.; Domenech-Abella, J.; Miret, M.; Ayuso-Mateos, J.L.; Haro, J.M. The impact of depression on the development of mild cognitive impairment over 3 years of follow-up: A population-based study. *Dement. Geriatr. Cogn. Disord.* **2017**, *43*, 155–169. [CrossRef] [PubMed]

12. Cherbuin, N.; Kim, S.; Anstey, K.J. Dementia risk estimates associated with measures of depression: A systematic review and meta-analysis. *BMJ Open* **2015**, *5*, e008853. [CrossRef] [PubMed]

13. Rushing, N.C.; Sachs-Ericsson, N.; Steffens, D.C. Neuropsychological indicators of preclinical Alzheimer's disease among depressed older adults. *Aging Neuropsychol. Cogn.* **2014**, *21*, 99–128. [CrossRef] [PubMed]

14. Perrino, T.; Mason, C.A.; Brown, S.C.; Spokane, A.; Szapocznik, J. Longitudinal relationships between cognitive functioning and depressive symptoms among Hispanic older adults. *J. Gerontol. Ser. B Psychol. Sci. Soc. Sci.* **2008**, *63*, 309–317. [CrossRef] [PubMed]

15. Bunce, D.; Batterham, P.J.; Christensen, H.; Mackinnon, A.J. Causal associations between depression symptoms and cognition in a community-based cohort of older adults. *Am. J. Geriatr. Psychiatry* **2014**, *22*, 1583–1591. [CrossRef]

16. Yuan, J.; Wang, Y.; Liu, Z. Temporal relationship between depression and cognitive decline in the elderly: A two-wave cross-lagged study in a Chinese sample. *Aging Ment. Health* **2023**, *27*, 2179–2186. [CrossRef]

17. Olaya, B.; Moneta, M.V.; Miret, M.; Ayuso-Mateos, J.L.; Haro, J.M. Course of depression and cognitive decline at 3-year follow-up: The role of age of onset. *Psychol. Aging* **2019**, *34*, 475–485. [CrossRef] [PubMed]

18. Mojtabai, R.; Olfson, M. Cognitive deficits and the course of major depression in a cohort of middle-aged and older community-dwelling adults. *J. Am. Geriatr. Soc.* **2004**, *52*, 1060–1069. [CrossRef] [PubMed]

19. Vinkers, D.J.; Gussekloo, J.; Stek, M.L.; Westendorp, R.G.; van der Mast, R.C. Temporal relation between depression and cognitive impairment in old age: Prospective population based study. *BMJ* **2004**, *329*, 881. [CrossRef]

20. Sanders, J.B.; Bremmer, M.A.; Comijs, H.C.; Deeg, D.J.; Lampe, I.K.; Beekman, A.T. Cognitive functioning and the natural course of depressive symptoms in late life. *Am. J. Geriatr. Psychiatry* **2011**, *19*, 664–672. [CrossRef]

21. Guan, Q.; Hu, X.; Ma, N.; He, H.; Duan, F.; Li, X.; Luo, Y.; Zhang, H. Sleep quality, depression, and cognitive function in non-demented older adults. *J. Alzheimer's Dis.* **2020**, *76*, 1637–1650. [CrossRef]

22. Lee, J.; Sung, J.; Choi, M. The factors associated with subjective cognitive decline and cognitive function among older adults. *J. Adv. Nurs.* **2020**, *76*, 555–565. [CrossRef] [PubMed]

23. Cheng, T.; Zhang, B.; Luo, L.; Guo, J. The influence of healthy lifestyle behaviors on cognitive function among older Chinese adults across age and gender: Evidence from panel data. *Arch. Gerontol. Geriatr.* **2023**, *112*, 105040. [CrossRef] [PubMed]

24. Barulli, D.; Stern, Y. Efficiency, capacity, compensation, maintenance, plasticity: Emerging concepts in cognitive reserve. *Trends Cogn. Sci.* **2013**, *17*, 502–509. [CrossRef] [PubMed]

25. Seidel, S.; Dal-Bianco, P.; Pablik, E.; Müller, N.; Schadenhofer, C.; Lamm, C.; Klösch, G.; Moser, D.; Klug, S.; Pusswald, G. Depressive symptoms are the main predictor for subjective sleep quality in patients with mild cognitive impairment—A controlled study. *PLoS ONE* **2015**, *10*, e0128139. [CrossRef] [PubMed]

26. Yu, J.; Rawtaer, I.; Mahendran, R.; Collinson, S.L.; Kua, E.-H.; Feng, L. Depressive symptoms moderate the relationship between sleep quality and cognitive function among the elderly. *J. Clin. Exp. Neuropsychol.* **2016**, *38*, 1168–1176. [CrossRef]

27. Pearlin, L.I.; Schieman, S.; Fazio, E.M.; Meersman, S.C. Stress, health, and the life course: Some conceptual perspectives. *J. Health Soc. Behav.* **2005**, *46*, 205–219. [CrossRef]

28. Pearlin, L.I.; Menaghan, E.G.; Lieberman, M.A.; Mullan, J.T. The stress process. *J. Health Soc. Behav.* **1981**, *22*, 337–356. [CrossRef] [PubMed]

29. Zhong, W.; Wang, F.; Chi, L.; Yang, X.; Yang, Y.; Wang, Z. Association between sleep duration and depression among the elderly population in China. *Exp. Aging Res.* **2022**, *48*, 387–399. [CrossRef]

30. Xue, K.; Nie, Y.; Wang, Y.; Hu, Z. Number of births and later-life depression in older adults: Evidence from China. *Int. J. Environ. Res. Public Health* **2022**, *19*, 11780. [CrossRef] [PubMed]

31. Lei, X.; Liu, H. Gender difference in the impact of retirement on cognitive function: Evidence from urban China. *J. Comp. Econ.* **2018**, *46*, 1425–1446. [CrossRef]

32. Dotson, V.M.; McClintock, S.M.; Verhaeghen, P.; Kim, J.U.; Draheim, A.A.; Syzmkowicz, S.M.; Gradone, A.M.; Bogoian, H.R.; Wit, L.D. Depression and cognitive control across the lifespan: A systematic review and meta-analysis. *Neuropsychol. Rev.* **2020**, *30*, 461–476. [CrossRef] [PubMed]

33. Zhou, L.; Ma, X.; Wang, W. Relationship between cognitive performance and depressive symptoms in Chinese older adults: The China Health and Retirement Longitudinal Study (CHARLS). *J. Affect. Disord.* **2021**, *281*, 454–458. [CrossRef]

34. Muhammad, T.; Meher, T. Association of late-life depression with cognitive impairment: Evidence from a cross-sectional study among older adults in India. *BMC Geriatr.* **2021**, *21*, 364. [CrossRef] [PubMed]

35. Zheng, F.; Zhong, B.; Song, X.; Xie, W. Persistent depressive symptoms and cognitive decline in older adults. *Br. J. Psychiatry* **2018**, *213*, 638–644. [CrossRef] [PubMed]

36. Rosenberg, P.B.; Mielke, M.M.; Xue, Q.-L.; Carlson, M.C. Depressive symptoms predict incident cognitive impairment in cognitive healthy older women. *Am. J. Geriatr. Psychiatry* **2010**, *18*, 204–211. [CrossRef] [PubMed]

37. Dotson, V.M.; Resnick, S.M.; Zonderman, A.B. Differential association of concurrent, baseline, and average depressive symptoms with cognitive decline in older adults. *Am. J. Geriatr. Psychiatry* **2008**, *16*, 318–330. [CrossRef]

38. Liu, Q.; Leng, P.; Gu, Y.; Shang, X.; Zhou, Y.; Zhang, H.; Zuo, L.; Mei, G.; Xiong, C.; Wu, T. The dose–effect relationships of cigarette and alcohol consumption with depressive symptoms: A multiple-center, cross-sectional study in 5965 Chinese middle-aged and elderly men. *BMC Psychiatry* **2022**, *22*, 657. [CrossRef]

39. Furihata, R.; Konno, C.; Suzuki, M.; Takahashi, S.; Kaneita, Y.; Ohida, T.; Uchiyama, M. Unhealthy lifestyle factors and depressive symptoms: A Japanese general adult population survey. *J. Affect. Disord.* **2018**, *234*, 156–161. [CrossRef]

40. Wiels, W.; Baeken, C.; Engelborghs, S. Depressive symptoms in the elderly—An early symptom of dementia? A systematic review. *Front. Pharmacol.* **2020**, *11*, 34. [CrossRef]

41. Peavy, G.M.; Lange, K.L.; Salmon, D.P.; Patterson, T.L.; Goldman, S.; Gamst, A.C.; Mills, P.J.; Khandrika, S.; Galasko, D. The effects of prolonged stress and APOE genotype on memory and cortisol in older adults. *Biol. Psychiatry* **2007**, *62*, 472–478. [CrossRef] [PubMed]

42. Kelley-Moore, J.A.; Ferraro, K.F. A 3-D model of health decline: Disease, disability, and depression among black and white older adults. *J. Health Soc. Behav.* **2005**, *46*, 376–391. [CrossRef] [PubMed]

Article

The Impact of the COVID-19 Pandemic on Depressive Disorder with Postpartum Onset: A Cross-Sectional Study

Livia Ciolac [1,2,3], Marius Lucian Craina [1,3,4], Virgil Radu Enatescu [5,6], Anca Tudor [7,8], Elena Silvia Bernad [1,3,4], Razvan Nitu [1,3], Lavinia Hogea [6,9], Lioara Boscu [2,10], Brenda-Cristiana Bernad [2,6,9,*], Madalina Otilia Timircan [1,3], Valeria Ciolac [11], Cristian-Octavian Nediglea [12] and Anca Laura Maghiari [13]

[1] Department of Obstetrics and Gynecology, Faculty of General Medicine, "Victor Babes" University of Medicine and Pharmacy Timisoara, 300041 Timisoara, Romania; livia.ciolac@umft.ro (L.C.); mariuscraina@hotmail.com (M.L.C.); bernad.elena@umft.ro (E.S.B.); nitu.dumitru@umft.ro (R.N.); timircan.madalina@yahoo.com (M.O.T.)

[2] Doctoral School, Faculty of General Medicine, "Victor Babes" University of Medicine and Pharmacy Timisoara, 300041 Timisoara, Romania; lioara.boscu@umft.ro

[3] Clinic of Obstetrics and Gynecology, "Pius Brinzeu" County Clinical Emergency Hospital, 300723 Timisoara, Romania

[4] Center for Laparoscopy, Laparoscopic Surgery and In Vitro Fertilization, "Victor Babes" University of Medicine and Pharmacy Timisoara, 300041 Timisoara, Romania

[5] Psychiatric Clinic, "Pius Brinzeu" Emergency County Hospital, 156 Liviu Rebreanu Blvd., 300723 Timisoara, Romania; enatescu.virgil@umft.ro

[6] Department of Neuroscience, "Victor Babes" University of Medicine and Pharmacy Timisoara, 2 Eftimie Murgu Square, 300041 Timisoara, Romania; hogea.lavinia@umft.ro

[7] Discipline of Computer Science and Medical Biostatistics, University of Medicine and Pharmacy, 300041 Timisoara, Romania; atudor@umft.ro

[8] Research Center in Dental Medicine Using Conventional and Alternative Technologies, School of Dental Medicine, "Victor Babes" University of Medicine and Pharmacy of Timisoara, 9 Revolutiei 1989 Ave., 300070 Timisoara, Romania

[9] Center for Neuropsychology and Behavioral Medicine, "Victor Babes" University of Medicine and Pharmacy, 300041 Timisoara, Romania

[10] Senate Office, "Victor Babes" University of Medicine and Pharmacy, Eftimie Murgu Square, No. 2, 300041 Timisoara, Romania

[11] Department of Sustainable Development and Environmental Engineering, University of Life Sciences "King Michael I", 300645 Timisoara, Romania; valeriaciolac@usvt.ro

[12] Clinic of Anaesthesia and Intensive Care, Emergency County Hospital "Pius Brinzeu", 325100 Timisoara, Romania; nedigleacristian@gmail.com

[13] Department of Anatomy and Embryology, "Victor Babes" University of Medicine and Pharmacy Timisoara, 300041 Timisoara, Romania; boscu.anca@umft.ro

* Correspondence: bernad.brenda@umft.ro; Tel.: +40-728-450-443

Citation: Ciolac, L.; Craina, M.L.; Enatescu, V.R.; Tudor, A.; Bernad, E.S.; Nitu, R.; Hogea, L.; Boscu, L.; Bernad, B.-C.; Timircan, M.O.; et al. The Impact of the COVID-19 Pandemic on Depressive Disorder with Postpartum Onset: A Cross-Sectional Study. *Healthcare* 2023, 11, 2857. https://doi.org/10.3390/healthcare11212857

Academic Editors: Carlos Laranjeira and Ana Querido

Received: 20 September 2023
Revised: 24 October 2023
Accepted: 25 October 2023
Published: 30 October 2023

Abstract: Background: COVID-19 has led to a global health crisis that is defining for our times and one of the greatest challenges to emerge since World War II. The potential impact of the pandemic on mental health should not be overlooked, especially among vulnerable populations such as women who gave birth during the COVID-19 pandemic. Materials and Methods: The study is a cross-sectional survey conducted from 1 March 2020 to 1 March 2023, during the period of the SARS-CoV-2 (COVID-19) pandemic, based on a retrospective evaluation of 860 postpartum women. The screening tool used to assess symptoms of postpartum depression was the Edinburgh Postnatal Depression Rating Scale (EPDS) questionnaire. The questionnaire was completed both in the Obstetrics and Gynaecology Clinical Sections I and II of the "Pius Brînzeu" County Emergency Hospital in Timisoara, Romania, and online using Google Forms. Results: The highest severity of postpartum depression symptoms was observed during the COVID-19 pandemic. The results of the study conducted during the period of the SARS-CoV-2 pandemic (COVID-19) showed that the prevalence of major postpartum depressive disorder (EPDS $\geq$ 13) was 54.2% (466 patients), while 15.6% (134) had minor depressive disorder (10 < EPDS $\leq$ 12) in the first year after delivery. Comparing these results with those obtained in research conducted before the onset of the pandemic period showed an alarming increase in the prevalence of postpartum depression. The risk factors associated with postpartum depression included the type of delivery, level of education, socio-economic conditions, health status, age, background, and personal obstetric history (number of abortions on demand, parity). Conclusions: The effects of the pandemic on mental health are of particular concern for women in the first year after childbirth. Observing these challenges and developing effective measures to prepare our health system early can be of great help for similar situations in the future. This will help and facilitate effective mental health screening for postpartum women, promoting maternal and child health.

Keywords: postpartum depression; screening; pandemic; COVID-19; Edinburgh Postnatal Depression Rating Scale (EPDS)

1. Introduction

The 2019 coronavirus disease (COVID-19), caused by severe acute respiratory syndrome coronavirus 2 (SARS-CoV-2), was first recognized in December 2019 in Wuhan, the capital of China's Hubei province. Since then, the disease has spread worldwide, leading to a coronavirus pandemic [1]. In January 2020, it was recognized by the World Health Organization (WHO) as a major public health concern [1]. COVID-19 has led to a global health crisis that is defining for our times and one of the greatest challenges to emerge since World War II. As a result of this sudden onset epidemic, governments and public health authorities urgently needed guidance and useful information on effective interventions to protect the health of the population [2].

Given the paucity of data in the existing academic literature on clinically manifested peripartum depression among women who gave birth during the COVID-19 pandemic, we set out to assess the risk of developing peripartum depressive disorder during the COVID-19 pandemic compared with the risk among women who gave birth before the onset of the COVID-19 pandemic.

The postpartum period represents a time of increased vulnerability for the development of psychiatric disorders [2]. The potential impact of the pandemic on mental health should not be overlooked, especially among vulnerable populations [3,4]. According to the World Health Organization, approximately 10% of parturients and 13% of postpartum women experience some form of mental disorder, mainly depression [5]. In the context of the COVID-19 pandemic, maternal distress may be exacerbated by concerns and fears about the risk of infection or hospitalization due to the coronavirus, given that perinatal morbidity and mortality associated with COVID-19 have been documented [6,7].

Postpartum depression is defined in psychiatric nomenclature as a major depressive disorder with a specific onset in the first month after parturition, with the possibility of extending the time interval up to one year [8]. It is considered a major public health problem as it affects both the mother and the child and has a high prevalence globally, ranging from 10% to 20% in most studies [9]. The term postnatal depression is generically used in the literature to designate the picture of depressive symptoms, with an onset in the period following childbirth and whose etiology is related to childbirth, as well as to hormonal (physiological), psychological, environmental, or social aspects, which occur in temporal proximity to the moment of birth [9].

Women who develop postnatal depression are at greater risk of relapsing during subsequent pregnancies and of developing major depressive disorder outside the perinatal period [8]. Studies in recent years have shown that the nature of the early mother–infant relationship in the context of postpartum depression is predictive of the child's cognitive, emotional, and social development [10].

Previous publications have noted an increased likelihood of depressive symptoms among pregnant or postpartum women, but these studies have been limited by the relatively small sample sizes of the patients included, as well as the individual particularities of each country in which they were conducted [11,12]. The extent to which parturients were emotionally affected by the pandemic remains unclear and problematic. It is therefore necessary to clarify which women are more at risk of being affected by peripartum depression. It is also important that the factors involved in the onset of mental distress are identified to help develop effective screening and prevention strategies among vulnerable populations.

The aim of this study was to assess the depressive symptomatology of postpartum women in pandemic times and to investigate potential associations between the symptoms of depressive disorder onset in close temporal proximity to the time of birth and the socio-demographic conditions, health status, and obstetric particularities of the patients.

2. Materials and Methods

2.1. Sample Description

The study is a cross-sectional survey conducted from 1 March 2020 to 1 March 2023, during the SARS-CoV-2 (COVID-19) pandemic, based on a retrospective evaluation of 860 postpartum women. This manuscript of the conducted observational study was prepared following STROBE guidelines [13]. We performed a G*Power (version 3.1.9.7) test for the chi-square test family, contingency tables as goodness of fit tests with 90% power, 0.05 level of significance, one degree of freedom, and 0.11 as an effect size. The estimated sample size was 853 respondents.

For the purpose of our study, the patients were meticulously screened and chosen based on a comprehensive set of inclusion and exclusion criteria, to ensure the specificity and uniformity of our participant pool. All 860 postpartum women were eligible participants incorporated in the investigation. The study was carried out in the Obstetrics and Gynaecology Clinical Sections I and II of the "Pius Brînzeu" County Emergency Hospital in Timisoara, Romania.

Inclusion and Exclusion Criteria

Participants were enrolled into the study if they met the following inclusion criteria:

- Delivered mothers within age group of 18–50 years;
- Women who had given birth in the last year from the date of completion of the survey;
- No history of psychiatric disorders;
- No history of peripartum depression in previous pregnancies;
- No past incidents or diagnoses of COVID-19 infection in the last year;
- Women who expressed an interest in this topic and who have provided informed consent to participate in the study.

Participants were excluded if they met any of the following conditions:

- Women with current and past use of psychotropic medications;
- Women with a high-risk pregnancy (including preeclampsia, gestational diabetes mellitus, chronic disease, intrauterine growth restriction, known fetal anomalies, or chromosomal aberrations);
- Women who had a history of psychiatric disorders or mental health issues.

Prior informed consent was obtained for each patient since it consisted of sensitive data and a vulnerable group of study participants.

The screening tool used to assess the symptoms of postpartum depression was the Edinburgh Postnatal Depression Rating Scale (EPDS) questionnaire. The questionnaire was completed both in the Obstetrics and Gynaecology Clinical Departments I and II and online using Google Forms. The percentages of participants that completed the EPDS questionnaire in the hospital was 93.26% (802), while 6.74% (58) completed it online.

In order to highlight the particularities of postpartum depression, as well as the factors favoring this pathology, in the patients included in this study, the following parameters were also taken into account: age; marital status; background; level of education; working conditions (risk at work); socio-economic conditions; health status; personal pathological history; parity; the method of obtaining pregnancy; type of birth, under the recommendation of the medical consultant; the mother's wishes regarding the type of birth; the number of miscarriages; and the number of abortions upon request.

2.2. Ethics Declarations

The Local Commission of Ethics for Scientific Research from the Timis County Emergency Clinical Hospital "Pius Brînzeu" in Timisoara, Romania, operates under the article 167 provisions of Law no. 95/2006, art. 28, chapter VIII of order 904/2006; with EU GCP Directives 2005/28/EC, International Conference of Harmonisation of Technical Requirements for Registration of Pharmaceuticals for Human Use (ICH); and with the Declaration of Helsinki—Recommendations Guiding Medical Doctors in Biomedical Research Involving Human Subjects. The current study was conducted according to the guidelines of the Declaration of Helsinki, followed the European Union General Data Protection Regulation (GDPR), and was approved by the Local Commission of Ethics for Scientific Research from the Timis County Emergency Clinical Hospital "Pius Brinzeu" in Timisoara, Romania, No. 184/10.02.2020.

2.3. Edinburgh Postnatal Depression Scale Questionnaire

Depression is a pathology where psychometric assessment is particularly useful in confirming the diagnosis.

The Edinburgh Postnatal Depression Rating Scale (EPDS) questionnaire is one of the most widely used screening tools for assessing symptoms of perinatal depression and anxiety [14]. The EPDS had been validated against clinical diagnoses in over 37 languages since its development, and it has been regarded as the most frequently used and well-validated screening tool for postpartum depression [15]. The EPDS is a 10-item questionnaire, which has been validated in different countries and populations, including Romania [16–18]. It is a simple instrument that assesses emotional experiences over the past seven days using 10 Likert scale questions, is easy to fill in and interpret, requires no specialist psychiatric expertise, and could easily be incorporated into the health care services offered to all women in the postnatal period [14].

The Edinburgh Postnatal Depression Scale (EPDS) has become a leading choice for identifying women at risk for the diagnosis [19]. This self-report tool was developed and tested in health centers in Edinburgh and Livingston (UK) by Cox, Holden, and Sagovsky in 1987 to help detect women suffering from postnatal depression [20]. Since its inception, the EPDS has been adapted for use in several countries and has become the most widely used tool for assessing postpartum depression [19].

The ratings of responses to the 10 questions are summed, and the resulting score can assess the likelihood that the patient has clinical depression. The EPDS scale is composed of three structural factors: the "depression" factor through questions 1, 2, and 8; the "anxiety" factor through questions 3, 4, and 5; and the "suicide" factor through question 10. A score >10 betrays a possible depression (minor depressive disorder), and a score ≥ 13 suggests a major depressive disorder (moderate to severe) [14,21,22].

The EPDS scale can provide stable results, especially when assessments are performed repeatedly. In comparison with a clinical diagnostic interview, the EPDS demonstrated the following psychometric properties: a specificity of 78%, a sensitivity of 86%, and a positive predictive value of 73% for women scoring >10 [20]. Validity studies show that the scale can correctly identify 92.3% of women with postpartum depression [20].

2.4. Statistical Assessment

The collected data were introduced in Excel format and statistically processed with the SPSSv.17 software package.

Nominal variables were represented as frequency tables for which percentage distributions (pie) were plotted and associated with the $\chi 2$ (Chi square) test of concordance. For the numeric variables, indicators of central tendency (mean and median) and dispersion (standard deviation and standard error of the mean) were calculated, and for the study of the association between them, a Spearman's nonparametric linear correlation analysis was carried out, with the help of which we calculated the correlation coefficients and probability values that provided us the significance of the correlation (p-values must be below 0. 05 for the association to be significant). For comparisons between two sets of numerical variables, the Mann–Whitney U nonparametric test was used, and for comparisons between more than two sets, the Kruskal–Wallis nonparametric test was applied.

3. Results

The study involved 860 women in their first year after childbirth. The distribution of socio-demographic characteristics and obstetric indicators among study participants are presented in Table 1.

Table 1. Distribution of socio-demographic characteristics and obstetric indicators among study participants.

Age	
Range	18–45
Mean ± SD	28.52 (4.93)
Marital status	
Married	86.7% (746)
Cohabiting	11.4% (98)
Single	1.9% (16)
Area of residence	
Urban	69.5% (598)
Rural	30.5% (262)
Level of education	
Higher education	59.2% (509)
High school	32.2% (277)
Primary education	8.6% (74)
Socio-economic conditions	
Very good standard of living	19.4% (167)
Good standard of living	59% (507)
Satisfactory conditions	18.6% (160)
Poor living conditions	3% (26)
Workplace hazard	
High	8.4% (72)
Medium	22.7% (195)
Low	69% (593)
Health status	
Good	79.4% (683)
Satisfactory	19.5% (168)
Poor	1% (9)
Parity	
Primiparous	65.2% (561)
Secondiparous	28.1% (242)
Tertiparous	5.2% (45)
Quarteparous	1.3% (11)
Quintiparous	0.1% (1)

Table 1. *Cont.*

Number of miscarriages in their personal obstetric history	
No miscarriage	80.1% (689)
One miscarriage	15.1% (130)
Two miscarriages	4% (34)
Three miscarriages	0.6% (5)
Four miscarriages	0.2% (2)
Number of abortions performed upon request in their personal obstetric history	
No abortion on request	87.2% (750)
One abortion on request	9% (77)
Two abortions on request	2.7% (23)
Three abortions on request	0.9% (8)
Four abortions on request	0% (0)
Five abortions on request	0.2% (2)
Method of achieving pregnancy	
Naturally	95.8% (824)
In vitro fertilization	1.2% (10)
With previous treatment	3% (26)

Following the assessment of the clinical status of the women, using the Edinburgh Psychiatric Postnatal Depression Rating Scale (EPDS), 54.2% (466 patients) had major depressive disorder, 15.6% (134) had minor depressive disorder, and 30.2% (260 patients) had no depressive disorder.

The recorded values of the Edinburgh score following the completion of the questionnaire ranged from 0 to 28, with a mean score of 13.06. The maximum possible Edinburgh score of 30 was not recorded in this study (Figure 1).

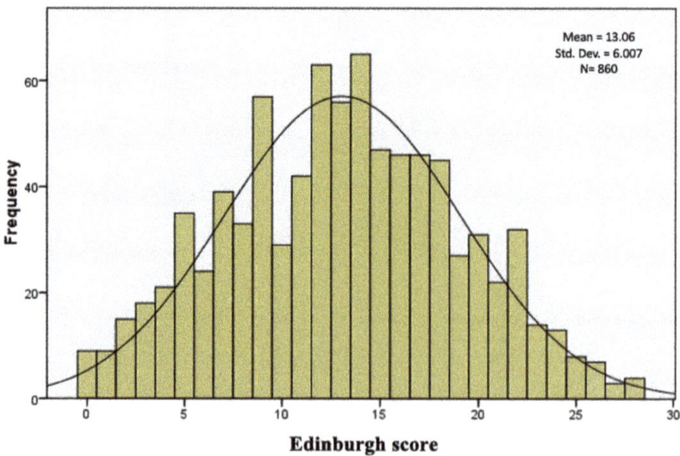

Figure 1. Histogram of Edinburgh scores.

Ratings of the responses to the 10 questions of the EPDS questionnaire are evidenced in Table 2.

Table 2. Percentage distribution of responses to the 10 questions of EPDS questionnaire.

Question Number	Item Question	Score 0	Score 1	Score 2	Score 3
Question 1	"I have been able to laugh and see the funny side of things."	57.2% (492)	34.1% (293)	7.3% (63)	1.4% (12)
Question 2	"I have looked forward with enjoyment to things."	65.9% (567)	22.4% (193)	9.4% (81)	2.2% (19)
Question 3	"I have blamed myself unnecessarily when things went wrong."	7.4% (64)	24% (206)	41.4% (356)	27.2% (234)
Question 4	"I have been anxious or worried for no good reason."	10% (86)	7.9% (68)	58.5% (503)	23.6% (203)
Question 5	"I have felt scared or panicky for no very good reason."	15.6% (134)	21% (181)	42.8% (368)	20.6% (177)
Question 6	"Things have been getting on top of me."	7.4% (64)	21% (181)	58.3% (501)	13.3% (114)
Question 7	"I have been so unhappy that I have had difficulty sleeping."	31.5% (271)	24.4% (210)	32.9% (283)	11.2% (96)
Question 8	"I have felt sad or miserable."	19.3% (166)	38.6% (332)	30.7% (264)	11.4% (98)
Question 9	"I have been so unhappy that I have been crying."	15% (129)	16.9% (145)	46.9% (403)	21.3% (183)
Question 10	"The thought of harming myself has occurred to me."	71.3% (613)	13.8% (119)	12.2% (105)	2.7% (23)

We aimed to analyze the prevalence of suicidal ideation in postpartum women in the context of the ongoing pandemic of the coronavirus disease. We found that 14.9% (247) of 860 mothers who completed the EPDS question related to suicidal ideation (question 10 of the questionnaire) reported some suicidal ideation: 2.7% (23) reported that the thought of harming themselves had occurred to them quite often, and 12.2% (105) reported that it sometimes occurred to them.

Of the 860 patients, 57.7% (496) gave birth by caesarean section, and 42.3% (364) gave birth naturally. As for the mother's wishes regarding the type of birth, 71.9% (618) would have liked to give birth naturally, and 28.1% (242) would have opted for caesarean section. A significant association was found between the mother's desire for future birth and the type of birth (Chi square, $p < 0.001$); in other words, the obstetrician took the mother's desire into account when determining the type of birth. The proportion of births by caesarean section, for mothers who wanted caesarean section, was significantly increased compared with the proportion of caesarean sections for mothers who wanted natural birth.

A significant association was established between type of delivery and depressive disorder (Chi square, $p = 0.003$). The proportion of mothers without depressive disorder was significantly increased among those who delivered naturally (Chi square, $p = 0.0012$), and the proportion of those with major depressive disorder was significantly decreased for mothers who delivered naturally (Chi square $p = 0.0041$) (Table 3). The occurrence of postpartum depressive disorder was significantly influenced by the type of delivery.

There was a significant association between the type of birth and marital status (Chi square, $p < 0.001$). Married mothers who gave birth by Caesarean section were significantly more numerous than those who gave birth naturally (Chi square, $p = 0.038$). Those living in cohabitation who gave birth by Caesarean section were significantly fewer than those who gave birth naturally (Chi square, $p = 0.0011$). Single mothers who gave birth by caesarean section were significantly more than those who gave birth naturally (Chi square, $p = 0.026$). (Table 3)

The association between the type of birth and the level of education was significant (Chi square, $p < 0.001$). The proportion of mothers without a high school education was significantly increased among those who gave birth naturally (Chi square, $p < 0.001$), while the proportion of mothers with higher education was significantly increased among those who gave birth by caesarean section (Chi square, $p = 0.012$). (Table 3)

The proportion of mothers who have a high school education was significantly increased among those with postpartum depression ($p = 0.010$), while the proportion of those with higher education was significantly increased among those without postpartum depressive disorder ($p = 0.028$) (Table 3). The association between the education level and postpartum depression was significant; higher education seemed to be a protective factor against the onset of depressive symptomatology (Chi square, $p = 0.028$).

Table 3. Comparative percentage representation of cases by socio-demographic variables, health status, type of birth, and occurrence of depressive disorder.

Association Variables		Type of Birth		*p*-Value	Postpartum Depression		*p*-Value
		Caesarean Section	Vaginal Delivery		Absence	Present	
Depressive disorder	Without	128 (25.80%)	132 (36.30%)	0.0012 *			
	Minor	78 (15.70%)	56 (15.40%)	0.98	-		
	Major	290 (58.50%)	176 (48.40%)	0.0041 *			
Marital status	Married	441 (88.90%)	305 (83.80%)	0.0382 *	229 (88.10%)	517 (86.17%)	0.511
	Cohabiting	41 (8.30%)	57 (15.70%)	0.0011 *	26 (10.00%)	72 (12.00%)	0.465
	Single	14 (2.80%)	2 (0.50%)	0.026 *	5 (1.90%)	11 (1.83%)	0.837
Education level	Less than high school	24 (4.80%)	50 (13.70%)	<0.001 *	24 (9.20%)	50 (8.30%)	0.764
	High school graduate	160 (32.30%)	117 (32.10%)	0.991	67 (25.8%)	210 (35.00%)	0.010 *
	Higher education	312 (62.90%)	197 (54.10%)	0.012 *	169 (65.00%)	340 (56.70%)	0.028 *
Socio-economic conditions	Good	290 (58.50%)	217 (59.60%)	0.799	157 (60.40%)	350 (58.30%)	0.617
	Very good	111 (22.40%)	56 (15.40%)	0.013 *	60 (23.10%)	107 (17.80%)	0.087
	Poor	10 (2.00%)	16 (4.40%)	0.067	4 (1.50%)	22 (3.70%)	0.131
	Satisfactory	85 (17.10%)	75 (20.60%)	0.224	39 (15.00%)	121 (20.20%)	0.088
Health status	Good	386 (77.80%)	297 (81.60%)	0.202	232 (89.20%)	451 (75.20%)	<0.001 *
	Poor	6 (1.20%)	3 (0.80%)	0.816	1 (0.40%)	8 (1.30%)	0.404
	Fair	104 (21.00%)	64 (17.60%)	0.248	27 (10.40%)	141 (23.50%)	<0.001 *

*—Significant difference.

The association between depressive disorder and health status was significant (Chi square, $p < 0.001$). The proportion of mothers with depressive disorder was significantly increased among those with fair or poor health status (Chi square, $p < 0.001$) (Table 3).

A direct, significant, and weak correlation was found between the number of miscarriages and the number of births (Spearman correlation coefficient r = 0.165, $p < 0.001$)—women who had an increased number of miscarriages also had an increased number of births.

A direct, significant and weak correlation was found between the number of abortions on demand and the number of births (Spearman correlation coefficient r = 0.142, $p < 0.001$)—women who had an increased number of abortions on demand also had an increased number of births.

A direct, significant, and weak correlation (Spearman correlation coefficient r = 0.138, $p = 0.001$) was found between the number of abortions on demand and the Edinburgh score—women who had an increased number of abortions on demand also had an increased Edinburgh score.

The maternal age was significantly lower for the respondents with depressive disorder (Mann–Whitney U nonparametric test, $p = 0.025$), indicating a possible association of younger age with the onset of postnatal depression (Table 4 and Figure 2).

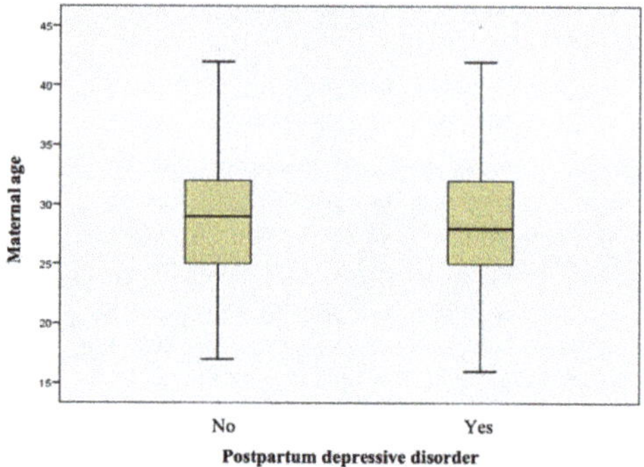

Figure 2. Boxplot representation for maternal age, comparative, according to depressive disorder.

Table 4. Descriptive statistics of numerical variables, comparatively, by depressive disorder.

Variable	Depressive Disorder	N	Mean	Std. Deviation	Std. Error Mean	Mean Rank
Maternal age	No	260	29.04	4.782	0.297	459.31
	Yes	600	28.29	4.982	0.203	418.01
Number of miscarriages	No	260	0.20	0.511	0.032	413.76
	Yes	600	0.28	0.602	0.025	437.75
Number of abortions on demand	No	260	0.09	0.328	0.020	409.48
	Yes	600	0.22	0.622	0.025	439.61
Number of births	No	260	1.43	0.639	0.040	437.21
	Yes	600	1.43	0.675	0.028	427.59
Number of miscarriages	Absence	260	0.20	0.511	0.032	413.76
	Minor	134	0.34	0.660	0.057	454.24
	Major	466	0.27	0.585	0.027	433.01
Number of abortions on demand	Absence	260	0.09	0.328	0.020	409.48
	Minor	134	0.23	0.612	0.053	443.26
	Major	466	0.22	0.625	0.029	438.56
Number of births	Absence	260	1.43	0.639	0.040	437.21
	Minor	134	1.54	0.752	0.065	462.49
	Major	466	1.39	0.648	0.030	417.56

Women with postnatal depressive disorder had a significantly increased number of abortions on demand (Mann–Whitney U nonparametric test, $p = 0.005$).

The number of abortions on demand was significantly increased among mothers with depressive disorder, both minor and major; therefore, the personal obstetric history of patients may be a factor in the development of postnatal depressive disorder (Kruskall–Wallis nonparametric test, $p = 0.018$) (Table 4 and Figure 3).

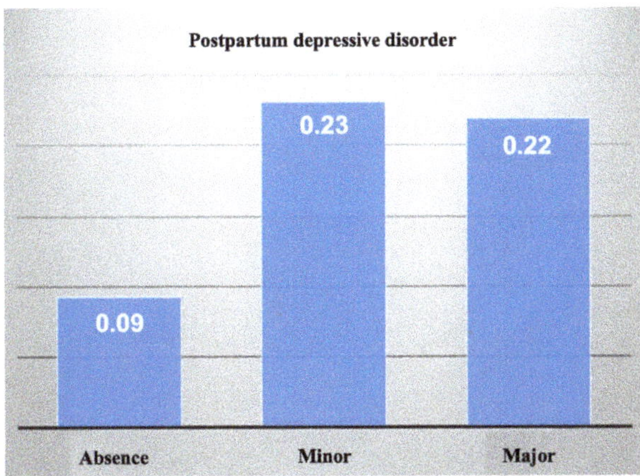

Figure 3. Mean values of the number of abortions on demand compared by severity of depressive disorder.

Mothers who gave birth by caesarean section had a significantly increased Edinburgh score (Mann–Whitney U nonparametric test, $p = 0.008$), which meant that they were more likely to develop depressive symptoms (Table 5).

Table 5. Descriptive statistics of the Edinburgh score, comparative, by type of birth.

Variable	Type of Birth	N	Mean	Std. Deviation	Std. Error Mean	Mean Rank
Edinburgh Score	Vaginal delivery	364	12.46	6.116	0.321	449.71
	Caesarean section	496	13.51	5.892	0.265	404.32

4. Discussion

The results of the study conducted during the period of the SARS-CoV-2 pandemic (COVID-19) showed that the prevalence of major postpartum depressive disorder was 54.2% (466 patients), while 15.6% (134) had minor depressive disorder in the first year after delivery. Comparing these results with those obtained in research conducted before the onset of the pandemic period showed an alarming increase in the prevalence of postpartum depression. The incidence of postpartum depressive disorder worldwide in the non-pandemic period was about 10% in developed countries and about 21–26% in developing countries [23,24]. Previous research has also found that during natural disasters that struck humanity, the prevalence rates of mental disorders among postpartum women were significantly higher than those among the general population [25]. Moreover, studies conducted in the pre-pandemic period have shown that in about 30% of patients with postnatal depressive disorder, the recovery or resolution of depressive symptoms may take more than 1 year, but it is not known to what extent the pandemic may influence the recovery time, and further research is needed.

It is known that postnatal depression has long-term consequences for both the mother and the infant, so identifying the risk factors involved could help to conduct targeted screening, as well as to design targeted intervention strategies to prevent the long-term impact of the pandemic on maternal mental health and infant development [26,27].

In terms of the mothers' backgrounds, 69.5% (598) were from urban areas, and 30.5% (262) were from rural areas. The implemented preventive measures, quarantine, home isolation, social distancing, aimed at stopping further spread of the virus, have increased the level of anxiety and stress among postpartum women [28,29], especially for mothers living in urban areas [30], regarded as high-incidence areas [31].

The occurrence of postpartum depressive disorder was significantly influenced by the type of birth. Of the 860 patients included in the study, 57.7% (496) gave birth by caesarean section, and 42.3% (364) gave birth naturally. In terms of the mother's wishes regarding the type of birth, 71.9% (618) would have preferred a natural birth, and 28.1% (242) would have opted for a caesarean birth. The obstetrician also considered the mother's wishes when deciding on the type of birth. The proportion of births by caesarean section, for mothers who wanted caesarean sections, was significantly increased compared with the proportion of caesarean sections, for mothers who wanted to give birth naturally during the period of our study.

Mothers who gave birth during the pandemic reported a higher level of perceived pain during labor [32] or during the recovery period secondary to the caesarean section. A significant association between the type of birth and depressive disorder was established in our own study; the proportion of mothers without depressive disorder was significantly increased among those who gave birth naturally, while the proportion of mothers with major depressive disorder was significantly decreased for mothers who gave birth naturally. In addition, patients who gave birth by caesarean section or those who experienced perinatal complications [33] tended to be more depressed, as the length of hospital stay increased [34]. It is important to mention that the presence of cardiovascular risk factors, with a negative impact on pregnancy, can also affect the mother's physical well-being; therefore, it is crucial to identify and properly treat these factors for optimal maternal health [35]. These aspects seem to be responsible for the increased risk of postpartum depression.

In contrast to the findings in most previous studies, single mothers did not show symptoms of postpartum depression to a significantly greater extent than mothers who were married or cohabiting [36]. By marital status, 86.7% (746) of the mothers included in the study were married, 11.4% (98) were cohabiting, and 1.9% (16) were single. Single-mother families often face structural disadvantages due to having lower income and less time together with their children [36]. In a Swedish study, children of single-parent households (90% women) were found to be at increased risk for childhood psychopathology, suicide attempts, and drug addiction [37]. Single mothers may face not only the non-shared care of a child but also economic problems resulting from discriminatory wage levels and the absence of a second income from a partner [38]. Paternal support is known to play an important role in the postnatal period but was not interpreted during this study.

In accordance with previous studies, the results indicated that giving birth at a young age was associated with symptoms of postpartum depression [36]. Teenage mothers were at increased risk for depression [39]. Several factors might be of importance for this finding. First, an adolescent mother faces the challenge of her own developmental tasks, in addition to the challenge of taking care of a newborn [36]. Second, early motherhood is associated with lower degrees of education and lower income [40]. Childbirth during adolescence is demanding as it takes place during an intense mental and physical developmental stage, challenging or forcing the transition from childhood to adulthood [36]. Children of teenage mothers have been shown to have delays in cognitive and language abilities [41].

The COVID-19 pandemic has increased women's insecurity from many perspectives. The present study also investigated the risk of postnatal depression according to the mothers' background, education level, and socio-economic conditions, and these variables were found to have a greater impact on depressive symptoms compared with younger age or unmarried status. Women with a good standard of living appeared to have a lower risk of experiencing postnatal depression. Research conducted between 2020 and 2021 highlighted an increased prevalence of food insecurity due to negative changes in food availability [42]. A lack of food security negatively influences quality of life and, thus, health status and is associated with poor nutrition for pregnant or breastfeeding women, obesity, depression, and even high mortality rates [43]. Post-partum depression, when associated with food insecurity found among families with poor living conditions, increases the risk of delayed early child development [44]. Families with poor living conditions should therefore be identified and supported to prevent maternal mental health problems.

The study showed that postpartum depression was influenced by parity. Of the patients, 65.2% were primiparous. Multiparous mothers were less likely to experience postpartum depression, compared with primiparous mothers. According to a study in the peer-reviewed literature, 50–60% of women experience postpartum depression after their first birth [45]. Also, a study conducted in Japan identified a positive correlation between perinatal depressive disorder and primiparity [46]. The experience of the first child may cause more fear and anxiety about childbirth or the immediate postpartum period; therefore, the support of these women could contribute to the prevention and timely diagnosis of depressive disorder with onset close to parturition.

The relationship between suicidal thoughts and suicidal acts in the postpartum period is not clear, but it is prudent to assume that suicidal thoughts are a marker of an increased risk of suicide [47]. We aimed to analyze the prevalence of suicidal ideation in postpartum women, in the context of the ongoing pandemic of coronavirus disease. Suicidal ideation was defined as an answer of "sometimes" or "yes, quite often" to question 10 of the EPDS, "The thought of harming myself has occurred to me"; no suicidal ideation was defined by answering "hardly ever" or "never" for question 10 [48]. We found that 14.9% (247) of 860 mothers who completed the EPDS question related to suicidal ideation (question 10 of the questionnaire) reported some suicidal ideation: 2.7% (23) reported that the thought of harming themselves had occurred to them quite often, and 12.2% (105) reported that it occurred to them sometimes. Like all previous studies, the main limitation of the EPDS suicidality measure is that it is about self-reported thoughts of self-harm. Self-reporting of suicidal thoughts may lead to under-reporting of them; also, it is possible that self-reporting of these ideations may more accurately reflect the truth problem than clinical interviews. However, healthcare professionals, using the EPDS, should be aware of the significant suicidality that is likely to be present in women endorsing "yes, quite often" to question 10 of the EPDS.

Although the original purpose of the EPDS was postpartum depression screening, recent studies revealed that subscales of the EPDS can be used in new ways, such as anxiety disorder screening [15,49]. Given that postpartum depression is often accompanied by anxiety [50], this means that the application of its subscales can help evaluate mothers' mental health conditions in greater detail [15]. Thus, the "anxiety" factor of the EPDS scale, known as a subscale of EPDS, consisting of items 3, 4, and 5 (EPDS-3A) of the questionnaire, may help to identify anxiety symptoms, expected to be accentuated in the context of the COVID-19 pandemic and the fact that the birth took place in the hospital, where patients that tested positive for COVID-19 infection were being treated, with the corresponding increased risk of infection for the rest of the hospitalized mothers. Table 2 shows the percentage distribution of the responses to questions 3, 4, and 5 of the EPDS questionnaire. It is noteworthy that the percentage distribution recorded for score 3 of questions 3 and 4 of the EPDS questionnaire was higher compared with the corresponding one for the rest of the questions. In addition to hospitalization, the fear of adverse effects of the virus and vaccines on the developing fetus; disruption of maternal–infant bonding; social isolation; less contact with friends, family, and social care services; and financial problems related to lockdown measures exerted additional anxiety in the pandemic context [51].

A cutoff of ≥ 5 on the EPDS-3A score was found to be efficient for identifying women experiencing clinical levels of anxiety (sensitivity: 70.9%; specificity: 92.2%) [52]. In settings where the EPDS is already implemented and where adding extra mental health screening instruments is not feasible, the EPDS-3A could be used as a resource-effective means of detecting mothers with possible anxiety disorder [52]. The vast majority of women screening positive on the EPDS-3A also screen positive on the total EPDS; using the EPDS-3A score along with the total EPDS score can indicate whether a mother may be suffering from anxiety either co-morbid with depression or as the primary problem [52].

The conducted research supports existing studies on the prevalence of mental health problems during the COVID-19 pandemic [53–55].

These findings indicated that the pandemic, caused by the spread of SARS-CoV-2 (COVID-19), as an acute public health problem, requires ongoing, comprehensive, and long-term health education to effectively mitigate panic and fear in women, thereby improving the ability of such a vulnerable population to respond in the future in a similar context.

Research on postnatal depression is challenging due to the complexity of the factors involved; therefore, some limitations of the study should also be considered. First, the questionnaire was also completed online via the Google Forms application and promoted via social media, a sampling technique that carries an inherent risk in terms of meeting the selection criteria of the study population. However, online surveys are also considered a good method of population recruitment for epidemiological research, especially in pandemic settings, and internet use is high among women of childbearing age in Europe [56,57]. Compared with national birth data, the enrolled participants were predominantly primiparous, highly educated, married, from urban areas, with a good standard of living, and without a significant personal pathological history. Given that women with higher education and the support of a partner tend to experience fewer symptoms of postpartum depression [58,59], while those with a lower level of education were susceptible to the appearance of postnatal suicide ideation [60], the high prevalence observed in our sample could shed light on the impact of the pandemic. It is also possible that mothers with marked anxiety and more severe symptoms of postnatal depressive disorder may not have completed the online questionnaire, and the likelihood of their being caught in the course of those approached during hospitalization is uncertain. Therefore, the high prevalence of postnatal depressive disorder observed in the study population may still reflect an underestimation of the pandemic situation in the general population. Second, the lack of a comparison group, as well as the cross-sectional study design, prevented us from drawing conclusions about the long-term consequences of the pandemic (whether the mental distress noted would subside in the short term or persist for a longer period of time). Such research requires longitudinal studies, conducted over several years, which are, therefore, more costly and time-consuming but which demonstrate the validity of the results. Third, recruiting patients in the postnatal period of pandemic times has been challenging because of the vulnerable terrain faced by new mothers, requiring extra involvement, patience, empathy, and extra attention from researchers. Ultimately, regression models only observed associations between socio-demographic conditions, health status, and a few obstetric features of the patients. While there may be some limitations in the study design and methods used, these inherent limitations by no means compromise the results reported. Although our data were based on a cross-sectional design, future studies may consider other methods, such as prospective designs, in-person interviews, or other objective measures, to avoid such limitations.

This study aimed to investigate the role of several maternal health status and socio-demographic factors in the risk of developing postpartum depression, yet the data were collected during the COVID-19 pandemic. We did not have information considering the impact of the pandemic on the risk of postpartum depression development among our study sample, as the design of the study was cross-sectional, and we did not have baseline data for the levels of postpartum depression risk in the pre-pandemic period. The COVID-19 pandemic may have played a major role in the risk of postpartum depression detected in this study.

We strongly encourage increased national attention to improve maternal mental health and reproductive health. Attention from primary care providers and other healthcare professionals and policymakers working in the reproductive and maternal health fields is also required [61]. Future studies should aim to examine what other variables may influence women's psychological well-being, including the impact of the COVID-19 pandemic. Postpartum depressive disorder is known to have multifactorial causes, with the substrate being generated by a combination of factors: biological, psychological, and social. Biological factors (hormonal changes, genetic predisposition, or neurochemical imbalances) interact with psychosocial factors (stress, lack of family support, or traumatic life events); this multifactorial nature causes it to be difficult to isolate specific causes or determine the relative contributions of each factor. It may also be of value to consider the major impact of a difficult birth experience (birth trauma) and the potential emotional impact of breastfeeding success and maternal wellness in the context of perinatal mental health issues.

Despite these difficulties, research efforts are aimed at improving the understanding of postpartum depressive disorder, the accuracy and precision of the diagnosis, and the development of effective interventions to reduce the stigma associated with the condition.

5. Conclusions

Women who gave birth during the COVID-19 pandemic are part of a susceptible, high-risk group that should be closely monitored to minimize the effects of possible undiagnosed postnatal depressive disorder. The conducted research indicated that the pandemic caused by the spread of SARS-CoV-2 (COVID-19), as an acute public health problem, shows an alarming increase in the prevalence of postpartum depression compared with studies conducted in the pre-pandemic period. As noted, the onset of the pandemic has generated major changes in postpartum care and created new challenges that could negatively impact maternal mental health. The effects of the pandemic on mental health are of particular concern for women in the first year after childbirth. Observing these challenges and developing effective measures to prepare our health system early can be of great help for similar situations in the future. This will help and facilitate effective mental health screening for postpartum women, promoting maternal and child health. Therefore, more research is needed to understand the relationship between COVID-19 and postnatal depressive disorder.

Author Contributions: Conceptualization, L.C. and M.L.C.; methodology, L.C. and E.S.B.; software, A.T.; validation, A.T., V.R.E. and R.N.; formal analysis, L.H., B.-C.B. and L.B.; investigation, M.O.T. and C.-O.N.; resources, A.L.M. and L.B.; data curation, L.C. and M.O.T.; writing—original draft preparation, L.C., V.C. and C.-O.N.; writing—review and editing, L.C., V.R.E., E.S.B. and V.C.; visualization, R.N. and B.-C.B.; supervision, M.L.C. and E.S.B.; project administration, M.L.C. All authors have read and agreed to the published version of the manuscript.

Funding: This research received no external funding.

Institutional Review Board Statement: The Local Commission of Ethics for Scientific Research from the Timis County Emergency Clinical Hospital "Pius Brinzeu" in Timisoara, Romania, operates under the article 167 provisions of Law no. 95/2006, art. 28, chapter VIII, of order 904/2006; with EU GCP Directives 2005/28/EC, International Conference of Harmonisation of Technical Requirements for Registration of Pharmaceuticals for Human Use (ICH); and with the Declaration of Helsinki—Recommendations Guiding Medical Doctors in Biomedical Research Involving Human Subjects. The current study was conducted according to the guidelines of the Declaration of Helsinki; followed the European Union General Data Protection Regulation (GDPR); and was approved by the Local Commission of Ethics for Scientific Research from the Timis County Emergency Clinical Hospital "Pius Brinzeu" in Timisoara, Romania, No. 184/10.02.2020.

Informed Consent Statement: Informed consent was obtained from all subjects involved in the study.

Data Availability Statement: The data presented in this study are available on request from the corresponding author.

Conflicts of Interest: The authors declare no conflict of interest.

References

1. do Nascimento, I.J.B.; Cacic, N.; Abdulazeem, H.M.; von Groote, T.C.; Jayarajah, U.; Weerasekara, I.; Esfahani, M.A.; Civile, V.T.; Marusic, A.; Jeroncic, A.; et al. Novel Coronavirus Infection (COVID-19) in Humans: A Scoping Review and Meta-Analysis. *J. Clin. Med.* **2020**, *9*, 941. [CrossRef] [PubMed]
2. Ceulemans, M.; Foulon, V.; Ngo, E.; Panchaud, A.; Winterfeld, U.; Pomar, L.; Lambelet, V.; Cleary, B.; O'Shaughnessy, F.; Passier, A.; et al. Mental Health Status of Pregnant and Breastfeeding Women during the COVID-19 Pandemic—A Multinational Cross-sectional Study. *Acta Obstet. Gynecol. Scand.* **2021**, *100*, 1219–1229. [CrossRef]
3. Thapa, S.B.; Mainali, A.; Schwank, S.E.; Acharya, G. Maternal Mental Health in the Time of the COVID-19 Pandemic. *Acta Obstet. Gynecol. Scand.* **2020**, *99*, 817–818. [CrossRef] [PubMed]
4. Pierce, M.; Hope, H.; Ford, T.; Hatch, S.; Hotopf, M.; John, A.; Kontopantelis, E.; Webb, R.; Wessely, S.; McManus, S.; et al. Mental Health before and during the COVID-19 Pandemic: A Longitudinal Probability Sample Survey of the UK Population. *Lancet Psychiatry* **2020**, *7*, 883–892. [CrossRef]
5. Maternal Mental Health. Available online: https://www.who.int/teams/mental-health-and-substance-use/promotion-prevention/maternal-mental-health (accessed on 28 August 2023).
6. Allotey, J.; Stallings, E.; Bonet, M.; Yap, M.; Chatterjee, S.; Kew, T.; Debenham, L.; Llavall, A.C.; Dixit, A.; Zhou, D.; et al. Clinical Manifestations, Risk Factors, and Maternal and Perinatal Outcomes of Coronavirus Disease 2019 in Pregnancy: Living Systematic Review and Meta-Analysis. *BMJ* **2020**, *370*, m3320. [CrossRef]
7. Westgren, M.; Pettersson, K.; Hagberg, H.; Acharya, G. Severe Maternal Morbidity and Mortality Associated with COVID-19: The Risk Should Not Be Downplayed. *Acta Obstet. Gynecol. Scand.* **2020**, *99*, 815–816. [CrossRef]
8. O'Connor, E.; Senger, C.A.; Henninger, M.; Gaynes, B.N.; Coppola, E.; Soulsby Weyrich, M. *Interventions to Prevent Perinatal Depression: A Systematic Evidence Review for the U.S. Preventive Services Task Force*; U.S. Preventive Services Task Force Evidence Syntheses, formerly Systematic Evidence Reviews; Agency for Healthcare Research and Quality (US): Rockville, MD, USA, 2019.
9. Yim, I.S.; Tanner Stapleton, L.R.; Guardino, C.M.; Hahn-Holbrook, J.; Schetter, C.D. Biological and Psychosocial Predictors of Postpartum Depression: Systematic Review and Call for Integration. *Annu. Rev. Clin. Psychol.* **2015**, *11*, 99–137. [CrossRef]
10. Yu, Y.; Liang, H.-F.; Chen, J.; Li, Z.-B.; Han, Y.-S.; Chen, J.-X.; Li, J.-C. Postpartum Depression: Current Status and Possible Identification Using Biomarkers. *Front. Psychiatry* **2021**, *12*, 620371. [CrossRef]
11. Davenport, M.H.; Meyer, S.; Meah, V.L.; Strynadka, M.C.; Khurana, R. Moms Are Not OK: COVID-19 and Maternal Mental Health. *Front. Glob. Women's Health* **2020**, *1*, 1. [CrossRef] [PubMed]
12. Ceulemans, M.; Hompes, T.; Foulon, V. Mental Health Status of Pregnant and Breastfeeding Women during the COVID-19 Pandemic: A Call for Action. *Int. J. Gynaecol. Obstet.* **2020**, *151*, 146–147. [CrossRef] [PubMed]
13. Cuschieri, S. The STROBE Guidelines. *Saudi J. Anaesth.* **2019**, *13*, S31. [CrossRef] [PubMed]
14. Cox, J.L.; Holden, J.M.; Sagovsky, R. Detection of Postnatal Depression. Development of the 10-Item Edinburgh Postnatal Depression Scale. *Br. J. Psychiatry* **1987**, *150*, 782–786. [CrossRef] [PubMed]
15. Matsumura, K.; Hamazaki, K.; Tsuchida, A.; Kasamatsu, H.; Inadera, H. Factor Structure of the Edinburgh Postnatal Depression Scale in the Japan Environment and Children's Study. *Sci. Rep.* **2020**, *10*, 11647. [CrossRef] [PubMed]
16. Bergant, A.M.; Nguyen, T.; Heim, K.; Ulmer, H.; Dapunt, O. German language version and validation of the Edinburgh postnatal depression scale. *Dtsch. Med. Wochenschr.* **1998**, *123*, 35–40. [CrossRef] [PubMed]
17. Vivilaki, V.G.; Dafermos, V.; Kogevinas, M.; Bitsios, P.; Lionis, C. The Edinburgh Postnatal Depression Scale: Translation and Validation for a Greek Sample. *BMC Public Health* **2009**, *9*, 329. [CrossRef]
18. Wallis, A.; Fernandez, R.; Florin, O.; Cherecheş, R.; Zlati, A.; Dungy, C. Validation of a Romanian Scale to Detect Antenatal Depression. *Cent. Eur. J. Med.* **2012**, *7*, 216–223. [CrossRef]

19. Zubaran, C.; Schumacher, M.; Roxo, M.; Foresti, K. Screening Tools for Postpartum Depression: Validity and Cultural Dimensions. *Afr. J. Psychiatry* **2010**, *13*, 357–365. [CrossRef]
20. Shrestha, S.D.; Pradhan, R.; Tran, T.D.; Gualano, R.C.; Fisher, J.R.W. Reliability and Validity of the Edinburgh Postnatal Depression Scale (EPDS) for Detecting Perinatal Common Mental Disorders (PCMDs) among Women in Low-and Lower-Middle-Income Countries: A Systematic Review. *BMC Pregnancy Childbirth* **2016**, *16*, 72. [CrossRef]
21. Lebel, C.; MacKinnon, A.; Bagshawe, M.; Tomfohr-Madsen, L.; Giesbrecht, G. Elevated Depression and Anxiety Symptoms among Pregnant Individuals during the COVID-19 Pandemic. *J. Affect. Disord.* **2020**, *277*, 5–13. [CrossRef]
22. Levis, B.; Negeri, Z.; Sun, Y.; Benedetti, A.; Thombs, B.D.; DEPRESsion Screening Data (DEPRESSD) EPDS Group. Accuracy of the Edinburgh Postnatal Depression Scale (EPDS) for Screening to Detect Major Depression among Pregnant and Postpartum Women: Systematic Review and Meta-Analysis of Individual Participant Data. *BMJ* **2020**, *371*, m4022. [CrossRef]
23. Dadi, A.F.; Miller, E.R.; Mwanri, L. Postnatal Depression and Its Association with Adverse Infant Health Outcomes in Low- and Middle-Income Countries: A Systematic Review and Meta-Analysis. *BMC Pregnancy Childbirth* **2020**, *20*, 416. [CrossRef] [PubMed]
24. Chen, J.; Cross, W.M.; Plummer, V.; Lam, L.; Tang, S. A Systematic Review of Prevalence and Risk Factors of Postpartum Depression in Chinese Immigrant Women. *Women Birth* **2019**, *32*, 487–492. [CrossRef] [PubMed]
25. Yan, H.; Ding, Y.; Guo, W. Mental Health of Pregnant and Postpartum Women during the Coronavirus Disease 2019 Pandemic: A Systematic Review and Meta-Analysis. *Front. Psychol.* **2020**, *11*, 617001. [CrossRef] [PubMed]
26. Wan Mohamed Radzi, C.W.J.B.; Salarzadeh Jenatabadi, H.; Samsudin, N. Postpartum Depression Symptoms in Survey-Based Research: A Structural Equation Analysis. *BMC Public Health* **2021**, *21*, 27. [CrossRef] [PubMed]
27. Cameron, E.E.; Joyce, K.M.; Delaquis, C.P.; Reynolds, K.; Protudjer, J.L.P.; Roos, L.E. Maternal Psychological Distress & Mental Health Service Use during the COVID-19 Pandemic. *J. Affect. Disord.* **2020**, *276*, 765–774. [CrossRef]
28. Stojanov, J.; Stankovic, M.; Zikic, O.; Stankovic, M.; Stojanov, A. The Risk for Nonpsychotic Postpartum Mood and Anxiety Disorders during the COVID-19 Pandemic. *Int. J. Psychiatry Med.* **2021**, *56*, 228–239. [CrossRef]
29. Zanardo, V.; Manghina, V.; Giliberti, L.; Vettore, M.; Severino, L.; Straface, G. Psychological Impact of COVID-19 Quarantine Measures in Northeastern Italy on Mothers in the Immediate Postpartum Period. *Int. J. Gynaecol. Obstet.* **2020**, *150*, 184–188. [CrossRef]
30. Baran, J.; Leszczak, J.; Baran, R.; Biesiadecka, A.; Weres, A.; Czenczek-Lewandowska, E.; Kalandyk-Osinko, K. Prenatal and Postnatal Anxiety and Depression in Mothers during the COVID-19 Pandemic. *J. Clin. Med.* **2021**, *10*, 3193. [CrossRef]
31. Matsushima, M.; Tsuno, K.; Okawa, S.; Hori, A.; Tabuchi, T. Trust and Well-Being of Postpartum Women during the COVID-19 Crisis: Depression and Fear of COVID-19. *SSM Popul. Health* **2021**, *15*, 100903. [CrossRef]
32. Mariño-Narvaez, C.; Puertas-Gonzalez, J.A.; Romero-Gonzalez, B.; Peralta-Ramirez, M.I. Giving Birth during the COVID-19 Pandemic: The Impact on Birth Satisfaction and Postpartum Depression. *Int. J. Gynaecol. Obstet.* **2021**, *153*, 83–88. [CrossRef]
33. Gupta, M.; Agarwal, N.; Agrawal, A.G. Impact of COVID-19 Institutional Isolation Measures on Postnatal Women in Level 3 COVID Facility in Northern India. *J. South Asian Fed. Obstet. Gynaecol.* **2021**, *13*, 50–54. [CrossRef]
34. Kumar, P.; Dhillon, P. Length of Stay after Childbirth in India: A Comparative Study of Public and Private Health Institutions. *BMC Pregnancy Childbirth* **2020**, *20*, 181. [CrossRef] [PubMed]
35. Abu-Awwad, S.-A.; Craina, M.; Gluhovschi, A.; Boscu, L.; Bernad, E.; Iurciuc, M.; Abu-Awwad, A.; Iurciuc, S.; Tudoran, C.; Bernad, R.; et al. Comparative Analysis of Neonatal Effects in Pregnant Women with Cardiovascular Risk versus Low-Risk Pregnant Women. *J. Clin. Med.* **2023**, *12*, 4082. [CrossRef] [PubMed]
36. Agnafors, S.; Bladh, M.; Svedin, C.G.; Sydsjö, G. Mental Health in Young Mothers, Single Mothers and Their Children. *BMC Psychiatry* **2019**, *19*, 112. [CrossRef] [PubMed]
37. Weitoft, G.; Hjern, A.; Haglund, B.; Rosén, M. Mortality, Severe Morbidity, and Injury in Children Living with Single Parents in Sweden: A Population-Based Study. *Lancet* **2003**, *361*, 289–295. [CrossRef]
38. Crosier, T.; Butterworth, P.; Rodgers, B. Mental Health Problems among Single and Partnered Mothers. *Soc. Psychiatry Psychiatr. Epidemiol.* **2007**, *42*, 6–13. [CrossRef]
39. Schmidt, R.M.; Wiemann, C.M.; Rickert, V.I.; Smith, E.O. Brian Moderate to Severe Depressive Symptoms among Adolescent Mothers Followed Four Years Postpartum. *J. Adolesc. Health* **2006**, *38*, 712–718. [CrossRef]
40. Geronimus, A.T.; Korenman, S.; Hillemeier, M.M. Does Young Maternal Age Adversely Affect Child Development? Evidence from Cousin Comparisons in the United States. *Popul. Dev. Rev.* **1994**, *20*, 585–609. [CrossRef]
41. Miller, C.L.; Miceli, P.J.; Whitman, T.L.; Borkowski, J.G. Cognitive Readiness to Parent and Intellectual-Emotional Development in Children of Adolescent Mothers. *Dev. Psychol.* **1996**, *32*, 533–541. [CrossRef]
42. Louie, S.; Shi, Y.; Allman-Farinelli, M. The Effects of the COVID-19 Pandemic on Food Security in Australia: A Scoping Review. *Nutr. Diet.* **2022**, *79*, 28–47. [CrossRef]
43. Ramsey, R.; Giskes, K.; Turrell, G.; Gallegos, D. Food Insecurity among Adults Residing in Disadvantaged Urban Areas: Potential Health and Dietary Consequences. *Public Health Nutr.* **2012**, *15*, 227–237. [CrossRef] [PubMed]
44. Pedroso, J.; Buccini, G.; Venancio, S.I.; Pérez-Escamilla, R.; Gubert, M.B. Maternal Mental Health Modifies the Association of Food Insecurity and Early Child Development. *Matern. Child Nutr.* **2020**, *16*, e12997. [CrossRef] [PubMed]
45. Dira, I.K.P.A.; Wahyuni, A.A.S. Prevalensi dan Faktor Risiko Depresi Postpartum di Kota Denpasar Menggunakan Edinburgh Postnatal Depression Scale. *E J. Med. Udayana* **2016**, *5*, 1–5.

46. Tamaki, R.; Murata, M.; Okano, T. Risk Factors for Postpartum Depression in Japan. *Psychiatry Clin. Neurosci.* **1997**, *51*, 93–98. [CrossRef]

47. Lindahl, V.; Pearson, J.L.; Colpe, L. Prevalence of Suicidality during Pregnancy and the Postpartum. *Arch. Womens Ment. Health* **2005**, *8*, 77–87. [CrossRef]

48. Howard, L.M.; Flach, C.; Mehay, A.; Sharp, D.; Tylee, A. The Prevalence of Suicidal Ideation Identified by the Edinburgh Postnatal Depression Scale in Postpartum Women in Primary Care: Findings from the RESPOND Trial. *BMC Pregnancy Childbirth* **2011**, *11*, 57. [CrossRef]

49. Matthey, S.; Fisher, J.; Rowe, H. Using the Edinburgh Postnatal Depression Scale to Screen for Anxiety Disorders: Conceptual and Methodological Considerations. *J. Affect. Disord.* **2013**, *146*, 224–230. [CrossRef]

50. Goodman, J.H.; Watson, G.R.; Stubbs, B. Anxiety Disorders in Postpartum Women: A Systematic Review and Meta-Analysis. *J. Affect. Disord.* **2016**, *203*, 292–331. [CrossRef]

51. Ozdil, M. Postpartum Depression among Mothers of Infants Hospitalized in the Neonatal Intensive Care Unit during the COVID-19 Pandemic. *Cureus* **2023**, *15*, e44380. [CrossRef]

52. Smith-Nielsen, J.; Egmose, I.; Wendelboe, K.I.; Steinmejer, P.; Lange, T.; Vaever, M.S. Can the Edinburgh Postnatal Depression Scale-3A Be Used to Screen for Anxiety? *BMC Psychol.* **2021**, *9*, 118. [CrossRef]

53. Wang, Z.; Liu, J.; Shuai, H.; Cai, Z.; Fu, X.; Liu, Y.; Xiao, X.; Zhang, W.; Krabbendam, E.; Liu, S.; et al. Mapping Global Prevalence of Depression among Postpartum Women. *Transl. Psychiatry* **2021**, *11*, 543. [CrossRef] [PubMed]

54. Wu, Y.; Zhang, C.; Liu, H.; Duan, C.; Li, C.; Fan, J.; Li, H.; Chen, L.; Xu, H.; Li, X.; et al. Perinatal Depressive and Anxiety Symptoms of Pregnant Women during the Coronavirus Disease 2019 Outbreak in China. *Am. J. Obstet. Gynecol.* **2020**, *223*, 240.e1–240.e9. [CrossRef] [PubMed]

55. Galletta, M.A.K.; Oliveira, A.M.d.S.S.; Albertini, J.G.L.; Benute, G.G.; Peres, S.V.; Brizot, M.d.L.; Francisco, R.P.V.; HC-FMUSP-Obstetric COVID19 Study Group. Postpartum Depressive Symptoms of Brazilian Women during the COVID-19 Pandemic Measured by the Edinburgh Postnatal Depression Scale. *J. Affect. Disord.* **2022**, *296*, 577–586. [CrossRef] [PubMed]

56. van Gelder, M.M.H.J.; Bretveld, R.W.; Roeleveld, N. Web-Based Questionnaires: The Future in Epidemiology? *Am. J. Epidemiol.* **2010**, *172*, 1292–1298. [CrossRef]

57. Statistics | Eurostat. Available online: https://ec.europa.eu/eurostat/databrowser/view/isoc_r_broad_h/default/table?lang=en (accessed on 28 August 2023).

58. Biaggi, A.; Conroy, S.; Pawlby, S.; Pariante, C.M. Identifying the Women at Risk of Antenatal Anxiety and Depression: A Systematic Review. *J. Affect. Disord.* **2016**, *191*, 62–77. [CrossRef]

59. Matsumura, K.; Hamazaki, K.; Tsuchida, A.; Kasamatsu, H.; Inadera, H.; Japan Environment and Children's Study (JECS) Group. Education Level and Risk of Postpartum Depression: Results from the Japan Environment and Children's Study (JECS). *BMC Psychiatry* **2019**, *19*, 419. [CrossRef]

60. Enătescu, I.; Craina, M.; Gluhovschi, A.; Giurgi-Oncu, C.; Hogea, L.; Nussbaum, L.A.; Bernad, E.; Simu, M.; Cosman, D.; Iacob, D.; et al. The Role of Personality Dimensions and Trait Anxiety in Increasing the Likelihood of Suicide Ideation in Women during the Perinatal Period. *J. Psychosom. Obstet. Gynecol.* **2021**, *42*, 242–252. [CrossRef]

61. Zedan, H.S.; Baattaiah, B.A.; Alashmali, S.; Almasaudi, A.S. Risk of Postpartum Depression: The Considerable Role of Maternal Health Status and Lifestyle. *Healthcare* **2023**, *11*, 2074. [CrossRef]

 *healthcare*

Article

Psychometric Properties of the Perceived Collective Family Efficacy Scale in Algeria

Aiche Sabah [1,*], Musheer A. Aljaberi [2], Kuo-Hsin Lee [3,4,*] and Chung-Ying Lin [5]

[1] Faculty of Human and Social Sciences, Hassiba Benbouali University of Chlef, Chlef 02076, Algeria
[2] Faculty of Medicine and Health Sciences, Taiz University, Taiz 6803, Yemen; musheer.jaberi@gmail.com
[3] Department of Emergency Medicine, E-Da Dachang Hospital, I-Shou University, Kaohsiung 824, Taiwan
[4] School of Medicine, College of Medicine, I-Shou University, No. 8, Yi-Da Road, Jiao-Su Village, Yan-Chao District, Kaohsiung 824, Taiwan
[5] Institute of Allied Health Sciences, College of Medicine, National Cheng Kung University, Tainan 701, Taiwan; cylin36933@gmail.com
* Correspondence: s.aiche@univ-chlef.dz (A.S.); peter1055@gmail.com (K.-H.L.)

Abstract: The Perceived Collective Family Efficacy Scale is a tool utilized to assess the effectiveness of a family as a functioning system. The scale has a single-factor structure with good validity and reliability. However, there is a shortage of psychometric evidence of the scale in an Arab context. This study aimed to assess the psychometric properties of the Perceived Collective Family Efficacy Scale among Algerian students. A cross-sectional study was conducted to recruit 300 students from Algerian universities. The students completed the 20-item Perceived Collective Family Efficacy Scale, Arabic version, to measure their beliefs regarding collective efficacy within families. Confirmatory factor analysis (CFA) and the Rasch model were employed to assess the psychometric properties and unidimensionality of the scale. Both CFA and Rasch findings supported the single-factor structure for the Perceived Collective Family Efficacy Scale. Specifically, the CFA indicated that the data aligned with a one-dimensional model. The Rasch analysis revealed favorable indicators of unidimensionality for the scale. Moreover, a thorough examination of the Principal Component Analysis of the Rasch residuals confirmed the existence of a single dimension, which is consistent with the original structure of the Perceived Collective Family Efficacy Scale. These findings provide scientific evidence for the validity and unidimensional nature of the Perceived Collective Family Efficacy Scale. Specifically, the satisfactory psychometric properties findings indicate that the Perceived Collective Family Efficacy Scale could be applied in an Arab context (i.e., in Algerian). The scale's unidimensional structure underscores its effectiveness in measuring beliefs in collective efficacy within families. These results enhance our understanding of family dynamics and provide a reliable measurement tool for assessing family efficacy in similar cultural contexts.

Keywords: self-efficacy; collective efficacy; psychological theory; family; factor analysis; Rasch model; scales

Citation: Sabah, A.; Aljaberi, M.A.; Lee, K.-H.; Lin, C.-Y. Psychometric Properties of the Perceived Collective Family Efficacy Scale in Algeria. *Healthcare* **2023**, *11*, 2691. https://doi.org/10.3390/healthcare11192691

Academic Editors: Carlos Laranjeira and Ana Querido

Received: 9 August 2023
Revised: 20 September 2023
Accepted: 27 September 2023
Published: 8 October 2023

1. Introduction

Family provides more than environments where individuals live; it also provides a complete and intricate social system for human development. Individuals interact within these systems, influencing each other's behavior. As a social system, the family is envisioned to possess unique characteristics, rules, roles, communication patterns, and power structures that extend beyond the individual [1–4]. The family systems theory asserts that family subsystems are closely interconnected, conceptualizing families as organized groups. It also suggests that understanding human behavior relies on the interactions between individuals within the family and between the family and its context, as the family is an integral part of its surrounding environment [5–7]. According to the family systems theory, family functioning encompasses task accomplishment, role performance, emotional involvement, control, values, standards, expression, and emotional communication. The concept of family functioning includes both the efficiency and style of the family. Family efficiency requires structure and the ability to adapt to changes over time, while family patterns refer to the quality of family interactions [8,9].

The social cognitive theory links behavior to four factors: goals, outcome expectations, self-efficacy, and social–structural variables [10–12]. The social cognitive theory, as proposed by Bandura, assumes an interaction between personal, behavioral, and social–environmental factors. The key point is that people strive to develop a sense of significant control over important events in their lives. The perceived efficacy of the group influences their aspirations, resource utilization, contribution to collective effort, resilience in the face of failed collective efforts or opposition, and adaptability when confronting challenging problems. Thus, the social cognitive theory establishes a central role for perceived efficacy in managing various relationships, interactions, and daily tasks within the family system [13,14]. Specifically, collective efficacy beliefs within the family refer to the judgments made by family members regarding the family's collective ability to accomplish necessary tasks for its functioning. Family collective efficacy focuses on the capabilities of family members to work together as a whole [15]. In order to better understand family collective efficacy, a validated instrument (i.e., the Perceived Collective Family Efficacy Scale in this study) should be used. However, the Perceived Collective Family Efficacy Scale does not have an Arabic version to assess Arab populations. Therefore, the present study translated the Collective Family Efficacy Scale into Arabic for further psychometric evaluation.

1.1. Collective Efficacy in Families

Self-efficacy is the belief in one's ability to perform complex life tasks successfully. It plays a crucial role in shaping a person's feelings, perceptions, motivational activities, and behaviors across various activities [16]. Collective efficacy is considered an extension of building self-efficacy and is a subsidiary model of the social cognitive theory proposed by Bandura. Bandura [17] defines collective efficacy as "a group's shared belief in its conjoint capabilities to organize and execute the actions required to produce given levels of attainment". Thus, perceived collective efficacy within the organization represents the group members' beliefs regarding the collective ability of the social system [18,19]. The dynamic characteristics of a group can encompass social support, solidarity, communication, collective participation, dialogue, trust, decision-making and sharing, group belongingness, and common goals. The willingness and ability to intervene for the benefit of the group depend on the level of solidarity, participation, and mutual trust among group members [20,21]. Beliefs about collective family efficacy reflect the judgments made by family members regarding the collective ability of the family as a whole to function as a complete system in accomplishing necessary tasks for the functioning of the family. Bandura, et al. [22] define perceived collective family efficacy as: "members' beliefs in the capabilities of their family to work together to promote each other's development and well-being, maintain beneficial ties to extrafamilial systems, and exhibit resilience to adversity".

While other self-efficacy beliefs primarily focus on dyadic relationships (e.g., between parent and child, husband and wife), collective family efficacy beliefs center around the perceived practical capabilities of the family as a whole [23]. Individual self-efficacy beliefs alone may be insufficient to achieve desired goals when focusing on family performance. Spouses, parents, and children cannot fulfill their roles independently of other family members' feelings, expectations, and behaviors. Many outcomes can only be achieved when all family members pool their resources and efforts together. This is because the family, as a social system, has a lasting impact on individual growth. Individuals face a variety of needs and challenges throughout life as part of an interconnected family system. Similar to any other social system, perceived collective efficacy influences the system's sense of purpose and message, the strength of members' commitment to its pursuit, their perception of their ability to fulfill mutual obligations, and the family's resilience in the face of adversity [17,23,24]. According to a study by Bandura, Caprara, Barbaranelli, Regalia and Scabini [22], a high sense of collective family efficacy is associated with open family communication and explicit disclosure by teenagers about their activities outside the home. Furthermore, family collective efficacy has contributed to the satisfaction of parents and teenagers with their family life. Another study by Kao, et al. [25] found that the perceived collective family efficacy of both teenagers and parents reduced the impact of parental and teenage depressive symptoms on risky health behaviors among teenagers. In fact, parents' and teenagers' perceived collective family efficacy protects against depressive symptoms and risky health behaviors.

1.2. Perceived Collective Family Efficacy Scale

The Perceived Collective Family Efficacy Scale, developed by Caprara [26] and Caprara, Regalia, Scabini, Barbaranelli and Bandura [24], is a measure used to assess the perceived effectiveness of families in accomplishing essential tasks and functioning as a complete system. It focuses on the family's practical capabilities and views it as a social system comprising interconnected and interactive relationships. This scale comprises 20 items that emphasize the family's ability to manage daily routines, reach consensus in decision-making and planning, cope with challenges, promote mutual agreement, provide emotional support during difficult times, engage in shared activities and relaxation despite multiple commitments, and maintain positive relationships with the community.

Several studies have been conducted on the Perceived Collective Family Efficacy Scale's psychometric properties. Caprara, Regalia, Scabini, Barbaranelli and Bandura [24] used a group of parents and adolescents to validate the scale's reliability and validity. The Principal Component Analysis (PCA) with oblimin rotation showed that collective family efficacy is unidimensional. The Cronbach's alpha coefficient for the collective family scale indicated high internal consistency, with values of 0.96 for boys and 0.97 for girls. The correlation coefficients between parents' and adolescents' family efficacy beliefs ranged from low to moderately high congruence, with same-sex dyads typically having stronger correlations than opposite-sex dyads.

Costa and colleagues conducted studies in Portuguese and Italian contexts to validate the cross-cultural stability of the Perceived Collective Family Efficacy Scale and its associations with communication, conflict management, and children's academic achievement [27]. The Perceived Collective Family Efficacy Scale's factor loadings were found to be robust in both samples, ranging from 0.71 to 0.89 in Portugal and 0.71 to 0.89 in Italy, indicating that cross-cultural invariance had been achieved in terms of configurable, metric, and scalar. The construct validity was supported by various correlations with internalized and externalized symptoms, close communication with parents, aggressive conflict styles, open communication, compromise in conflict styles, and children's academic achievement.

Pepe, et al. [28] conducted validation studies in Spanish adolescents and found that all items displayed factor loadings exceeding 0.40, indicating a robust relationship with the underlying factor. The Cronbach's alpha coefficient achieved a value of 0.92, meeting the standard criteria for internal consistency. The construct validity was supported by various correlations, including positive correlations between perceived collective family efficacy and parental affection, the promotion of autonomy, and productive coping strategies and negative correlations with psychological control exerted by parents. The scale also exhibited positive correlations with certain non-productive coping strategies (e.g., worry, wishful thinking) and negative correlations with others (e.g., tension reduction). Additionally, adolescents with a higher family efficacy tended to use fewer drugs.

The psychometric properties of the Perceived Family Collective Efficacy revised scales were also evaluated in the Iranian population by Panaghi and colleagues [29]. Exploratory factor analysis revealed a two-factor solution, while confirmatory factor analysis provided support for both the two-factor and one-factor models, with a preference for the two-factor model due to its superior fit. The calculated Cronbach's alpha coefficient was 0.92, and the test–retest reliability score was 0.83, highlighting high internal consistency and stability. These findings suggest that the Perceived Family Collective Efficacy Scale has robust psychometric properties suitable for research and family counseling endeavors within the Iranian context.

Overall, the literature evidence indicates that the Perceived Collective Family Efficacy Scale is reliable and valid in assessing family effectiveness and functioning. Its utility spans different cultural contexts and age groups, making it a valuable tool for research and psychological assessment.

1.3. Purpose of the Present Study

The Perceived Collective Family Efficacy Scale has yet to be validated in Arab populations. Because of the lack of the Perceived Collective Family Efficacy Scale, studies conducted in Arabic could not investigate in-depth research on family collective efficacy. Introducing this new research tool in this field provides a consistent and reliable way to assess family efficacy, particularly in the quickly evolving social, economic, and political landscape of Arab societies, with a focus on Algeria. These changes significantly impact family life, social upbringing processes, marital harmony, and stability [30], including the psychological long-term effects of COVID-19 and quarantine, as well as future pandemics [4,31–33]. Given developing countries' unique challenges, a trustworthy method for measuring family efficacy is crucial. The study examines the scale's psychometric properties and expands its applicability to include various Arab cultures, making it a valuable addition to research tools for the wider Arab community.

The Perceived Collective Family Efficacy Scale has yet to be validated in Arab populations. Considering that the widespread use of this scale within families represents a valuable tool for assessing collective efficacy, this current study aims to verify the psychometric properties of the Perceived Collective Family Efficacy Scale among Algerian university students. Confirmatory factor analysis (CFA) and the Rasch model were the primary psychometric methods to assess the Perceived Collective Efficacy Scale's psychometric properties.

The objectives of this study are as follows: (1) To verify the psychometric properties of the translated and adapted version of the Perceived Collective Family Efficacy Scale through CFA on Algerian university students. (2) To examine the psychometric properties of the translated and adapted version of the Perceived Collective Family Efficacy Scale within the framework of Rasch modeling on Algerian university students.

2. Materials and Methods

2.1. Study Design and Participants

A cross-sectional research design was used in this study to create general models that relate groups of variables under specific conditions [34–36]. The data were manually collected through the Perceived Collective Family Efficacy Scale self-administered questionnaire to students from various disciplines at the University of Chlef. The response process was voluntary, and students were informed about the scale's purpose, with participation being optional. Students who did not have siblings at home were excluded from the study. The participants in this study completed the Arabic-translated version of the Perceived Collective Family Efficacy Scale. The research was conducted as a part of the research conducted by Projects on University Training and Research (PRFU), Chlef University, and the research was approved by the Ministry of Higher Education and Scientific Research, Hassiba Benbouali University of Chlef, Algeria, Faculty of Humanities and Social Sciences, Department of Social Sciences with ethical approval reference number (I05L03UN020120200002).

2.2. Study Sample Size

Based on research by Hair et al., a sample size of at least 100 is necessary to conduct a structural equation modeling (SEM) analysis using covariance [37]. Given that confirmatory factor analysis (CFA) is a type of SEM, the sample size of 300 students in the present study was thus deemed suitable for utilizing CFA to achieve research objectives [38,39]. The importance of sample size in modeling and CFA studies has been emphasized by various researchers, with a consensus that SEM (including CFA) studies require a minimum of 200 participants [40,41]. However, it is worth noting that this recommended sample size may be insufficient for complex models, non-normal data distributions, or when using different estimation methods than maximum likelihood (ML), as Kline has pointed out [38,40]. Our study's model is not complex as it only included one latent variable with 20 observed variables; therefore, we adopted the N: q rule introduced by Jackson [42,43]. With the use of 10:1 sample size ratio to the number of parameters to be estimated in SEM [40], the minimum sample size required is 200. In brief, a sample size of 200 is often considered reasonable for relatively small and simple models [44,45], such as the model tested in the present study.

2.3. Instrument

The Perceived Collective Family Efficacy was assessed using a 20-item scale developed by Caprara [26] to measure beliefs regarding the family's effectiveness in functioning as a complete system and accomplishing essential tasks for family functioning. The scale covered various aspects of the family's capabilities, including managing daily routines, reaching consensus in decision-making and planning, dealing with challenges, promoting mutual agreement, providing emotional support during difficult situations, enjoying and relaxing together despite multiple responsibilities, and maintaining positive relationships with the community as a whole. Participants rated each item on a 5-point scale ranging from 1 for "Not at all" to 5 for "Very well". All items are positive, and there are no negative items. The scale spans from 20 to 100, with higher scores indicating a higher family self-efficacy. To calculate family self-efficacy, the mean of all items is computed. The scale also exhibits excellent internal consistency, with Cronbach's alpha coefficients ranging from 0.96 to 0.97 [28].

Translation Procedures for the Scale

To translate the scale into Arabic, back-translation is the preferred translation technique. This method involves a group of interpreters and experts who translate the items from the source language to the target language and then back-translate them into the source language, ensuring agreement on meaning and word choice for each item [46,47]. Afterward, a small test group of participants is used to confirm that the target population easily understands the tool. To ensure cultural appropriateness, investigators should use commonly used words by the target population [48–50].

Therefore, the scale was translated from English to Arabic through a series of steps [45,46,51,52]. Firstly, permission was obtained from the developers (i.e., Caprara, Regalia, Scabini, Barbaranelli and Bandura [24]) to translate the scale from the English version (see Supplementary Materials, Table S1) [28] into Arabic.

A team of proficient Arab researchers and an English language expert with a good command of Arabic conducted a preliminary translation. The two translations were harmonized to create an initial Arabic scale version (Supplementary Materials, Table S2). During this step, a comparative analysis was conducted between the preliminary translation and the original scale to select clear vocabulary and phrases that are closely aligned with the English version.

In the second stage, a "Back Translation" process was carried out. An English language expert, who had not seen the scale's English or Arabic versions, translated the proposed Arabic version back into English (Supplementary Materials, Table S3). The back-translated English version of the instrument was compared to the original English version, which increased confidence in the proposed Arabic translation. The comparison revealed a near-perfect match between the English translation and the original, particularly concerning the scale's items. Some items varied slightly in wording but did not significantly impact their intended meaning. Notable item translations included:

Item 1: "Set aside leisure time with your family when other things press for attention" became "Allocate free time for the family when there are other things that require attention."

Item 3: "Resolve conflicts when family members feel they are not being treated fairly" became *"Resolve conflicts when other people feel like they are not being treated fairly."*
Item 15: "Celebrate family traditions even in difficult times" became *"Celebrate family occasions even during hard times."*
Item 17: "Face up to difficulties without excessive tension" became *"Face difficulties effortlessly."*

After confirming the accuracy of the back-translation, the scale was administered to a small sample of university students to ensure clarity of vocabulary, suitability of items, and comprehensibility of instructions for the target age group. It was found that the scale items were clear and free from ambiguity.

Following these steps, researchers were confident that the Arabic version of the questionnaire was ready for implementation. The scale was subsequently administered in Arabic to a sample of students at the University of Chlef.

2.4. Data Analysis

Descriptive statistics were used to summarize the item properties, providing information on central tendencies such as skewness, kurtosis, mean, and standard deviation. According to Hair Jr, et al. [53], skewness values between -1 and $+1$ are considered excellent, while kurtosis values should fall within the range of -2 to $+2$. These statistics offer a concise overview of the distribution and characteristics of the items in the scale [39,54,55].

To assess the factor structure of the Perceived Collective Family Efficacy Scale, CFA with the maximum likelihood estimation method was conducted to test if the scale has a single-factor structure. The following fit indices were used to examine data–model fit: the p-value of the chi-squared statistic (non-significant), comparative fit index (CFI) (≥ 0.90), standardized root-mean-squared residual (SRMR) (≤ 0.08), and root-mean-squared error of approximation (RMSEA) (≤ 0.08) [33,39,56–59]. By examining the relationship between the observed data and the expected factor structure, the CFA provided insights into how well the items were related to the measured latent construct [31,56,60–62]. In addition, Cronbach's alpha, Composite Reliability, and MaxR(H) were used to estimate the scale's internal reliability.

The Rasch model was also employed to confirm the unidimensionality of the Perceived Collective Family Efficacy Scale. Rasch analysis is a statistical method used to examine the properties of items (questions) and individuals on a measurement scale. It aims to assess the extent to which the items in a scale function together to measure a latent trait or construct accurately. The Rasch model is a widely used statistical model in psychometrics that assesses how well the observed responses align with the expected response patterns based on the underlying construct [56,63]. This analysis helps ensure that the scale is unidimensional, meaning that all items effectively contribute to measuring the intended construct. In Rasch analysis, we used Outfit mean square (MnSq) and Infit MnSq through Winsteps software version 3.72.3. The first step in Rasch analysis was to exclude individuals whose data did not fit the model, meaning their fit exceeded a threshold of 2. The acceptable fit range for individuals is typically between 0.60 and -1.40, as suggested by Bond and Fox [64]. Additionally, according to Linacre [65], several conditions should be considered when assessing the fit of individuals and items to the Rasch model. These conditions include examining Outfit before Infit prioritizing mean squares before ZSTD, prioritizing high mean squares before low mean squares, considering positive ZSTD before negative ZSTD, and starting with the worst item or person. After excluding the "worst" item or person, there will always be another item or person that may appear as the "worst" in the newly adjusted context, which is more suitable for the model. Therefore, it is important not to mechanically remove items as this may result in no remaining items or persons. In other words, the ideal range for fitting suitable individuals should fall within the required values [66,67]. Those individuals who have statistically exceeded the acceptable threshold, either by correctly answering items that are more difficult than their abilities or by failing to answer correctly to items that require lower abilities than their own, might have relied on guessing, lacked in seriousness, or provided inaccurate responses [66,67]. By employing these data analysis techniques, the researcher aimed to validate the factor structure and unidimensionality of the Perceived Collective Family Efficacy Scale in the Arab context. These rigorous analyses contribute to evaluating the scale's psychometric properties and establish its suitability for assessing perceived collective family efficacy among Arab populations. The statistical analyses were performed using AMOS 24.0 (for CFA), Winsteps (for Rasch), and SPSS 24.0 (for other analyses).

3. Results

3.1. Sociodemographic Characteristics of the Sample

The sociodemographic characteristics of the sample are summarized in Table 1. The number of females was 255 (85%). Regarding the number of siblings, 49% of participants had one to four siblings, followed by 43% having five to eight siblings, and 8% having more than nine siblings. Notably, the majority of the sample were single individuals, with a percentage of 92.7%, while the percentage of married individuals was low, estimated at 7.3%. A significant proportion of the sample reported that their parents lived together, accounting for 84.3%, while the percentage of individuals with divorced parents was 3.7%, and 12% had one or both parents deceased. The economic level of the majority of the sample's families was moderate, with a percentage of 85.3%, followed by a low economic level of 6%, and a small percentage of 8.7% had a high economic level. The most common field of study among the students was humanities and social sciences, accounting for 84.7%, followed by natural sciences with a percentage of 9.3%. Finally, the percentage of students in the arts and languages field was 6%.

Table 1. Characteristics of the study sample.

Variables	Groups	N	%
Gender	Boys	45	15.0
	Girls	255	85.0
Number of siblings	1–4	147	49.0
	5–8	129	43.0
	>9	24	8.0
Marital status	Single	278	92.7
	Married	22	7.3
Parental status	Live together	253	84.3
	Divorced	11	3.7
	One or both of them is dead	36	12.0
Family economic status	Lower	18	6.0
	Middle	256	85.3
	Upper	26	8.7
Specialties	Social and human sciences	254	84.7
	Natural sciences	28	9.3
	Literature and language	18	6.0

3.2. Descriptive Statistics

The descriptive statistics for the scale items are presented in Table 2, which includes the skewness, kurtosis, mean, and standard deviation (SD) for each item individually. These statistics summarize the central tendencies and variability of each item in the scale.

The skewness values ranged from −0.915 to −0.05, indicating a normal distribution of the items. The kurtosis values ranged from −0.741 to 0.371, also indicating a normal distribution. The mean values ranged from 3.18 to 4.01. The item "Serve as a positive example for the community" had the highest mean, while the rest of the items had means above 3.

3.3. Confirmatory Factor Analysis

Table 3 presents the CFA fit indices. Overall, the results of the fit indices indicate a good model fit after modification. The initial model fit was unsatisfactory, such as the CFI value at 0.878. After making two modifications (i.e., deleting item 17 and linking up the residual correlation between items 18 and 19), the fit indices were acceptable: CFI = 0.912, SRMR = 0.04, and RMSEA = 0.05. Regarding the loadings of the items after conducting CFA, as shown in Figure 1, they ranged from 0.418 (item 5) to 0.756 (item 11), all of which are acceptable loadings.

Table 2. Descriptive statistics for The Perceived Collective Family Efficacy Scale.

Items	Skewness	Kurtosis	Mean	SD
Set aside leisure time with your family when other things press for attention.	−0.189	−0.300	3.18	1.06
Agree to decisions that require some sacrifice of personal interests.	−0.208	−0.326	3.28	1.01
Resolve conflicts when family members feel they are not being treated fairly.	−0.308	−0.174	3.42	0.99
Prevent family disagreements from turning into heated arguments.	−0.560	−0.181	3.71	1.06
Get family members to share household responsibilities.	−0.604	−0.278	3.71	1.10
Support each other in times of stress.	−0.531	−0.464	3.71	1.09
Help each other to achieve their personal goals.	−0.617	−0.223	3.75	1.07
Help each other with work demands.	−0.362	−0.448	3.68	1.00
Build respect for each other's particular interests.	−0.481	−0.395	3.58	1.09
Get family members to carry out their responsibilities when they neglect them.	−0.550	−0.331	3.71	1.07
Build trust in each other.	−0.764	0.124	3.78	1.08
Figure out what choices to make when the family faces important decisions.	−0.403	−0.235	3.48	1.05
Find community resources and make good use of them for the family.	−0.312	−0.193	3.37	1.04
Get the family to keep close ties to their larger family.	−0.469	−0.151	3.64	1.01
Celebrate family traditions even in difficult times.	−0.144	−0.677	3.38	1.12
Cooperate with schools to improve their educational practices.	−0.094	−0.741	3.18	1.17
Face up to difficulties without excessive tension.	−0.050	−0.349	3.22	0.97
Remain confident during difficult times.	−0.398	−0.548	3.65	1.06
Accept each member's need for independence.	−0.261	−0.364	3.40	1.02
Serve as a positive example for the community.	−0.915	0.371	4.01	1.01

Table 3. Model fit.

Model Fit	Without Modifications	With Modifications
χ^2	390.211	299.780
χ^2/df	2.295	1.985
CFI	0.878	0.912
SRMR	0.05	0.04
RMSEA	0.06	0.05

Model Validity

The model's Cronbach's alpha value was 0.898, indicating high internal consistency, and the Composite Reliability value was 0.896. These values are considered good, suggesting a strong reliability of the model. Furthermore, the MaxR(H) value of 0.907 exceeded the CR value, which indicates the establishment of discriminant validity. This implies that the constructs in the model measure different aspects of the Perceived Collective Family Efficacy under investigation. Overall, these findings provide further support for the validity and reliability of the model.

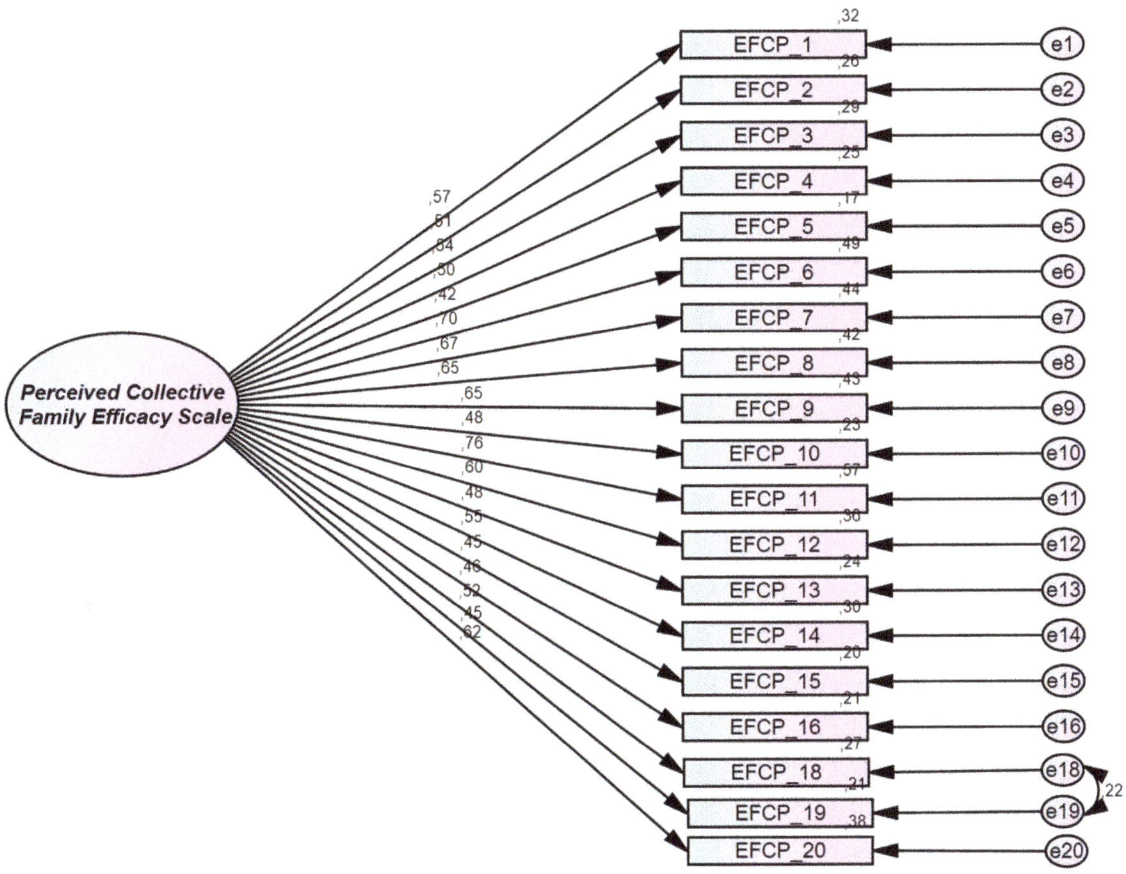

Figure 1. Perceived Collective Family Efficacy Scale in its final form after modification through confirmatory factor analysis.

3.4. Rasch Analysis

3.4.1. Fit of Individuals and Items to Rasch Analysis

The participants' Infit mean-square (IN.MSQ) values ranged from 0.1132 to 2.8103, and Outfit mean square ranged from 0.1141 to 2.8711, indicating a general fit of persons to the Rasch model. Regarding the fit of items to the Rasch analysis (see Table 4), the Outfit MnSq ranged from 0.79 to 1.38, while the Infit MnSq ranged from 0.81 to 1.33. These values align well with the Rasch model, as the items do not exceed the fit boundaries of 0.60 and 1.40.

Table 4. Item statistics: Measure order.

Items	Measure	Error	IN.MSQ	IN.ZSTD	OUT.MSQ	OUT.ZSTD
EFCP_1	0.48	0.06	0.9079	−1.1991	0.9471	−0.6691
EFCP_2	0.35	0.07	0.9252	−0.9591	0.9585	−0.519
EFCP_3	0.17	0.07	0.8721	−1.6891	0.9195	−1.0291
EFCP_4	−0.22	0.07	1.109	1.3711	1.1048	1.2911
EFCP_5	−0.22	0.07	1.3305	3.8513	1.3847	4.3314

Table 4. *Cont.*

Items	Measure	Error	IN.MSQ	IN.ZSTD	OUT.MSQ	OUT.ZSTD
EFCP_6	−0.22	0.07	0.9137	−1.1091	0.8859	−1.4591
EFCP_7	−0.28	0.07	0.9627	−0.449	0.9328	−0.8291
EFCP_8	−0.18	0.07	0.8118	−2.5392	0.7902	−2.8192
EFCP_9	−0.05	0.07	0.9331	−0.8491	0.9118	−1.1291
EFCP_10	−0.22	0.07	1.1817	2.2112	1.1312	1.6011
EFCP_11	−0.31	0.07	0.8222	−2.3792	0.79	−2.7792
EFCP_12	0.09	0.07	0.9259	−0.9491	0.9064	−1.2091
EFCP_13	0.24	0.07	1.054	0.7111	1.0503	0.6611
EFCP_14	−0.13	0.07	0.9164	−1.0691	0.8909	−1.3991
EFCP_15	0.22	0.07	1.2206	2.6812	1.2222	2.6812
EFCP_16	0.47	0.06	1.2196	2.6812	1.1984	2.4312
EFCP_17	0.43	0.06	0.923	−0.9991	0.9421	−0.7391
EFCP_18	−0.13	0.07	1.0245	0.341	1.0081	0.131
EFCP_19	0.19	0.07	0.9865	−0.149	1.0368	0.491
EFCP_20	−0.66	0.07	1.0108	0.161	0.945	−0.6291

In this study, the item difficulty values ranged from −0.66 to 0.48, as shown in the table above. The logit value (0) was not observed in items with moderate difficulty, indicating the absence of items with moderate difficulty. However, the logit values were positive for items with higher-than-moderate difficulty and deviated from zero. Specifically, the items with positive logits were 12, 3, 19, 15, 13, 2, 17, 16, and 1. On the other hand, items with lower difficulty had negative logits, as represented by the following items: 20, 11, 7, 6, 4, 10, 5, 8, 14, 18, and 9 (see Figure 2). The average logit difficulty score was 0, with a standard deviation of 0.30. The average score and standard deviation in item difficulty logit suggest homogeneity and proximity to the mean (0) logit, indicating item consistency and uniformity.

Through the grading map in Figure 2, we observe that it illustrates the order of items, ranging from (−1 to +1). Furthermore, the map reveals the presence of the ceiling effect, which means that individuals with high abilities do not encounter items that challenge their proficiency beyond a certain level. However, the map does not measure high proficiency accurately (meaning that we need items that match the abilities of individuals with high capabilities).

3.4.2. Empirical Item Characteristic Curves (ICCs)

We checked the fit of the items using Empirical Item Characteristic Curves (ICCs). We found that all items fit within the two-sided 95% confidence bands, except for item 17, which showed a misfit. Figure 3 shows that item 17's empirical data fell outside the confidence bands.

3.4.3. Unidimensionality of the Perceived Collective Family Efficacy Scale

Due to the assumption of unidimensionality in the Rasch model, it should be noted that unidimensionality is not absolute. Unidimensionality should not be equated with factor analysis, as their goals differ. Factor analysis aims to identify the factors that make up the test, while item response theory aims to identify deviations from the measured trait and determine whether they constitute an independent factor. Therefore, the software provides Rasch residual-based Principal Component Analysis (PCAR) to analyze the underlying dimensions, as shown in Table 5. This analysis reveals differences between dimensions and allows for an assessment of unidimensionality based on the following criteria:

(a) The variance explained by measures should be greater than or equal to 20% to 80% (in our study, the variance explained was 36.3%, which is good).

(b) The raw variance explained by items (36.3%) is larger than the raw variance explained by persons (14.4%).

(c) At most, five contrasts are reported, and in our model, there are five variances.

(d) All conditions for the unidimensionality of Rasch are acceptable, as shown in the table above, except for the unexplained variance in the first contrast, which is 2.1, slightly higher than the recommended 2.0.

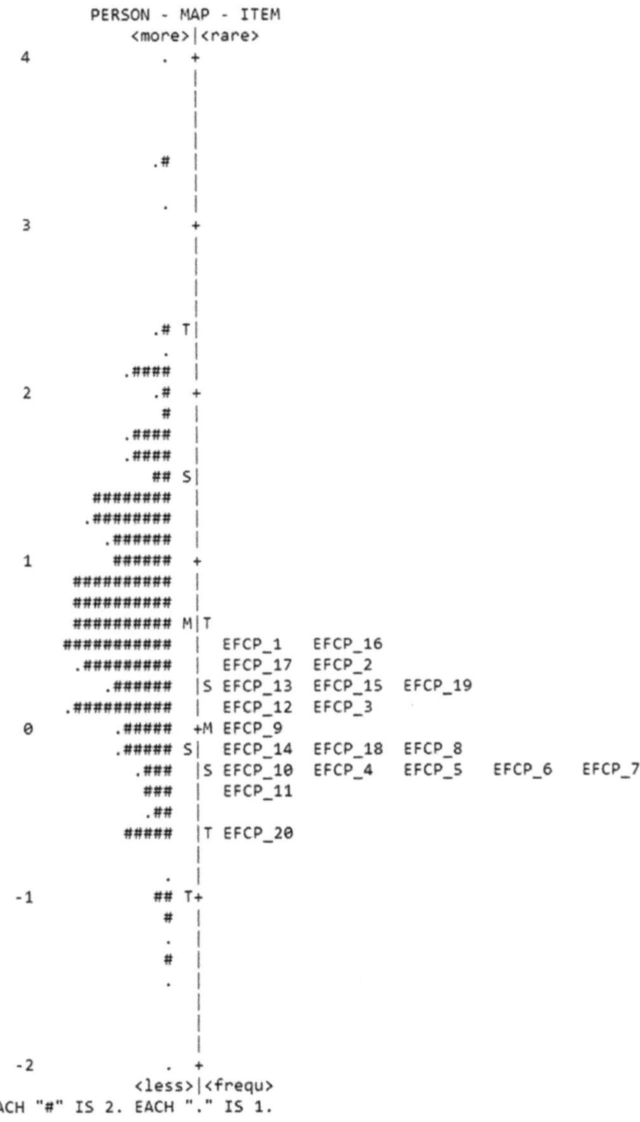

Figure 2. Grading the difficulty of items and assessing individuals' abilities based on the distribution map.

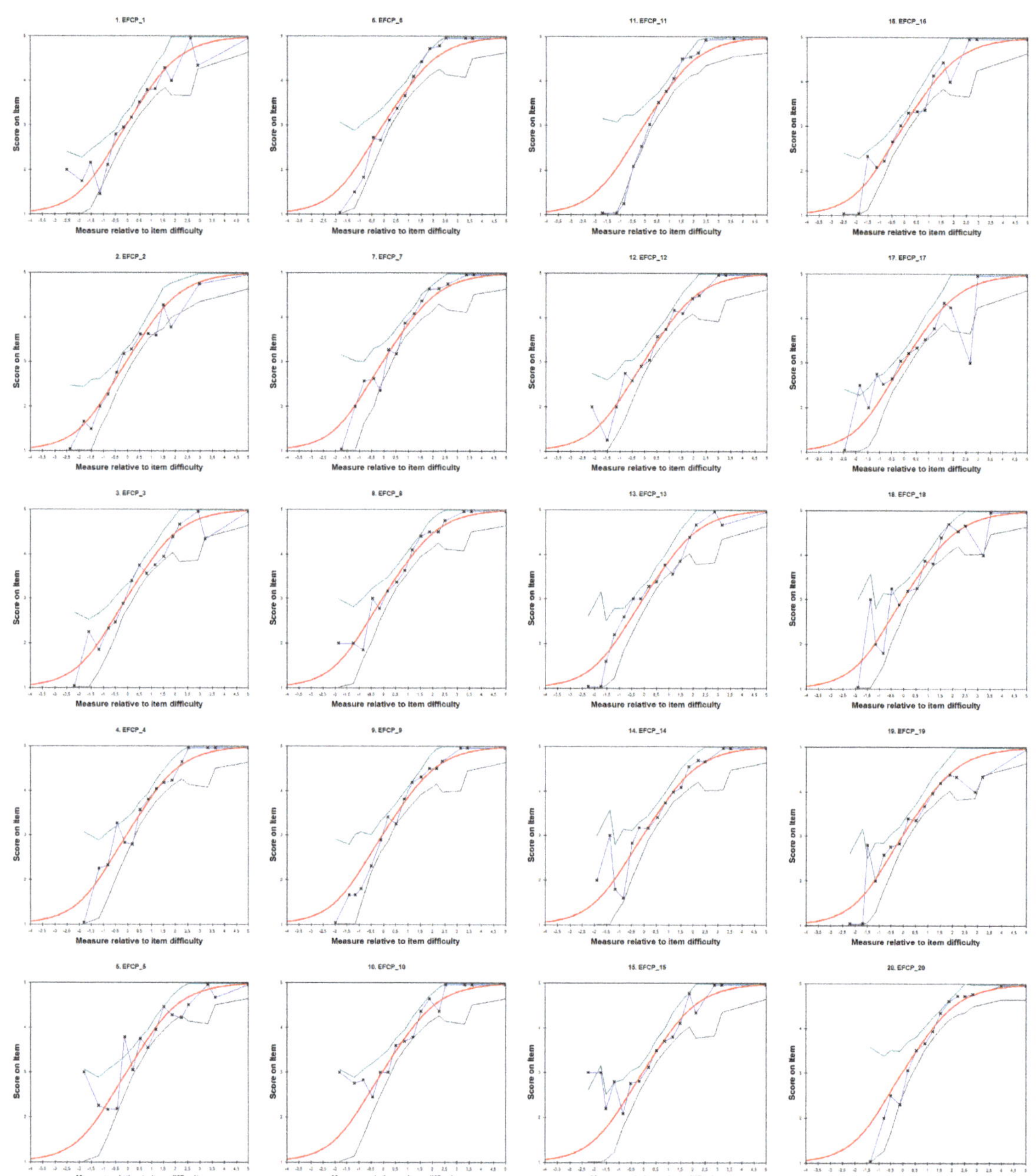

Figure 3. Empirical Item Characteristic Curves (ICCs) of the Perceived Collective Family Efficacy Scale. Note: The red line represents the item characteristic curve, as predicted by the Rasch model. It shows the average score that students at different levels of the latent variable (*x*-axis) would obtain on the item (*y*-axis) according to the Rasch model. The blue line depicts the empirical ICC. Each "x" on the *x*-axis summarizes the responses of students whose measurements are near that particular point. The green-gray lines represent the two-sided 95% confidence bands.

Table 5. Unidimensionality of the Perceived Collective Family Efficacy Scale.

	Empirical		Modeled
Total raw variance in observations	31.4	100.0%	100.0%
Raw variance explained by measures	11.4	36.3%	36.5%
Raw variance explained by persons	4.5	14.4%	14.5%
Raw variance explained by items	6.9	21.8%	22.0%
Raw unexplained variance (total)	20.0	63.7%	63.5%
Unexplained variance in 1st contrast	2.1	6.6%	10.4%
Unexplained variance in 2nd contrast	1.7	5.4%	8.4%
Unexplained variance in 3rd contrast	1.5	4.8%	7.5%
Unexplained variance in 4th contrast	1.4	4.3%	6.8%
Unexplained variance in 5th contrast	1.3	4.2%	6.6%

3.4.4. Reliability

The Rasch model provides an overall test reliability coefficient, as in classical measurement theory, and reliability coefficients for items and persons. It is evident that the item separation coefficient for the test was estimated at 4.23, which exceeds 2. This confirms the hierarchical ordering of the scale items based on item difficulty. The item reliability value was 0.95, indicating high reliability. The person separation coefficient was 2.62, which is greater than 2. Moreover, the person reliability was 0.87, a good value indicating scale stability. This suggests that individuals can effectively differentiate between the items, accurately defining the targeted trait.

3.4.5. Response Category Functioning of the Perceived Collective Family Efficacy Scale

The analysis of category performance under the Rasch measurement requirements is presented in Table 6 and illustrated in Figure 4. It shows the category probability curves for the Perceived Collective Family Efficacy Scale, ranging from 1 (not at all well) to 5 (very well). As shown in Figure 4, the graphs demonstrate the likelihood of individuals selecting various categories for Perceived Collective Family Efficacy. The horizontal axis shows the measured variable, while the vertical axis displays the probability of choosing a category between 1 and 5. Each curve represents responses on a five-point Likert scale, with 'Never' represented in red, 'Rarely' in blue, 'Sometimes' in pink, 'Often' in gray, and 'Very often' in green.

Table 6. Summary of category structure. Model = "R".

Category		Observed		Observed	Sample	Infit	Outfit	Structure	Category
Label	Score	Count	%	Average	Expect	Mnsq	Mnsq	Calibration	Measure
1	1	265	4	−0.54	−0.61	1.08	1.11	NONE	(−2.69)
2	2	662	11	−0.08	−0.07	0.99	0.98	−1.25	−1.21
3	3	1914	32	0.38	0.40	0.94	0.93	−0.89	−0.06
4	4	1849	31	0.87	0.86	0.94	0.92	0.66	1.19
5	5	1310	22	1.40	1.39	1.04	1.03	1.47	(2.82)

The perfect graph would exhibit a peak for each category. An analysis was conducted to verify the effectiveness of the five-category Likert response format. The results demonstrated a consistent distribution of responses, logit measures increased as categories increased, and outfit statistics within the range (<2.0). The sequential arrangement of category thresholds suggested that the 5-category rating scale performed optimally.

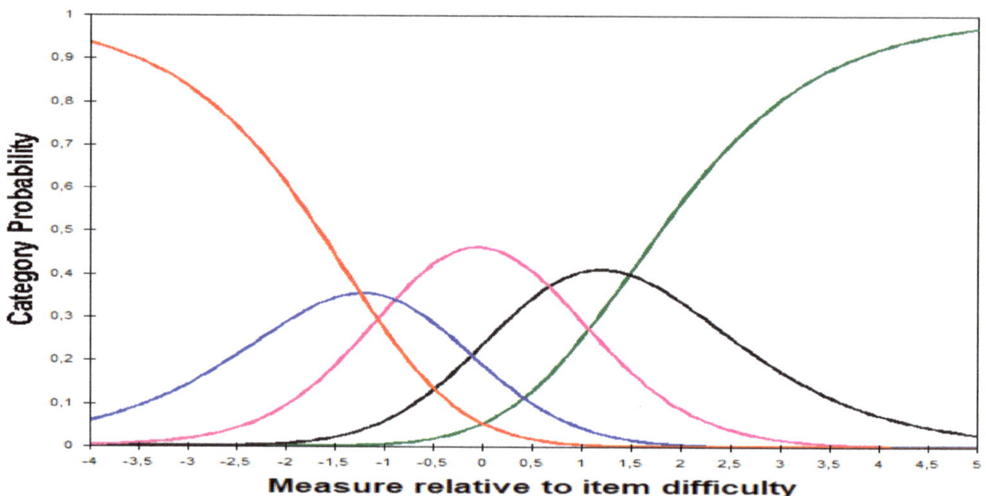

Figure 4. Measure relative to item difficulty: five categories. Note: The horizontal axis shows the measured variable, while the vertical axis displays the probability of choosing a category between 1 and 5. Each curve represents responses on a five-point Likert scale, with 'Never' represented in red, 'Rarely' in blue, 'Sometimes' in pink, 'Often' in gray, and 'Very often' in green.

Both Infit and Outfit MnSq ranged from 0.6 to 1.4, which are considered acceptable for rating scale measurement [66]. Table 6 confirms that none of the values exceeded 1.40 or fell below 0.60. The observed average person measures for respondents endorsing each category progressed monotonically with the categories: $-0.54 < -0.08 < 0.38 < 0.87 < 1.40$. This pattern indicates that individuals with higher abilities endorse higher categories, while those with lower abilities support lower categories [66]. Regarding the thresholds between categories, it is optimal for the Andrich threshold step values to have a minimum difference in step difficulty of 1.4 logits for an optimum response category performance. From Table 6, it is found that the width between the Andrich thresholds for categories 1 and 2 is -1.25 logits, categories 2 and 3 is $(-1.25) + (-0.89) = 2.14$ logits, categories 3 and 4 is 1.55 logits, and categories 4 and 5 is 2.13 logits.

4. Discussion

The study's results demonstrated that the proposed single-factor structure for the Perceived Collective Family Efficacy Scale exhibited an acceptable fit in both CFA and Rasch results. The CFA results supported the single-factor structural model, and it was found that the scale possesses good validity and reliability after being analyzed through the Rasch model. This finding is consistent with previous studies [24,27–29]. The study extended the psychometric properties of the Perceived Collective Family Efficacy Scale to the Arab context (i.e., in Algeria). The findings revealed that the scale exhibited acceptable indicators of quality, supported by confirmatory factor analysis. This suggests that the scale effectively measures the Algerian population's perceived collective family efficacy construct. The single-factor structural model was validated, confirming the theoretical framework of the scale. These results align with previous studies [24,27,28] conducted in different cultural contexts, indicating the generalizability of the scale's psychometric properties. Our study in Algeria has echoed well with recent findings reported in Italian and Portuguese participants [27]. Other prior results suggest that the Perceived Collective Family Efficacy Scale is applicable across different cultures: Caprara's study in 2004 supported the scale's unidimensionality and good reliability among family participants from

Genzano, a residential community near Rome, and from Milan and its surroundings [24], Pepe et al.'s study in 2008 also supported the scale's unidimensionality and good reliability among Spanish participants [28]. Our study findings extended the psychometric evidence of the Perceived Collective Family Efficacy Scale to Algeria with a reliability of 0.898. Overall, our study enhances the scale's applicability in Arab cultures, with Algeria serving as a model for its validity and unidimensional structure.

The scale demonstrated good levels of validity and reliability in the Algerian context, implying that it could accurately measure the intended construct and produce results consistent with prior findings [24]. This finding further enhances the scale's utility and applicability in an Arab context. The study's findings contribute to the existing body of literature on perceived collective family efficacy and provide valuable insights into its measurement and psychometric properties. Researchers and practitioners can confidently employ the Perceived Collective Family Efficacy Scale in the Algerian context to assess and understand the collective efficacy beliefs within families. The scale's validity and reliability establish a foundation for future research and interventions to promote family efficacy and well-being in similar cultural settings. The findings support its applicability in an Arab cultural context and contribute to the broader understanding of family dynamics and functioning. Future research can build upon these findings by exploring the scale's associations with other relevant variables and examining its effectiveness in intervention programs to strengthen family efficacy and resilience.

4.1. Implications

4.1.1. Theoretical Implications

The study establishes the validity and reliability of the Perceived Collective Family Efficacy Scale in the Algerian context. This provides a foundation for future research and ensures the credibility of findings based on the scale. That is, the scale can be used in diverse populations and cultural settings, enhancing its utility as a cross-cultural measurement tool. The validation of the single-factor structural model and alignment with theoretical assumptions (unidimensionality, reliability, validity) support the conceptual understanding of perceived collective family efficacy. The study highlights the applicability of the Perceived Collective Family Efficacy Scale in an Arab cultural context, contributing to the broader understanding of family dynamics and functioning within Arab societies.

4.1.2. Practical Implications

The study findings could assist healthcare practitioners in assessing and understanding collective efficacy beliefs within families. Additionally, the findings hold practical implications for healthcare practitioners involved in designing intervention programs aimed at promoting family efficacy and well-being within the Algerian context. The scale's established validity and reliability ensure its effective utilization for evaluating the impact of interventions on enhancing family efficacy. Practitioners can thus customize their interventions based on the scale's measurements and pinpoint areas for improvement within families. This underscores the importance of considering cultural factors when assessing and promoting family efficacy, highlighting the necessity for culturally sensitive interventions. The Perceived Collective Family Efficacy Scale is culturally appropriate for use with other Arabic-speaking populations. The scale's formulation and items align with the cultural values of various Arab countries. This suitability was demonstrated through the scale's validity and reliability within the Algerian Arab population in the present study. For instance, Cronbach's alpha coefficient for the collective family scale demonstrated a high internal consistency.

5. Limitations and Recommendations

5.1. Limitations

There are a few limitations to this study that should be addressed. Firstly, it is important to note that the sample used for this study was drawn from only one university, so caution should be taken when trying to generalize the results to other student populations. Secondly, this study focused only on a sample of students with a low representation of males. Previous studies have shown that gender differences in Algeria's education system are clear [68,69], with girls consistently performing better than boys in secondary school and university. This gender gap has widened over the years, with women making up nearly 60% of university students. It is worth noting that women tend to dominate areas such as education, humanities, social sciences, and health and welfare, while men display a stronger inclination toward pursuing STEM (science, technology, engineering, and mathematics) disciplines. Lastly, the study's reliance solely on confirmatory factor analysis and the Rasch model for assessing validity may represent a limitation. Using additional instruments to evaluate the study's concurrent, convergent, and divergent validity would be beneficial.

5.2. Recommendations for Future Research

In order to improve the generalizability of the study's findings, we propose that similar research be conducted in other Algerian and Arab universities from different specializations with random sampling, as they speak the same language. It would be valuable to explore the psychometrics of this scale across a variety of sample sizes and age groups, including adolescents and adults with equal gender representation. Additionally, we suggest using different statistical methods such as retesting, Exploratory Factor Analysis (EFA), and Differential Item Functioning (DIF) analysis to evaluate the scale's validity and reliability comprehensively.

To improve future research in this field, it is recommended to use multiple measurement tools to ensure data accuracy. To establish concurrent validity, we recommend comparing the results with established measures of the same concept, as well as examining relationships with related and unrelated concepts to assess convergent and divergent validity. Test–retest assessments should also be conducted to ensure measurement reliability over time. These methodological improvements will strengthen the validity of the study's findings and contribute to a comprehensive validation process.

6. Conclusions

In conclusion, this study examined the psychometric properties of the Arabic version of the Perceived Collective Family Efficacy Scale by employing CFA and the Rasch model with a sample of university students. The study's findings demonstrate that the Perceived Collective Family Efficacy Scale exhibits satisfactory validity and reliability within the Algerian context. The established validity and reliability of the scale provide a foundation for future investigations and interventions aimed at promoting family efficacy and well-being in similar cultural contexts. Further research can explore its associations with other variables and assess its effectiveness in family efficacy and resilience intervention programs.

Supplementary Materials: The following supporting information can be downloaded at: https://www.mdpi.com/article/10.3390/healthcare11192691/s1, Table S1: English version of The Perceived Collective Family Efficacy Scale; Table S2: Arabic version of The Perceived Collective Family Efficacy Scale; Table S3: Back-translation version of The Perceived Collective Family Efficacy Scale.

Author Contributions: Conceptualization, A.S. and M.A.A.; methodology, A.S. and M.A.A.; software, A.S. and C.-Y.L.; validation, A.S., M.A.A., K.-H.L. and C.-Y.L.; formal analysis, A.S. and C.-Y.L.; investigation, A.S., M.A.A., K.-H.L. and C.-Y.L.; resources, A.S., M.A.A., K.-H.L. and C.-Y.L.; data curation, A.S.; writing—original draft preparation, A.S., M.A.A., K.-H.L. and C.-Y.L.; writing—review and editing, A.S., M.A.A., K.-H.L. and C.-Y.L.; visualization, A.S., M.A.A. and C.-Y.L.; supervision, A.S.; project administration, A.S. All authors have read and agreed to the published version of the manuscript.

Funding: This research received no external funding.

Institutional Review Board Statement: The study was conducted in accordance with the Declaration of Helsinki, and approved by the Department of Social Sciences, Faculty of Humanities and Social Sciences, Hassiba Ben Bouali University, Chlef, Algeria, with reference number I05L03UN020120200002/2022/02.

Informed Consent Statement: Informed consent was obtained from all subjects involved in the study.

Data Availability Statement: The dataset supporting this study's findings is not openly available and will be available from the corresponding author upon reasonable request.

Conflicts of Interest: The authors declare no conflict of interest.

References

1. Gilbertson, S.; Graves, B.A. Chapter 4—Heart Health and Children. In *Lifestyle in Heart Health and Disease*; Watson, R.R., Zibadi, S., Eds.; Academic Press: Cambridge, MA, USA, 2018; pp. 35–46.
2. McGinnis, H.A.; Wright, A.W. Adoption and child health and psychosocial well-being. In *Encyclopedia of Child and Adolescent Health*, 1st ed.; Halpern-Felsher, B., Ed.; Academic Press: Oxford, UK, 2023; pp. 582–598.
3. Aiche, S. Building a measure of functional family performance: A field study on a sample of respondents in Algeria. *J. Educ. Qual. Res.* **2021**, *8*, 167–182. [CrossRef]
4. Taresh, S.M.; Ahmad, N.A.; Roslan, S.; Ma'rof, A.M. Preschool Teachers' Beliefs towards Children with Autism Spectrum Disorder (ASD) in Yemen. *Children* **2020**, *7*, 170. [CrossRef] [PubMed]
5. Schermerhorn, A.C.; Mark Cummings, E. Transactional Family Dynamics: A New Framework for Conceptualizing Family Influence Processes. In *Advances in Child Development and Behavior*; Kail, R.V., Ed.; JAI: Beijing, China, 2008; Volume 36, pp. 187–250.
6. Watson, W.H. Family Systems. In *Encyclopedia of Human Behavior*, 2nd ed.; Ramachandran, V.S., Ed.; Academic Press: San Diego, CA, USA, 2012; pp. 184–193.
7. Aiche, S.; Hammadi, F. The factorial structure of the satisfaction with family life scale on a sample of Algerians. *Al-Qabas J. Psychol. Soc. Stud.* **2023**, *5*, 69–79.
8. Caporino, N.E. Chapter 14—Involving family members in exposure therapy for children and adolescents. In *Exposure Therapy for Children with Anxiety and OCD*; Peris, T.S., Storch, E.A., McGuire, J.F., Eds.; Academic Press: Cambridge, MA, USA, 2020; pp. 323–357.
9. Coulacoglou, C.; Saklofske, D.H. Chapter 8—The Assessment of Family, Parenting, and Child Outcomes. In *Psychometrics and Psychological Assessment*; Coulacoglou, C., Saklofske, D.H., Eds.; Academic Press: San Diego, CA, USA, 2017; pp. 187–222.
10. Conner, M. Health Behaviors. In *International Encyclopedia of the Social & Behavioral Sciences*, 2nd ed.; Wright, J.D., Ed.; Elsevier: Oxford, UK, 2015; pp. 582–587.
11. Johnson, D.W.; Johnson, R.T. Cooperation and Competition. In *International Encyclopedia of the Social & Behavioral Sciences*, 2nd ed.; Wright, J.D., Ed.; Elsevier: Oxford, UK, 2015; pp. 856–861.
12. Schunk, D.H.; DiBenedetto, M.K. Learning from a social cognitive theory perspective. In *International Encyclopedia of Education*, 4th ed.; Tierney, R.J., Rizvi, F., Ercikan, K., Eds.; Elsevier: Oxford, UK, 2023; pp. 22–35.
13. Procentese, F.; Gatti, F.; Di Napoli, I. Families and Social Media Use: The Role of Parents' Perceptions about Social Media Impact on Family Systems in the Relationship between Family Collective Efficacy and Open Communication. *Int. J. Environ. Res. Public Health* **2019**, *16*, 5006. [CrossRef] [PubMed]
14. Bandura, A. Social cognitive theory: An agentic perspective. *Annu. Rev. Psychol.* **2001**, *52*, 1–26. [CrossRef] [PubMed]
15. Urdan, T.; Pajares, F. *Self-Efficacy Beliefs of Adolescents*; IAP: Cape Canaveral, FL, USA, 2006.
16. Butler, J. Self-Efficacy. In *Encyclopedia of Behavioral Medicine*; Gellman, M.D., Ed.; Springer International Publishing: Cham, Switzerland, 2020; pp. 1983–1985.
17. Bandura, A. *Self-Efficacy: The Exercise of Control*; W H Freeman/Times Books/Henry Holt & Co: New York, NY, USA, 1997; pp. 604–609.
18. Donohoo, J.; Hattie, J.; Eells, R. The power of collective efficacy. *Educ. Leadersh.* **2018**, *75*, 40–44.
19. Pietrantoni, L. Collective Efficacy. In *Encyclopedia of Quality of Life and Well-Being Research*; Michalos, A.C., Ed.; Springer: Dordrecht, The Netherlands, 2014; pp. 987–989.
20. Dehingia, N.; Dixit, A.; Heskett, K.; Raj, A. Collective efficacy measures for women and girls in low- and middle-income countries: A systematic review. *BMC Women's Health* **2022**, *22*, 129. [CrossRef] [PubMed]
21. Aiche, S. Family Sacrifice: A Theoretical Approach. *Soc. Empower. J.* **2021**, *3*, 148–159.
22. Bandura, A.; Caprara, G.V.; Barbaranelli, C.; Regalia, C.; Scabini, E. Impact of family efficacy beliefs on quality of family functioning and satisfaction with family life. *Appl. Psychol.* **2011**, *60*, 421–448. [CrossRef]
23. Scabini, E.; Marta, E.; Lanz, M. *The Transition to Adulthood and Family Relations: An Intergenerational Approach*; Psychology Press: London, UK, 2007.
24. Caprara, G.V.; Regalia, C.; Scabini, E.; Barbaranelli, C.; Bandura, A. Assessment of Filial, Parental, Marital, and Collective Family Efficacy Beliefs. *Eur. J. Psychol. Assess.* **2004**, *20*, 247–261. [CrossRef]

25. Kao, T.-S.A.; Ling, J.; Dalaly, M.; Robbins, L.B.; Cui, Y. Parent–Child Dyad's Collective Family Efficacy and Risky Adolescent Health Behaviors. *Nurs. Res.* **2020**, *69*, 455–465. [CrossRef] [PubMed]
26. Caprara, G.V. *La Valutazione Dell'autoefficacia. Costrutti e Strumenti*; Edizioni Erickson: Trento, Italy, 2001.
27. Costa, M.; Faria, L.; Alessandri, G.; Caprara, G.V. Measuring parental and family efficacy beliefs of adolescents' parents: Cross-cultural comparisons in Italy and Portugal. *Int. J. Psychol.* **2016**, *51*, 421–429. [CrossRef] [PubMed]
28. Pepe, S.; Sobral, J.; Gómez-Fraguela, J.A.; Villar-Torres, P. Spanish adaptation of the Adolescents' perceived collective family efficacy scale. *Psicothema* **2008**, *20*, 148–154. [PubMed]
29. Panaghi, L.; Mokhtarnai, I.; Kalantary, F. A Preliminary Study of Psychometric Properties of the Adolescents' Perceived Family Collective Efficacy Scale in Adolescent. *J. Fam. Res.* **2016**, *11*, 531–550.
30. Sabah, A.; Khalaf Rashid Al-Shujairi, O.; Boumediene, S. The Arabic Version of the Walsh Family Resilience Questionnaire: Confirmatory Factor Analysis of a Family Resilience Assessment Among Algerian and Iraq Families. *Int. J. Syst. Ther.* **2021**, *32*, 273–290. [CrossRef]
31. Aljaberi, M.A.; Al-Sharafi, M.A.; Uzir, M.U.H.; Sabah, A.; Ali, A.M.; Lee, K.-H.; Alsalahi, A.; Noman, S.; Lin, C.-Y. Psychological Toll of the COVID-19 Pandemic: An In-Depth Exploration of Anxiety, Depression, and Insomnia and the Influence of Quarantine Measures on Daily Life. *Healthcare* **2023**, *11*, 2418. [CrossRef] [PubMed]
32. Sabah, A.; Boumediene, S.; Zineb, D. Adverse Life Events and Family Distress During the Coronavirus Pandemic: A Field Study in Algeria. *Arab. J. Psychiatry* **2021**, *32*, 35–42.
33. Aljaberi, M.A.; Alareqe, N.A.; Alsalahi, A.; Qasem, M.A.; Noman, S.; Uzir, M.U.H.; Mohammed, L.A.; Fares, Z.E.A.; Lin, C.-Y.; Abdallah, A.M.; et al. A cross-sectional study on the impact of the COVID-19 pandemic on psychological outcomes: Multiple indicators and multiple causes modeling. *PLoS ONE* **2022**, *17*, e0277368. [CrossRef]
34. Hunziker, S.; Blankenagel, M. Cross-Sectional Research Design. In *Research Design in Business and Management: A Practical Guide for Students and Researchers*; Springer Fachmedien Wiesbaden: Wiesbaden, Germany, 2021; pp. 187–199.
35. Daniels, C.D. Cross-Sectional Research. In *Encyclopedia of Child Behavior and Development*; Goldstein, S., Naglieri, J.A., Eds.; Springer: Boston, MA, USA, 2011; p. 444.
36. Voelkle, M.C.; Hecht, M. Cross-Sectional Research Designs. In *Encyclopedia of Personality and Individual Differences*; Zeigler-Hill, V., Shackelford, T.K., Eds.; Springer International Publishing: Cham, Switzerland, 2017; pp. 1–4.
37. Hair, J.; Black, W.; Babin, B.; Anderson, R. *Multivariate Data Analysis*, 8th ed.; Cengage Learning EMEA: Hampshire, UK, 2019.
38. Kline, R.B. *Principles and Practice of Structural Equation Modeling*, 5th ed.; Guilford Publications: New York, NY, USA, 2023.
39. Alareqe, N.A.; Hassan, S.A.; Kamarudin, E.M.E.; Aljaberi, M.A.; Nordin, M.S.; Ashureay, N.M.; Mohammed, L.A. Validity of Adult Psychopathology Model Using Psychiatric Patient Sample from a Developing Country: Confirmatory Factor Analysis. *Ment. Illn.* **2022**, *2022*, 9594914. [CrossRef]
40. Kline, R.B. *Principles and Practice of Structural Equation Modeling*, 4th ed.; Guilford Press: New York, NY, USA, 2016; pp. 534–617.
41. Selig, J.P.; Card, N.A.; Little, T.D. Latent variable structural equation modeling in cross-cultural research: Multigroup and multilevel approaches. In *Multilevel Analysis of Individuals and Cultures*; Psychology Press Ltd.: London, UK, 2014; pp. 93–119.
42. Jackson, D.L. The effect of the number of observations per parameter in misspecified confirmatory factor analytic models. *Struct. Equ. Model. A Multidiscip. J.* **2007**, *14*, 48–76. [CrossRef]
43. Jackson, D.L. Revisiting sample size and number of parameter estimates: Some support for the N:q hypothesis. *Struct. Equ. Model.* **2003**, *10*, 128–141. [CrossRef]
44. Keith, T.Z. *Multiple Regression and Beyond: An Introduction to Multiple Regression and Structural Equation Modeling*; Routledge: Oxfordshire, UK, 2014.
45. Nichols, A.L.; Edlund, J. *The Cambridge Handbook of Research Methods and Statistics for the Social and Behavioral Sciences: Volume 1: Building a Program of Research*; Cambridge University Press: Cambridge, UK, 2023.
46. Chapman, D.W.; Carter, J.F. Translation Procedures for the Cross Cultural Use of Measurement Instruments. *Educ. Eval. Policy Anal.* **1979**, *1*, 71–76. [CrossRef]
47. Brislin, R.W. The wording and translation of research instruments. In *Field Methods in Cross-Cultural Research*; Cross-Cultural Research and Methodology Series; Sage Publications, Inc.: Thousand Oaks, CA, USA, 1986; Volume 8, pp. 137–164.
48. Bravo, M.; Canino, G.J.; Rubio-Stipec, M.; Woodbury-Fariña, M. A cross-cultural adaptation of a psychiatric epidemiologic instrument: The diagnostic interview schedule's adaptation in Puerto Rico. *Cult. Med. Psychiatry* **1991**, *15*, 1–18. [CrossRef] [PubMed]
49. Zaid, S.M.R.; Jamaluddin, S.; Baharuldin, Z.; Taresh, S.M. Psychometric properties of an adapted Yemeni version of rejection sensitivity questionnaire. *Pertanika J. Soc. Sci. Humanit.* **2020**, *28*, 2149–2165.
50. Kelley, K.; Clark, B.; Brown, V.; Sitzia, J. Good practice in the conduct and reporting of survey research. *Int. J. Qual. Health Care* **2003**, *15*, 261–266. [CrossRef] [PubMed]
51. Nadhiroh, S.; Nurmala, I.; Pramukti, I.; Tivany, S.; Tyas, L.; Zari, A.; Poon, W.; Siaw, Y.-L.; Kamolthip, R.; Chirawat, P.; et al. Weight stigma in Indonesian young adults: Validating the indonesian versions of the weight self-stigma questionnaire and perceived weight stigma scale. *Asian J. Soc. Health Behav.* **2022**, *5*, 169–179. [CrossRef]
52. World Health Organization. Process of Translation and Adaptation of Instruments. Available online: http://www.who.int/substance_abuse/research_tools/translation/en/ (accessed on 18 August 2023).

53. Hair, J.F., Jr.; Hult, G.T.M.; Ringle, C.M.; Sarstedt, M. *A Primer on Partial Least Squares Structural Equation Modeling (PLS-SEM)*; Sage Publications: Thousand Oaks, CA, USA, 2021.
54. Aljaberi, M.A.; Juni, M.H.; Al-Maqtari, R.A.; Lye, M.S.; Saeed, M.A.; Al-Dubai, S.A.R.; Kadir Shahar, H. Relationships among perceived quality of healthcare services, satisfaction and behavioural intentions of international students in Kuala Lumpur, Malaysia: A cross-sectional study. *BMJ Open* **2018**, *8*, e021180. [CrossRef] [PubMed]
55. Mohammed, L.A.; Aljaberi, M.A.; Amidi, A.; Abdulsalam, R.; Lin, C.-Y.; Hamat, R.A.; Abdallah, A.M. Exploring Factors Affecting Graduate Students’ Satisfaction toward E-Learning in the Era of the COVID-19 Crisis. *Eur. J. Investig. Health Psychol. Educ.* **2022**, *12*, 1121–1142. [CrossRef]
56. Aljaberi, M.A.; Lee, K.-H.; Alareqe, N.A.; Qasem, M.A.; Alsalahi, A.; Abdallah, A.M.; Noman, S.; Al-Tammemi, A.A.B.; Mohamed Ibrahim, M.I.; Lin, C.-Y. Rasch Modeling and Multilevel Confirmatory Factor Analysis for the Usability of the Impact of Event Scale-Revised (IES-R) during the COVID-19 Pandemic. *Healthcare* **2022**, *10*, 1858. [CrossRef]
57. Abiddine, F.Z.E.; Aljaberi, M.A.; Gadelrab, H.F.; Lin, C.-Y.; Muhammed, A. Mediated effects of insomnia in the association between problematic social media use and subjective well-being among university students during COVID-19 pandemic. *Sleep Epidemiol.* **2022**, *2*, 100030. [CrossRef]
58. Alareqe, N.A.; Roslan, S.; Taresh, S.M.; Nordin, M.S. Universality and Normativity of the Attachment Theory in Non-Western Psychiatric and Non-Psychiatric Samples: Multiple Group Confirmatory Factor Analysis (CFA). *Int. J. Environ. Res. Public Health* **2021**, *18*, 5770. [CrossRef] [PubMed]
59. Fekih, L. Bullying among high school students and their relationship with diligence at school Field study on a sample of secondary school students in Algeria. *Qatar Found. Annu. Res. Conf. Proc.* **2018**, *2018*, SSAHPP129. [CrossRef]
60. Sabah, A.; Aljaberi, M.A.; Lin, C.-Y.; Chen, H.-P. The Associations between Sibling Victimization, Sibling Bullying, Parental Acceptance–Rejection, and School Bullying. *Int. J. Environ. Res. Public Health* **2022**, *19*, 16346. [CrossRef] [PubMed]
61. Alareqe, N.A.; Roslan, S.; Nordin, M.S.; Ahmad, N.A.; Taresh, S.M. Psychometric Properties of the Millon Clinical Multiaxial Inventory–III in an Arabic Clinical Sample Compared With American, Italian, and Dutch Cultures. *Front. Psychol.* **2021**, *12*, 562619. [CrossRef] [PubMed]
62. Sabah, A.; Al-Shujairi, O.K.R. The Confirmatory Factor Analysis of the Spiritual Wellbeing Scale of a Sample of Students from Iraq and Algeria. *J. STEPS Humanit. Soc. Sci.* **2022**, *1*, 1. [CrossRef]
63. Kimberlin, C.L.; Winterstein, A.G. Validity and reliability of measurement instruments used in research. *Am. J. Health-Syst. Pharm.* **2008**, *65*, 2276–2284. [CrossRef] [PubMed]
64. Bond, T.G.; Fox, C.M. *Applying the Rasch Model: Fundamental Measurement in the Human Sciences*, 2nd ed.; Psychology Press: New York, NY, USA, 2013.
65. Linacre, J.M. *WINSTEPS Rasch Measurement Computer Program*; WINSTEPS.com: Chicago, IL, USA, 2006.
66. Bond, T.G.; Fox, C.M. *Applying the Rasch Model: Fundamental Measurement in the Human Sciences*, 3rd ed.; Routledge: Oxfordshire, UK, 2015.
67. Ariffin, S.R.; Omar, B.; Isa, A.; Sharif, S. Validity and reliability multiple intelligent item using rasch measurement model. *Procedia-Soc. Behav. Sci.* **2010**, *9*, 729–733. [CrossRef]
68. Zahia, O.-B. Gender inequity in education in Algeria: When inequalities are reversed. *J. Educ. Soc. Policy* **2018**, *5*, 84–105. [CrossRef]
69. AHMAID, I. What Stands behind the Balanced Ratio of Male/Female Students in the Algerian STEM Education Despite the Country's Low Gender Equity? Master's Thesis, University of Agder, Kristiansand, Norway, 2021.

 *healthcare*

Review

An Ethical Analysis Regarding the COVID-19 Pandemic Impact on Oral Healthcare in Patients with Mental Disorders

Oana-Maria Isailă [1,*], Eduard Drima [2,*] and Sorin Hostiuc [1]

[1] Department of Legal Medicine and Bioethics, Faculty of Dentistry, "Carol Davila" University of Medicine and Pharmacy, 020021 Bucharest, Romania; sorin.hostiuc@umfcd.ro

[2] Medical Clinical Department, Dunărea de Jos University, 800201 Galați, Romania

* Correspondence: oana_maria.isaila@yahoo.com (O.-M.I.); drima_eduardpolea@yahoo.com (E.D.)

Abstract: During the COVID-19 pandemic, restrictive measures were imposed that significantly impacted the healthcare system in general, and the dental healthcare system in particular. The literature cites a possible association between mental and oral health, as psychiatric patients have decreased awareness of their oral health and, therefore, poor dental status. Moreover, several studies have found a positive association between SARS-CoV-2 infection and oral health conditions, as well as between SARS-CoV-2 infection and mental health status. This context generated multiple ethical dilemmas in the case of persons with mental health disorders who require dental treatment because they are more vulnerable in this respect. This article aims to analyze the ethical issues in dental care for patients with mental disorders concerning the COVID-19 restrictive measures. The ethical aspects involved here are the basic principles of bioethics and the related elements of accessibility, equity, consent, and confidentiality.

Keywords: oral health; COVID-19; mental disorders; justice; vulnerable; stigma

Citation: Isailă, O.-M.; Drima, E.; Hostiuc, S. An Ethical Analysis Regarding the COVID-19 Pandemic Impact on Oral Healthcare in Patients with Mental Disorders. *Healthcare* **2023**, *11*, 2585. https://doi.org/10.3390/healthcare11182585

Academic Editors: Carlos Laranjeira and Ana Querido

Received: 21 August 2023
Revised: 17 September 2023
Accepted: 18 September 2023
Published: 19 September 2023

1. Introduction

During the COVID-19 pandemic, restrictive social healthcare-related measures were imposed, which were required to properly manage a large number of infected patients, but also to protect citizens from disease transmission. The main route of transmission of the virus has been found to be airborne via small infected fluid particles (including aerosols) from the sick person when coughing, talking, sneezing, or breathing [1,2]. Thus, given the high risk of contamination with potentially dire consequences, many social, economic, or even medical services have been temporarily halted, leading to changes in living and working conditions. One of the medical areas that has been severely impacted by the pandemic, mainly due to its significant exposure to aerosols, has been outpatient dental care [3].

In this context, at the level of the general population, amid uncertainty and restrictive measures, common psychological reactions such as stress, anxiety, depression, frustration, panic, hopelessness, and despair, including self-harming behavior, were observed [4,5]. The impact of COVID-19 was disproportionate, with vulnerability being higher among persons with low socio-economic status and pre-existing somatic conditions [6]. A significant burden was identified in patients with pre-existing psychiatric conditions, both due to the overall pandemic context, which included lower accessibility to health services [6], higher risk of infection, and the SARS-CoV-2 infection itself, which has been shown to cause various short- and long-term mental health issues [7]. The literature cites a possible association between mental and oral health, as psychiatric patients have a decreased awareness of their oral health, a situation that was intensified by the limitations imposed by the pandemic context [8]. The interconnection between mental and oral health, following unhealthy behaviors and neglect of oral health, involves bacterial translocation and systemic inflammation against the backdrop of microbiome dysregulation, thereby observing a precarious dental status in conditions such as Alzheimer's disease, depression, Parkinson's disease, bipolar disorder, and Schizophrenia [9].

In the same manner, studies have found a positive association between SARS-CoV-2 infection and oral health conditions such as xerostomia, aphthous lesions [10], orofacial pain, and periodontal symptoms. Additionally, a positive association was observed between the severity of the infection and periodontal symptoms [11]. In the same sense, the specialized literature reveals possible positive associations between temporomandibular joint disease and the context of the SARS-CoV-2 pandemic, given the impact on the mental health of the population through the prism of isolation, social distancing, quarantine, and the infection itself. These increased the level of stress, anxiety, and depression, with stress being an aggravating factor for bruxism [12].

Considering the above-stated facts, a complex interconnection between COVID-19, mental health, and oral health exists, which is summarized in Figure 1.

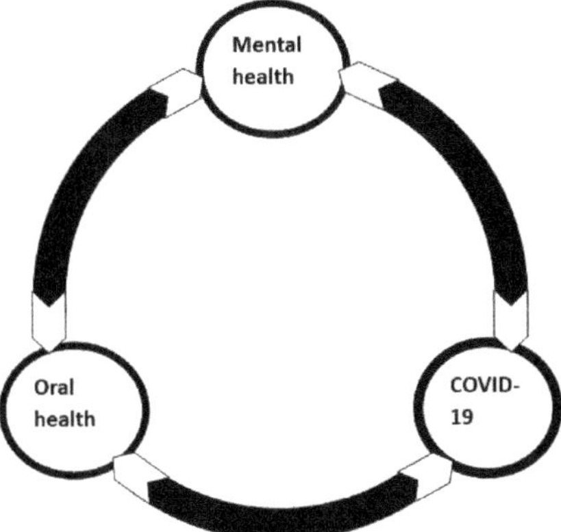

Figure 1. The interlinking association between COVID-19, mental health, and oral health.

The main ethical issues in the dental care of patients with neurocognitive disorders before the COVID-19 pandemic, identified according to the literature on this topic, for example purposes, are listed in Table 1.

Table 1. The main ethical issues in dental care for patients with neurocognitive disorders prior to the COVID-19 pandemic.

Authors, Year	Ethical Issues Identified
Nordenram et al., 1994 [13]	The treatment decision and treatment task
Nordenram and Norberg, 1998 [14]	The uncertainty regarding the provision of adequate medical treatment Respect for human integrity
Shuman 1999 [15]	Valid informed consent, ensuring patient safety and dignity
Gitto et al., 2001 [16]	Achieving the ethical norms without discrimination
Yao and MacEntee, 2014 [17]	Equity, accessibility
Delwel et al., 2017 [18]	The need to improve the training of health personnel for patient-centered dental treatment
Geddis-Regan et al., 2018 [19]	Medical personnel education for patient-centered dental treatment

The purpose of our article is to analyze the ethical implications of the COVID-19 pandemic on oral healthcare in patients with mental illness.

2. Materials and Methods

We analyzed the scientific literature on moral principles and values related to oral healthcare for psychiatric patients during the COVID-19 pandemic. The search was conducted on Web of Science, Scopus, PubMed, and Google Scholar for peer-reviewed research that explores the dental approach to patients with psychiatric symptoms or disorders during the COVID-19 pandemic from a clinical and ethical perspective. The included literature focused on oral health structures in relation to the aforementioned populations with special needs, through the impact of the restrictive measures implemented during the pandemic, with their repercussions on the medical act and the patient. We also took into account scenarios centered on particular patient cases exposed to the general public, as well as the reactions and medical methods of adaptation generated.

3. Results

3.1. Beneficence and Nonmaleficence

As a general rule, during the early stages of the COVID-19 pandemic, dental beneficence has been deprioritized in relation to nonmaleficence, as the health-related risks associated with disease transmission, both for patients and physicians, have been considered higher than the consequences of postponing dental care [20]. Dental care has generally been limited to emergency cases during the pandemic to reduce the usage of scarce protective equipment [21]. Some authors have even recommended new frameworks to identify dental emergencies during the pandemic, based on a multilevel evaluation (oral, general, and psychological), associated with a risk assessment score to manage them [22]. Since the availability of protective equipment and vaccines, along with a better understanding of disease transmission and mechanisms, the deprioritization of beneficence has decreased. This has made oral care more accessible to patients regardless of their emergency status. Many patients with mental disorders who required dental consultation were institutionalized, lived in poorer communities, often lacked the material resources or protective equipment to travel to a dental practice or were unable to tolerate them, did not have proper access to newer technologies that were shown to facilitate dental assessment (such as telemedicine), were exposed to being infected as well as exposed their caregivers to being infected, were unable to respect general rules designed to minimize transmission, etc. [23–26]. Numerous measures have been implemented to prevent the transmission of infections and enable dental procedures. However, these measures were not always viable for individuals with mental health concerns. Dental organizations, for instance, devised protocols that called for conducting virtual consultations or telephonic conversations with patients before commencing treatment. This helped limit the number of physical interactions required [27]. Nonetheless, this course of action was not always a feasible option for patients with mental health conditions, as they either lacked the technical know-how or were disinterested in engaging with these modalities. The use of plexiglass shields, which were employed to separate the medical staff from patients to minimize direct aerosol exposure and limit personal contact, potentially increased patient anxiety and therefore decreased compliance [28]. These specific issues, associated with an overall reported increased difficulty in treating mental health patients by dental practitioners [29–33], have led to a longer deprioritization of dental beneficence in this population.

3.2. Veracity

Veracity is one of the pillars of the physician–patient relationship, involving therapeutic premises and patient expectations. Truthfulness confers respect for the autonomy of the patient or the parental/legal guardian authority in the decision-making process, with the patient's best medical interests as its objective. To achieve the goal of veracity, physicians have a duty to prevent the promotion of misinformation, and therefore, speculative and misleading therapies. In the initial pandemic phase, external anxiety and caution prevailed in the absence of sufficient definitive evidence regarding COVID-19 to underpin medical practice in this regard [34]. In psychiatry, in the case of patients without decision-making capacity, one can speak of the physician–patient–legal representative relationship, a situation of increased complexity, when a person requires dental care against the background of the COVID-19 pandemic. Effective communication between physicians and patients is a pivotal component of medical care. However, several factors can impede this process. This may encompass challenges in accurately diagnosing and assessing the urgency of a patient's condition when they are unable to articulate their discomfort. Additionally, prioritizing patients based on valid and ethical criteria, maintaining a solid physician–patient relationship while utilizing new equipment that limits nonverbal communication, and promoting patient adherence to treatment during physician–patient interactions can also pose significant obstacles [35]. Nonverbal communication is an important element in interpersonal relationships, even more so in a medical context, when patients with certain cognitive impairments are easily influenced by certain clothing and gestures. Trust and respect are largely inspired by nonverbal communication. For example, in the study by Lavelle et al., it was revealed that the nonverbal behavior of patients with schizophrenia can denote the avoidance of social interaction or the desire for social interaction, and this is associated with the behavior of their psychiatrist as well as the quality of the therapeutic relationship [36].

Furthermore, the COVID-19 pandemic has created uncertainty surrounding the accuracy of medical information as well as the effectiveness of certain treatments and investments in nonstandardized bio-decontamination equipment [37]. Studies in this regard have found varying levels of knowledge among the medical staff. For example, Wahed et al. found a knowledge level of 80.4% in respondents with higher education, and in those working directly with COVID-19-positive patients, a significantly increased knowledge score [38]. Similarly, Zhang et al. found that 89% of medical staff respondents had sufficient knowledge about COVID-19, and 89.7% followed correct practices regarding COVID-19 [39].

It is crucial for dental professionals to prioritize providing accurate medical information to patients [40]. However, the approach may need to be tailored to the patient's physical and mental health conditions, social factors, and individual traits. Although truthful communication is typically expected by healthcare providers, conflicts may arise when certain values are at odds [41]. One such scenario that can be difficult for dentists is when a patient lacks the ability to make decisions, which may occur in those experiencing mental health disorders [42]. In such cases, uncertainty about the reliability of information may lead to anxiety for both the patient and the practitioner [43].

3.3. Consent and Confidentiality

Triage and informing patients of the risks of viral infection is mandatory, and confidentiality can be breached regarding the patient's COVID-19 status with the maintenance of standards of care [44]. In particular, in patients with mental disorders, the possibility of absent/altered decisional capacity should be considered. For patients with impaired decision-making capacity, alternative, more suggestive methods of information can be used, with the minimization of medical terms, the use of beneficence centered physician–patient relationship models to decrease anxiety about making a complex decision, and the inclusion of the family in the decision-making process, with the explicit consent of the patient. Once the decision-making algorithm has been completed, the informed consent form is the sole responsibility of the patient, and the procedure is performed solely on the patient's signature. In the case of patients whose decision-making capacity is absent, decisions are made by legal representatives, based on the patient's previous wishes, concerning the patient's current biological status and current dental needs [45]. The COVID-19 pandemic posed significant challenges to providing interventional dental treatment. Given the limited time for physician–patient interaction and the risk of contamination for the patient's legal representative, the prioritization of dental emergencies was critical. Consent information also evolved due to the pandemic's impact on acceptable procedures and available interventions. Drug treatment and tooth extraction emerged as the primary practices, while procedures involving aerosol use were avoided. Staged interventions that required constant monitoring were deferred until safe conditions were established [46]. Additionally, certain dentists have suggested implementing a written consent form to inform patients of the risk of COVID-19 transmission during dental appointments [47].

3.4. Justice, Accessibility, and Equity

The oral health of mental health patients and persons with special needs is frequently overlooked due to their lack of motivation and/or lack of awareness, dental treatment phobias, and poor economic situation. Common barriers such as decreased awareness, decreased expression of physical discomfort, and dependence on family members to reach the dental office or to provide treatment expenses are well known [48]. Pathological conditions, including oral health, can be assessed clinically, objectively, and strictly based on local examinations. The concept of "oral health-related quality of life" refers to the patient's perception of the disease and its impact on their life. This is a multidimensional parameter that involves biopsychosocial elements. It can be influenced by cultural background, age, and other coexisting pathologies including mental disorders, with a major impact on the patient's subjective perceptions [3,49]. In the context of the COVID-19 pandemic, these have been extended and augmented by restrictive measures, with reduced outpatient dental care and increased economic instability [50]. A scenario that reveals the amplification of the patient's vulnerabilities is that of a person with paranoid schizophrenia with severe dental pain who has a panic attack when trying to visit the dentist's office because is afraid of people with protective masks seen on the street [51]. If in the pre-pandemic period, the simple phobia of the dentist prevailed, during the pandemic, new phobias and new adjacent anxiety reasons emerged, making it more difficult for the patient to access treatment, in addition to the restrictive measures. Ensuring tailored accessibility for vulnerable populations with special needs is essential to avoid serious scenarios with irrevocable consequences and can also be a key factor in improving patient cooperation and compliance.

In the same sense, the absence of professional training for dentists regarding patient care for patients with special needs, the risk of stigmatizing this vulnerable population group, and the lack of training for psychiatrists on oral health screening [48], combined with the novelty of the insufficiently elucidated SARS-CoV-2 viral pathology, have increased barriers to dental care. Aljabri et al. identified anxiety, difficulty in keeping appointments, and treatment costs as the main barriers to accessing dental care among psychiatric patients [52]. In the context of the COVID-19 pandemic, Phadraig et al. found a decrease in the provision of dental care for people with disabilities during isolation, with the pandemic having an additional impact on their ability to access already scarce oral healthcare services, with dentists reporting significant decreases in pharmacological support for these patients [53]. Access to medical care was also further compromised in the case of people with dental emergencies who, due to neuropsychological impairments, were unable to express or become aware of their dental pathological status. Given the new working conditions, with the appropriate protective measures imposed, additional costs were incurred, which increased the cost of dental services [54]. A study conducted by Wolf et al. among dentists in Switzerland and Lichtenstein revealed that most dentists had to limit their dental practice to a minimum. By the end of 2020, 1.4% of dentists were forced to close their practice either permanently or temporarily due to a severely reduced economic situation [55]. Corollary to the aforementioned barriers, the pandemic has escalated these issues by increasing the incidence or worsening of mental disorders [7,56,57]. An obstacle frequently encountered by individuals with mental disorders seeking COVID-19 triage is the standardized protocol. To contain the spread of the virus, individuals were required to disclose whether they have been in contact with infected individuals or exhibit symptoms of the illness. However, this can pose a challenge to individuals with certain mental conditions. Knowledge and application of general and pandemic legislative frameworks, empathy, professionalism, and tact in treating these patients, with the establishment of psychiatrist–dentist collaborations were required [58], but at the same time, under the burden of the "duty to treat", and the fear of contamination, physicians have experienced anxiety, which has led to reservations in approaching patients [59].

3.5. Autonomy and Phronesis: Patient versus Community?

Phronesis is based on practical experience and determines a person's ability to perceive a situation to deliberate and act appropriately in given circumstances [60]. The pandemic has imposed caution and wisdom to minimize risk. Thus, public health measures took precedence over individual liberty to protect people with increased vulnerability and increased risk of death [61]. However, vulnerability is a hypothetical notion given the novelty of the etiological agent and insufficiently understood mechanisms of action. For example, in the study by Bain et al. among children with cystic fibrosis, SARS-CoV-2 infection was found to be clinically mild [62]. In pandemic management, moral resistance has played an important role in managing the ethical dilemmas generated in medical practice [63] by highlighting the physician as a public health agent at risk of compromising loyalty to the potential vector patient [64]. Delaying dental treatment in nonemergencies to control viral spread, according to some authors, has been associated with an "ethical barter for the greater good of the community" [46]. Patient autonomy and the need for regular dental care were undermined by an approach in which maximum caution dominated amid limited resources and incompletely understood infection mechanisms. This was more pronounced in people with mental disorders, in whom concern about oral health is much lower. According to a survey carried out by Knights et al. amidst the COVID-19 pandemic, dental health professionals expressed their primary concerns to be the potential of being accused of subpar treatment by patients, the sense of remorse for failing to provide adequate information to patients, postponed or missed diagnoses, and the disappointment of being unable to offer oral hygiene assistance to vulnerable patients with additional requirements [65].

For people with mental disorders, stigma can be extrinsic (public) but also intrinsic (self-stigma) [66]. In addition, SARS-CoV-2 infection has generated stigma and discrimination [67], given the exacerbation of negative attitudes toward already marginalized groups. This has created discussions about prioritizing the allocation of scarce resources, highlighting inequities in healthcare delivery among disadvantaged population groups [68].

The pandemic has caused a decrease in the number of dental procedures, leading to financial struggles in the dental profession. This is concerning, and there is a need for the profession to recover by prioritizing preventative services and ensuring equal access to oral health services [69].

3.6. Teledentistry and Its Implications

In the COVID-19 pandemic, telemedicine, which was already perceived as a promising tool for the prevention and promotion of oral health [70,71], gained momentum and became the preferred method given the background of protective measures. The main consideration behind this modern dentistry approach has been overcoming the limited personnel and infrastructure resources [72]. Teledentistry is considered reliable and feasible for patients with special needs for triage and treatment planning [73]. Remote oral examination via telecommunication is less stressful as it occurs in an environment familiar to the patient. At the same time, in the pandemic context, it has become preferable to avoid direct physician–patient interactions that could have favored the spread of the virus [74]. Rahman et al.'s research on the use of teledentistry among patients revealed that they were highly satisfied with the service. This study focused on factors such as the ease of use, increased accessibility to healthcare, reliability, and usefulness to patients. Over 90% of the respondents had a positive response to all parameters [75]. In their study among dentists on teledentistry, Cheuk et al. found that 49.3% of dentists use teledentistry, 36% since the pandemic. The three major uses of telestomatology by the respondents were in the areas of patient consultation and education. The mixed perception of this type of dentistry approach is derived from the lack of resources, lack of interest, and a limited dental curriculum to optimize its use [76]. In a study conducted among dentists and patients, Menhadji et al. revealed that 75.7% of the patients found the teleconsultation option more comfortable, and most of them showed a better understanding of their dental status, although they were initially skeptical about telemedicine. As far as dentists are concerned, those specializing in dental restorative processes considered telemedicine to be of little use in cases requiring specific in-person investigations for differential diagnosis. The barriers to teledentistry were technological because of the poor quality of service in the telecommunication network [77].

Although seemingly salutary given the particularity of the situation, teledentistry involves certain ethical challenges, including confidentiality, which may be compromised by breaches in information security [78] and informed consent, noting the risk of incomplete/inaccurate diagnosis given the remote physician–patient interaction. The validity of informed consent can be problematic in patients with mental disorders, as a person's decision-making capacity is more difficult to assess via telecommunication [71,79].

3.7. Treatment Refusal vs. Duty to Treat

The duty to treat infected patients is based on four considerations: duty as an intrinsic moral obligation, duty as a response to a given trust, duty as a professional norm, and duty as an employer-imposed norm [45].

The provision of dental care involved medical risks for dentists in the initial pandemic phase with limited protective materials. Additionally, through the lens of neurocognitive disorders, a patient may be noncompliant or exhibit negative behaviors, which gives the clinician the right to refuse to initiate the therapeutic alliance as long as the patient's dental status does not represent a dental emergency. Treatment refusal may be due to multiple causes, including cultural differences, inadequate (poor/excessive) information, information from different, contradictory sources (Internet, TV, magazines, relatives, etc.), physicians' verbal and nonverbal communication, and medical conditions that may alter the decision-making process and implicitly lead to refusal of treatment [45]. According to Picciani et al., in the pandemic context, there has been an increase in dental emergencies in which uncooperative patients with neurocognitive deficits refuse dental care because of anxiety and fear, which would require dedicated dental sedation services as well as the behavioral management of anxiety and pain [80]. At the same time, this also implies an assessment of the patient's decisional capacity to validate treatment refusal [45]. Pharmacological sedation techniques have been found to increase patient cooperation, create less traumatic therapeutic contexts, and allow for more personalized care. This approach is particularly beneficial for individuals with mild intellectual disability [81,82]. In the current pandemic situation, Alharbi et al. have categorized dental treatments into different groups. The first category is emergencies which include maxillofacial fractures, infections, and bleeding. The second category includes emergencies that can be treated with minimal invasion and without the production of aerosols, such as dental pain, dental fractures, and periodontitis. The third category includes emergencies that require invasive treatment and/or the use of aerosols. Nonemergencies such as asymptomatic pathologies fall into the fourth category. The fifth category includes elective treatment [20].

In the COVID-19 scenario, the refusal or interruption of the physician–patient relationship was questionable, with the physician having the obligation to treat in the situation of dental emergencies. In support of this obligation came the further obligation of the physician to have and use protective materials (mask, goggles, gloves, face shield, etc.) as well the patient's obligation to provide the necessary information during the triage procedure about COVID-19 status, personal contacts, and symptoms. The drastic measures took into account the possible unfavorable consequences generated by the potential contamination of the physician, which would have required the interruption of his activity and the accentuation of the staff shortage, as well as the scenario of the physician becoming a vector of transmission of the SARS-CoV-2 virus both professionally (patients and colleagues) as well as on a social level [83], which, on the background of the epidemic, led to the premises for deprioritization in the provision of dental care for people with special needs, including patients with mental disorders with deficiencies in awareness of their dental status or even their pain and reluctance to seek specialist services. The survey conducted by Riguzzi and Gashi in June–July 2020 highlighted the marked concern of healthcare staff about patients, the elderly, and family members, and less about their own health. Increased levels of stress were detected owing to the lack of protective materials, staff shortages, and absence at the time of emergency structures and plans [84].

3.8. Future Perspectives

Although a higher prevalence of dental diseases among people with mental disorders is known, the issue of dental treatment is rarely addressed in the literature, which can be considered a limitation of this review. Suggestions to improve oral health in people with special needs include prophylactic dental programs in the context of interdisciplinary psychiatrist–dentist collaboration, the implementation of appropriate preventive dental measures, and easily accessible regular check-ups [85]. The implementation of protocols tailored to people with neurocognitive impairments that reduce anxiety and pain and the association of dental treatment with traumatic experiences is also required [80]. In light of the pandemic, there has been renewed focus on pre-existing challenges, and telemedicine has emerged as a reliable screening option for patients. Nonetheless, dentists must receive adequate training, and additional measures must be in place to uphold ethical standards and prioritize patients' well-being [71]. As psychiatry and dentistry are two distinct medical fields, it would be beneficial to offer training for dentists to better understand the challenges that patients with neurocognitive disorders face when seeking and adhering to treatment.

4. Conclusions

The global impact of COVID-19 has rendered individuals with mental disorders more susceptible, and consequently, their dental healthcare has been adversely affected. It is crucial to acknowledge this reality and implement measures to prevent future circumstances that could inflict irreversible harm upon patients and the healthcare system. Safe and regulated means of obtaining dental services must be prioritized, while also analyzing the ethical implications of medical practices and their effects on vulnerable groups.

Author Contributions: Conceptualization, O.-M.I.; methodology, O.-M.I., S.H. and E.D.; software, O.-M.I., E.D. and S.H.; validation, O.-M.I., S.H. and E.D.; formal analysis: O.-M.I., E.D. and S.H.; investigation, O.-M.I., S.H. and E.D.; resources, E.D., O.-M.I. and S.H.; data curation, O.-M.I., E.D. and S.H.; writing—original draft preparation, O.-M.I., E.D. and S.H.; writing—review and editing, O.-M.I., E.D. and S.H.; visualization, O.-M.I., E.D. and S.H.; supervision, O.-M.I. and S.H.; project administration, O.-M.I.; funding acquisition, O.-M.I., E.D. and S.H. All authors have read and agreed to the published version of the manuscript.

Funding: This research received no external funding.

Institutional Review Board Statement: Not applicable.

Informed Consent Statement: Not applicable.

Data Availability Statement: Not applicable.

References

1. World Health Organization. Coronavirus Disease (COVID-19). Available online: https://www.who.int/health-topics/coronavirus#tab=tab_1 (accessed on 11 June 2023).
2. Rothan, H.A.; Byrareddy, S.N. The epidemiology and pathogenesis of coronavirus disease (COVID-19) outbreak. *J. Autoimmun.* **2020**, *109*, 102433. [CrossRef]
3. Ciardo, A.; Simon, M.M.; Sonnenschein, S.K.; Büsch, C.; Kim, T.-S. Impact of the COVID-19 pandemic on oral health and psychosocial factors. *Sci. Rep.* **2022**, *12*, 4477. [CrossRef] [PubMed]

4. Serafini, G.; Parmigiani, B.; Amerio, A.; Aguglia, A.; Sher, L.; Amore, M. The psychological impact of COVID-19 on the mental health in the general population. *QJM Int. J. Med.* **2020**, *113*, 531–537. [CrossRef] [PubMed]
5. Duan, L.; Zhu, G. Psychological interventions for people affected by the COVID-19 epidemic. *Lancet Psychiatry* **2020**, *7*, 300–302. [CrossRef] [PubMed]
6. Penninx, B.W.J.H.; Benros, M.E.; Klein, R.S.; Vinkers, C.H. How COVID-19 shaped mental health: From infection to pandemic effects. *Nat. Med.* **2022**, *28*, 2027–2037. [CrossRef]
7. Nersesjan, V.; Amiri, M.; Lebech, A.-M.; Roed, C.; Mens, H.; Russell, L.; Fonsmark, L.; Berntsen, M.; Sigurdsson, S.T.; Carlsen, J.; et al. Central and peripheral nervous system complications of COVID-19: A prospective tertiary center cohort with 3-month follow-up. *J. Neurol.* **2021**, *268*, 3086–3104. [CrossRef]
8. Persson, K.; Olin, E.; Östman, M. Oral health problems and support as experienced by people with severe mental illness living in community-based subsidised housing-a qualitative study. *Health Soc. Care Community* **2010**, *18*, 529–536. [CrossRef] [PubMed]
9. Skallevold, H.E.; Rokaya, N.; Wongsirichat, N.; Rokaya, D. Importance of oral health in mental health disorders: An updated review. *J. Oral Biol. Craniofac. Res.* **2023**, *13*, 544–552. [CrossRef]
10. Aragoneses, J.; Suárez, A.; Algar, J.; Rodríguez, C.; López-Valverde, N.; Aragoneses, J.M. Oral Manifestations of COVID-19: Updated Systematic Review with Meta-Analysis. *Front. Med.* **2021**, *8*, 726753. [CrossRef]
11. Qi, X.; Northridge, M.E.; Hu, M.; Wu, B. Oral health conditions and COVID-19: A systematic review and meta-analysis of the current evidence. *Aging Health Res.* **2022**, *2*, 100064. [CrossRef]
12. Rokaya, D.; Koontongkaew, S. Can Coronavirus Disease-19 Lead to Temporomandibular Joint Disease? *Open Access Maced. J. Med. Sci.* **2020**, *8*, 142–143. [CrossRef]
13. Nordenram, G.; Norberg, A.; Bischofberger, E. Ethical aspects of dental care for demented patients. *Methodol. Consid. Swed. Dent. J.* **1994**, *18*, 155–164.
14. Nordenram, G.; Norberg, A. Ethical issues in dental management of patients with severe dementia: Ethical reasoning by hospital dentists. A narrative study. *Swed. Dent. J.* **1998**, *22*, 61–76. [PubMed]
15. Shuman SK, Doing the Right Thing: Resolving Ethical Issues in Geriatric Dental Care. *J Calif. Dent. Assoc.* **1999**, *27*, 693–698. [CrossRef]
16. Gitto, C.A.; Moroni, M.J.; Terezhalmy, G.T.; Sandu, S. The patient with Alzheimer's disease. *Quintessence Int.* **2001**, *32*, 221–231.
17. Yao, C.S.; Macentee, M.I. Inequity in Oral Health Care for Elderly Canadians: Part 2. Causes and Ethical Considerations. *J. Can. Dent. Assoc.* **2014**, *80*, e10.
18. Delwel, S.; Binnekade, T.T.; Perez, R.S.G.M.; Hertogh, C.M.P.M.; Scherder, E.J.A.; Lobbezoo, F. Oral hygiene and oral health in older people with dementia: A comprehensive review with focus on oral soft tissues. *Clin. Oral Investig.* **2017**, *22*, 93–108. [CrossRef]
19. Geddis-Regan, A.R.; Stewart, M.; Wassall, R.R. Orofacial pain assessment and management for patients with dementia: A meta-ethnography. *J. Oral Rehabil.* **2018**, *46*, 189–199. [CrossRef]
20. Alharbi, A.; Alharbi, S.; Alqaidi, S. Guidelines for dental care provision during the COVID-19 pandemic. *Saudi Dent. J.* **2020**, *32*, 181–186. [CrossRef]
21. Cohen, D.F.; Kurkowski, M.A.; Wilson, R.J.; Jonke, G.J.; Patel, O.R.; Pappas, R.P.; Hall, D.W.; Pandya, A. Ethical practice during the COVID-19 pandemic. *J. Am. Dent. Assoc.* **2020**, *151*, 377–378. [CrossRef]
22. Spicciarelli, V.; Marruganti, C.; Viviano, M.; Baldini, N.; Franciosi, G.; Tortoriello, M.; Ferrari, M.; Grandini, S. A new framework to identify dental emergencies in the COVID-19 era. *J. Oral Sci.* **2020**, *62*, 344–347. [CrossRef] [PubMed]
23. Søraa, R.A.; Manzi, F.; Kharas, M.W.; Marchetti, A.; Massaro, D.; Riva, G.; Serrano, J.A. Othering and deprioritizing older adults' lives: Ageist discourses during the COVID-19 pandemic. *Eur. J. Psychol.* **2020**, *16*, 532–541. [CrossRef]
24. Dintrans, P.V.; Valenzuela, P.; Castillo, C.; Granizo, Y.; Maddaleno, M. Bottom-up innovative responses to COVID-19 in Latin America and the Caribbean: Addressing deprioritized populations. *Rev. Panam. Salud. Publica* **2023**, *47*, e92. [CrossRef] [PubMed]
25. Allwood, L.; Bell, A. COVID-19: Understanding inequalities in mental health during the pandemic. *Cent. Ment. Health* **2020**, *18*, 2020-07.
26. Ettinger, R.; Marchini, L.; Zwetchkenbaum, S. The Impact of COVID-19 on the Oral Health of Patients with Special Needs. *Dent. Clin. N. Am.* **2022**, *66*, 181–194. [CrossRef] [PubMed]
27. Ren, Y.; Rasubala, L.; Malmstrom, H.; Eliav, E. Dental Care and Oral Health under the Clouds of COVID-19. *JDR Clin. Transl. Res.* **2020**, *5*, 202–210. [CrossRef]
28. Pinsonnault-Skvarenina, A.; de Lacerda, A.B.M.; Hotton, M.; Gagné, J.-P. Communication with Older Adults in Times of a Pandemic: Practical Suggestions for the Health Care Professionals. *Public Health Rev.* **2021**, *42*, 1604046. [CrossRef]
29. Scrine, C.; Durey, A.; Slack-Smith, L. Enhancing oral health for better mental health: Exploring the views of mental health professionals. *Int. J. Ment. Health Nurs.* **2017**, *27*, 178–186. [CrossRef]
30. Slack-Smith, L.; Hearn, L.; Scrine, C.; Durey, A. Barriers and enablers for oral health care for people affected by mental health disorders. *Aust. Dent. J.* **2017**, *62*, 6–13. [CrossRef]
31. Scrine, C.; Durey, A.; Slack-Smith, L. Providing oral care for adults with mental health disorders: Dental professionals' perceptions and experiences in Perth, Western Australia. *Community Dent. Oral Epidemiol.* **2018**, *47*, 78–84. [CrossRef]
32. Teng, P.-R.; Lin, M.-J.; Yeh, L.-L. Utilization of dental care among patients with severe mental illness: A study of a National Health Insurance database. *BMC Oral Health* **2016**, *16*, 87. [CrossRef]

33. Wallace, B.B.; MacEntee, M.I. Access to Dental care for Low-Income Adults: Perceptions of Affordability, Availability and Acceptability. *J. Community Health* **2011**, *37*, 32–39. [CrossRef] [PubMed]

34. Graf, W.D.; Epstein, L.G.; Pearl, P.L. Practical Bioethics during the Exceptional Circumstances of a Pandemic. *Pediatr. Neurol.* **2020**, *108*, 3–4. [CrossRef]

35. Grzybowski, S.C.; Stewart, M.A.; Weston, W.W. Nonverbal communication and the therapeutic relationship: Leading to a better understanding of healing. *Can. Fam. Physician* **1992**, *38*, 1994–1998. [PubMed]

36. Lavelle, M.; Dimic, S.; Wildgrube, C.; McCabe, R.; Priebe, S. Non-verbal communication in meetings of psychiatrists and patients with schizophrenia. *Acta Psychiatr. Scand.* **2014**, *131*, 197–205. [CrossRef] [PubMed]

37. Maart, R.; Mulder, R.; Khan, S. COVID-19 in dentistry—Ethical considerations. *S. Afr. Dent. J.* **2020**, *75*, 396–399. [CrossRef]

38. Wahed, W.Y.A.; Hefzy, E.M.; Ahmed, M.I.; Hamed, N.S. Assessment of Knowledge, Attitudes, and Perception of Health Care Workers Regarding COVID-19, A Cross-Sectional Study from Egypt. *J. Community Health* **2020**, *45*, 1242–1251. [CrossRef]

39. Zhang, M.; Zhou, M.; Tang, F.; Wang, Y.; Nie, H.; Zhang, L.; You, G. Knowledge, attitude, and practice regarding COVID-19 among healthcare workers in Henan, China. *J. Hosp. Infect.* **2020**, *105*, 183–187. [CrossRef]

40. American Dental Association. The ADA Principle of Ethics and Code of Conduct. 2023. Available online: https://www.ada.org/-/media/project/ada-organization/ada/ada-org/files/about/ada_code_of_ethics.pdf?rev=ba22edfdf1a646be9249fe2d870d7d31&hash=CCD76FCDC56D6F2CCBC46F1751F51B96 (accessed on 3 September 2023).

41. Olsen, K. Ross and the particularism/generalism divide. *Can. J. Philos.* **2014**, *44*, 56–75. [CrossRef]

42. Khorshidian, A.; Parsapoor, A.; Gooshki, E.S. Conflicts and Challenges of Truth-Telling in Dentistry: A Case-Based Ethical Analysis. *Front. Dent.* **2022**, *19*, 1. [CrossRef]

43. Aboujaoude, E. Ethics Commentary: Ethical Challenges in the Treatment of Anxiety. *Focus* **2011**, *9*, 289–291. [CrossRef]

44. Kathree, B.A.; Khan, S.B.; Ahmed, R.; Maart, R.; Layloo, N.; Asia-Michaels, W. COVID-19 and its impact in the dental setting: A scoping review. *PLoS ONE* **2020**, *15*, e0244352. [CrossRef] [PubMed]

45. Hostiuc, S. Tratat de Bioetică Medicală și Stomatologică. In *Tratat de Bioetică Medicală și Stomatologică*; Editura C.H Beck: București, Romania, 2021; pp. 495–503.

46. Ramanarayanan, V.; Sanjeevan, V.; Janakiram, C. The COVID-19 uncertainty and ethical dilemmas in dental practice. *Indian J. Med. Ethics* **2021**, *6*, 1–3. [CrossRef] [PubMed]

47. Merivale, J. Adapting patient consent in response to COVID-19. *BDJ Pract.* **2020**, *33*, 26–27. [CrossRef]

48. Narayan, V. Management of Dental Patients with Mental Health Problems in Special Care Dentistry: A Practical Algorithm. *Cureus* **2023**, *15*, e34809. [CrossRef]

49. Campos, L.A.; Peltomäki, T.; Marôco, J.; Campos, J.A.D.B. Use of Oral Health Impact Profile-14 (OHIP-14) in Different Contexts. What Is Being Measured? *Int. J. Environ. Res. Public Health* **2021**, *18*, 13412. [CrossRef]

50. Butt, R.T.; Janjua, O.S.; Qureshi, S.M.; Shaikh, M.S.; Guerrero-Gironés, J.; Rodríguez-Lozano, F.J.; Zafar, M.S. Dental Healthcare Amid the COVID-19 Pandemic. *Int. J. Environ. Res. Public Health* **2021**, *18*, 11008. [CrossRef] [PubMed]

51. Singh, K.; Kaur, J. Dental Anxiety and PTSD During the COVID-19 Pandemic. *Prim. Care Companion J. Clin. Psychiatry* **2021**, *23*, 21cr03104. [CrossRef]

52. Aljabri, M.K.; Gadibalban, I.Z.; Kalboush, A.M.; Sadek, H.S.; Abed, H.H. Barriers to special care patients with mental illness receiving oral healthcare. *Saudi Med. J.* **2018**, *39*, 419–423. [CrossRef]

53. Phadraig, C.M.G.; van Harten, M.; Diniz-Freitas, M.; Posse, J.L.; Faulks, D.; Dougall, A.; Dios, P.D.; Daly, B. The impact of COVID-19 on access to dental care for people with disabilities: A global survey during the COVID-19 first wave lockdown. *Med. Oral Patol. Oral Cir. Bucal.* **2021**, *26*, e770–e777. [CrossRef]

54. Farooq, I.; Ali, S. COVID-19 outbreak and its monetary implications for dental practices, hospitals and healthcare workers. *Postgrad. Med. J.* **2020**, *96*, 791–792. [CrossRef] [PubMed]

55. Wolf, T.G.; Zeyer, O. Guglielmo Campus COVID-19 in Switzerland and Liechtenstein: A Cross-Sectional Survey among Dentists' Awareness, Protective Measures and Economic Effects. *Int. J. Environ. Res. Public Health* **2020**, *17*, 9051. [CrossRef] [PubMed]

56. Cullen, W.; Gulati, G.; Kelly, B.D. Mental health in the COVID-19 pandemic. *QJM Int. J. Med.* **2020**, *113*, 311–312. [CrossRef]

57. Leung, C.M.C.; Ho, M.K.; Bharwani, A.A.; Cogo-Moreira, H.; Wang, Y.; Chow, M.S.C.; Fan, X.; Galea, S.; Leung, G.M.; Ni, M.Y. Mental disorders following COVID-19 and other epidemics: A systematic review and meta-analysis. *Transl. Psychiatry* **2022**, *12*, 205. [CrossRef]

58. Sarac, Z.; Zovko, R.; Curlin, M.; Filakovic, P. Dental Medicine and Psychiatry: The Need for Collaboration and Bridging the Professional Gap. *Psychiatr. Danub.* **2020**, *32*, 151–158. [CrossRef] [PubMed]

59. De Haro, J.C.; Rosel, E.M.; Salcedo-Bellido, I.; Leno-Durán, E.; Requena, P.; Barrios-Rodríguez, R. Psychological Impact of COVID-19 in the Setting of Dentistry: A Review Article. *Int. J. Environ. Res. Public Health* **2022**, *19*, 16216. [CrossRef]

60. Aristotle. *The Nicomachean Ethics*; Oxford University Press: Oxford, UK, 1980.

61. Benedictis, F.M.; Bush, A. Science, medicine and ethics during COVID-19 pandemic. *Acta Paediatr.* **2021**, *111*, 213–214. [CrossRef]

62. Bain, R.; Cosgriff, R.; Zampoli, M.; Elbert, A.; Burgel, P.-R.; Carr, S.B.; Castaños, C.; Colombo, C.; Corvol, H.; Faro, A.; et al. Clinical characteristics of SARS-CoV-2 infection in children with cystic fibrosis: An international observational study. *J. Cyst. Fibros.* **2020**, *20*, 25–30. [CrossRef] [PubMed]

63. Delgado, J.; Siow, S.; de Groot, J.; McLane, B.; Hedlin, M. Towards collective moral resilience: The potential of communities of practice during the COVID-19 pandemic and beyond. *J. Med. Ethic* **2021**, *47*, 374–382. [CrossRef]

64. Jones-Bonofiglio, K.; Nortjé, N. A policy and decision-making framework for South African doctors during the COVID-19 pandemic. *S. Afr. Med. J.* **2020**, *110*, 613–616.

65. Knights, J.; Beaton, L.; Young, L.; Araujo, M.; Yuan, S.; Clarkson, J.; Freeman, R. Uncertainty and Fears around Sustainability: A Qualitative Exploration of the Emotional Reactions of Dental Practitioners and Dental Care Professionals during COVID-19. *Front. Oral Health* **2022**, *2*, 799158. [CrossRef] [PubMed]

66. Ahmedani, B.K. Mental Health Stigma: Society, Individuals, and the Profession. *J. Soc. Work Values Ethics* **2011**, *8*, 41–416. [PubMed]

67. Bagcchi, S. Stigma during the COVID-19 pandemic. *Lancet Infect. Dis.* **2020**, *20*, 782. [CrossRef] [PubMed]

68. Chaimowitz, G.A.; Upfold, C.; Géa, L.P.; Qureshi, A.; Moulden, H.M.; Mamak, M.; Bradford, J.M.W. Stigmatization of psychiatric and justice-involved populations during the COVID-19 pandemic. *Prog. Neuro-Psychopharmacol. Biol. Psychiatry* **2020**, *106*, 110150. [CrossRef]

69. Witton, R.; Plessas, A.; Wheat, H.; Baines, R.; Delgado, M.B.; Mills, I.; Paisi, M. The future of dentistry post-COVID-19: Perspectives from Urgent Dental Care centre staff in England. *Br. Dent. J.* **2021**, 1–5. [CrossRef]

70. Fernández, C.; Maturana, C.; Coloma, S.; Carrasco-Labra, A.; Giacaman, R. Teledentistry and mHealth for Promotion and Prevention of Oral Health: A Systematic Review and Meta-analysis. *J. Dent. Res.* **2021**, *100*, 914–927. [CrossRef]

71. Islam, R.R.; Islam, R.; Ferdous, S.; Watanabe, C.; Yamauti, M.; Alam, M.K.; Sano, H. Teledentistry as an Effective Tool for the Communication Improvement between Dentists and Patients: An Overview. *Healthcare* **2022**, *10*, 1586. [CrossRef]

72. Al-Muzian, L.; Farsi, D.; Hiremath, S. Global Scenario of Teledentistry during COVID-19 Pandemic: An Insight. *Int. J. Clin. Pediatr. Dent.* **2021**, *14*, 426–429. [CrossRef]

73. Omezli, M.M.; Torul, D.; Yilmaz, E.B. Is Teledentistry a Feasible Alternative for People Who Need Special Care? *Disaster Med. Public Health Prep.* **2023**, *17*, e129. [CrossRef]

74. Wakhloo, T.; Reddy, G.S.; Chug, A.; Dhar, M. Relevance of teledentistry during the COVID-19 pandemic. *J. Fam. Med. Prim. Care* **2020**, *9*, 4494–4495. [CrossRef]

75. Rahman, N.; Nathwani, S.; Kandiah, T. Teledentistry from a patient perspective during the coronavirus pandemic. *Br. Dent. J.* **2020**, 1–4. [CrossRef] [PubMed]

76. Cheuk, R.; Adeniyi, A.; Farmer, J.; Singhal, S.; Jessani, A. Teledentistry use during the COVID-19 pandemic: Perceptions and practices of Ontario dentists. *BMC Oral Health* **2023**, *23*, 72. [CrossRef] [PubMed]

77. Menhadji, P.; Patel, R.; Asimakopoulou, K.; Quinn, B.; Khoshkhounejad, G.; Pasha, P.; Sanchez, R.G.; Ide, M.; Kalsi, P.; Nibali, L. Patients' and dentists' perceptions of tele-dentistry at the time of COVID-19. A questionnaire-based study. *J. Dent.* **2021**, *113*, 103782. [CrossRef] [PubMed]

78. Schrimshaw, E.W.; Siegel, K.; Wolfson, N.H.; Mitchell, D.A.; Kunzel, C. Insurance-related barriers to accessing dental care among African American adults with oral health symptoms in Harlem, New York City. *Am. J. Public Health* **2011**, *101*, 1420–1428. [CrossRef]

79. Nutalapati, R.; Jampani, N.; Dontula, B.S.K.; Boyapati, R. Applications of teledentistry: A literature review and update. *J. Int. Soc. Prev. Community Dent.* **2011**, *1*, 37–44. [CrossRef]

80. Picciani, B.L.S.; Bausen, A.G.; dos Santos, B.M.; Marinho, M.A.; Faria, M.B.; Bastos, L.F.; Dziedzic, A. The challenges of dental care provision in patients with learning disabilities and special requirements during COVID-19 pandemic. *Spéc. Care Dent.* **2020**, *40*, 525–527. [CrossRef]

81. Collado, V.; Faulks, D.; Nicolas, E.; Hennequin, M. Conscious sedation procedures using intravenous midazolam for dental care in patients with different cognitive profiles: A prospective study of effectiveness and safety. *PLoS ONE* **2013**, *8*, e71240. [CrossRef]

82. Lim, G.X.D.; Boyle, C.A. Conscious sedation service for geriatric and special-care dentistry: A health policy brief. *Proc. Singap. Health* **2020**, *29*, 126–138. [CrossRef]

83. Isailă, O.-M.; Hostiuc, S. The duty to treat in the context of the COVID-19 pandemic. *Rev. Bioét.* **2020**, *28*, 426–431. [CrossRef]

84. Riguzzi, M.; Gashi, S. Lessons from the First Wave of COVID-19: Work-Related Consequences, Clinical Knowledge, Emotional Distress, and Safety-Conscious Behavior in Healthcare Workers in Switzerland. *Front. Psychol.* **2021**, *12*, 628033. [CrossRef]

85. Gowda, E.; Bhat, P.; Swamy, M. Dental Health Requirements for Psychiatric Patients. *Med. J. Armed Forces India* **2007**, *63*, 328–330. [CrossRef] [PubMed]

healthcare

Brief Report

Differences in Improvement of Physical Function in Older Adults with Long-Term Care Insurance with and without Falls: A Retrospective Cohort Study

Masahiro Kitamura [1,*], Junichi Umeo [2], Kyohei Kurihara [2], Takuji Yamato [2], Takayuki Nagasaki [1], Katsuhiko Mizota [1], Haruki Kogo [1], Shinichi Tanaka [1] and Takashi Yoshizawa [1]

[1] School of Physical Therapy, Faculty of Rehabilitation, Reiwa Health Sciences University, 2-1-12 Wajirooka, Higashi-ku, Fukuoka 811-0213, Japan
[2] Kizuna Daycare Center, 1399-1 Imai, Yukuhashi 824-0018, Japan; junichi-umeo@reha-kizuna.com (J.U.)
* Correspondence: pt_masa0808@yahoo.co.jp; Tel.: +81-92-607-6701

Abstract: (1) Background: This study examined the differences in changes in physical function with and without falls after daycare use among frail older adults with long-term care insurance (LTCI). (2) Methods: In this retrospective cohort study, 82 of 96 consecutive daycare center users met the inclusion criteria. The participants were divided into two groups based on the presence or absence of falls 6–12 months after use. Participant characteristics in the fall and non-fall groups and physical function at baseline and six months in each group were compared. Using analysis of covariance, we analyzed physical function and its changes between the two groups, and cut-off values were calculated using receiver operating characteristic curves. (3) Results: Gait speed, timed up-and-go test, and 30 s chair stand test (CS30) improved significantly over six months in the no-fall group ($n = 70$) and all participants ($n = 82$) ($p < 0.01$). Gait speed in the fall group ($n = 12$) improved significantly over six months ($p = 0.04$). The fall group had significantly lower adjusted ΔCS30 scores than the no-fall group ($p = 0.03$), with a cutoff value of 2 ($p = 0.024$). (4) Conclusions: In older adults with LTCI, physical function with and without falls after daycare use differed by ΔCS30, with a cutoff value of 2.

Keywords: physical function; fall; older adults; long-term care insurance

Citation: Kitamura, M.; Umeo, J.; Kurihara, K.; Yamato, T.; Nagasaki, T.; Mizota, K.; Kogo, H.; Tanaka, S.; Yoshizawa, T. Differences in Improvement of Physical Function in Older Adults with Long-Term Care Insurance with and without Falls: A Retrospective Cohort Study. *Healthcare* **2023**, *11*, 2558. https://doi.org/10.3390/healthcare11182558

Academic Editors: José Carmelo Adsuar Sala, Carlos Laranjeira and Ana Querido

Received: 28 June 2023
Revised: 4 September 2023
Accepted: 13 September 2023
Published: 15 September 2023

1. Introduction

Among falls and fractures in older adults worldwide, at least one fall per year occurs in 30% of those over 65 years of age [1], and 10% of these falls result in fractures that cause difficulties in the activities of daily living (ADL) [2,3]. The annual fall rate among community-dwelling older adults in Japan is approximately 20% [4], and fractures due to falls often require long-term care insurance (LTCI) because of ADL difficulties [5]. Moreover, an increase in healthcare and long-term care expenditures associated with hospitalization and the use of LTCI pose serious problems [6]. Furthermore, older adults with LTCI have a fall rate of 25% and are more likely to fall than those without [7]. Repeated falls cause a significant decline in ADL and serious illness due to refracture and fear; therefore, measures to reduce future falls are extremely important [8–10]. Among fall risk factors, physical function deterioration, such as deterioration in lower extremity muscle strength, walking ability, and balance, has been reported [2,11].

Exercise has been shown to be effective in improving physical function in older adults, including muscle strength, balance, and walking ability [12]. Improvements in physical function have also been reported in older adults with frailty after a six-month exercise program that included resistance, balance, and gait training [13–18]. Furthermore, the effectiveness of lower-extremity strength, gait, and balance training in preventing falls in older adults has been reported [12,19–22].

However, the rate of falls among older adults with LTCI after using daycare is problematic, as high as 18.3% over a 3-month period [23]. In addition, there is a lack of research on the relationship between improvement in their physical function and falls. Furthermore, although the presence or absence of falls after an exercise program is important, very few studies have investigated the difference in change in physical function between the presence and absence of falls. We hypothesized that the fall group of older adults with LTCI would show less improvement in physical function with an exercise program in daycares than the no-fall group would. If we can identify those indicators of physical function that are less likely to change, we can revise the exercise program to improve those physical functions, and possibly prevent falls after daycare use. Therefore, this study aimed to retrospectively examine the differences in physical function and its changes in LTCI older adults with and without falls after using daycare.

2. Materials and Methods

2.1. Design and Participants

This was a retrospective cohort study. The participants were 96 consecutive new users of the daycare center between April 2018 and December 2020. We included participants who could walk with or without aid, were aged 65 years or older, had used LTCI, and had been using the center for 12 months. We excluded those with dementia who scored less than 24 on the mini-mental state examination [24]. Age, sex, weight, LTCI level [25], comorbidities, history of falls [26], and physical functions such as gait speed [27], timed up-and-go test (TUG) [28], and 30 s chair stand test (CS30) were also investigated [29]. This study complied with the Declaration of Helsinki and was approved by the Reiwa Health Sciences University Ethics Committee (approval no. 22-009) and obtained the informed consent of each participant.

2.2. LTCI Levels

The LTCI system in Japan was introduced to meet the demands of older adults, based on the social insurance system [25]. The LTCI level is determined by the LTCI committee of the community in which the participants live and is available to persons aged 65 years and older and 40–64 years with certain diseases. Support level 1 is for people who are independent in ADL but need supervision in activities such as shopping. Support level 2 is for people whose ability to walk is impaired because of lower-limb muscle weakness. Care level 1 refers to individuals requiring simple care for ADL. Care level 2 refers to people who need care for ADL such as eating, urinating, and bathing. Care level 3 refers to people who use walking aids or wheelchairs for mobility and require considerable care for ADL. Care level 4 refers to those who use a wheelchair for mobility and cannot perform ADL without care. Care Level 5 refers to patients who are bedridden, unable to communicate, or unable to eat alone. The LTCI level was investigated by two physical therapists based on clinical data from the participants.

2.3. History of Falls

An older adult's fall was determined by "unintentionally coming to the ground or at a lower level for reasons other than as a consequence of sustaining a violent blow, loss of consciousness, sudden onset of paralysis as in stroke, or an epileptic seizure" [30]. The presence of falls in daily life was investigated by two physical therapists, one nurse, and one caregiver during each weekly daycare center visit. If a fall occurred, a questionnaire was administered to investigate the date and time of the fall and its causes.

2.4. Physical Function

The main outcomes of the study were changes in physical function between baseline and 6 months later: Δgait speed, ΔTUG, and ΔCS30. Gait speed, a measure of walking ability, was measured by a physical therapist using a stopwatch and defined as the time required to walk 5 m at a normal speed [27]. A 1 m acceleration and deceleration walkway was set before and after the 5 m walkway. The TUG test was used to assess balance and mobility [28]. In this test, the measurer asked the participant to "get up from the chair, walk 3 m back and forth at a normal pace, and then sit back down on the chair," and measured the time with a stopwatch. The CS30 was used as a measure of lower extremity muscle strength [29]. The measurer prepared a chair with a seat height of 40 cm, asked the participants to "stand and sit on a chair for 30 s as many times as they could without using their arms," and measured the number of times they could stand. These physical function measurements were performed by two physical therapists based on the measurement manual.

2.5. Exercise Program

The exercise program at the daycare center consisted of multicomponent exercises: (1) warm-ups such as stretching and hot packs; (2) lower extremity resistance training using body weight and machines; (3) aerobic exercise using an ergometer; (4) balance training using a ball; and (5) repetitive standing exercises [30]. These exercises were performed in groups. Initial and periodic evaluations, exercise program planning, and exercise prescriptions were set by the physical therapist with confirmation from a family doctor. Regarding exercise intensity, for aerobic exercise, the rating of perceived exertion (RPE) was "fairly light to somewhat hard" [31], and for resistance training, "approximately 60% intensity of maximal load $\times$ 10–15 times $\times$ three sets" [32]. Regarding time and frequency, the total time was 120 min, with 20 min for each practice session plus rest, and the frequency was 1–2 times per week, with the assistance of two physical therapists, one nurse, and one caregiver. The exercise program was implemented for a period of at least 6 months. Also, the purpose of the program and participation rules were explained at the beginning of use, and the staff recorded attendance for each use.

2.6. Sample Size

This study's sample size used G-Power 3.1 software. The total sample size was at least 36, calculated with an effect size of 0.96, α error of 0.05, and power of 0.8 [33].

2.7. Statistical Analysis

Participant characteristics and clinical parameter values were reported as mean $\pm$ standard deviation for continuous variables and as percentages for categorical variables. Statistical analysis was performed after assessing the normal distribution of the data using the Shapiro–Wilk test. Categorical variables are reported as numbers (%) and variables as mean ($\pm$standard deviation) in the participant characteristics and evaluated parameter values. The participants were divided into two groups based on the presence or absence of falls 6–12 months after the start of the daycare center. Unpaired *t*-tests, Mann–Whitney U-tests, and chi-square tests were used to compare participant characteristics and clinical parameters between the fall and no-fall groups. Paired *t*-tests and Wilcoxon tests were used to compare physical function at baseline and six months in the fall group, no-fall group, and all participants. Analysis of covariance was used to compare the differences in physical function and changes between the two groups. The covariates used were age, sex, and LTCI level between the two groups. A receiver operating characteristic (ROC) curve was used for fall identification and the area under the curve (AUC) was calculated. The Youden index determines the cut-off value for physical function in the presence of falls. AUC values > 0.9 indicate high precision, 0.7–0.9 indicate medium precision, and <0.7 indicate low precision [34]. Statistical significance was set at p value < 0.05. Statistical analyses were performed using the IBM SPSS 25.0 J statistical software (IBM SPSS Japan, Inc., Tokyo, Japan).

3. Results

3.1. Participant Flow

The flowchart of the participants included in this study is shown in Figure 1. Of the 96 participants, 83 met the inclusion criteria for this study; of the 83 participants, those with dementia were excluded, resulting in a final number of 82 participants (fall group, $n = 12$; no-fall group, $n = 70$; fall rate, 14.6%). The six-month exercise program participation rate for the participants analyzed was 83%.

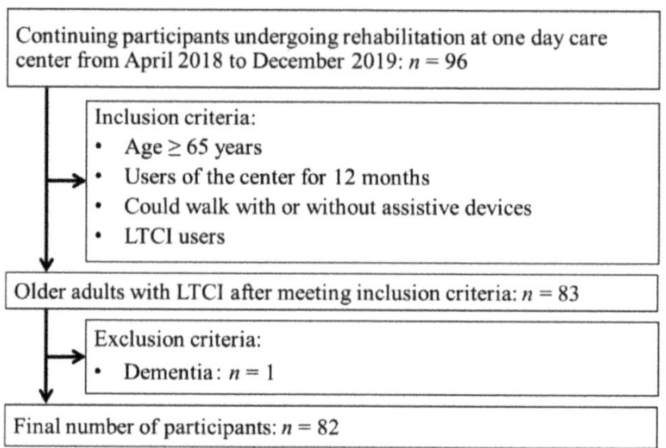

Figure 1. Participant flow.

3.2. Characteristics of Fall Group and No-Fall Group

Table 1 presents the characteristics such as age, sex, weight, LTCI, and comorbidity of the fall and no-fall groups. There were no significant differences between the two groups.

Table 1. Characteristics of fall group and no-fall group.

	Fall, $n = 12$			No-Fall, $n = 70$			Z or χ^2 Value	p Value
Fall rate, %	14.6			0				
Age, years	84.8	±	8.1	81.0	±	7.8	1.5 [b]	0.14
Sex, male, %	41.7			34.3			0.2 [a]	0.62
Weight, kg	57.3	±	13.3	54.7	±	13.7	0.2 [b]	0.84
LTCI level, support level 1/2, care level 1/2/3, %	50.0/33.3/16.7/0/0			48.6/41.4/12.9/5.7/1.4			0.9 [a]	0.91
Comorbidity, %								
Hypertension	41.7			50.0			0.3 [a]	0.59
Diabetes	16.7			25.7			0.5 [a]	0.50
Orthopedic disease	41.7			35.7			0.2 [a]	0.69
Cerebrovascular disease	25.0			30.0			0.1 [a]	0.73
Parkinson's disease	16.7			5.7			0.8 [a]	0.18
Heart disease	8.3			24.3			1.5 [a]	0.21
Cancer disease	8.3			6.1			<0.1 [a]	0.89

Values are presented as mean ± standard deviation or %, [a]: χ^2 value, [b]: Z value. Abbreviations: LTCI, long-term care insurance.

3.3. Pre- and Post-Comparison of Physical Function for Each Group

Results are presented below for a before and after comparison of each group's physical function. Comparing physical function at the start and six months later, there were significant improvements in gait speed (0.93 ± 0.33 m/s vs. 1.14 ± 0.34 m/s; $p < 0.001$, 0.94 ± 0.32 m/s vs. 1.17 ± 0.36 m/s; $p < 0.001$), TUG (13.3 ± 8.07 s vs. 11.4 ± 6.1 s; $p = 0.001$, 13.1 ± 7.8 s vs. 11.1 ± 5.8 s; $p < 0.001$), and CS30 (12.8 ± 7.4 times vs. 15.3 ± 6.1 times; $p < 0.001$, 13.3 ± 7.5 times vs. 15.3 ± 6.1 times; $p = 0.001$) in the no-fall group and all participants, respectively. The fall group showed a significant improvement in gait speed six months after starting at the daycare center (0.99 ± 0.32 m/s vs. 1.34 ± 0.41 m/s; $p = 0.04$), with no significant improvement in TUG and CS30 (10.9 ± 4.8 s vs. 9.6 ± 3.7 s; $p = 0.06$, 17.2 ± 6.6 times vs. 15.7 ± 6.2 times; $p = 0.29$).

3.4. Two-Group Comparison of Physical Function: Start, 6 Months, and Changes during That Period

A comparison of the physical function and the main outcome, Δphysical function, between the fall and non-fall groups is presented in Table 2. The comparison reflected that the fall group had a significantly lower ΔCS30 value than the no-fall group ($p = 0.02$). After adjusting for age, sex, and LTCI, ΔCS30 was significantly different between the two groups ($p = 0.03$).

Table 2. Two-group comparison of physical function: at the start, at 6 months, and changes during that period.

	Fall, $n = 12$			No-Fall, $n = 70$			t or Z Value	p Value
Physical function								
Gait speed at start, m/s	0.99	$\pm$	0.32	0.93	$\pm$	0.33	0.6	0.57
Gait speed at 6 months, m/s	1.34	$\pm$	0.41	1.14	$\pm$	0.34	1.8	0.07
ΔGait speed, m/s	0.35	$\pm$	0.51	0.21	$\pm$	0.36	1.2	0.25
TUG at start, s	10.9	$\pm$	4.8	13.3	$\pm$	8.1	-1.1 [b]	0.28
TUG at 6 months, s	9.6	$\pm$	3.7	11.4	$\pm$	6.1	-1.4 [b]	0.17
ΔTUG, m/s	-1.3	$\pm$	2.5	-2.0	$\pm$	6.9	-0.3 [b]	0.77
CS30 at start, times	17.2	$\pm$	6.6	12.8	$\pm$	7.4	2.0	0.053
CS30 at 6 months, times	15.7	$\pm$	6.2	15.3	$\pm$	6.1	0.2	0.84
ΔCS30, times	-1.6	$\pm$	4.9	2.5	$\pm$	4.8	-2.3 [b]	0.02
Physical function after adjustment								
Gait speed at start, m/s	0.99	$\pm$	0.10	0.93	$\pm$	0.04	0.6	0.57
Gait speed at 6 months, m/s	1.35	$\pm$	0.10	1.14	$\pm$	0.05	1.9	0.06
ΔGait speed, m/s	0.36	$\pm$	0.1	0.21	$\pm$	0.05	1.2	0.10
TUG at start, s	11.3	$\pm$	2.3	13.2	$\pm$	0.9	-0.8	0.43
TUG at 6 months, s	9.7	$\pm$	1.7	11.3	$\pm$	0.7	-0.9	0.39
ΔTUG, m/s	-1.6	$\pm$	1.9	-1.9	$\pm$	0.8	0.2	0.88
CS30 at start, times	16.7	$\pm$	2.1	12.8	$\pm$	0.9	0.7	0.10
CS30 at 6 months, times	15.7	$\pm$	1.8	15.2	$\pm$	0.7	0.2	0.81
ΔCS30, times	-1.0	$\pm$	1.4	2.4	$\pm$	0.5	-2.3	0.03

Values are presented as mean $\pm$ standard deviation or %, [b]: Z value. Abbreviations: CS30, 30 s chair stand test; TUG, timed up-and-go test. Analysis of covariance adjustment: age, sex, long-term care insurance.

3.5. Cut-Off Value for the Presence of Falls

The cut-off value of ΔCS30 for the presence of a fall is shown in Figure 2. The cut-off value was calculated two times (AUC, 0.704; $p = 0.024$).

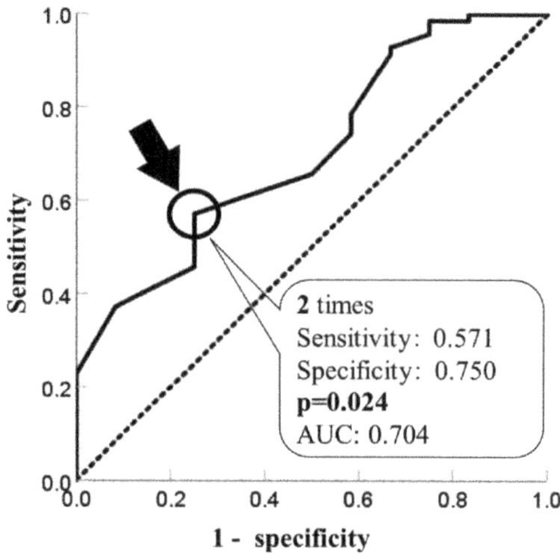

Figure 2. Cut-off value of ΔCS30 for the presence of falls.

4. Discussion

To the best of our knowledge, this is the first report to show differences in changes in physical function depending on the presence or absence of falls in older Japanese adults with LTCI after using daycare. The results showed that the rate of falls among older adults with LTCI after 6–12 months of daycare center use was 14.6% and that the fall group had a lower ΔCS30, the change in lower limb muscle strength, than the no-fall group. In addition, the cut-off value for the presence of falls was two times.

4.1. Fall Rate and Improvement in Physical Function

The fall rate of 14.6% per six months in this study was lower than the 44.5% per year for home-discharged adults after stroke [35] but higher than the 10–20% per year for community-dwelling older adults [4,36]. The fall rate in older adults with LTCI was 25.3% per year and 18.3% per 3 months [7,23], similar to the results of our study.

Regarding the significant improvement in gait speed, TUG, and CS30 for the no-fall group and all participants, and gait speed for the fall group, physical function improvement in older adults with LTCI has been reported with the use of a daycare center for six months [16–18]. Thus, the exercise program at the daycare center is expected to improve the physical function of most participants. However, regarding the failure to show improvement in TUG and CS30 tests in the fall group, older adults at risk of falling reported higher TUG and lower CS30 values than those at no risk [33].

4.2. Cut-Off Value for the Presence of Fall

To the best of our knowledge, there have been no reports on the differences in improvements in physical function between LTCI older adults with and without falls after using daycare. In this study, Δgait speed and ΔTUG showed no difference in change in physical function between the fall and no-fall groups. In a reported randomized controlled trial (RCT) of older adults with LTCI, an exercise program increased walking ability and balance [37], and it is possible that the present study had similar effects on gait speed and TUG in both groups. On the other hand, the ΔCS30 was shown as the difference in physical function in the presence of a fall, and its cutoff value was two times, which was a novelty of this study. CS30 is a screening test for fall risk in older adults [38,39]. The fall risk group of pre-frail older adults reported a lower CS30 than the no-fall risk group [33]. In community-dwelling older adults, the known CS30 cut-off for fall risk is 14.5 times [40]. Furthermore, the reference value for CS30 in the 80s was 15 times [41]. The CS30 at six months in the no-fall group was 15.3 times by exercise, exceeding the cut-off value of a previous study and reaching the reference value [40,41]. Also, the minimal clinically important difference (MCID) for CS30 in older adults with orthopedic disease was two times, similar to the cutoff value of two times in this study [42]. ΔCS30 may predict the risk of falling, which was not captured by the cut-off value of CS30. Therefore, in older adults with LTCI, failure to improve ΔCS30 more than twice with exercise indicated the possibility of future falls. In clinical practice, an increase in CS30 of two or more times from the start through an exercise program may be important in preventing falls.

4.3. Limitations

This study had some limitations. This study was conducted in a daycare center with a small sample size. In addition, we were unable to investigate confounding factors related to fall risk, such as educational history, medications, nutrition, smoking, and environmental modifications. Thus, we were unable to consider the impact of these factors on fall risk [43]. Moreover, because participants were not investigated for their use of LTCI services other than those at the center, we were unable to examine the impact of their use of these services on falls and physical function. The compliance rate for the exercise program has not been investigated and there are some missing data, and interpretation of this study should take this into account.

5. Conclusions

The fall rate for older adults with LTCI during 6–12 months of daycare center use was 14.6%. The difference in physical function with and without falls after using daycare showed a ΔCS30, with a cut-off value of two times. Failure to improve ΔCS30 more than twice after a daycare-based exercise program may result in a future fall.

Author Contributions: Conceptualization, M.K. and J.U.; methodology, M.K.; software, M.K.; validation, M.K.; formal analysis and investigation, M.K., J.U., K.K. and T.Y. (Takuji Yamatoand); resources, writing—original draft preparation, and writing—review and editing, M.K., J.U., K.K., T.Y. (Takuji Yamatoand), T.N., K.M., H.K., S.T. and T.Y. (Takashi Yoshizawa); visualization, supervision, and project administration, M.K. and J.U.; funding acquisition, M.K. All authors have read and agreed to the published version of the manuscript.

Funding: This work was supported by JSPS KAKENHI Grant Number JP23K16629.

Institutional Review Board Statement: This study was conducted in accordance with the Declaration of Helsinki, and approved by Reiwa Health Sciences University Ethics Committee (approval no. 22-009) for studies involving humans.

Informed Consent Statement: Written informed consent was obtained from each participant in this study.

Data Availability Statement: Not applicable.

Acknowledgments: This study benefited from the support and encouragement of the staff of the Daycare Center, Kizuna.

Conflicts of Interest: The authors declare no conflict of interest.

References

1. NICE Guideline CG161. Falls in Older People: Assessing Risk and Prevention. Available online: https://www.nice.org.uk/guidance/cg161/resources/falls-in-older-people-assessing-risk-and-prevention-35109686728645 (accessed on 4 January 2023).
2. Tinetti, M.E.; Speechley, M.; How, C.H. Risk factors for falls among elderly persons living in the community. *N. Engl. J. Med.* **1988**, *319*, 1701–1707. [CrossRef] [PubMed]
3. Campbell, A.J.; Borrie, M.J.; Spears, G.F.; Jackson, S.L.; Brown, J.S.; Fitzgerald, J.L. Circumstances and consequences of falls experienced by a community population 70 years and over during a prospective study. *Age Ageing* **1990**, *19*, 136–141. [CrossRef] [PubMed]

4. Suzuki, T.; Sugiura, M.A.; Furuna, T.; Nishizawa, S.; Yoshida, H.; Ishizaki, T.; Kim, H.; Yukawa, H.; Shibata, H. Association of physical performance and falls among the community elderly in Japan in a five year follow-up study. *Nihon Ronen Igakkai Zasshi* **1999**, *36*, 472–478. [CrossRef]

5. Cabinet Office. White Paper on the Elderly Society. 2021. Available online: https://www8.cao.go.jp/kourei/whitepaper/w-2021/html/zenbun/s1_2_2.html (accessed on 4 January 2023). (In Japanese)

6. Mori, T.; Tamiya, N.; Jin, X.; Jeon, B.; Yoshie, S.; Iijima, K.; Ishizaki, T. Estimated expenditures for hip fractures using merged healthcare insurance data for individuals aged ≥ 75 years and long-term care insurance claims data in Japan. *Arch. Osteoporos.* **2018**, *13*, 37. [CrossRef] [PubMed]

7. Suzukawa, M.; Shimada, H.; Makizako, H.; Watanabe, S.; Suzuki, T. Incidence of falls and fractures in disabled elderly people utilizing long-term care insurance. *Nihon Ronen Igakkai Zasshi* **2009**, *46*, 334–340. (In Japanese) [CrossRef]

8. Koga, R.; Yagi, H.; Togami, K.; Tominaga, T.; Kido, K. The study of factors that influence the fear of falling among the elderly patients with fractured bones. *Jpn. J. Occup. Med. Traumatol.* **2014**, *62*, 23–26. (In Japanese)

9. Aiba, I.; Saito, Y.; Yoshioka, M.; Matsuo, H.; Fujimura, H.; Inui, T.; Kawai, M.; Tobita, M.; Chida, K.; Kaneko, M.; et al. Incidence and characteristics of fall-related serious injuries in patients under long-term care insurance: Japanese prospective fall study in elderly patients under home nursing care (J-FALLS). *Jpn. J. Fall Prev.* **2015**, *2*, 19–33. (In Japanese) [CrossRef]

10. Trevisan, C.; Bedogni, M.; Pavan, S.; Shehu, E.; Piazzani, F.; Manzato, E.; Sergi, G.; March, A. The impact of second hip fracture on rehospitalization and mortality in older adults. *Arch. Gerontol. Geriatr.* **2020**, *90*, 104175. [CrossRef]

11. Panel on Prevention of Falls in Older Persons; American Geriatrics Society and British Geriatrics Society. Summary of the Updated American Geriatrics Society/British Geriatrics Society clinical practice guideline for prevention of falls in older persons. *J. Am. Geriatr. Soc.* **2011**, *59*, 148–157. [CrossRef]

12. Arai, H. *Kaigoyobogaido*; Medical View Co., Ltd.: Tokyo, Japan, 2019; pp. 90–116. Available online: https://www.ncgg.go.jp/ri/topics/documents/cgss1.pdf (accessed on 1 March 2023). (In Japanese)

13. Treacy, D.; Hassett, L.; Schurr, K.; Fairhall, N.J.; Cameron, I.D.; Sherrington, C. Mobility training for increasing mobility and functioning in older people with frailty. *Cochrane Database Syst. Rev.* **2022**, *6*, CD010494. [CrossRef]

14. Takai, I. Effect of balance exercise on functional reach test and body sway for the frail elderly. *Nihon Ronen Igakkai Zasshi* **2008**, *45*, 505–510. (In Japanese) [CrossRef] [PubMed]

15. Takami, N.; Suina, S.; Eguchi, K.; Komuro, H.; Okuyama, T.; Sekiwa, R.; Ishii, T.; Omori, Y. Effectiveness of physical therapy in older people with long-term care insurance certification: Practice at a day rehabilitation for care prevention institution. *Rigakuryoho-Gizyutsu Kenkyu* **2018**, *46*, 67–73. (In Japanese)

16. Kugota, H.; Ikeda, T.; Sagara, M.; Ishida, K. The effect of frequency of exercise on muscle strength and balance abilities for the user of outpatient rehabilitation for preventive long-term care. *Rigakuryoho-Gizyutsu Kenkyu* **2014**, *42*, 47–52. (In Japanese)

17. Shinoda, M.; Arai, T.; Osaki, M.; Kurihara, D.; Yonezawa, M.; Matida, Y.; Sugita, R.; Kobuti, M.; Otuka, Y. The end of use and cure effect of exercise in ambulatory rehabilitation. *J. Saitama Compr. Rehabil.* **2015**, *1*, 2–5. (In Japanese)

18. Wakida, M.; Asai, T.; Kubota, R.; Kuwabara, T.; Fukumoto, Y.; Sato, H.; Nakano, J.; Mori, K.; Ikezoe, T.; Hase, K. Longitudinal effects of physical exercise on health-related outcomes based on frailty status in community-dwelling older adults. *Geriatr. Gerontol. Int.* **2022**, *22*, 213–218. [CrossRef]

19. Sherrington, C.; Tiedemann, A.; Fairhall, N.; Close, J.C.; Lord, S.R. Exercise to prevent falls in older adults: An updated meta-analysis and best practice recommendations. *New South Wales Public Health Bull.* **2011**, *22*, 78–83. [CrossRef]

20. Gillespie, L.D.; Robertson, M.C.; Gillespie, W.J.; Sherrington, C.; Gates, S.; Clemson, L.M.; Lamb, S.E. Interventions for preventing falls in older people living in the community. *Cochrane Database Syst. Rev.* **2012**, *2012*, CD007146. [CrossRef]

21. Sherrington, C.; Fairhall, N.J.; Wallbank, G.K.; Tiedemann, A.; Michaleff, Z.A.; Howard, K.; Clemson, L.; Hopewell, S.; Lamb, S.E. Exercise for preventing falls in older people living in the community. *Cochrane Database Syst. Rev.* **2019**, *1*, CD012424. [CrossRef]

22. Sherrington, C.; Fairhall, N.; Wallbank, G.; Tiedemann, A.; Michaleff, Z.A.; Howard, K.; Clemson, L.; Hopewell, S.; Lamb, S. Exercise for preventing falls in older people living in the community: An abridged Cochrane systematic review. *Br. J. Sports Med.* **2020**, *54*, 885–891. [CrossRef]

23. Shimada, H.; Tiedemann, A.; Lord, S.R.; Suzukawa, M.; Makizako, H.; Kobayashi, K.; Suzuki, T. Physical factors underlying the association between lower walking performance and falls in older people: A structural equation model. *Arch. Gerontol. Geriatr.* **2011**, *53*, 131–134. [CrossRef]

24. Mitchell, A.J. A meta-analysis of the accuracy of the Mini-Mental State Examination in the detection of dementia and mild cognitive impairment. *J. Psychiatr. Res.* **2009**, *43*, 411–431. [CrossRef] [PubMed]

25. Ministry of Health, Labour and Welfare. Long-Term Care, Health and Welfare Services for the Elderly. Available online: https://www.mhlw.go.jp/english/policy/care-welfare/care-welfare-elderly/dl/ltcisj_e.pdf (accessed on 4 January 2023).

26. Gibson, M.J.; Andres, R.O.; Isaacs, B.; Radebaugh, T.; Worm-Petersen, J. The prevention of falls in later life. A report of the Kellogg International Work Group on the Prevention of Falls by the Elderly. *Dan. Med. Bull.* **1987**, *34* (Suppl. 4), 1–24.

27. Nagasaki, H.; Itoh, H.; Furuna, T. The structure underlying physical performance measures for older adults in the community. *Aging* **1995**, *7*, 451–458. [CrossRef]

28. Podsiadlo, D.; Richardson, S. The timed "UP & Go": A test of basic functional mobility for frail elderly persons. *J. Am. Geriatr. Soc.* **1991**, *39*, 142–148. [CrossRef] [PubMed]

29. Jones, C.J.; Rikli, R.E.; Beam, W.C. A 30-s chair-stand test as a measure of lower body strength in community-residing older adults. *Res. Q. Exerc. Sport* **1999**, *70*, 113–119. [CrossRef]

30. Cadore, E.L.; Rodríguez-Mañas, L.; Sinclair, A.; Izquierdo, M. Effects of different exercise interventions on risk of falls, gait ability, and balance in physically frail older adults: A systematic review. *Rejuvenation Res.* **2013**, *16*, 105–114. [CrossRef]

31. Scherr, J.; Wolfarth, B.; Christle, J.W.; Pressler, A.; Wagenpfeil, S.; Halle, M. Associations between Borg's rating of perceived exertion and physiological measures of exercise intensity. *Eur. J. Appl. Physiol.* **2013**, *113*, 147–155. [CrossRef]

32. Barajas-Galindo, D.E.; González Arnáiz, E.; Ferrero Vicente, P.; Ballesteros-Pomar, M.D. Effects of physical exercise in sarcopenia. A systematic review. *Endocrinol. Diabetes Nutr.* **2021**, *68*, 159–169. [CrossRef]

33. Lee, Y.C.; Chang, S.F.; Kao, C.Y.; Tsai, H.C. Muscle Strength, Physical Fitness, Balance, and Walking Ability at Risk of Fall for Prefrail Older People. *Biomed. Res. Int.* **2022**, *7*, 4581126. [CrossRef]

34. Akobeng, A.K. Understanding diagnostic tests 3: Receiver operating characteristic curves. *Acta Paediatr.* **2007**, *96*, 644–647. [CrossRef]

35. Walsh, M.E.; Galvin, R.; Williams, D.J.; Harbison, J.A.; Murphy, S.; Collins, R.; McCabe, D.J.; Crowe, M.; Horgan, N.F. Falls-Related EvEnts in the first year after StrokE in Ireland: Results of the multi-centre prospective FREESE cohort study. *Eur. Stroke J.* **2018**, *3*, 246–253. [CrossRef] [PubMed]

36. Yasumura, S.; Haga, H.; Nagai, H.; Suzuki, T.; Amano, H.; Shibata, H. Rate of falls and the correlates among elderly people living in an urban community in Japan. *Age Ageing* **1994**, *23*, 323–327. [CrossRef] [PubMed]

37. Shinohara, H.; Mikami, Y.; Kuroda, R.; Asaeda, M.; Kawasaki, T.; Kouda, K.; Nishimura, Y.; Ohkawa, H.; Uenishi, H.; Shimokawa, T.; et al. Rehabilitation in the long-term care insurance domain: A scoping review. *Health Econ. Rev.* **2022**, *12*, 59. [CrossRef]

38. Cho, K.H.; Bok, S.K.; Kim, Y.J.; Hwang, S.L. Effect of lower limb strength on falls and balance of the elderly. *Ann. Rehabil. Med.* **2012**, *36*, 386–393. [CrossRef] [PubMed]

39. Roongbenjawan, N.; Siriphorn, A. Accuracy of modified 30-s chair-stand test for predicting falls in older adults. *Ann. Phys. Rehabil. Med.* **2020**, *63*, 309–315. [CrossRef]

40. Kawabata, Y.; Hiura, M. The CS-30 test is a useful assessment tool for predicting falls in community-dwelling elderly people. *Rigakuryoho Kagaku* **2008**, *23*, 441–445. (In Japanese) [CrossRef]

41. Nakatani, T.; Nadamoto, M.; Miura, K.I.; Itoh, M. Validation of a 30-sec chair-stand test for evaluating lower extremity muscle strength in Japanese elderly adults. *Jpn. J. Phys. Educ. Health Sport. Sci.* **2002**, *47*, 451–461. (In Japanese) [CrossRef]

42. Wright, A.A.; Cook, C.E.; Baxter, G.D.; Dockerty, J.D.; Abbott, J.H. A comparison of 3 methodological approaches to defining major clinically important improvement of 4 performance measures in patients with hip osteoarthritis. *J. Orthop. Sports Phys. Ther.* **2011**, *41*, 319–327. [CrossRef]

43. Xu, Q.; Ou, X.; Li, J. The risk of falls among the aging population: A systematic review and meta-analysis. *Front. Public Health* **2022**, *10*, 902599. [CrossRef]

Article

"Feeling Trapped in Prison" Due to the COVID-19 Pandemic: Perceptions and Practices among Healthcare Workers and Prison Staff from a Brazilian Maximum Security Unit

Wanessa Cristina Baccon [1], Maria Aparecida Salci [1], Lígia Carreira [1], Adriana Martins Gallo [1], Francielle Renata Danielli Martins Marques [1] and Carlos Laranjeira [2,3,4,*]

[1] Departamento de Pós-Graduação em Enfermagem, Universidade Estadual de Maringá, Avenida Colombo, 5790—Campus Universitário, Maringá 87020-900, PR, Brazil; wanessabaccon@hotmail.com (W.C.B.); masalci@uem.br (M.A.S.); ligiacarreira.uem@gmail.com (L.C.); adriana.gallo@ifpr.edu.br (A.M.G.); franrenata.martins@gmail.com (F.R.D.M.M.)

[2] School of Health Sciences, Polytechnic of Leiria, Campus 2, Morro do Lena, Alto do Vieiro, Apartado 4137, 2411-901 Leiria, Portugal

[3] Centre for Innovative Care and Health Technology (ciTechCare), Campus 5, Polytechnic University of Leiria, Rua de Santo André-66-68, 2410-541 Leiria, Portugal

[4] Comprehensive Health Research Centre (CHRC), University of Évora, 7000-801 Évora, Portugal

* Correspondence: carlos.laranjeira@ipleiria.pt

Abstract: The COVID-19 pandemic had several repercussions on prison staff, but the currently available evidence has mainly ignored these effects. This qualitative study aimed to understand the impact of COVID-19 on the prison system through the narratives of health and security professionals, using the methodological framework of the constructivist grounded theory proposed by Charmaz. The sample included 10 healthcare workers and 10 security professionals. Data collection took place between October and November 2022 through individual in-depth interviews. The data were analyzed using the MaxQDA software. Three categories of interrelated data emerged: (1) "Confrontation and disruption" caused by the COVID-19 pandemic in the prison system; (2) "Between disinfodemic and solicitude" referring to the tension between information management and the practice of care centered on the needs of inmates; and, finally, (3) "Reorganization and mitigation strategies during the fight against COVID-19". Continuous education and the development of specific skills are essential to enable professionals to face the challenges and complex demands that arise in prison contexts. The daily routines professionals had previously taken for granted were disrupted by COVID-19. Thus, investing in adequate training and emotional support programs is crucial to promote the resilience and well-being of these professionals, ensuring an efficient and quality response to critical events.

Keywords: prisons; prison personnel; Brazil; COVID-19 pandemic; occupational health

Citation: Baccon, W.C.; Salci, M.A.; Carreira, L.; Gallo, A.M.; Marques, F.R.D.M.; Laranjeira, C. "Feeling Trapped in Prison" Due to the COVID-19 Pandemic: Perceptions and Practices among Healthcare Workers and Prison Staff from a Brazilian Maximum Security Unit. *Healthcare* **2023**, *11*, 2451. https://doi.org/10.3390/healthcare11172451

Academic Editor: Philippe Gorce

Received: 26 July 2023
Revised: 24 August 2023
Accepted: 30 August 2023
Published: 1 September 2023

1. Introduction

The COVID-19 pandemic has expanded and emphasized the multiple threats to public health and human rights that have long been associated with the prison system [1]. Prisons around the world were identified as high-risk environments for the spread of the disease [2]. According to estimates, incarcerated individuals have an infection rate that is 5.5 times higher than that of the general population [3]. This rate stems from several factors, including overcrowding of prison facilities, ineffective ventilation, unsanitary conditions, shortage of human resources, and high professional turnover [2,4]. Moreover, prisons house people with several chronic comorbidities [5].

With more than 11.5 million people imprisoned worldwide [6], the overcrowded conditions in prisons make it difficult to apply health control measures, in many cases violating human rights. Brazil has one of the largest prison populations worldwide, with over 835,000 imprisoned people [7]. Individuals are housed in filthy conditions, with poor provision of basic necessities, along with inadequate delivery of basic and essential health care, contributing to the rapid spread of the disease [8,9]. To protect people who live or work in detention facilities, the World Health Organization [10] has highlighted the importance of implementing strategies to control the spread of the coronavirus in these institutions.

The system's response to the COVID-19 crisis reflects the Foucauldian "principles of segregation, segmentation, and surveillance" [11]. The measures implemented by prison authorities to address the COVID crisis illustrate the transformation of prisoners into 'docile bodies' [12] (p. 136). These measures include physical distancing between inmates, adequate isolation of suspected and confirmed cases, quarantine, compassionate release (based on their offense background), suspension of visits, and transfers of detainees between units [12]. Foucault argues that disciplinary power operates through the control and regulation of bodies and individual conduct. In this sense, fear, perceived risk, and safety concerns regarding the virus seem to be directly linked to engagement in preventive behaviors [13,14].

Other measures to prevent the spread of the virus were also fundamental, such as protocols for the regular disinfection of the prison's physical spaces and the use of personal protective equipment, carrying out large-scale tests, providing information and psychological support, and prioritizing the vaccination campaign [15–17]. According to Vella et al. [18], the implementation of the vaccination plan was effective in preventing and protecting both the inmates and healthcare staff. Although inmates, the prison healthcare professionals and other workers were priority targets for vaccination, the former received less timely access.

Prison staff were essential during the pandemic, having to adapt to changes in the prison system and assume new responsibilities [19]. This can cause confusion and distress, as there are different expectations and demands to be met [20]. Due to their difficult jobs, hostile working environments, and exposure to both direct and indirect stress, prison staff have poor physical and mental health results. This may make them susceptible populations [21,22].

Likewise, prison staff is disproportionately at high risk for COVID-19 consequences in prison facilities since they move between correctional institutions and their communities [23,24], and this may be exacerbated by additional occupational concerns. Many professionals reported signs of burnout, as distress in the work environment can negatively affect the team's attitudes and, thus, affect the quality of health care [22]. Prison staff faced additional barriers to adequately meeting the needs of inmates due to "legacy factors" in the prison system [25], including the state of prisoners' health before the pandemic, and their vulnerability to COVID-19. The fact that they deal with a high-risk population generates greater fear of the impact of SARS-CoV-2 compared to other groups [26]. In addition to their responsibility to preserve the health and safety of detainees, who are considered a vulnerable group [27,28], they should ensure compliance with preventive measures inherent to their function [17], generating high pressure and professional wear.

The main health concerns and dangers related to COVID-19 from a transmission and containment perspective are covered in the most recent evidence on COVID-19 and prisons [29]. However, there are additional, less common discussions regarding the effects of disadvantages and inequities faced by people working in prison settings, which go beyond those intrinsic challenges of prison as a "total institution" dealing with the pandemic [30].

Although there is some qualitative research available [22,31,32], there is a dearth of studies in Brazil looking specifically at the impact of the pandemic on the prison system through the eyes of healthcare and security personnel.

The present study aimed to understand qualitatively the impact of COVID-19 on the prison system through the perceptions and practices of health and security professionals. In this sense, the research question was: How do healthcare and prison staff perceive and deal with the impact of COVID-19 in a Brazilian high-security prison? We hope this study creates a robust field of knowledge and generates guidelines for the implementation of effective measures capable of dealing with future catastrophe scenarios.

2. Materials and Methods

2.1. Study Design

This study is part of a larger research project aimed at assessing the effects of the pandemic on the prison system in the state of Paraná, Brazil. The present study used a Constructivist Grounded Theory design (CGT) based on the principles of Kathy Charmaz [33]. This approach seeks to understand how individuals construct their realities and focuses on the interpretation of the meanings attributed by the subjects. It values the interpretive perspective of the participants and seeks to explore the processes by which meanings are constructed and shared [33].

This study was conducted and reported following the Consolidated Criteria for Reporting Qualitative Research (COREQ) checklist [34].

2.2. Setting, Participants and Recruitment

The study was carried out in a maximum-security prison with a capacity for 960 male persons located in the municipality of Maringá—northwest of the State of Paraná—in Brazil. The facility was built with the initial objective of only receiving individuals who were awaiting trial, but, due to the shortage of vacancies in other prisons, it ended up receiving convicted individuals. This choice was made because this is one of the largest prison facilities in one of the Brazilian states most affected by the COVID-19 pandemic [35].

The participants were included incrementally, according to the principles of CGT [33], first by purposeful selection and then through a focused theoretical sampling approach.

A maximum variation purposive sample was used to guarantee a heterogeneous sample in terms of age and professional experience. Participants were healthcare and security professionals with a strategic role in the prisons during the pandemic. Study eligibility depended on the following criteria: (a) work experience in a prison before the onset of the COVID-19 pandemic; (b) preset format work during the data collection period; and (c) ability to understand and communicate in Portuguese.

The theoretical sample included 10 health professionals and 10 security professionals (prison officers) who voluntarily participated in the in-depth qualitative interviews. This number was determined by theoretical saturation, as recommended by Charmaz [30].

2.3. Data Collection

Data collection was conducted over 2 months (from October to November 2022), for reasons of organization and functioning of the unit. We identified target participants through the research team's professional network. Before signing a consent form, those who manifested interest in the research were given a participant information leaflet.

The interviews were carried out by only one nurse, with large clinical activity in prison facilities (W.C.B.). A semi-structured interview guide prepared by the researchers and based on available evidence was used as an instrument for data collection [17,21]. The interview guide addressed the following topics: perceptions and experiences regarding work during the pandemic period; barriers encountered in the performance of their functions; strategies used to deal with these barriers; the availability and access of prisoners to health services; and, finally, how professionals have changed moving forwards.

Before starting data collection, the instrument was validated by three experts with PhDs and experience in qualitative research, who validated the content, form, and clarity of the questions, obtaining an agreement level of greater than 90%.

Given that data collection was concomitant with data analysis, the interview script was modified according to emerging themes, in particular, to improve understanding of the phenomenon under study [33]. Field notes were taken during and after the interviews, and memos were written by the researcher to guide the analysis, following the principles of CGT.

The interviews were carried out in a specific place at the institution, which ensured the privacy of the interviews. All interviews were conducted in person. There were no repeat interviews. The interviews lasted between 40 min and 90 min, with an average duration of 60 min, and were recorded in audio and transcribed in full.

In order to ensure the accuracy of the citation translations, they were first translated into English and then back-translated into Portuguese. The sample excerpts were numbered according to the role of the participant, being HealthCare Professionals (HCP) and Security Professionals (SP).

2.4. Data Analysis

According to the approach suggested by CGT, an abductive approach was adopted in the analysis and codification of qualitative data [33]. Data analysis was conducted with the support of MaxQDA® Software–version 2018 (VERBI Software, Berlin, Germany) to facilitate data management as well as store it [36].

The data collection process was carried out simultaneously, based on three stages [33]. The first stage was the initial coding, which involved open coding carried out incident by incident. The second stage was focused coding, in which categories and subcategories were formed, and grouped based on their similarities. In this process, the researcher's perception became relevant to identify the central category, representing a significant step in the research area [37]. The theoretical links between the categories will be more apparent after the selective coding step. The theoretical coding phase of a grounded theory is the next stage, during which more, carefully chosen data is gathered in order to advance the emergent theory and elucidate the key categories that make it up [33].

2.5. Study Rigor

To guarantee the rigor and trustworthiness of the study, the criteria proposed by Charmaz and Thornberg [38] were adopted: reliability, originality, resonance, and usefulness. Credibility was achieved through various approaches. The interviews were meticulously transcribed, and the field notes were carefully compared and verified, ensuring the fidelity of the collected data. In addition, additional interviews were carried out to confirm the emerging theoretical categories, ensuring the reliability of the findings.

Likewise, triangulation was used as a strategy to validate the evidence obtained. Several sources of data, such as interviews, observation, and review of the existing literature, were used to corroborate the results and strengthen the reliability of the research.

Another relevant aspect was the prolonged involvement in the field. This commitment allowed establishing a relationship of trust with the participants, ensuring that they felt comfortable authentically sharing their experiences. Peer debriefing was carried out throughout the research process [39]. This practice involved members of the research team, which enabled in-depth discussions and constant reflection on the findings and interpretations. In addition, brief memos were prepared together with the project team, allowing for a critical and innovative approach [33,37]. To ensure the resonance and usefulness criteria, during the analysis process, schematic representations were also created to visually describe the study findings.

The researcher's reflexivity has become increasingly relevant in the context of GT, which values the joint construction of knowledge [33]. This reflective approach contributed to increasing the transparency and credibility of the research process. Throughout the different stages of the study, from topic selection and research question formulation to study design, data collection and analysis, and even the writing process, reflexivity was used to reflect on the researcher's interactions.

The research team involved a nurse with experience in caring for the prison population and undertaking a Ph.D. in nursing at the moment of data collection (W.C.B.). In addition, the team had researchers with expertise in qualitative health research from a constructivist stance (M.A.S., L.C., A.M.G., F.R.M.M., and C.L.).

2.6. Ethics

The research was conducted under the assumptions of the Declaration of Helsinki and was duly authorized by the Research Ethics Committee of the State University of Maringá—UEM (Opinion No. 3,211,746). Before each interview, written informed consent was obtained from participants, including approval for audio recording. They were duly informed that they had the right to withdraw from the research at any time, without any consequences. It should be noted that participants did not receive any monetary incentive for their participation.

3. Results

3.1. Sample Description

The sample consisted of 20 participants, 10 of whom are health professionals, with an average age of 44.7 years. The majority of these health professionals are white females (60%), and they have an average tenure in the prison unit of 14.9 years. All participants had a history of COVID-19.

The second sample group is comprised of 10 male security professionals, with an average age of 49.3 years. The majority are white (80%), and they have about 11.4 years of professional activity in prison. Only one prison officer reported having no history of COVID-19. Table 1 shows the characterization of the sample.

Table 1. Participant characteristics (n = 20).

Variables	Healthcare Professionals (n = 10)	Security Professionals (n = 10)
Age (years) Mean ± SD (range)	44.7 ± 4.92 (37–52)	49.3 ± 8.24 (30–58)
Race		
White	6	8
Yellow	2	–
Black/African descent	2	2
Sex		
Female	6	–
Male	4	10
Education		
≤8 years	–	–
≥9 years	10	10
Profession		
Clinical Social Worker	1	–
Dentist	1	–
Nurse	4	–
Nurse technician	4	–
Prison Officer	–	10
Length of service (years) Mean ± SD (range)	14.9 ± 4.77 (9–27)	11.4 ± 6.29 (3–17)
History of COVID-19		
Yes	10	9
No	–	1

3.2. Findings from the Analysis

The findings were related through an analytical process formed by three categories: (1) Confrontation and disruption; (2) between disinfodemic and solicitude; and (3) reorganization and mitigation strategies during the fight against COVID-19. The relationships and interactions between the categories were analyzed and a central element that permeates the entire data set was identified. The core category revealed was "Feeling trapped in prison" (Figure 1). Working in a highly restrictive environment, where actions and movements are strictly controlled, can create a feeling of restriction even for those who are responsible for ensuring the health and safety of prisoners.

According to the participants, the changes in routines and the functioning of work activities due to the COVID-19 pandemic resulted in a series of challenges. The fear generated by the spread of the virus combined with the misinformation that was circulating exacerbated the feeling of being trapped in the workplace. The separation between personal and professional life has become increasingly difficult, leading to unregulated schedules and an imbalance between the two domains.

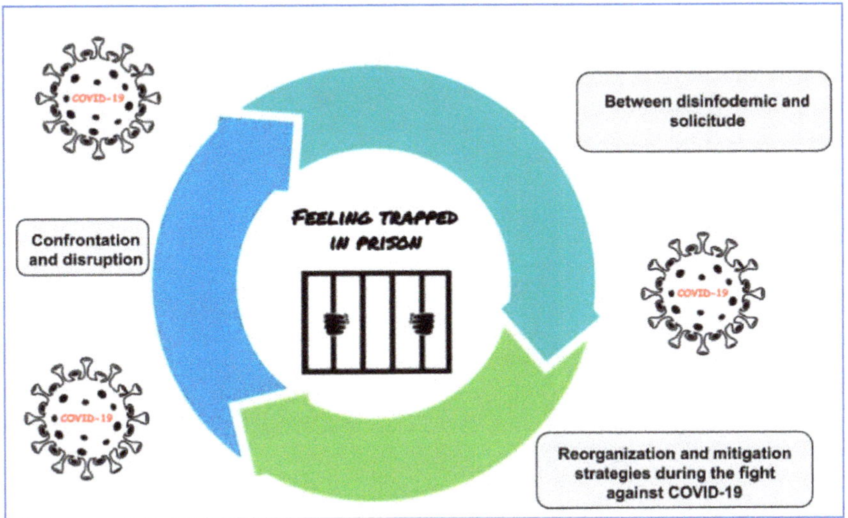

Figure 1. Model of repercussions of COVID-19 on the prison system.

3.2.1. Confrontation and Disruption

This category reveals the confrontation of health and security professionals in the prison system with an unknown reality, which generates manifestations such as fear and emotional dysregulation. The participants faced a complex and restrictive environment, with difficult working conditions, with special emphasis on the initial phase of the pandemic, given the high degree of unpredictability and uncertainty. However, they acknowledge the greater pressure experienced by other structures, such as hospitals. As stated,

[...] Its start was frightening because it was a new disease with an uncertain evolution, long hospitalization periods, and many cases progressing to death. (HCP4)

It was emotionally draining [...] dealing with fear during the pandemic was one of the hardest things. The fact that we don't know much about the virus and all the resulting complications; deaths happening daily due to the various difficulties health teams faced everywhere. (HCP3)

At first, there was a bit of despair, but then I started following other health professionals, and I realized that our front line wasn't really that much of a front line. Compared to my friends working in the ICU (Intensive Care Units), hospitals, and UPAS (Emergency Care Units), so I think that, relatively, it was not as difficult as what was faced at the entrance doors. (HCP6)

Prior knowledge about infectious diseases made the participants fear both personal contamination and the possibility of transmitting the virus to their loved ones, reducing contact with the outside world.

Knowing the impact that an infectious disease has, the biggest concern was taking it to our families, so we reduced the number of trips home [. . .] when the initial fear subsided, there was a concern about elaborating or trying to minimize the impact of the pandemic. (HCP2)

It was difficult, I won't say it was easy, I had to control what I would take home, it was complicated, I was afraid, I had already been infected and I was afraid. In September, I had a bad time, I didn't go to the hospital, I was afraid to take it back home, and it was complicated. (SP5)

Given the unprecedented character of the pandemic and the structural limitations, the participants underline the difficulty of identifying the chains of infection within the prison. Their share of responsibility for the transmissibility of the disease was recognized, as they were the ones who entered and left the prison unit daily.

The agents of the penal system were the main vehicles of transmission because they left the prison daily, so when cases arose, it was because they brought it from the community. (SP9)

The prison officers were the ones who brought the disease into the prison. (SP4)

Some security professionals perceive the prison context as more dangerous concerning the spread of the virus compared to other community contexts. In this regard, SP9 mentions: "[. . .] *the virus was contagious, but it didn't seem as aggressive outside as it did inside the system* [. . .]". This perception may be related to the specific challenges experienced within the prison system, such as the difficulty of adequate physical distancing, the scarcity of resources, and limited access to health care.

[. . .] Human resources are needed. We don't have enough human resources to provide a quality health service. (HCP2)

The health area is very precarious [. . .] we have no voice, we are not recognized or even remembered. (HCP9)

[. . .] we have a big shortage of employees, [. . .] (the institution) should hire more people. (HCP1)

The difficulty in finding cells inside the prison for the isolation of positive cases was identified as a challenge for the participants. Given that overcrowding is a recurring problem, there is a regular demand for cells to accommodate a large number of inmates. In this sense, finding space for isolation cells created significant disruption.

[. . .] all incoming prisoners, even those without COVID-19, had to be quarantined, and we didn't have space for that, and that created a certain operational difficulty. (SP7)

With the implementation of protocols for the care of suspected or confirmed cases of COVID-19, participants reported a sense of unpredictability in the face of the evolution of the pandemic in the prison system. This fact determined that new information and guidelines were constantly emerging. Added to this is the lack of prior experience in dealing with a pandemic of this magnitude and the need to adapt quickly to changes. At the same time, the complexity of the prison environment also exacerbated feelings of uncertainty among all those involved on the front lines. In several cases, working hours were extended to respond to needs, and sometimes feelings of detachment from the outside reality arose.

We made decisions according to protocol [. . .], but always with a feeling of uncertainty about everything we did. Whether it was good or not, I don't know [. . .] but at the same time, it relieves me, seeing that other places have gone through the same situation, where the problem was the same. (HCP6)

Non-health professionals were always asking questions, wanting answers [. . .] some new behavior raised questions, everything was new. Behaviors changed from one day to the next and we had to adapt. (HCP5)

We had to work long hours, which was very complicated because we didn't see the family. (SP1)

3.2.2. Between Disinfodemic and Solicitude

During the COVID-19 pandemic, fighting the disinfodemic and showing concern were recognized as key strategies to ensure the safety and well-being of inmates, as well as the prison staff involved.

A disinfodemic describes the large-scale spread of false, unreliable, or misleading information. Health professionals faced the challenge of dealing with a significant volume of information about COVID-19. Therefore, they had to adopt strategies such as the active search for reliable sources and the constant updating of scientific knowledge. In the initial phase of the pandemic, there was concern among workers about the misinformation associated with COVID-19, which generated high skepticism. In parallel, participants report having responsibility for filtering and disseminating accurate and up-to-date information, with the aim of promoting an adequate understanding of the disease and preventive measures among inmates and their families.

In the beginning, it was a new experience, but as the days went by, it brought fear and insecurity, culminating in skepticism [. . .] I didn't believe in anything anymore. (HCP10)

We had a large amount of information, much of it inaccurate. We serve as multipliers of information not only in our functions but also to the prison population and staff. We experience difficulties with several misguided guidelines without scientific basis that have been released. (HCP4)

Health professionals point out a limitation, the lack of communication during the pandemic. Several factors may have had an impact on the capacity to respond and manage the health crisis in prisons, causing delays in the dissemination of information, such as health guidelines, safety protocols, and updates on COVID-19 cases.

There are no meetings to discuss or talk about assistance, standardization of conduct, or procedures with the security team and the health sector. (HCP4)

The exchange of information and moments of debriefing between colleagues was evaluated as a positive aspect of managing day-to-day situations in the pandemic context. This collaboration allowed professionals to share experiences, coping strategies, and up-to-date knowledge on disease management in a challenging environment. This interaction and collaboration between colleagues contributed to improving the care provided to prisoners and to strengthening the health team in facing the pandemic in a prison context.

I can say that crucial facilitating aspects in this context were the support and information shared with/among co-workers. They were fundamental! (HCP9)

The moments of sharing between peers were important, I see more in that sense, the positive points. (HCP2)

We stayed up late, discussing what could be done even more in an environment that is conducive to the proliferation of these microorganisms and we work directly with confined people, a higher-risk environment. Our concern was mainly to try to reduce the effects here. (HCP2)

In many cases, prisoners' access to the health sector within the prison units went through the scrutiny of prison agents. These professionals usually act as intermediaries between prisoners and health professionals. In this context, solicitude emerges as a possibility of openness and acceptance of the other through an intersubjective relationship that provides attention and care.

Our role is to be the link because one thing they (inmates) *usually say is that we are their hands and feet, so we try to have a good relationship. When there is good contact and mutual respect, this favors the relationship.* (SP2)

The participants' concern to promote monitoring and the creation of an environment conducive to dialogue is highlighted, particularly at times when inmates adopt extreme measures to make themselves heard. Different forms of protest were mentioned, such as making noise, shouting, banging on the railings, or even starting riots to draw the attention of the authorities and demand medical attention.

At certain times they (inmates) *start to get excited, so the tendency is to knock, shout, kick, to knock on doors. They think that by doing this, they will speed up the process, but sometimes they don't [. . .] that's how they do it. (SP2)*

3.2.3. Reorganization and Mitigation Strategies during the Fight against COVID-19

The reorganization and implementation of COVID-19 mitigation strategies in prisons was a complex process due to the specific characteristics of this environment. The closed nature of prisons and close contact among inmates has posed challenges in preventing the spread of the virus.

Health professionals highlight the concern with the underreporting of cases in the prison system. Additionally, they mention that there is a lack of adequate and accurate records of COVID-19 cases within prisons, which may lead to an underestimation of the real extent of the disease in these institutions.

[. . .] The doctor did not frequently ask for the screening test. In the UPA (Emergency care unit) all people are tested. She (the doctor) tests one case or another, the most serious, but it is not transverse conduct for all symptomatic inmates. (HCP6)

Participants evaluate virus mitigation strategies as positive points in prisons. They further recognize that the implementation of these measures was crucial to reducing the risks of spreading COVID-19 and ensuring the safety of both inmates and staff.

There was a greater demand among employees for prevention and this was a positive point. (SP10)

The changes were effective, I believe so [. . .] they made it much easier for the virus not to spread. (SP3)

I believe that some measures were effective in reducing exposure to the SARS-CoV-2 virus, however, the administration should have provided masks with greater protection capacity. (HCP2)

The compassionate release of prisoners was also a measure to mitigate the virus, particularly for individuals with greater vulnerability to COVID-19, but always depending on the type of crime committed. HCP6 mentions: *[. . .] one of the strategies was the departure of inmates who had some comorbidity. It was very complicated [. . .] because there are crimes that could not go home.*

Awareness of maintaining hygiene standards even after the COVID-19 pandemic is an important aspect highlighted by the participants. Although the pandemic has increased the need for hygiene measures and precautions to prevent the spread of the virus, it is necessary to maintain these practices in the long term.

It is important to continue using a mask, especially when there is a risk of infection, just adjust the use of a mask, alcohol gel, and gloves. (SP8)

We've already lost a lot of people [. . .], we have to learn from what happened. And we must be careful, wear a mask, and get vaccinated, or we run the risk of coming back again. (HCP1)

The implementation of stricter screening measures for the early identification of suspected cases, increased testing of inmates and staff, increased access to Personal Protective Equipment (PPE), and specific training on prevention and response to COVID-19, in addition to establishing an efficient flow of communication, were strategies adopted to mitigate the virus.

We had several strategies, we adopted protocols within the prison units, and we had a lower mortality rate because, through DEPEN (National Penitentiary Department), some ordinances determined the carrying out of screening tests and prohibited, for example, visits, the delivery of food and hygiene goods. (HCP2)

We produced fabric masks for the inmates and the agents, aprons and caps for us and for the hospitals [. . .] the PPE was in short supply everywhere, all the stock was running out, and that was one of the strategies that we also had to invent. (HCP6)

We implemented continuous education to create moments of discussion and training with prison agents. In the first phase of the pandemic, we were concerned with clarifying the doubts of security professionals. (HCP2)

Some professionals noticed improvements in resources and processes within the prison system resulting from the measures implemented due to COVID-19. Emphasis on the availability of gel alcohol dispensers in strategic locations, the expansion of the supply of hygiene products, the implementation of cleaning protocols, and awareness of the importance of personal hygiene. These changes were seen as positive, as they contributed to the promotion of a healthier environment.

The placement of alcohol dispensers in the corridors, in strategic locations, and in the toilet, to wash hands, has increased significantly. It was even good because we have always wanted it, but we didn't have it. Then, liquid soap and paper towels to dry hands started to appear, which was something very scarce. (HCP6)

The prison environment is somewhat unsanitary, so hygiene measures ended up making the environment cleaner. (SP7)

Likewise, vaccination against COVID-19 brought peace of mind to the professionals involved, considerably reducing the risk of contamination and the possible serious effects of the disease.

Encouraging people to take the vaccine and booster doses became my main focus [. . .] receiving materials for the rapid COVID tests and vaccine doses has positively eased the policy of social isolation. (HCP10)

[. . .] and after the vaccine things got easier [. . .] and the prisoners joined in. (HCP1)

Professionals also mentioned that by restricting visits it was possible to reduce the likelihood of the virus entering the prison environment.

Regarding facilitating actions, the cancellation of visits, quarantines after leaving the unit for care, and the creation of a sentinel unit in suspected or confirmed cases of COVID-19. (PS4)

The inmate was left without a visit and isolated and that was positive to control and not turn the pandemic (in prison). (PS7)

In this sense, virtual visits were instituted as an alternative to the suspension of face-to-face visits. Perceptions about virtual visits varied, ranging from recognizing their importance in guaranteeing the proximity of family ties to expediting the holding of court hearings, despite the lack of training and digital literacy in their use.

Online visits were positive, it worked well. The system already existed and was perfected. Before, the connection was precarious, but it has improved, court hearings with video were also a gain. (SP5)

[. . .] The system of visits by videoconference was initiated without any training or IT support being offered to the professionals who would carry out this task. This was detrimental to overworked professionals and also to visitors and prisoners. (HCP8)

Videoconferencing visits helped a lot because there was no contact between people. (SP8)

Participants have the perception that prison is a complex environment permeated by prejudices and stigma, which determines the need for a holistic and humanized approach, capable of considering individual particularities, with a view to their reintegration and resocialization.

> *[...] here is not a bunch of people [...] society thinks that the person is arrested and disappears [...] the prisoner will continue eating, living, needing medical attention, that's why the State has to wake up and see that one-day people will get out... whether they will leave "worse" or "better", depends on the available staff and structure. (SP9)*

> *[...] We value doing our best, we refrain from passing judgment. As representatives of the State, we go and find solutions. (SP8)*

4. Discussion

In this study, we aimed to understand the impact of COVID-19 on the prison system through the narratives of health and security personnel. These narratives cast doubt on widely held notions about the effective control of COVID-19 in prison environments and clarify the problems and obstacles the pandemic has presented prison staff and inmates. Like previous studies, participants showed fear of contracting COVID-19 and of being vehicles of transmission, putting their family members at risk [8,40].

Fear can have an undesired effect, generating emotions of powerlessness that can lead to defensive responses instead of an active approach [17]. Additionally, excessive fear can be harmful to mental health, manifesting in symptoms such as anxiety, stress, depression, and other psychological problems [41].

Participants in our study pointed out uncertainties regarding COVID-19, which according to Liu Ye [9] reflects the need for more protective measures to ensure safety in the workplace. The existing concern highlights the need to improve precautions to provide a safer work environment.

Although prison staff and inmates are subject to measures of varying complexity instituted to control COVID-19, they were all required to follow the same guidelines in the use of Personal Protective Equipment (PPE) and social distancing, reflecting a regulatory uniformity imposed by institutional power. This observation is related to the ideas of Michel Foucault, who discussed the concept of biopolitics [13], which refers to the use of state power to regulate and control the life of the population, including in terms of health and well-being. In this sense, health guidelines during the pandemic can be understood as a biopower mechanism, exercising control over the lives of prisoners and staff [13]. The findings highlight how COVID-19 disrupted previous roles and identities, challenging the power dynamics established in the prison environment as discussed by Foucault [42,43]. This situation has led to the emergence of new relationships and interactions between staff and inmates, with potential changes in the power structure.

During the pandemic, important challenges were highlighted by prison staff due to the high labor demands and pressures [4,44]. Managing and promoting a positive work environment plays an important role in improving burnout and compassion fatigue, as they offer an empathetic lens to help employees deal with stressors associated with the health crisis. In this regard, studies reveal moderate levels of burnout and compassion fatigue, as well as difficulties in maintaining satisfaction with compassion for professionals working in a prison context [44]. This situation was aggravated by the pandemic crisis, where levels of anxiety and burnout increased [45].

Although participants generally recognized they were capable of protecting themselves against COVID-19 in the workplace, some mentioned working in unhealthy conditions, perceiving a negative impact on mental health brought about by issues related to the pandemic, and frequent concerns about the possibility of becoming infected in the workplace and taking the virus from the prison into their personal environment [23].

Foucault argues that power relations are not limited to the scope of law or violence, nor are they limited to contractual or merely repressive aspects. According to him, power operates as a network in constant circulation, where individuals are not only passive or consenting targets but also centers imparting this power [13,46]. This conception of networked power is relevant to understanding how prison staff dealt with the challenges of the pandemic. Amid constant changes in the environment to prevent the spread of the virus, they came together and adopted safety strategies, demonstrating an evident concern with the unique consequences of this experience [4].

In this context, the shortage of PPE, the difficulties in maintaining physical distance, and the lack of consolidated information about the necessary care in relation to the disease, reveal the limitations and challenges faced by the staff. This situation generated anxiety among the staff since they were working in an environment where most people were advised to stay at home selves [4]. The interaction between power, care, and life control strategies in the prison environment during the pandemic highlights the relevance of Foucauldian concepts to understanding the dynamics present in this context.

Foucault [47] highlighted the interconnection between power and self-care, emphasizing how power structures can influence the way individuals perceive and care for themselves. In this context, prison staff faced obstacles to ensuring their health and well-being while facing the demands and pressures of the prison environment during the COVID-19 crisis [4].

Health and security professionals faced difficulties when handling the detainees' demands. Prison officers perceived themselves as spokespersons for the detainees within the prison, but often the dialogue between the parties was not satisfactory. The findings of this study must be interpreted considering the broader power dynamics that occur in prison institutions [43]. The employee-inmate division and the perceived distance between these two roles are challenged when individuals assumed different roles or when similarities between groups are identified [42]. The COVID-19 pandemic fostered this division between staff and detainees, upsetting the previously existing balance between these two groups [42].

As in our findings, the literature indicates the possibility of underreported COVID-19 cases in prisons [9,48,49], due to the lack of adequate testing among inmates and prison staff. Evidence underlines that mass testing has proven to be highly effective in containing the spread of COVID-19 in a prison environment [16]. Participants in the current study revealed concern about the lack of prisoner testing, a situation also identified by Liu [9]. This perception seems to reflect the existing distrust in institutions or healthcare professionals, hindering the pursuit of proper care and adoption of preventive measures, such as vaccination.

The inmates' distrust of the health care they receive can be understood as a manifestation of the disciplinary power that permeates the prison environment, where the surveillance and control exercised by the staff can result in a perception of negligence or lack of care [43]. In contrast, professionals had a higher level of exposure to the virus, as they maintained contact between the prison environment and the community [50]. This exposure justified subjecting them to mandatory tests. Evidence revealed a significantly higher prevalence of infection among professionals when compared to inmates [16,23].

Although the need to implement measures to control the virus within prisons is widely discussed, professionals reported difficulties in following these guidelines and implementing protocols. These guidelines include structural changes, such as reducing the number of inmates, cells for preventive isolation, and improving ventilation to reduce the potential for transmission in prisons. In addition, efforts to increase vaccination coverage were considered important, especially among prison officers [51].

Given the resource constraints, some participants expressed concerns about the acquisition of personal protective and hygiene materials, such as masks, gloves, and other essential items. Faced with this situation, professionals took the initiative and the inmates dedicated themselves to the production of protective equipment for local prison institutions, in addition to supplying materials to hospitals. Other studies have reported this difficulty faced by inmates, who often improvise their own personal protective equipment, highlighting the importance of solicitude and mutual care among these people [52].

To mitigate the impacts of the pandemic on the prison system, drastic measures were implemented to eliminate or reduce COVID-19 infections in prisons. These measures included the early release of incarcerated individuals, especially those at higher risk of serious complications due to COVID-19; the suspension of social or legal visits to detained persons; the quarantine of newly incarcerated individuals; and the reinforcement of hand hygiene [23]. This study's participants also mentioned these measures.

It is necessary to direct resources toward alternatives to incarceration and approaches that prioritize the health and well-being of individuals, recognizing that the prison system poses significant challenges to the protection of public health and guaranteeing respect for human rights [5].

Our findings suggest that the shortage of qualified professionals is a significant obstacle, representing one of the greatest challenges to containing the spread of COVID-19 in prisons [5]. In line with other studies [23,53], the lack of a primary healthcare team in prison institutions was identified as the main difficulty, resulting in unnecessary escorts and security threats.

Misinformation about COVID-19 has also been prolific, threatening not just individuals but societies as a whole. This leads people who ignore scientific advice to put themselves in harm's way, widens the lack of confidence in politicians and governments, and diverts the media's efforts, which work reactively to refute untruths rather than proactively producing information from new data. In this sense, and similarly to other studies [54], our findings underscore the need for monitoring and countering misinformation responses (peer debriefings, dissemination of information from credible sources) that help identify, demystify, and report misinformation about COVID-19. In contrast with policies focused on punishing or rehabilitating inmates, policies focused on lowering COVID-19 risks do not explicitly aim to reduce crime. Instead, they represent compassionate criminal justice policies, whose main goal is to enhance the welfare of detainees [55].

The performance of health professionals in the elaboration of mitigation strategies is important since their knowledge and experience are fundamental for the collective construction of knowledge during the COVID-19 pandemic [56]. In this sense, Foucault [43] asserts there is an interconnection between power and knowledge, wherein power is not limited to hierarchical structures but is also present in knowledge relations and in the production, distribution, and use of knowledge.

4.1. Strengths and Study Limitations

One of the strengths of the present study was the use of a sample that includes health and safety professionals with varied experiences, providing a comprehensive view of perceptions and practices in the prison context. Furthermore, the qualitative approach adopted in the study allowed an in-depth analysis of the qualitative data so that it resonates with the voices of the participants. Lastly, the study fills a knowledge gap, given that the findings have the potential to inform the implementation of measures that promote better health and safety for incarcerated individuals.

However, some limitations must be considered. One of the limitations of the study is that it only included data from a prison in the south of Brazil, which limits the transferability of the results to other prison contexts. The interviews took place in prisons during the times when correctional personnel were on duty. Due to time constraints, it was not possible to guarantee that the interviews would go without pause. It is important to consider that the study was carried out in a certain period of time (the third year of the COVID-19 era), reflecting the perceptions and practices of professionals during this specific period of the pandemic. Given that the evolution of COVID-19 had different trajectories in different Brazilian states, the findings may not reflect the reality of the prison system at the national level.

It is essential to highlight the need for more research with quantitative and qualitative approaches of a longitudinal nature to deepen the understanding of the impact of the pandemic in the prison context. Future research should also validate our findings including samples with maximum variation in terms of geographic location and type of prison unit.

4.2. Implications for Practice

The study presents the challenges faced by health and security professionals in the performance of their duties during the pandemic. Based on these perceptions, it was possible to identify the support needs of these professionals, such as emotional support, additional resources, or specialized training [57].

The current study has implications of great relevance for health practice in the prison context. By understanding the complexities and challenges faced by health professionals in this specific environment, it is possible to identify measures and strategies that help to reduce the impacts of the pandemic and ensure the provision of adequate care to incarcerated individuals. This deeper understanding contributes toward improving health practices and allows for a more effective and targeted approach to health promotion and disease prevention within the prison system, where the use of digital technologies can facilitate bringing services closer to the community.

Another important implication of this study is the possibility of developing targeted interventions to address the specific challenges of the prison system during the pandemic. This would allow the development of more effective programs and policies, such as mental health programs, infection prevention education, and improved training in biosecurity measures. Meanwhile, these measures can also reduce other contagious diseases (such as tuberculosis, influenza, or mumps), highly prevalent in prisons [58].

The leaders of a prison organization must consider implementing comprehensive and regular testing policies and protocols, as well as prevention and control measures, to ensure the safety and well-being of inmates and prison staff and prepare for future disaster scenarios.

Finally, carrying out this in-depth qualitative study contributes to the advancement of scientific knowledge about the pandemic in the prison context. By filling existing knowledge gaps, the study provides a broader understanding of the challenges faced and best practices in coping with the pandemic in prisons. This is fundamental to support future research and guide interventions and policies in the prison system.

5. Conclusions

As far as we know, this is one of the first attempts to explore the impact of COVID-19 on the prison system, providing important insights through the lens of prison staff. In the context of the pandemic, research in prison units is a rarely explored field due to the novelty and complexity of the disease.

Prison is, by nature, an environment where one is likely to feel confined. Moreover, qualitative analysis revealed a sense of "feeling trapped" due to several changes associated with the pandemic and the legacy factors that still prevail in prison settings. One of the main aspects highlighted by participants was the fear of contracting the disease and subsequently transmitting it to their loved ones. This fear reflects concern for their own health and the responsibility to protect those in close contact with them. The unavailability of clear and up-to-date guidelines on disease prevention and management protocols in the prison environment made it difficult to make appropriate decisions and effectively implement preventive measures. This generated uncertainties and difficulties in adopting practices that could minimize the spread of the virus among inmates and health and security workers. Another challenge was the shortage of personal protective materials and equipment, essential to guarantee the safety of prison professionals. Likewise, staff clearly struggled to navigate humanity in the prison and solicitude was a necessary condition to respect inmate vulnerability inmates.

Author Contributions: Conceptualization, C.L., M.A.S. and W.C.B.; methodology, C.L., A.M.G. and W.C.B.; software, W.C.B. and F.R.D.M.M.; validation, C.L. and M.A.S.; formal analysis, C.L., M.A.S., L.C. and W.C.B.; investigation, C.L., A.M.G., M.A.S., L.C., W.C.B. and F.R.D.M.M.; resources, C.L. and M.A.S.; data curation, L.C., C.L. and M.A.S.; writing—original draft preparation, W.C.B., M.A.S., L.C., F.R.D.M.M., A.M.G. and C.L.; writing—review and editing, C.L., M.A.S., L.C. and W.C.B.; visualization, W.C.B., C.L., A.M.G., M.A.S., L.C. and F.R.D.M.M.; supervision, C.L. and M.A.S.; project administration, M.A.S. and C.L.; funding acquisition, C.L. All authors have read and agreed to the published version of the manuscript.

Funding: This work is supported by FCT—Fundação para a Ciência e a Tecnologia, I.P. (UIDB/05704/2020 and UIDP/05704/2020) and under the Scientific Employment Stimulus—Institutional Call—[CEECINST/00051/2018].

Institutional Review Board Statement: The study was conducted according to the guidelines of the Declaration of Helsinki and approved by the Research Ethics Committee of the State University of Maringá—UEM (approval no 3.211.746).

Informed Consent Statement: Informed consent was obtained from all subjects involved in the study. Participation in the study was completely voluntary and anonymous. Participants received no compensation.

Data Availability Statement: All data generated or analyzed during this study are included in this article. This article is based on the first author's doctoral thesis in Nursing at the State University of Maringá—Brazil.

Acknowledgments: We would like to thank the prison staff for sharing their valuable experiences and knowledge with us.

Conflicts of Interest: The authors declare no conflict of interest. The funders had no role in the design of the study, in the collection, analyses, or interpretation of data, in the writing of the manuscript, or in the decision to publish the results.

References

1. Pont, J.; Enggist, S.; Stöver, H.; Baggio, S.; Gétaz, L.; Wolff, H. COVID-19: The case for rethinking health and human rights in prisons. *Am. J. Public Health* **2021**, *111*, 1081–1085. [CrossRef]
2. Johnson, L.; Czachorowski, M.; Gutridge, K.; McGrath, N.; Parkes, J.; Plugge, E. The mental wellbeing of prison staff in England during the COVID-19 pandemic: A cross-sectional study. *Front. Public Health* **2023**, *11*, 1049497. [CrossRef]
3. Saloner, B.; Parish, K.; Ward, J.A.; DiLaura, G.; Dolovich, S. COVID-19 cases and deaths in federal and state prisons. *JAMA* **2020**, *324*, 602–603. [CrossRef] [PubMed]
4. Kothari, R.; Forrester, A.; Greenberg, N.; Sarkissian, D.; Tracy, D. COVID-19 and prisons: Providing mental health care for people in prison, minimizing moral injury and psychological distress in mental health staff. *Med. Sci. Law* **2020**, *60*, 165–168. [CrossRef] [PubMed]
5. LeMasters, K.; Ranapurwala, S.; Maner, M.; Nowotny, K.M.; Peterson, M.; Brinkley-Rubinstein, L. COVID-19 community spread and consequences for prison case rates. *PLoS ONE* **2022**, *17*, e0266772. [CrossRef] [PubMed]
6. Penal Reform International. *Global Prison Trends 2022*; Penal Reform International: London, UK, 2022. Available online: www.penalreform.org/global-prison-trends-2022/ (accessed on 10 July 2022).
7. World Prison Brief: Brazil. Institute for Crime & Justice Policy Research. Available online: https://www.prisonstudies.org/coun-713try/brazil (accessed on 10 May 2023).
8. Andrade, R.O. COVID-19: Prisons exposed in Brazil's crisis. *BMJ* **2020**, *370*, m2884. [CrossRef] [PubMed]
9. Liu, Y.E.; Oto, J.; Will, J.; LeBoa, C.; Doyle, A.; Rens, N.; Aggarwal, S.; Kalish, I.; Rodriguez, M.; Sherif, B.; et al. Factors associated with COVID-19 vaccine acceptance and hesitancy among residents of Northern California jails. *Prev. Med. Rep.* **2022**, *27*, 101771. [CrossRef]
10. World Health Organization. Preparedness, Prevention and Control of COVID-19 in Prisons and Other Places of Detention: Interim Guidance 15 March 2020. Available online: www.who.int/docs/default-source/coronaviruse/who-china-jointmission-on-covid-19-final-report.pdf (accessed on 5 July 2023).
11. Foucault, M. *Discipline and Punish: The Birth of the Prison*; Vintage Books: New York, NY, USA, 1995.
12. Raghavan, V. Prisons and the pandemic: The panopticon plays out. *J. Soc. Econ. Dev.* **2021**, *23*, 388–397. [CrossRef]
13. Foucault, M. *Microfísica do Poder*, 8th ed.; Graal: Rio de Janeiro, Brazil, 1989.
14. Harper, C.A.; Satchell, L.P.; Fido, D.; Latzman, R.D. Functional fear predicts public health compliance in the COVID-19 pandemic. *Int. J. Ment. Health Addict.* **2020**, *19*, 1875–1888. [CrossRef]

15. Braithwaite, I.; Edge, C.; Lewer, D.; Hard, J. High COVID-19 death rates in prisons in England and Wales, and the need for early vaccination. *Lancet Respir. Med.* **2021**, *9*, 569–570. [CrossRef]
16. Franchi, C.; Rossi, R.; Malizia, A.; Gaudio, P.; Di Giovanni, D. Biological risk in Italian prisons: Data analysis from the second to the fourth wave of COVID-19 pandemic. *Occup. Environ. Med.* **2023**, *80*, 273–279. [CrossRef] [PubMed]
17. Gonçalves, L.C.; Baggio, S.; Schnyder, N.; Zaballa, M.E.; Baysson, H.; Guessous, I.; Rossegger, A.; Endrass, J.; Wolff, H.; Stringhini, S.; et al. COVID-19 Fears and Preventive Behaviors among Prison Staff. *Vict. Offenders* **2023**, *18*, 673–690. [CrossRef]
18. Vella, R.; Giuga, G.; Piizzi, G.; Alunni Fegatelli, D.; Petroni, G.; Tavone, A.M.; Potenza, S.; Cammarano, A.; Mandarelli, G.; Marella, G.L. Health Management in Italian Prisons during COVID-19 Outbreak: A Focus on the Second and Third Wave. *Healthcare* **2022**, *10*, 282. [CrossRef] [PubMed]
19. O'Moore, É. *Briefing Paper: Interim Assessment of Impact of Various Population Management Strategies in Prisons in Response to COVID-19 Pandemic in England*; HMPPS: London, UK, 2020. Available online: https://assets.publishing.service.gov.uk/government/uploads/system/uploads/attachment_data/file/882622/covid-19-population-management-strategy-prisons.pdf (accessed on 4 January 2022).
20. Boaron, F.; Gerocarni, B.; Fontanesi, M.G.; Zulli, V.; Fioritti, A. Across the walls: Treatment pathways of mentally ill offenders in Italy, from prisons to community care. *J. Psychopathol.* **2021**, *27*, 34–39. [CrossRef]
21. Testoni, I.; Nencioni, I.; Arbien, M.; Iacona, E.; Marrella, F.; Gorzegno, V.; Selmi, C.; Vianello, F.; Nava, A.; Zamperini, A.; et al. Mental health in prison: Integrating the perspectives of prison staff. *Int. J. Environ. Res. Public Health* **2021**, *18*, 11254. [CrossRef]
22. Wainwright, L.; Senker, S.; Canvin, K.; Sheard, L. It was really poor prior to the pandemic. It got really bad after: A qualitative study of the impact of COVID-19 on prison healthcare in England. *Health Justice* **2023**, *11*, 6. [CrossRef]
23. Wallace, M.; Hagan, L.; Curran, K.G.; Williams, S.P.; Handanagic, S.; Bjork, A.; Davidson, S.L.; Lawrence, R.T.; Joseph McLaughlin, J.; Butterfield, M.; et al. COVID-19 in Correctional and Detention Facilities: United States, February–April 2020. *MMWR Morb. Mortal. Wkly. Rep.* **2020**, *69*, 587–590. [CrossRef]
24. Costa, F.A.; Neufeld, M.; Hamad, M.; Carlin, E.; Ferreira-Borges, C. Response measures to COVID-19 in prisons and other detention centers. *Int. J. Prison. Health* **2021**, *17*, 351–358. [CrossRef]
25. Canvin, K.; Sheard, L. *Impact of COVID-19 on Healthcare in Prisons in England: Early Insights*; University of York: York, UK, 2021. Available online: https://qualporg.files.wordpress.com/2021/11/canvin-and-sheard-2021-impact-of-c19-on-healthcare-in-prisons-1.pdf (accessed on 10 July 2023).
26. Oladeru, O.T.; Tran, N.-T.; Al-Rousan, T.; Williams, B.; Zaller, N. A call to protect patients, correctional staff and healthcare professionals in jails and prisons during the COVID-19 pandemic. *Health Justice* **2020**, *8*, 17. [CrossRef]
27. Montoya-Barthelemy, A.G.; Lee, C.D.; Cundiff, D.R.; Smith, E.B. COVID-19 and the Correctional Environment: The American Prison as a Focal Point for Public Health. *Am. J. Prev. Med.* **2022**, *58*, 888–891. [CrossRef]
28. Mendes, R.; Baccon, W.C.; Laranjeira, C. Fear of COVID-19, Mental Health and Resilient Coping in Young Adult Male Inmates: A Portuguese Cross-Sectional Study. *Int. J. Environ. Res. Public Health* **2023**, *20*, 5510. [CrossRef]
29. Esposito, M.; Salerno, M.; Di Nunno, N.; Ministeri, F.; Liberto, A.; Sessa, F. The Risk of COVID-19 Infection in Prisons and Prevention Strategies: A Systematic Review and a New Strategic Protocol of Prevention. *Healthcare* **2022**, *10*, 270. [CrossRef] [PubMed]
30. Maycock, M. COVID-19 has caused a dramatic change to prison life: Analysing the impacts of the COVID-19 pandemic on the pains of imprisonment in the Scottish Prison Estate. *Br. J. Criminol.* **2022**, *62*, 218–233. [CrossRef]
31. Testoni, I.; Francioli, G.; Biancalani, G.; Libianchi, S.; Orkibi, H. Hardships in Italian Prisons During the COVID-19 Emergency: The Experience of Healthcare Personnel. *Front. Psychol.* **2021**, *12*, 619687. [CrossRef] [PubMed]
32. Baccon, W.C.; Salci, M.A.; Carreira, L.; Gallo, A.M.; Marques, F.R.D.M.; Paiano, M.; Baldissera, V.D.A.; Laranjeira, C. Meanings and Experiences of Prisoners and Family Members Affected by the COVID-19 Pandemic in a Brazilian Prison Unit: A Grounded Theory Analysis. *Int. J. Environ. Res. Public Health* **2023**, *20*, 6488. [CrossRef] [PubMed]
33. Charmaz, K. *Constructing Ground Theory: A Pratical Guide through Qualitative Analysis*; Sage Publication: London, UK, 2014.
34. Tong, A.; Sainsbury, P.; Craig, J. Consolidated criteria for reporting qualitative research (COREQ): A 32-item checklist for interviews and focus groups. *Int. J. Qual. Health Care* **2007**, *19*, 349–357. [CrossRef] [PubMed]
35. Statista. Number of Confirmed Cases of Coronavirus (COVID-19) in Brazil as of July 7, 2023, by State. Available online: https://www.statista.com/statistics/1103791/brazil-coronavirus-cases-state/ (accessed on 10 July 2023).
36. Rädiker, S.; Kuckartz, U. *Analyse Qualitativer Daten Mit MAXQDA*, 1st ed.; Springer: Wiesbaden, Germany, 2019.
37. Santos, J.L.G.; Cunha, K.S.D.; Adamy, E.K.; Backes, M.T.S.; Leite, J.L.; Sousa, F.G.M.D. Data analysis: Comparison between the different methodological perspectives of the Grounded Theory. *Rev. Esc. Enferm. USP* **2018**, *52*, 1–8. [CrossRef]
38. Charmaz, K.; Thornberg, R. The pursuit of quality in grounded theory. *Qual. Res. Psychol.* **2021**, *18*, 305–327. [CrossRef]
39. Amin, M.E.K.; Nørgaard, L.S.; Cavaco, A.M.; Witry, M.J.; Hillman, L.; Cernasev, A.; Desselle, S.P. Establishing trustworthiness and authenticity in qualitative pharmacy research. *Res. Social. Adm. Pharm.* **2020**, *16*, 1472–1482. [CrossRef]
40. Hagan, L.M.; Williams, S.P.; Spaulding, A.C.; Toblin, R.L.; Figlenski, J.; Ocampo, J.; Ross, T.; Bauer, H.; Hutchinson, J.; Lucas, K.D. Mass testing for SARS-CoV-2 in 16 Prisons and Jails—Six jurisdictions, United States, April–May 2020. *MMWR Morb. Mortal. Wkly. Rep.* **2020**, *69*, 1139–1143. [CrossRef]

41. Koçak, O.; Koçak, Ö.E.; Younis, M.Z. The psychological consequences of COVID-19 fear and the moderator effects of individuals' underlying illness and witnessing infected friends and family. *Int. J. Environ. Res. Public Health* **2021**, *18*, 1836. [CrossRef] [PubMed]
42. Song, M.; Kramer, C.T.; Sufrin, C.B.; Eber, G.B.; Rubenstein, L.S.; Beyrer, C.; Saloner, B. It was like you were being literally punished for getting sick: Formerly incarcerated people's perspectives on liberty restrictions during COVID-19. *AJOB Empir. Bioeth.* **2023**, *14*, 155–166. [CrossRef] [PubMed]
43. Foucault, M. *Vigiar e Punir: Nascimento da Prisão*, 42nd ed.; Vozes: Rio de Janeiro, Brazil, 2014.
44. Bell, S.; Hopkin, G.; Forrester, A. Exposure to traumatic events and the experience of burnout, compassion fatigue and compassion satisfaction among prison mental health staff: An exploratory survey. *Issues Ment. Health Nurs.* **2019**, *40*, 304–309. [CrossRef] [PubMed]
45. Memon, A.; Hunt, A.; Thelwall, E.; Hardwick, N.; Johnson, S.U. The mental health of staff working in UK prisons during the COVID-19 pandemic. *Howard J. Crim. Justice* **2023**, *62*, 1–17. [CrossRef]
46. Costa, R.; Souza, S.S.; Ramos, F.R.S.; Padilha, M.I. Foucault e sua utilização como referencial na produção científica em enfermagem. *Texto Contexto-Enferm.* **2008**, *17*, 629–637. [CrossRef]
47. Foucault, M. *História da Sexualidade I: A Vontade de Saber*, 1st ed.; Graal: Rio de Janeiro, Brazil, 1988.
48. Tyagi, E.; Marquez, N.; Manson, J. A Crisis of Undertesting: How Inadequate COVID-19 Detection Skews the Data Costs Lives. UCLA Law COVID behind Bars Data Project, Oct. 2021. Available online: https://uclacovidbehindbars.org/undertesting (accessed on 8 April 2022).
49. Moyer, J.W. DC Inmates Reported Lack of Access to Coronavirus Tests, Medical Care as Pandemic Unfolded, Report Says. *The Washington Post*. 2021. Available online: https://www.washingtonpost.com/local/public-safety/dc-prison-covid-tests-medical-care/2021/07/20/3f7d9afc-e975-11eb-84a2-d93bc0b50294_story.html (accessed on 4 June 2023).
50. Kinner, S.A.; Young, J.T.; Snow, K.; Southalan, L.; Lopez-Acuña, D.; Ferreira-Borges, C. Prisons and custodial settings are part of a comprehensive response to COVID-19. *Lancet Public Health* **2020**, *5*, e188–e189. [CrossRef]
51. KhudaBukhsh, W.R.; Khalsa, S.K.; Kenah, E.; Rempała, G.A.; Tien, J.H. COVID-19 dynamics in an Ohio prison. *Front. Public Health* **2023**, *11*, 1087698. [CrossRef]
52. Riback, L.R.; Dickson, P.; Ralph, K.; Saber, L.B.; Devine, R.; Pett, L.A.; Clausen, A.J.; Pluznik, J.A.; Bowden, C.J.; Sarrett, J.C.; et al. Coping with COVID in corrections: A qualitative study among the recently incarcerated on infection control and the acceptability of wastewater-based surveillance. *Health Justice* **2023**, *11*, 5. [CrossRef]
53. Ely, K.Z.; Schwarzbold, P.; Ely, G.Z.; Vendrusculo, V.G.; Dotta, R.M.; Rosa, L.R.; Krug, S.B.F.; Valim, A.R.M.; Possuelo, L.G. A Educação Permanente em Saúde e os atores do sistema prisional no cenário pandêmico. *Trab. Educ. Saúde* **2023**, *21*, e01224207. [CrossRef]
54. Kincaid, R. Mass incarceration and misinformation: The COVID-19 infodemic behind bars. *Univ. St. Thomas Law J.* **2023**, *19*, 323–363. Available online: https://ir.stthomas.edu/cgi/viewcontent.cgi?article=1565&context=ustlj (accessed on 9 July 2023).
55. Hickert, A.; Shi, L.; Silver, J.R. Is compassion the flip side of punitiveness? Incorporating COVID-19 crisis in experimental vignettes to examine support for visitation and vaccination in prison. *J. Exp. Criminol.* **2022**, 1–22. [CrossRef] [PubMed]
56. Neves, V.N.S.; Machado, C.J.S.; Fialho, L.M.F.; Sabino, R.N. Utilização de lives como ferramenta de educação em saúde durante a pandemia pela COVID-19. *Educ. Soc.* **2021**, *42*, e240176. [CrossRef]
57. Ryan, C.; Brennan, F.; McNeill, S.; O'Keeffe, R. Prison Officer Training and Education: A Scoping Review of the Published Literature. *J. Crim. Justice Educ.* **2020**, *33*, 110–138. [CrossRef]
58. Beaudry, G.; Zhong, S.; Whiting, D.; Javid, B.; Frater, J.; Fazel, S. Managing outbreaks of highly contagious diseases in prisons: A systematic review. *BMJ Glob. Health* **2020**, *5*, e003201. [CrossRef] [PubMed]

Article

Effects of Psychotherapy on the Problem Behaviors of Humidifier Disinfectant Survivors: The Role of Individual Characteristics and Adaptive Functioning

Min Joo Lee [1], Yubin Chung [1], Soeun Hong [1], Hun-Ju Lee [2], Gippeum Park [1] and Sang Min Lee [1,*]

[1] Department of Education, Korea University, Seoul 02841, Republic of Korea; mjbravo@korea.ac.kr (M.J.L.); yubin90@gmail.com (Y.C.); soeunhong@gmail.com (S.H.); joy3151@naver.com (G.P.)
[2] University Industry Foundation, Yonsei University, Seoul 03722, Republic of Korea; hunjulee@yonsei.ac.kr
* Correspondence: leesang@korea.ac.kr; Tel.: +82-2-3290-2306

Abstract: This study aimed to examine group differences in the survivors of humidifier damage and the effect of individual psychotherapy on the psychological symptoms of the survivor groups, using the single group pre–post study design. A series of Wilcoxon–Mann–Whitney tests were conducted to investigate the level of psychological problems before and after psychotherapy, as well as the main and interaction effects of demographic characteristics and adaptive functioning on the treatment effects in 69 humidifier disinfectant survivors. The results demonstrated significant differences in problems with socioeconomic status (SES), life functioning, friendships, family relationships, and job adjustment in the survivor groups. Groups with high SES, low life functioning, and poor friend relationships had more problem behaviors than other groups. Problem behaviors related to friendship levels were different before and after psychotherapy. After psychotherapy, individuals with limited social connections exhibited a greater decrease in problem behaviors compared to those with strong friendships. This paper extends the international literature on the long-term consequences of environmental health hazards and the importance of tailored mental health interventions.

Keywords: adaptive function; effects of psychotherapy; humidifier disinfectant survivors; problem behaviors; psychological symptoms

Citation: Lee, M.J.; Chung, Y.; Hong, S.; Lee, H.-J.; Park, G.; Lee, S.M. Effects of Psychotherapy on the Problem Behaviors of Humidifier Disinfectant Survivors: The Role of Individual Characteristics and Adaptive Functioning. *Healthcare* **2023**, *11*, 2179. https://doi.org/10.3390/healthcare11152179

Academic Editors: Carlos Laranjeira and Ana Querido

Received: 6 April 2023
Revised: 26 July 2023
Accepted: 26 July 2023
Published: 1 August 2023

1. Introduction

It is necessary to look at not only the physical but also the psychological damage of the humidifier disinfectant survivors. This can be an important foundation for developing and organizing psychological interventions based on their psychological problems. To do this, it is necessary to examine the damage caused by humidifier disinfectants in South Korea. It was proposed that humidifier disinfectants, which have been sold since 1994, could prevent diseases caused by bacteria. People believed that this messaging, promoted by a large company, was true and that humidifier disinfectant would help prevent diseases and protect their health and that of their family members [1]. Families with newborn children also became attracted to the fact that they could effectively eliminate bacterial diseases.

In April 2011, the Asan Medical Center in Seoul reported the occurrence of a strange disease in six mothers, the symptoms of which were respiratory failure and pulmonary fibrosis [2]. These cases were distributed throughout all areas in South Korea and were not concentrated in a particular region. The Korea Centers for Disease Control and Prevention (K-CDC) conducted an epidemiological investigation into the reasons behind the occurrence of these symptoms. The K-CDC found that humidifier disinfectant, biocide water placed in humidifiers along with water, was associated with the particular symptoms being displayed by the six women.

In December 2012, a Lung Injury Investigation Committee (LIIC) was formed to investigate the damage caused by these humidifiers, and in July 2013, investigations on humidifier disinfectants and their damage status began [3]. In-depth investigations have found that the chemical products used in humidifiers are associated with a wide range of lung injuries, including interstitial pneumonitis [4–6]. The chemicals contained in humidifier disinfectants include substances that are fatal to the human body, such as chloromethylisothiazolinone (CMIT), methylisothiazolinone (MIT), hexamethylene guanidine phosphate (PHMG), and oligo 2-(2-ethoxy) ethoxy ethyl guanidine chloride (PGH). In particular, products containing CMIT and MIT were sold from 1994 to 2011, and approximately 9.98 million units of humidifier disinfectants were sold [7]. It is inferred that humidifier disinfectant products may have affected a random selection of people over an extended period of time [8]. Toxic chemicals were absorbed into the body through the nose, mouth, and skin via nebulizers and began to penetrate the lungs, causing irreversible damage to lung cells and widespread health damage [6,9–11]. As reported in a study conducted on 94 adults in Gwangmyeong City in Gyeonggi Province, 37.2% used a humidifier, and 18.1% used a humidifier disinfectant [12]. In an epidemiological survey of 1144 pregnant women, the rate of humidifier use was 28.2% [13]. Yoon et al. [8] argued that 75.6% of children in Korea used humidifiers, and 31.1% of children were exposed to humidifier disinfectants. Considering that humidifier disinfectant was also used in families with children and/or newborns, it can be inferred that the amount of damage to survivors and their families will increase. However, it is difficult to accurately estimate the number of victims [14]. As of April 2022, the number of people who applied for damages was 7685, including 1751 deaths.

Additionally, the problems of humidifier disinfectant survivors appeared not only as physical problems but also as psychological problems. Yoo et al. [15] analyzed the psychological problems faced by 26 survivors and 92 survivors' families (45 bereaved). They found that survivors still experience anxiety and fear, even after the event has passed, and bereaved families often display alcohol abuse and insomnia. According to Leem et al. [16], 57.5% of humidifier disinfectant survivors reported depression, 55.1% reported guilt and self-blame, 54.3% reported anxiety, 27.6% reported suicidal thoughts, and 11% reported suicide attempts. Ko et al. [17] compared the mental health status of humidifier disinfectant survivors from 2018 to 2021 in a general population of 228, and the survivors reported more psychological problems, such as anxiety, depression, atrophy, and thinking problems, compared to the general population.

In 2017, the Ministry of Environment created an organization to provide psychological psychotherapy interventions for humidifier disinfectant survivors and their families. The organization, named Health Monitoring of Humidifier Disinfectant Victims (Mental Health), provides various psychological interventions, such as group psychotherapy and social improvement programs; individual psychotherapy is one of the most essential interventions that can be provided to the survivors and their families. Therefore, it is necessary to examine the effectiveness of individual psychotherapy for humidifier disinfectant survivors to alleviate their psychological distress. Several studies have shown that psychological interventions are effective for people suffering from various psychological problems, such as PTSD, stress, anger, and depression [18,19]. Howard et al. [20] argued that psychotherapy was more significant than spontaneous recovery, based on the results of a meta-analysis of the effects of psychotherapy on 2400 clients over a 30 year study period. On the other hand, while Cuijpers et al. [21] did not deny the effectiveness of psychotherapy, they reported that the effectiveness of psychotherapy was somewhat exaggerated.

The purpose of this study is to examine the effectiveness of individual therapy on psychological symptoms. Additionally, this study examines the moderating role of demographic characteristics in treatment effect to identify the characteristics of survivors who can most benefit from individual psychotherapy. By analyzing the psychological symptoms of humidifier disinfectant survivors and the effectiveness of psychotherapy on them, this study can contribute to the understanding of what constitutes an appropriate psychological intervention for humidifier disinfectant survivors.

The following are the specific research questions for this study: First, what characteristics did survivors who had significant psychological problems describe? We specifically expect that there could be statistical differences in the extent of psychological symptoms by SES, according to the findings of the Ko et al. [17] study. Second, is individual psychotherapy effective for survivors of humidifier disinfectants? According to the results of previous research [18–21], we assume that individual psychotherapy has a considerable impact in relieving psychological symptoms. Third, which survivor groups benefited the most from psychotherapy? We predict that several traits of survivors (such as the quality of their friendships) will moderate the effectiveness of individual psychotherapy in reducing psychiatric symptoms. Several scholars [22,23] note that the effectiveness of psychotherapy is a complex interaction of several traits and that what works best for one person may not work the same way for another. Tailoring treatment to the individual's unique characteristics and needs is essential for achieving the best possible outcomes.

2. Materials and Methods

2.1. Participants

The ethical approval for this study was obtained from the Institutional Review Board of Yonsei University (No. 7001988-202104-HR-1178-02l) in South Korea. This study used data from a total of 69 individuals who survived exposure to toxic humidifier disinfectants and sought individual psychotherapy through a government support program administrated by the National Institute of Environmental Research (NIER) in 2021. Among the survivors aged 13 years and older, a total of 224 individuals received psychotherapy. However, only 69 of them voluntarily participated in this study with informed consent. The response rate among the participants was 30.8%. The age range of the survivors was 13 to 60 years (M = 38.13, SD = 14.75), and 55.1% (n = 38) were female. The survivors received treatment at one of the psychotherapy centers officially approved by the government program, which stipulates the conditions for a psychotherapy center to be eligible for the support program. The centers had certified therapists from the Korean Counseling Psychological Association (KCPA) or the Korean Counseling Association (KCA). The participants completed the survey of Achenbach System of Empirically Based Assessment (ASEBA) provided by HUNO Inc. (ASEBA Provider Company in Seoul, Republic of Korea), before and after treatment.

2.2. Treatment

Therapists who held level 2 or higher psychotherapy-related certificates issued by the KCPA or the KCA provided individual psychotherapy to survivors. One therapist conducted psychotherapy with only one or two clients. The therapists' theoretical orientations mainly included cognitive behavioral therapy, psychodynamic therapy, interpersonal psychotherapy, and integrative therapy. The default number of sessions provided was a total of 10 sessions, with each session lasting 50 min. However, sessions could be extended based on the client's status. The average number of sessions was 15 (SD = 7.52), with a range from 4 to 41. In each session, the psychotherapy process was monitored using the Outcome Rating Scale (ORS) and the Session Rating Scale (SRS) completed by the survivors [24]. Therapists submitted weekly session evaluation reports and completed psychotherapy records and online psychological assessments. The quality of treatment was steadily managed through regular monitoring by the government support program.

2.3. Measures

2.3.1. Problem Behavior Scale

The Adult Self-Report (ASR) [25,26] and the Youth Self-Report (YSR) [27] were used to measure the psychological problem behaviors of participants before and after treatment. Adolescents under the age of 19 responded to the YSR, and adults responded to the ASR. In the YSR and ASR, the total scores of the problem behavior scale consist of the internalizing scale, externalizing scale, thought problems, attention problems, social problems (only for YSR), and other similar problems. Anxious/depressed, withdrawn/depressed, and somatic complaints correspond to the internalizing scale, whereas aggressive behavior, rule-breaking behavior, and intrusive behavior (only for ASR) correspond to the externalizing scale. The reliability of the problem behavior scale was 0.95 for youth and 0.97 for adults, respectively. The Korean versions of the YSR and ASR were provided by HUNO Inc., and standardized T-scores were reported based on representative samples of adolescents (11–18 years old) and adults (19–60 years old) in South Korea. The higher the score, the more serious the mental health problems: $T \geq 64$ indicates clinical, $T = 63–60$ is borderline clinical, and $T < 60$ is non-clinical mental health problems, meaning that participants in a clinical or borderline clinical group are more likely to exhibit emotional and behavioral problems in adapting to situations than other participants in the same age and gender groups.

2.3.2. Adaptive Functioning Scale

The degree of adaptive functioning of the participants was measured using the YSR and ASR, before and after treatment. For ASR, the total score of adaptive functioning was calculated using the scales of friends, spouse/partner, family, job, and education. For the YSR, sociality and academic performance were considered to calculate the total score of adaptive functioning. The reliability of the adaptive functioning scale was 0.56 for youth and 0.70 for adults, respectively. The lower the score, the lower the adaptive level: $T \leq 36$ indicates clinical, $T = 37–40$ is borderline clinical, and $T > 40$ is non-clinical. The clinical standards of the sub-scale of adaptive functioning, including friendship (friends in ASR and sociality in YSR), family, and job, are clinical ($T \leq 30$), borderline clinical ($T = 31–35$), and non-clinical ($T > 35$). For example, clinical group participants in the case of friendships were less likely to adapt to having meaningful relationships with friends than non-clinical group participants.

2.4. Data Analysis

Data analyses were performed using IBM SPSS version 21. First, we conducted descriptive statistics analysis to provide an overview of the demographic and life functioning variables before and after treatments. Regarding missing data, we addressed it by imputing the missing values with the mode score, following the rules of ASEBA. Second, we conducted the Mann–Whitney test for two independent groups, the Kruskal–Wallis test for three or more independent groups, and the Wilcoxon signed-rank test for paired samples to assess treatment effects (i.e., whether the total problem level changed significantly before and after treatment) [28–30]. Although the power analysis using G*power software with an effect size of 0.40 [31] indicated that 60 participants were sufficient to run a dependent t-test and one-way ANOVA to examine the treatment effects, we chose to perform a series of nonparametric tests to compare outcomes between multiple groups in order to obtain clearer results. Lastly, for practical significance, effect sizes using Cohen's d were calculated and interpreted as $d = 0.20$ (small effect), $d = 0.50$ (medium effect), and $d = 0.80$ (large effect [32]).

3. Results

3.1. Mean Differences in Problem Behavior Scores

Mann–Whitney tests and Kruskal–Wallis tests were conducted for the group differences of each variable at the pre-treatment stage to identify groups that reported more problem behaviors than the other groups. The median and interquartile of each variable are listed in Table 1. There were no significant differences between survivors with and without compensation, between self and family members, and between male and female participants. There were significant differences among age groups; the age group of 40–50 ($T_{Md} = 62.50$, 54.75–72.75) displayed more problem behaviors than the age group under 19 ($T_{Md} = 57.00$, 48.25–63.50). There were significant differences in problem behaviors regarding SES, life functioning, friendships, family relationships, and job adjustment. For example, the SES group with the highest level ($T_{Md} = 63.50$, 57.00–86.25) reported more problem behaviors than the middle ($T_{Md} = 56.00$, 49.00–62.75) and the lowest ($T_{Md} = 55.00$, 48.00–63.00) levels. For life functioning, the group with the lower level

(T_{Md} = 63.00, 57.00–82.50) showed more problem behaviors than the group with higher levels (T_{Md} = 53.50, 47.25–61.75). In the case of friendship, the group with poor relationships (T_{Md} = 63.00, 57.00–80.75) displayed more problem behaviors than the group with good relationships (T_{Md} = 57.00, 48.00–63.00). Similarly, in family relationships, the group with poor relationships (T_{Md} = 59.00, 57.00–89.25) reported more problem behaviors than the one with good relationships (T_{Md} = 53.50, 47.00–62.25).

3.2. Treatment Effects and Interaction with the Demographic and Life Functioning Variables

The Wilcoxon Signed Ranks Test was conducted with problem behaviors at treatment time points (pre- and post-intervention) to determine whether there were significant treatment effects. Problem behaviors significantly decreased with time ($Z = -2.955$, $p = 0.003$), with medium to large effect sizes. The results of the Wilcoxon Signed Ranks Test showed a significant main effect for time, but no significant interaction effect between time and groups of variables: survivors with and without compensation, self and family member, male and female, age, SES, family, and job adjustment. Additionally, there was a significant main effect of time and an interaction effect ($Z = -2.342$, $p = 0.019$) on friendship. No significant interaction effects of time and family relationships, time, life functioning, or job adjustment were found.

Because there were some significant differences in problem behaviors before and after treatment according to the levels of friendship, a post hoc analysis (simple effect comparisons) was conducted. In other words, we examined whether the level of friendship influenced the treatment effects. The quality of friendship was divided into two groups: the poor relation group (T $\leq$ 35), which includes standard scores of clinical and borderline clinical provided by ASEBA and the good relation group (T > 35). There was no notable difference in problem behaviors in the good relation group before (T_{Md} = 57.00, 48.00–63.00) and after (T_{Md} = 55.00, 43.00–65.00) treatment. In the case of the poor relation group, problem behaviors significantly decreased after (T_{Md} = 55.50, 50.75–67.25) treatment, compared to before (T_{Md} = 63.00, 57.00–80.75) (see Figure 1). The poor relation group reported fewer problem behaviors after treatment than the good relation group.

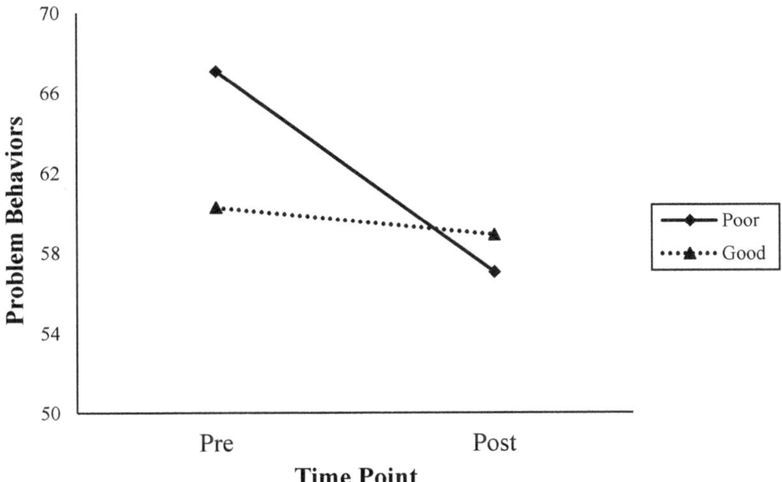

Figure 1. Interaction effect between quality of friendship (poor or good) and problem behaviors by time point.

Table 1. Descriptive statistics of problem behaviors according to research variables at pre- and post-treatment stages.

Variable		*n*	Pre-Test		Post-Test		Cohen'*d*
			M(SD)	*Md(IQR)*	*M(SD)*	*Md(IQR)*	
Compensation Presence	With Compensation	26	63.77(17.138)	59.50(52.75–80.75)	57.88(12.854)	55.00(49.50–65.25)	0.39
	Without Compensation	43	61.81(13.512)	60.00(52.75–80.75)	58.53(15.906)	54.00(45.00–66.00)	0.22
Compensation Subject	Self	44	64.36(14.989)	61.50(55.00–71.25)	60.86(15.200)	58.50(49.25–67.75)	0.23
	Family member	25	59.36(14.454)	57.00(49.50–66.50)	53.76(12.950)	52.00(43.50–62.50)	0.41
Gender	Male	31	59.65(16.390)	58.00(48.00–68.00)	54.97(12.828)	52.00(44.00–65.00)	0.32
	Female	38	64.92(13.294)	60.50(56.75–69.75)	61.00(15.774)	55.50(50.75–70.50)	0.27
Age	Under 19	12	54.92(9.885)	57.00(48.50–63.50)	48.17(8.100)	45.00(42.00–53.75)	0.75
	20 to 39	15	62.47(15.793)	58.00(49.00–79.00)	57.07(13.956)	52.00(47.00–65.00)	0.36
	40 to 59	42	64.76(15.303)	62.50(54.75–72.75)	61.62(15.300)	61.50(50.75–68.50)	0.21
SES	High	38	67.50(15.866)	63.50(57.00–86.25)	61.05(16.319)	59.00(50.00–68.75)	0.40
	Middle	24	56.46(11.077)	56.00(49.00–62.75)	55.83(12.345)	54.50(45.00–62.00)	0.053
	Low	7	56.57(12.012)	55.00(48.00–63.00)	51.71(10.547)	47.00(43.00–65.00)	0.43
Life functioning	High	28	54.50(12.816)	53.50(47.25–61.75)	53.64(14.952)	50.50(42.24–61.75)	0.062
	Low	41	68.05(13.769)	63.00 (57.00–82.50)	61.46(13.884)	59.00(51.50–68.00)	0.48
Friendship	Good	35	60.28(15.156)	57.00(48.00–63.00)	58.91(16.220)	55.00(43.00–65.00)	0.087
	Poor	34	67.09(13.527)	63.00(57.00–80.75)	57.04(11.424)	55.50(50.75–67.25)	0.80
Family Relationship	Good	14	61.07(14.143)	53.00(47.00–62.25)	58.98(14.455)	50.00(43.50–63.75)	0.15
	Poor	14	71.36(15.174)	59.00 (57.00–89.25)	62.21(14.110)	64.50(54.75–71.75)	0.62
Job Adjustment	High	20	55.60(10.713)	57.00(49.50–60.75)	53.50(10.511)	54.00(44.75–63.75)	0.20
	Low	8	77.88(18.019)	89.00(59.25–91.50)	71.75(16.334)	68.00(57.50–89.00)	0.36

4. Discussion

This study aimed to closely examine group differences in the survivors of humidifier disinfectant damage and the effect of individual psychotherapy on the survivor groups. For this purpose, the results are as follows: First, the differences in problem behaviors at the pre-treatment stage based on socio-demographic and psychological variables were examined to identify groups that reported more problem behaviors than others. There were no significant differences between survivors with and without compensation, between self and family members, between male and female participants, and between the three age groups. However, survivors from high socioeconomic status (SES) backgrounds reported more problem behaviors compared to those from low- and middle-SES backgrounds. The higher family economic status group's psychological symptoms may be influenced by social comparison. Ko et al. [17] also reported the results that the high-SES group among humidifier disinfectant survivors experienced more psychological problems than the other groups. Humidifier disinfectant survivors who reported a high level of life functioning, friendships, family relationships, and job adjustment were less likely to report problem behaviors. The results of this study are consistent with those of several studies reporting that good functioning in life and good interpersonal relationships are closely related to a low level of psychological problems. Furthermore, many studies have found that individuals who function well in life and have positive relationships tend to experience better physical health, increased life satisfaction, higher confidence, and lower stress and anxiety levels [33]. In this way, the correlation between psychological problems and life function appears to be that they influence each other and accelerate their impact.. The previously mentioned study findings allow us to confirm the requirement for psychological intervention in a person whose life function has been negatively affected by psychological problems.

Second, the treatment effects of individual psychotherapy and interactions with demographic and life functioning variables were analyzed. The results indicated a significant main effect for time, but no significant interaction effect between time and groups of variables: survivors with and without compensation, self and family members, male and female, age, SES, life functioning, family, and job adjustment. Specifically, the effects of individual psychotherapy were more pronounced in survivors with poor friendship, as their problem behavior scores significantly decreased following treatment.

The results of this study can be explained by the study of the life adaptation of people who have achieved psychological recovery through individual psychotherapy [33]. Research on disaster survivors has shown that some survivors recover well psychologically, while others experience psychological difficulties to the point of experiencing trauma from the disaster [34–40]. Through psychotherapy, individuals achieve psychological recovery by gaining social support, forming a sense of solidarity, and accepting their pain [41–43]. Survivors can learn to cope with stress, enhance life satisfaction, reduce psychological symptoms, function effectively at work and home, and cultivate positive relationships through the psychotherapy process. Finally, it was found that psychological problems were significantly reduced after psychotherapy in the group with poor friendships. This result is in line with the research findings that posit that social support helps psychological recovery after an accident or disaster [44–47]. In addition, this is consistent with the results of studies reporting that happiness and social support are closely related to life satisfaction [48–50]. Mexico, which is exceptional in the correlation between the average income level, crime rate, and happiness index, has a lower income level and higher crime rate than the United States but a higher happiness index. This index can be attributed to the cultural characteristics of Mexico, where there are stronger family bonds and a higher frequency of contact among family members than in the United States [51]. Individuals with social support are less susceptible to stress and experience fewer psychological difficulties. The group with poor friendships may improve their ability to seek social support through psychotherapy, which leads to a reduction in psychological problems. In summary, this indicates that treatment may enhance the survivors' ability to seek social support.

The limitations of this study are as follows. Firstly, the subjects were a unique group of survivors of humidifier disinfectant disasters. Therefore, it is necessary to be careful when interpreting the research results for the general public. Secondly, there is a limitation in verifying statistical significance because the number of subjects was relatively small. Although G*Power provided a minimum sample size of 60, it was small to divide the sample into adults and minors and perform all the analyses. Thus, it is necessary to replicate these results by increasing the number of subjects in the future. Thirdly, we considered demographic characteristics (e.g., age, SES, compensation presence, etc.) as moderators for treatment effectiveness. However, other potentially critical moderation variables such as the severity of the health damages by the humidifier disinfectant use, losses of participating family members, the duration of victimization, taking psychotropic drugs, and the number of psychotherapy sessions should also be included as moderation variables in future studies. Fourthly, this study focused on verifying the effectiveness of a distinctive group of social disaster survivors who immediately needed psychotherapy support. Due to the nature of these subjects, it is ethically problematic to randomly assign survivors to the control group, effectively suspending them from psychotherapy. To explore the effect of psychotherapy in this situation, where a control group could not be secured due to ethical issues, this study observed changes before and after psychotherapy through a time-series design. Because this study has limitations in that it cannot control factors for internal validity from the experimental design, attention should be paid to the interpretation of the results of the study. Lastly, all items were self-reported; therefore, the response could be biased due to faking negatively or positively (e.g., social desirability). In future studies, the researchers need to utilize objective measures such as significant others' observations.

Based on the findings and limitations of this study, the following recommendations can be made for practice, research, and management. Tailored interventions addressing these specific concerns of survivors may be beneficial. Future studies should replicate this research with larger sample sizes to enhance statistical power and allow for subgroup analyses, such as comparing adult and minor survivors separately. In addition to demographic variables, researchers should consider including other critical factors as moderation or mediation variables, such as the severity of health damages, losses of family members, duration of victimization, and number of psychotherapy sessions, to better understand their influence on treatment effectiveness. Objective measures, such as observations from significant others, should be incorporated alongside self-reported measures to minimize response biases and enhance the validity of the research findings. Public health authorities and policymakers should consider implementing and promoting accessible mental health services for survivors of environmental disasters. Adequate resources and support should be allocated to ensure timely and effective psychological interventions are available to those in need. Managers of mental health facilities and organizations should encourage interdisciplinary collaboration and research partnerships to further explore the psychological impacts of environmental disasters. This can help develop evidence-based practices and interventions tailored to the specific needs of different survivor groups.

It is important to acknowledge the limitations of this study, including the small sample size, the specificity of the population studied, the lack of a control group, and the reliance on self-reported measures. These limitations should be taken into account when interpreting our findings, and recommendations should be considered in light of these limitations. Further research and replication studies are needed to validate the findings and address the identified limitations. Despite all the limitations, this study has meaning: (1) It examined the effect of psychotherapy on survivors of social disasters. (2) Survivors were classified according to differences in their ability to adapt to life, even though they experienced the same disaster. The differences in the effectiveness of psychotherapy according to the classified group were examined. (3) It is especially meaningful that we explored which group showed the greatest change due to psychotherapy. Overall, this study provides insights into the group differences among survivors of humidifier disinfectant damage and underscores the positive impact of individual psychotherapy on psychological recovery, particularly in individuals with poor friendship networks.

In conclusion, this study aimed to investigate group differences among the survivors of humidifier disinfectant damage and the effects of individual psychotherapy on their psychological symptoms. The findings of this study contribute to our understanding of the psychological impact of such environmental disasters from an international perspective. The results demonstrate the effectiveness of individual psychotherapy in reducing psychological symptoms among survivors of humidifier disinfectant damage. The treatment significantly decreased problem behaviors over time, with medium to large effect sizes. Significant interaction effects were found between treatment effects and friendship levels on problem behaviors. This study highlights the importance of considering the quality of friendships in psychological interventions for survivors. Individuals with poor friendships showed significant improvements in problem behaviors following psychotherapy, emphasizing the role of social support in psychological recovery after disasters or accidents. Overall, this study contributes to the scientific and civil communities' understanding of the psychological consequences of environmental disasters, underscores the positive impact of individual psychotherapy on survivors' psychological recovery, and highlights the importance of addressing social support and relational approaches in interventions for disaster survivors.

5. Conclusions

The paper aimed to investigate the psychological impact of environmental disasters on survivors and the effects of individual psychotherapy on their psychological symptoms. The study found that individual psychotherapy was effective in reducing psychological symptoms among survivors of humidifier disinfectant damage. The treatment significantly decreased problem behaviors over time, with medium to large effect sizes. This study also highlighted the importance of considering the quality of friendships in psychological interventions for survivors, as individuals with poor friendships showed significant improvements in problem behaviors following psychotherapy. The paper suggested tailoring interventions to address survivors' specific issues, replicating the study with a larger sample size, and incorporating objective measures alongside self-report measures to increase the validity of the findings. Public health authorities and policymakers should consider implementing and promoting accessible mental health services for survivors of environmental disasters, and adequate resources and support should be allocated to ensure timely and effective psychological interventions are available to those in need. Managers of mental health facilities and organizations should encourage interdisciplinary collaboration and research partnerships to further explore the psychological impacts of environmental disasters.

Author Contributions: Conceptualization, S.M.L.; methodology, S.H.; formal analysis, Y.C., S.H. and G.P.; investigation, M.J.L. and H.-J.L.; data curation, Y.C.; writing—original draft preparation, M.J.L., Y.C., S.H., H.-J.L. and G.P.; writing—review and editing, S.M.L., M.J.L., Y.C., S.H., H.-J.L. and G.P.; supervision, S.M.L. All authors have read and agreed to the published version of the manuscript.

Funding: This research was supported by the National Institute of Environmental Research, Ministry of Environment (ME) of the Republic of Korea (NIER-2021-04-03-001).

Institutional Review Board Statement: This research was approved by the Institutional Review Board of Yonsei University (No. 7001988-202104-HR-1178-02l). Informed consent was obtained from all the individual participants included in this study. Informed consent was also obtained from all the parents/guardians of the minor participants (under 19 years old). All methods were carried out in accordance with relevant guidelines and regulations (Declaration of Helsinki).

Informed Consent Statement: Written informed consent has been obtained from the participants to publish this paper.

Data Availability Statement: Data sharing not available due to restrictions of privacy.

Acknowledgments: The norm data was provided by HUNO, an ASEBA Provider Company in Republic of Korea. We appreciate the sharing of the data for this research.

Conflicts of Interest: The authors declare that there are no conflict of interest.

References

1. Park, H.S.; Kwon, S.Y. A study on the life experiences of family of social disaster victim: Focusing on spouse of victim of humidifier disinfectants. *Korean Assoc. Christ. Counsel. Psychol.* **2020**, *31*, 175–221. [CrossRef]
2. Kim, J. Life after "Humidifier Disinfectant": Parents' Becoming Victims in a Risk Society. Master's Thesis, Seoul National University, Seoul, Republic of Korea, 2018. Available online: https://hdl.handle.net/10371/142138 (accessed on 1 February 2023).
3. Choi, J.E.; Hong, S.B.; Do, K.H.; Kim, H.J.; Chung, S.; Lee, E.; Choi, J.; Hong, S.J. Humidifier disinfectant lung injury, how do we approach the issues? *Environ. Health Toxicol.* **2016**, *31*, 147–157. [CrossRef] [PubMed]
4. Lee, E.; Seo, J.H.; Kim, H.Y.; Yu, J.; Jhang, W.K.; Park, S.J.; Kwon, J.W.; Kim, B.J.; Do, K.H.; Cho, Y.A.; et al. Toxic inhalational injury-associated interstitial lung disease in children. *J. Korean Med. Sci.* **2013**, *28*, 915–923. [CrossRef]
5. Kim, H.J.; Lee, M.S.; Hong, S.B.; Huh, J.W.; Do, K.H.; Jang, S.J.; Lim, C.M.; Chae, E.J.; Lee, H.; Jung, M.; et al. A cluster of lung injury cases associated with home humidifier use: An epidemiological investigation. *Thorax* **2014**, *69*, 703–708. [CrossRef] [PubMed]
6. Park, D.U.; Lee, S.; Lim, H.K.; Kim, S.Y.; Kim, J.; Park, J.; Zoh, K.E. Comprehensive review on humidifier disinfectant (HD) products, focusing on the number of products and their disinfectant type. *J. Environ. Health Sci.* **2020**, *46*, 481–494. [CrossRef]
7. Choi, Y.; Lee, I. A comparative study of the humidifier disinfectant disaster and Minamata disease. *J. Environ. Health Sci.* **2019**, *45*, 326–339. [CrossRef]
8. Yoon, J.; Cho, H.J.; Lee, E.; Choi, Y.J.; Kim, Y.H.; Lee, J.L.; Lee, Y.J.; Hong, S.J. Rate of humidifier and humidifier disinfectant usage in Korean children: A nationwide epidemiologic study. *Environ. Res.* **2017**, *155*, 60–63. [CrossRef]
9. Yoon, J.; Kang, M.; Jung, J.; Ju, M.J.; Jeong, S.H.; Yang, W.; Choi, Y.H. Humidifier disinfectant consumption and humidifier disinfectant-associated lung injury in South Korea: A nationwide population-based study. *Int. J. Environ. Res. Public Health* **2021**, *18*, 6136. [CrossRef]
10. Park, J.H.; Ko, M.H. A Study on the Damage Relief in Humidifier Sterilizer Case—Focused on Expanding the Role of Epidemiological Causality for Presumption of Causal Relationship. *Law J.* **2020**, *48*, 265–306. [CrossRef]
11. Ryu, S.H.; Park, D.U.; Lee, E.; Park, S.; Lee, S.Y.; Jung, S. Humidifier disinfectant and use characteristics associated with lung injury in Korea. *Indoor Air* **2019**, *29*, 735–747. [CrossRef]
12. Jeon, B.H.; Park, Y.J. Frequency of humidifier and humidifier disinfectant usage Gyeonggi provine. *Environ. Health Toxicol.* **2012**, *27*, e2012002. [CrossRef]
13. Chang, M.H.; Park, H.; Ha, M.; Kim, Y.; Hong, Y.C.; Ha, E.H. Characteristics of humidifier use in Korean pregnant women: The Mothers and Children's Environmental Health (MOCEH) study. *Environ. Health Toxicol.* **2012**, *27*, 123–137. [CrossRef]
14. Choi, Y.; Ryu, H.; Yoon, J.; Lee, S.; Kwak, J.; Han, B.; Chu, Y.H.; Kim, P.G.; Yang, W. Demographic Characteristics and Exposure Assessment for Applicants Who Have Been Injured by Humidifier Disinfectant: Focusing on 4-1 and 4-2 Applicants. *J. Korean Soc. Environ. Health* **2018**, *44*, 301–314. [CrossRef]
15. Yoo, S.; Sim, M.; Choi, J.; Jeon, K.; Shin, J.; Chung, S.; Hong, S.B.; Lee, S.Y.; Hong, S.J. Psychological responses among humidifier disinfectant disaster victims and their families. *J. Korean Med. Sci.* **2019**, *34*, e29. [CrossRef] [PubMed]
16. Leem, J.H.; Joh, J.S.; Hong, Y.S.; Kim, J.; Park, S.; Lim, S.; Kim, Y. Characteristics of a new respiratory syndrome associated with the use of a humidifier disinfectant: Humidifier disinfectant-related respiratory syndrome (HDRS). *Int. J. Occup. Med. Environ. Health* **2020**, *33*, 829–839. [CrossRef]
17. Ko, H.-Y.; Ryu, S.-H.; Lee, M.-J.; Lee, H.-J.; Kwon, S.-Y.; Kim, S.-M.; Lee, S.-M. Exploring Socio-Demographic Factors Affecting Psychological Symptoms in Humidifier Disinfectant Survivors. *Int. J. Environ. Res. Public Health* **2021**, *18*, 11811. [CrossRef] [PubMed]
18. Matliwala, K. The effect of psychological counseling on mental health. *J. Psychol. Clin. Psychiatry* **2017**, *7*, 4–36. [CrossRef]
19. Shea, M.T.; Stout, R.L.; Reddy, M.K.; Sevin, E.; Presseau, C.; Lambert, J.; Cameron, A. Treatment of anger problems in previously deployed post-911 veterans: A randomized controlled trial. *Depress. Anxiety* **2022**, *39*, 274–285. [CrossRef]
20. Howard, K.I.; Kopta, S.M.; Krause, M.S.; Orlinsky, D.E. The dose–effect relationship in psychotherapy. *Am. Psychol.* **1986**, *41*, 159–164. [CrossRef]
21. Cuijpers, P.; Karyotaki, E.; Reijnders, M.; Ebert, D.D. Was Eysenck right after all? A reassessment of the effects of psychotherapy for adult depression. *Epidemiol. Psychiatr. Sci.* **2019**, *28*, 21–30. [CrossRef]
22. Wang, M.-C.; Nyutu, P.; Tran, K.; Spears, A. Finding resilience: The mediation effect of sense of community on the psychological well-being of military spouses. *J. Ment. Health Couns.* **2015**, *37*, 164–174. [CrossRef]
23. Humphreys, K.; Mankowski, E.S.; Moos, R.H.; Finney, J.W. Do enhanced friendship networks and active coping mediate the effect of self-help groups on substance abuse? *Ann. Behav. Med.* **1999**, *21*, 54–60. [CrossRef]
24. Miller, S.D.; Duncan, B.L. *The Outcome and Session Rating Scales: Administration and Scoring Manuals*; Institute for the Study of Therapeutic Change: Chicago, IL, USA, 2004.
25. Achenbach, T.M.; Rescorla, L.A. *Manual for the ASEBA Adult Forms & Profiles*; ASEBA: Burlington, VT, USA, 2003.

26. Kim, M.Y.; Kim, Y.A.; Lee, J.; Kim, H.J.; Oh, K.J. A Validity Study on the Korean Version of the Adult Self Report. *Korean J. Clin. Psychol.* **2014**, *33*, 615–632.

27. Achenbach, T.M.; Rescorla, L.A. *Manual for the ASEBA School-Age Forms & Profiles*; ASEBA: Burlington, VT, USA, 2001.

28. Mann, H.B.; Whitney, D.R. On a test of whether one of two random variables is stochastically larger than the other. *Ann. Math. Stat.* **1947**, *18*, 50–60. [CrossRef]

29. Norman, B. A generalized Kruskal-Wallis test for comparing K samples subject to unequal patterns of censorship. *Biometrika* **1970**, *57*, 579–594. [CrossRef]

30. Wilcoxon, F. Individual comparisons by ranking methods. *Biometrics* **1945**, *1*, 80–83. [CrossRef]

31. Faul, F.; Erdfelder, E.; Lang, A.G.; Buchner, A. G*Power 3: A flexible statistical power analysis program for the social, behavioral, and biomedical sciences. *Behav. Res.* **2007**, *39*, 175–191. [CrossRef] [PubMed]

32. Cohen, J. *Statistical Power Analysis for the Behavioral Science*, 2nd ed.; Routledge: New York, NY, USA, 1988; pp. 25–27.

33. McGraw, A.P.; Mellers, B.A.; Tetlock, P.E. Expectations and emotions of Olympic athletes. *J. Exp. Soc. Psychol.* **2005**, *41*, 438–446. [CrossRef]

34. Ford, M.T.; Cerasoli, C.P.; Higgins, J.A.; Decesare, A.L. Relationships between psychological, physical, and behavioural health and work performance: A review and meta-analysis. *Work Stress* **2011**, *25*, 185–204. [CrossRef]

35. Brooks, S.K.; Dunn, R.; Amlôt, R.; Rubin, G.J.; Greenberg, N. Social and occupational factors associated with psychological wellbeing among occupational groups affected by disaster: A systematic review. *J. Ment. Health* **2017**, *26*, 373–384. [CrossRef]

36. Lee, M.J.; Lee, H.J.; Ko, H.; Ryu, S.H.; Lee, S.M. Multi-layer relationships between psychological symptoms and life adaptation among humidifier disinfectant survivors. *Front. Psychiatry* **2022**, *13*, 890122. [CrossRef]

37. Wiseman, T.; Foster, K.; Curtis, K. Mental health following traumatic physical injury: An integrative literature review. *Injury* **2013**, *44*, 1383–1390. [CrossRef]

38. Lee, J.Y.; Kim, S.W.; Kim, J.M. The Impact of Community Disaster Trauma: A Focus on Emerging Research of PTSD and Other Mental Health Outcomes. *Chonnam Med. J.* **2020**, *56*, 99–107. [CrossRef] [PubMed]

39. O'Donnell, M.L.; Lau, W.; Fredrickson, J.; Gibson, K.; Bryant, R.A.; Bisson, J.; Burke, S.; Busuttil, W.; Coghlan, A.; Creamer, M.; et al. An open label pilot study of a brief psychosocial intervention for disaster and trauma survivors. *Front. Psychiatry* **2020**, *11*, 483. [CrossRef] [PubMed]

40. Levers, L.L. An introduction to counseling survivors of trauma: Beginning to understand the historical and psychosocial implications of trauma, stress, crisis, and disaster. In *Trauma Counseling*; Springer Publishing Company: New York, NY, USA, 2022. [CrossRef]

41. Norris, F.H.; Friedman, M.J.; Watson, P.J. 60,000 disaster victims speak: Part II. Summary and implications of the Disaster Mental Health Research. *Psychiatry Interpers. Biol. Process.* **2002**, *65*, 240–260. [CrossRef]

42. Kaniasty, K. Predicting social psychological well-being following trauma: The role of postdisaster social support. *Psychol. Trauma Theory Res. Pract. Policy* **2012**, *4*, 22–33. [CrossRef]

43. Watson, P.J.; Brymer, M.J.; Bonanno, G.A. Postdisaster psychological intervention since 9/11. *Am. Psychol.* **2011**, *66*, 482–494. [CrossRef]

44. Stellman, J.M.; Smith, R.P.; Katz, C.L.; Sharma, V.; Charney, D.S.; Herbert, R.; Moline, J.; Luft, B.J.; Markowitz, S.; Udasin, I.; et al. Enduring mental health morbidity and social function impairment in world trade center rescue, recovery, and cleanup workers: The psychological dimension of an environmental health disaster. *Environ. Health Perspect.* **2008**, *116*, 1248–1253. [CrossRef]

45. Lê, F.; Tracy, M.; Norris, F.H.; Galea, S. Displacement, County Social Cohesion, and Depression After a Large-Scale Traumatic Event. *Soc. Psychiatry Psychiatr. Epidemiol.* **2013**, *48*, 1729–1741. [CrossRef]

46. Huang, Y.; Wong, H. Impacts of Sense of Community and Satisfaction with Governmental Recovery on Psychological Status of the Wenchuan Earthquake Survivors. *Soc. Indic. Res.* **2014**, *117*, 421–436. [CrossRef]

47. Manesi, Z.; Van Lange, P.A.; Van Doesum, N.J.; Pollet, T.V. What Are the Most Powerful Predictors of Charitable Giving to Victims of Typhoon Haiyan: Prosocial Traits, Socio-Demographic Variables, or Eye Cues? *Personal. Individ. Differ.* **2019**, *146*, 217–225. [CrossRef]

48. Guilaran, J.; de Terte, I.; Kaniasty, K.; Stephens, C. Psychological Outcomes in Disaster Responders: A Systematic Review and Meta-Analysis on the Effect of Social Support. *Int. J. Disaster Risk Sci.* **2018**, *9*, 344–358. [CrossRef]

49. Strine, T.W.; Kroenke, K.; Dhingra, S.; Balluz, L.S.; Gonzalez, O.; Berry, J.T.; Mokdad, A.H. The associations between depression, health-related quality of life, social support, life satisfaction, and disability in community-dwelling US adults. *J. Nerv. Ment. Dis.* **2009**, *197*, 61–64. [CrossRef] [PubMed]

50. Young, K.W. Social support and life satisfaction. *Int. J. Psychosoc. Rehabil.* **2005**, *10*, 155–164.

51. Helliwell, J.F.; Layard, R.; Sachs, J.D.; Neve, J.E.D. *World Happiness Report 2021*; Sustainable Development Solutions Network: New York, NY, USA, 2021.

 *healthcare*

Article

The Prospective Effects of Coping Strategies on Mental Health and Resilience at Five Months after HSCT

Maya Corman [1,*], Michael Dambrun [1,*], Marie-Thérèse Rubio [2], Aurélie Cabrespine [3], Isabelle Brindel [4], Jacques-Olivier Bay [3] and Régis Peffault de La Tour [4]

[1] LAPSCO UMR CNRS 6024, Université Clermont Auvergne (UCA), 34 Avenue Carnot, 63000 Clermont-Ferrand, France
[2] Service D'Hématologie, CHRU Nancy-Hôpitaux de Brabois, 54511 Vandoeuvre-les-Nancy, France
[3] CHU de Clermont-Ferrand, Site Estaing, Service de Thérapie Cellulaire et D'hématologie Clinique Adulte, 63000 Clermont-Ferrand, France
[4] Service D'Hématologie, Hôpital Saint-Louis, Greffe de Moelle, 75010 Paris, France
*   Correspondence: maya.corman@uca.fr (M.C.); michael.dambrun@uca.fr (M.D.)

Abstract: Objectives: Hematopoietic stem cell transplantation (HSCT) is a stressful event that engenders psychological distress. This study examines the prospective effects of coping strategies during hospitalization on resilience and on various mental-health dimensions at five months after transplantation. Methods. One hundred and seventy patients (M_{age} = 52.24, SD = 13.25) completed a questionnaire assessing adjustment strategies during hospitalization, and 91 filled out a questionnaire five months after HSCT (M_{age} = 51.61, SD = 12.93). Results: Multiple regression analyses showed that a fighting spirit strategy positively predicted resilience ($p < 0.05$), whereas anxious preoccupations predicted anxiety ($p < 0.05$), poorer mental QoL ($p < 0.01$), and were associated with an increased risk of developing PTSD (OR = 3.27, $p < 0.01$; 95% CI: 1.36, 7.84) at five months after transplantation. Hopelessness, avoidance, and denial coping strategies were not predictive of any of the mental health outcomes. Finally, the number of transplantations was negatively related to a fighting spirit ($p < 0.01$) and positively related to hopelessness-helplessness ($p < 0.001$): Conclusions: These results highlight the importance of developing psychological interventions focused on coping to alleviate the negative psychological consequences of HSCT.

Keywords: hematopoietic stem cell transplant (HSCT); adjustment coping strategies; mental health; quality of life; post-traumatic stress disorder; resilience

Citation: Corman, M.; Dambrun, M.; Rubio, M.-T.; Cabrespine, A.; Brindel, I.; Bay, J.-O.; Peffault de La Tour, R. The Prospective Effects of Coping Strategies on Mental Health and Resilience at Five Months after HSCT. *Healthcare* **2023**, *11*, 1975. https://doi.org/10.3390/healthcare11131975

Academic Editors: Carlos Laranjeira and Ana Querido

Received: 31 May 2023
Revised: 3 July 2023
Accepted: 5 July 2023
Published: 7 July 2023

1. Background

Hematopoietic Stem Cell Transplantation (HSCT) can significantly impair several psychological health aspects of the individuals who benefit from this treatment. It alters short- and long-term physical and mental Quality of Life (QoL) [1–3] and engenders symptoms of psychological distress, such as anxiety [1,4], depression [1,5], and post-traumatic stress disorder (PTSD) [1,6]. The pre-graft period is a source of psychological distress because of both the treatment and the anticipation of hospitalization [7]. Hospitalization, particularly isolation in a protective area, is recognized as distressing for patients [8–13]. Regarding these issues, studies are necessary to explore the psychological factors involved in the nature and quality of the psychological and physical health of patients during the early stage of hospitalization and its impact several months later.

This question can be addressed through the exploration of coping strategies used during HSCT hospitalization. These coping strategies include behavioral, cognitive, and emotional responses used to reduce or manage psychological distress engendered by a stressful and threatening situation that exceeds the patient's external and/or internal personal resources [14]. According to the transactional model [14], the coping process is dynamic and depends on the individual's perception and assessment of his/her relationship to his/her environment in terms of perceived stress, control, and social support. Generally, studies in oncology show that the perception of good social support [15,16], a situation perceived as controllable [17], and lower or moderate perceived stress [18,19] are associated with active coping and better mental health and QoL. There are no bad or good coping strategies per se, but some are considered to be linked to better mental health and QoL than others [20,21]. Therefore, coping strategies can be conceptualized as adaptive (e.g., problem solving, positive reappraisal) or maladaptive (e.g., negative rumination or emotional suppression) and as two independent processes that can interact [21,22]. More precisely, adaptive coping seems to mediate the relationship between positive psychological resources or dispositions (e.g., resilience, optimism) and mental health in the sense of protecting and buffering psychological distress in the short and long term [21,23,24]. Conversely, negative psychological dispositions, such as neurotic personality traits (i.e., the tendency to experience negative emotions such as sadness, anxiety, nervousness, and tension), and both mental health and QoL impairment are associated with maladaptive coping strategies [25].

Studies testing the prospective effects of coping strategies on mental health and resilience after HSCT are scarce. In the case of hematological malignancies, a meta-analysis [26] showed that a fighting spirit (i.e., an active and engaged coping strategy) is positively linked with a better QoL [27], whereas hopelessness-helplessness, fatalism, or anxious preoccupations (i.e., passive and disengagement strategies) are positively related to psychological distress [28]. In HSCT, to our knowledge, four prospective studies examined the relationships between depression and coping [29–32]. They mainly reveal that the greater the depression symptomatology, the more often an individual will use maladaptive coping strategies such as avoidance or fatalism, and that such an individual will rarely use adaptive coping strategies (i.e., fighting spirit, problem solving). The results of this study revealed that high scores of depression symptomatology are associated with a more frequent use of maladaptive coping strategies such as avoidance or fatalism and a less frequent use of adaptive strategies such as a fighting spirit or problem-solving. Finally, lower social support and higher use of an avoidance coping strategy one month prior to transplantation predicted greater PTSD symptomatology seven months after transplantation [33].

Our study explored the prospective effects of five coping strategies during HSCT on several markers of mental health (i.e., QoL, depression, anxiety, happiness, and PTSD symptomatology) and on resilience in a five-month follow-up. First, we examined the relationships between perceived stress, control, perceived social support and adjustment coping strategies with a sample of people undergoing HSTC. The objective was to verify that our measures correlated with each other in the expected direction, in accordance with the existing literature, thus attesting to the validity of our scales. Second, we tested the prospective effects of coping strategies on various dimensions of mental health and resilience.

In sum, hospitalization in a protected area is relatively specific to the case of bone marrow transplantation (isolation, risk of infection, aplasia, side effects of treatment during transplantation, etc.). This stage generates stress, which in turn generates specific coping strategies. We know that coping strategies partly explain certain aspects of mental health, such as anxiety, depression, post-traumatic stress symptoms and quality of life. The main objective of this research was to identify the coping strategies that predict mental health and resilience five months after HSCT.

The research will enable us to identify the most suitable strategies midway through the HSCT process, so that we can implement more targeted and effective psychotherapeutic interventions at a later stage.

We predicted that maladaptive coping strategies during hospitalization (i.e., avoidance, hopelessness-helplessness, anxious preoccupations, and denial) would positively predict PTSD, anxiety, and depression and negatively predict QoL, happiness, and resilience at five months after transplantation. We anticipated the opposite for the adaptive coping strategies (i.e., fighting spirit). The prospective effects of coping strategies on several medical outcomes (e.g., acute graft versus host disease [GvHD], relapse, death) were also explored.

2. Methods

2.1. Participants

Two hundred and fifty-seven participants were invited to participate in the "Psy-Greffe" protocol between November 2017 and September 2020. Among them, 70 declined to participate or could not participate for various reasons. The recruited sample filled out three questionnaires: one before hospitalization, one during transplantation, and one five months after the allograft. One hundred and eighty-seven participants filled out the first questionnaire (M_{age} = 52.07, *SD* = 13.22, age range from 19 to 72 years old) and 170 filled out the second one (M_{age} = 52.24, *SD* = 13.25, age range from 19 to 72 years old). Finally, 91 completed the third questionnaire at the five-month follow-up (M_{age} = 51.61, *SD* = 12.93, age range from 23 to 70 years old). They came from three hospital centers in France and were candidates for an allogenic Hematopoietic Stem Cell Transplantation (HSCT) after diagnoses of hematologic malignancies. (See Figure 1). We estimated the required sample size for sufficient correlation power (90%). On the basis of the correlations between coping strategies and QoL reported by O'Connor et al. [28] (i.e., in absolute value, r between 0.34 and 0.67), the minimum required sample size was 87 with r = 0.34.

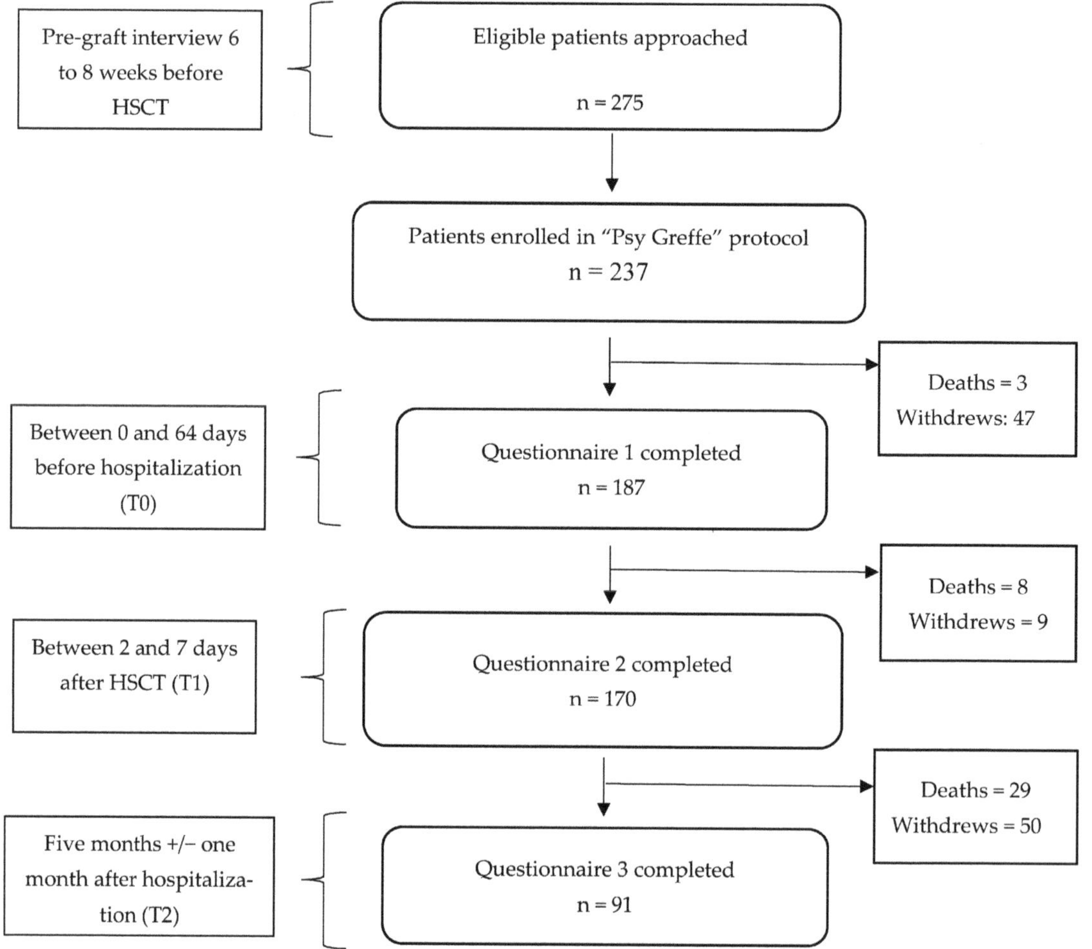

Figure 1. Flow diagram of protocol.

2.2. Procedures

The study population concerned all patients proposed for an allograft, aged over 18 years, who did not object to answering the various questionnaires over a period of approximately seven months post-transplant. The participants came from the hematology departments in Paris, Clermont-Ferrand, and Nancy. During the pre-graft interview with the doctor, the study was proposed to each eligible patient by reading an information note about the protocol. After a 15-day cooling-off period, patients who agreed to take part in the study filled out informed consent forms and provided sociodemographic and medical information. Their levels of QoL, anxiety, depression, and happiness were also assessed at this time (Time 0). Next, a questionnaire assessing coping, perceived control, stress, and social support was given during hospitalization, between day one and day seven after transplantation (Time 1). The third questionnaire was proposed five months (+/− one month) after the allograft (Time 2). This final questionnaire measured QoL, anxiety, depression, happiness, PTSD, and resilience. The relevant medical data were extracted from the ProMISe (Project Manager Internet Server) database. Participants knew they were not retributed for their participation. All participants were volunteers.

2.3. Measures

Adjustment strategies (Time 1). Mental adjustment to cancer scale (MACs) [34,35]. This French version contains 45 items measuring five different coping strategies used by patients with a cancer diagnosis. This is a 4-point Likert-type scale (from 1 "not at all" to 4 "completely"). The scale is divided into five subscales for each meta-coping strategy: fighting spirit ($\alpha = 0.81$), hopelessness-helplessness ($\alpha = 0.80$), anxious preoccupations ($\alpha = 0.89$), avoidance ($\alpha = 0.75$) and denial ($\alpha = 0.80$). To obtain a global score of maladaptive adjustment strategies, we summed up the four means of hopelessness-helplessness, anxious preoccupations, avoidance, and denial scores. Some statements were adapted to match with the HSCT conditions of patients.

Perceived social support (Time 1). Social Provisions Scale-Revised (SPS-R) [36,37]. Twenty-four items measure the perceived social support on a four-point scale from 1 "strongly disagree" to 4 "strongly agree." Six subscales assess the different dimensions of social support: emotional support (attachment), social integration, reassurance of value (reassurance of worth), material assistance (reliable alliance), advice and information (guidance), and the need to feel useful (opportunity for nurturance). The total score is the sum of all subscale scores and varies from 0 to 96 ($\alpha = 0.84$). The higher the score, the more people perceive good social support.

Perceived stress (Time 1). The Perceived Stress Scale (PSS) [38,39] assesses perceived stress regarding 10 items on a five-point scale from 0 "never" to 4 "very often." The total score is the sum of all items. A score between 0 and 13 means low stress; a score between 14 and 26 indicates medium stress, and a score between 27 and 40 means that people report a high level of perceived stress. The internal validity is very satisfying ($\alpha = 0.87$).

Perceived control (Time 1). Internal, Powerful others, and Chance Locus of Control (IPC) [40]. This is a Canadian version of Levenson's (1973) [41] three-dimensional locus of control assessment of 24 items. This is a 6-point Likert-type scale (from 1 "totally agree" to 6 "totally disagree"). Eight items measure the "Internality" dimension ($\alpha = 0.64$), eight the "Powerful others" dimension (externality), and eight the "Chance" dimension (externality). We added both "Powerful others" and "Chance" dimensions to obtain an externality score ($\alpha = 0.77$).

Quality of Life (Time 0 and Time 2). We used the SF-12 [42], a short version of the SF-36 which assessed both mental/social (i.e., vitality, social functioning, role-emotional, mental health) and physical (physical functioning, role-physical, bodily pain, general health) QoL. Calculation of Cronbach's alpha is not possible because all items are weighted. Each item score is converted into standardized data.

Mental Health/Illness (Time 0 and Time 2). Anxiety and Depression symptomatology was measured with the Hospital Anxiety and Depression scale (HAD) [43]. Seven items estimate the anxiety symptomatology ($\alpha_{t0} = 0.76$, $\alpha_{t2} = 0.72$) and seven items assess symptoms of depression ($\alpha_{t0} = 0.70$, $\alpha_{t2} = 0.80$). Happiness was assessed with the Subjective Authentic-Durable Happiness scale (SA-DHS) [44] through 13 items ($\alpha_{t0} = 0.95$, $\alpha_{t2} = 0.97$).

Post-Traumatic Stress Disorder (Time 2). Post-Traumatic Stress Disorder Checklist Scale (PCLS) [45,46]. This scale is used to detect post-traumatic stress disorder through 17 items assessing the severity of 17 symptoms of PTSD listed in the DSM-V. For each item, individuals indicate how much they have experienced these symptoms during the last month from 1 ("Not at all") to 5 ("Very often"). This scale has a very good internal consistency ($\alpha = 0.91$).

Resilience (Time 2). The Connor-Davidson Resilience scale (CD-RISC) [47,48] comprises 25 items, each rated on a 5-point scale (from 0 "not at all" to 4 "almost all the time"), with higher scores reflecting greater resilience. The internal consistency in our sample is very satisfying ($\alpha = 0.87$).

Medical and socio-demographic variables. Patients provided some information about their sex, age, educational level, marital status, and socio-professional category. Controlled medical variables included alcohol consumption, smoking, physical activity, body mass index (BMI), sleeping hours, type of disease, number of transplantations, the latency time between disease diagnosis and transplantation, myeloablative conditioning, the type of donor and chronic GvHD. Whether or not there was acute GvHD, latency between transplantation and engraftment, a relapse, a number of infections, or death during/following the hospitalization were included as dependent variables in our study with a mean follow-up of seven months ($SD = 3.96$).

Statistical Analyses. Data were expressed in numbers and percentages for categorical variables and as mean $\pm$ standard deviation (SD) for quantitative variables. Statistics were computed using Jamovi 2.3.24.0 and in this paper, p-value was chosen as an indicator to compare significant values through all tests used [49]. In a first step, we examined the relationships between the various variables using Pearson correlations. Then, in a second step, prospective effects were investigated using multiple linear regression (MLR) analyses. In these analyses, we calculated the z-score for all independent variables as well as for the covariates. For each MLR, we entered the different coping strategies that were significantly correlated with the selected health outcome (in step 1) as independent variables and the same health outcome as the dependent variable. Thus, an MLR analysis was realized with each health outcome (i.e., QoL, anxiety, depression, happiness, resilience, and PTSD), resulting in a series of six multiple regression analyses. Whenever available, we included the health outcome at time 0 as a covariate. Taking QoL as an example, our independent variables were hopelessness at time 1, anxious preoccupations at time 1 and quality of life at time 0. Our dependent variable was quality of life at time 2. None of the socio-demographic and medically controlled variables correlated significantly with our main dependent variables (i.e., QoL, anxiety, depression, happiness, resilience and PTSD) except the socio-professional category for anxiety, and gender for happiness. Consequently, we included the socio-professional category as a covariate in the MLR analysis with anxiety as the dependent variable, and we included participants' gender as a covariate in the MLR analysis with happiness as the dependent variable. Multicollinearity was adequate (i.e., all VIFs < 3). Homoscedasticity was checked using a scatterplot with residuals against the dependent variable. No extreme values influenced the results of the multiple regression analyses (all Cook's distance < 0.025). Except for depression as a DV ($p < 0.01$), the normality tests did not reject the normality hypothesis (all $p > 0.10$). A square root transformation was applied to the measure assessing depression at T2. As this transformation recovered normality ($p > 0.43$), we used it as the dependent variable in the correlation and multiple regression analyses. In all analyses, we used all available data without any imputations for missing data.

3. Results

3.1. Descriptive Statistics

The characteristics of our sample are available in Table 1. Among the patients in the sample who provided information, 42.7% were women, 46.4% were married, 46.3% had graduated, and 69.6% were employed. Thirty-six percent were allogenic HSCT candidates for acute leukemias, 17.4% for myelodysplastic syndromes, 10.1% for myeloproliferative neoplasia, and 11.8% for non-Hodgkin's lymphomas. For 38% of patients, the graft came from a matched unrelated donor.

Table 1. Descriptive Statistics for Socio Demographic and Medical Variables (*n* = 220).

	% (Excluding Missing Values)	Mean (*SD*)	*n*
Controlled socio demographic variables			
Age		52.03 (13.28)	217
Sex (*women*)	42.7		221
Marital Status (*married*)	46.4		181
Educational Level (*post-graduate*)	46.3		175
Socioprofessional Category (*employed*)	69.6		151
Follow-up (in months) Controlled medical variables		6.58 (4.04)	
Disease Status			178
Acute Leukemia	36		
Myelodysplastic Syndrome	17.4		
Myeloproliferative Neoplasia	10.1		
Non Hodgkin Lymphoma	11.8		
Alcohol consumption (*yes*)	30.8		172
Smoking (*yes*)	15.8		177
Physical Activity (*yes*)	45.3		172
Body Mass Index		24.92 (4.61)	176
Sleeping hours		7.42 (1.15)	161
Number of transplantations		1.07 (0.3)	178
Latency between disease diagnostic and transplantation (*in years*)		2.61 (4.41)	178
Myeloablative conditioning	25.8		
Chronic GvHD	16.5		178
Donor type			164
Identical sibling	25.7		179
Mismatched unrelated	8.9		
Mismatched relative	12.8		
Matched unrelated	38		
Unrelated	14		
Matched other relative	0.6		
Dependant Medical Variables			
Latency engrafment (*in days*)		20.24 (6.95)	161
Acute GvHD	51.5		171
Relapse	14.8		162
Number of infections		2.14 (1.8)	170
Death	16.4		177

Note: *n* = number of observations; *SD* = Standard Deviation.

Perceived stress in our sample (*M* = 14.25, *SD* = 6.8) was moderate because it ranged between 14 and 26 on scale (The mean for PTSD symptomatology in our sample was 31.26 (*SD* = 11.86), with 16.3% of patients meeting the criteria for PTSD (i.e., a score above or equal to 44) and 32.6% meeting the criteria for psychological distress regardless of their post-traumatic condition (i.e., score above 34). In order to identify how our sample compared to the norm (through a representative sample), a t-test was used to compare the quality of life score of our sample at time 2 with an average score of the general population in France (n = 2743) obtained from an SF-12 validation study. Results showed that the score for quality of life was significantly lower in the sample than the average score of the general French population [42] (mental QoL: M = 45.60 instead of 51.2, SD = 8.47, t = -3.15, $p < 0.01$, M_{diff} = -2.79; physical QoL: M = 41.04 instead of 48.4, SD = 8.52, t = -11.38, $p < 0.001$, M_{diff} = -10.16). In the sample, 59.3% had a score below the mean of the general population for the mental component of QoL, whereas 87.9% had a level of physical QoL below the mean of a non-clinical sample. Among the patients, 19.1% had a symptomatology of depression (i.e., a score above 7; 21.1% of women and 17.6% of men) and 37.1% of patients (44.7% of women and 31.4% of men) had a score above the threshold level for clinical anxiety (i.e., a score above 7).

3.2. Relationships between Perceived Stress, Perceived Social Support, Perceived Control, and Adjustment Coping Strategies at t1

We used the correlation analyses between all measures at t1 to verify accordance with the transactional model of stress in the case of HSCT. Pearson correlations demonstrated a significant relationship between perceived social support and all five coping strategies (see Table 2). Greater social support was associated with a higher fighting spirit and with lower maladaptive strategies. Anxious preoccupations, hopelessness-helplessness, and a low fighting spirit were negatively and significantly related to perceived stress but not avoidance and denial. Finally, while only fighting spirit was positively related to internality, externality was positively associated with all four maladaptive strategies (i.e., hopelessness-helplessness, avoidance, denial, and anxious preoccupations).

Table 2. Pearson's Correlation Coefficients Between Perceived Stress, Perceived Social Support, Perceived Control and Adjustment Coping Strategies.

	SPS-R	PSS	IPC Internality	IPC Externality	FS	H/H	A	A/P	D
Social Provisions Scale-Revised (SPS-R)	-	−0.22 **	0.14	−0.19 *	−0.30 ***	−0.41 ***	−0.19 *	−0.28 ***	−0.23 **
Perceived Stress Scale (PSS)		-	0.01	0.34 ***	−0.46 ***	0.57 ***	0.04	0.60 ***	0.08
IPC internality			-	0.23 **	0.24 **	0.01	0.08	0.08	−0.02
IPC externality				-	−0.15	0.43 ***	0.29 ***	0.47 ***	0.27 ***
Fighting Spirit (FS)					-	−0.47 ***	0.10	−0.29 ***	0.04
Hopelessness/ Helplessness (H/H)						-	0.14	0.61 ***	0.21 *
Avoidance (A)							-	0.32 ***	0.46 ***
Anxious Preoccupations (A/P)								-	0.29 ***

Note: *** $p < 0.001$; ** $p < 0.01$; * $p < 0.05$.

3.3. The Prospective Effects of Coping Strategies on Quality of Life, Mental Health, PTSD, and Resilience at Five Months after Transplantation

Table 3 presents the Pearson correlations between coping strategies and health outcomes (see the column zero-ordered effects; i.e., "Z-O"). To examine the prospective effect of coping strategies, a series of multiple regression analyses was performed (see Table 3; columns "Adjusted β").

Table 3. Prospective Effects of Coping Strategies (at t1) on Mental Quality of Life, Mental Health, Post-Traumatic Stress Disorder (PTSD) and Resilience at Five Months Post-Transplantation (at t2).

	Fighting Spirit		Hopelessness/Helplessness		Avoidance		Anxious Preoccupations		Denial	
	Z-O	Adjusted β	Z-O	Adjusted β	Z-O	Adjusted β	Z-O	Adjusted β	Z-O	Adjusted β
Health Outcomes:										
Mental QoL [a]	0.20	-	−0.35 **	0.01	−0.15	-	−0.52 ***	−0.42 **	−0.14	-
Anxiety [b]	−0.27 *	−0.01	0.46 ***	−0.04	−0.02	-	0.62 ***	0.34 *	0.16	-
Depression [c]	−0.30 *	0.10	0.34 **	0.10	−0.06	-	0.35 **	0.07	0.11	-
Happiness [d]	0.53 ***	0.19 +	−0.45 ***	0.04	0.05	-	−0.47 ***	−0.17	0.07	-
PTSD [a]	−0.06	-	0.30 *	−0.25 +	0.08	-	0.59 ***	0.72 ***	0.20	-
Resilience [a]	0.40 ***	0.33 *	−0.36 **	−0.16	−0.08	-	−0.34 **	−0.05	−0.22	-

Note: The column Z-O depicts the zero-ordered effects of the variable, with other variables not included in the model. [a] Adjusted for Mental QoL at Time 0; [b] Adjusted for anxiety at Time 0 and occupation, [c] Adjusted for depression at Time 0; [d] Adjusted for happiness at Time 0 and sex. *** $p < 0.001$; ** $p < 0.01$; * $p < 0.05$; + $p < 0.10$.

Concerning mental QoL (as a DV), adjusting for mental QoL at Time 0 (i.e., prior to hospitalization), only the anxious preoccupations coping strategy ($\beta = -0.42$, $p < 0.01$) significantly and negatively predicted QoL at Time 2 (at 5-month follow-up).

Concerning anxiety as a DV, while controlling for anxiety and the socio-professional category at Time 0, the analyses showed that the anxious preoccupations coping strategy ($\beta = 0.34$, $p < 0.05$) was the only robust predictor of anxiety symptomatology at the follow-up.

As shown in Table 3, depression was not significantly robustly predicted by any adjustment strategies. The results were similar with happiness. Adjusting for happiness at Time 0 and for sex, happiness at Time 2 was only marginally predicted in the expected direction by a fighting spirit ($\beta = 0.19$, $p < 0.10$).

Because both resilience and PTSD were not assessed at Time 0, we adjusted them for mental QoL at Time 0. Resilience was only significantly predicted by a fighting spirit ($\beta = 0.33$, $p < 0.05$). Finally, concerning PTSD symptomatology, the anxious preoccupations coping strategy ($\beta = 0.72$, $p < 0.001$) was predictive of a higher level of PTSD at five months post-transplantation. Transforming the PTSD scores into a binary variable (PTSD: 0 = no; 1 = yes), and adjusting for mental QoL at Time 0, the anxious preoccupations coping style was associated with an increased risk of developing PTSD five months after HSCT (OR = 3.27, $p < 0.01$; 95% CI: 1.36, 7.84).

Finally, the prospective effects of perceived stress, perceived control, and perceived social support on QoL, PTSD, depressive and anxious symptomatology, happiness, and resilience were also explored. A series of regression analyses revealed that adjusting for mental QoL at Time 0, only perceived stress ($\beta = 0.31$, $p < 0.05$) significantly predicted PTSD symptomatology at five months. However, this effect disappeared when the anxious preoccupations coping strategy was included in the model ($\beta = -0.08$, $p = 0.62$).

3.4. Relationships between Coping Strategies and Medical Outcomes

We proceeded to a bivariate correlation analysis to explore the relationships between coping strategies and medical outcomes. We found that only the number of transplantations was negatively related to a fighting spirit ($r = -0.25$, $p < 0.01$) and positively related to hopelessness-helplessness ($r = 0.30$, $p < 0.001$). The higher the number of previous transplantations, the less frequently patients used the fighting spirit coping strategy, and the more often they used the hopelessness/helplessness strategy.

4. Discussion

The aim of this study was to test the prospective effects of coping strategies on QoL, PTSD, depressive and anxious symptomatology, happiness and, finally, resilience at five months after HSCT. There are few studies examining these relationships in cases of threatening diseases or treatments.

First, we examined the relationships between coping and perceived stress, control, and social support. Having good social support, low perceived stress, and a low external locus of control were related to less maladaptive coping strategies and more adaptive ones. Social support is well-recognized as a protective factor for adjustment in HSCT [32]. Despite the negative effect of perceived stress on the adjustment process [18,19] in the present study, both avoidance and denial were not related to the perceived stress scale. This result may suggest that such strategies protect in the short term against the perception of being stressed, even if the long-term adverse effects of these strategies are well-known [50]. In addition, our sample had a low to moderate amount of perceived stress. It may be that a certain level of perceived stress must be reached to trigger more denial or avoidance coping strategies. It would also be relevant to separately consider each adaptive/maladaptive coping strategy according to their function at a specific step or condition of the disease [51].

Finally, during HSCT, a lack of control was related to maladaptive coping strategies. The effects of internality and externality is not clear in the literature. Perceived internal control is linked to a fighting spirit [23], but internal control can increase the patient's feeling of responsibility for the disease process and thus be deleterious in terms of health [23,52]. Interestingly, the higher the number of transplantations, the more frequently patients used the hopelessness-helplessness strategy and the less often they used the fighting spirit coping strategy. The fact that some patients had already been through this event seemed to increase this maladaptive coping, which highlights that multiple hospitalizations for HSCT represent a challenge for the quality of adaptation.

The results concerning the prospective effects of coping strategies reveal several interesting findings. First, avoidance strategy was not related to any of the mental health outcomes. It is not consistent with other studies highlighting the relationship between avoidance coping and psychological distress [29,32] or PTSD [33].

To our knowledge, no study has put forward the predictive effect of coping on QoL five months post-transplantation. The main factors studied, and involved in impairment of QoL, after auto- and allo-HSCT are medical (i.e., GvHD, regimen conditioning); socio-demographic (being young and female); environmental (lack of social support); and psychological (pre-existing psychological distress) [53]. Anxiety preoccupation was the only strategy we found to be significantly and negatively related to QoL at follow-up. Targeting factors such as perceived stress or anxiety, which are related to this maladaptive strategy, would reduce the deleterious effects of these coping strategies on QoL several months after hospitalization.

Interestingly, a fighting spirit was only related to resilience, a factor characterizing a positive recovery. For example, the degree of resilience seems to be a relevant factor that distinguishes patients who coped better with post-HSCT. Indeed, patients with high levels of resilience reported higher levels of QoL and lower anxiety and depression symptoms [54]. These results on the effect of a fighting spirit suggest that other factors potentially involved in resilience should be considered and that the focus should not only be on maladaptive/passive/disengagement coping strategies but also on adaptive/active/engaged strategies to promote recovery and improve well-being [54,55]. In this regard, acceptance coping, a strategy not explored in this research, appears to be predictive of positive outcomes in the case of incurable cancer [56], whereas a lack of acceptance predicts poorer long-term psychological adjustment to breast cancer [57]. In a meta-analysis [51], the results of prospective studies about coping in breast cancer indicated that secondary-control coping (i.e., positive reappraisal, acceptance, and fighting spirit), aimed at facilitating the adaptation to stress without modifying the stressor or the related emotions, were associated with more positive effects and less negative ones. On the other hand, primary-control coping strategies (i.e., planning, social-support seeking, direct action), focusing on the modification of the stressor or related emotions, were associated with more positive effects but not with negative ones. As indicated by Cousson-Gélie (2019) [58] and Folkman (2008) [59], such results suggest a focus on positive and engagement coping strategies to promote positive effects and reduce negative ones.

Surprisingly, depression was predicted by any of the coping strategies, which is congruent with the results obtained by Barata et al. (2018) [29] showing that relationships between depression and coping disappear after a few weeks and that depression prior to hospitalization explains part of the symptomatology of depression after a few months. However, PTSD and anxiety are predicted by anxious preoccupations, which is not surprising given the high level of stress, and that maladaptive coping strategies tend to elicit a PTSD symptomatology [60]. Hopelessness-helplessness did not predict any of the psychological outcomes. However, such a coping strategy is related to a lack of control and self-efficacy, and a less optimistic view of the future prevents has a negative impact on the capacity for resilience, i.e., the ability to restore psychological functioning despite the stressful events, and this engenders higher levels of stress [54].

4.1. Study Limitations

The loss of part of our sample due to death and decreased desire to continue participating in this study between Time 1 (i.e., one week during transplantation) and Time 2 (i.e., five months after transplantation) necessarily limit the results of this study. Despite the heterogeneity of the medical centers selected to conduct this study, the study is not representative of the different bone marrow transplantation centers in France. For reasons of anonymity, it was not possible to know which center the patients came from. The results of the regression analyses may therefore be biased because our observations were not independent, which is the first assumption of a regression analysis. This type of study must also be replicated by combining psychological and medical variables, especially immunological variables (e.g., natural killer cells, lymphocytes, cytokines), which are rare in the case of HSCT [61]. For example, some studies reveal promising results concerning the effect of emotion-regulating coping on inflammatory biomarkers in the case of prostate cancer [62]. In addition, the mediation effect of coping between dispositional factors (e.g., dispositional mindfulness, acceptance, experiential avoidance, or optimism) and adjustment outcomes should be explored in further studies, given that some dispositions, such as experiential avoidance [63] or self-efficacy [58], seem to influence the process of coping and hence the subsequent psychological and physical health outcomes. Patients' adjustment strategies were assessed at the precise moment of hospitalization and not several days or weeks after, which prevented an understanding of the effect of extended hospitalization and isolation. Change scores on coping strategies assessed at two different times would also be relevant for the testing of our prospective effects [1]. Only a fighting spirit, as an adaptive coping strategy, was explored in this study. However other secondary-control coping strategies identified by Kvillemo and Bränström (2014) [51], such as acceptance, should be explored as predictors of psychological health outcomes [24]. Such research would lead to the identification of the most appropriate engagement coping strategy to promote recovery and also target an appropriate intervention after HSCT, particularly with regard to the symptomatology of PTSD, which was particularly prevalent even five months after transplantation.

4.2. Clinical Implications

A focus on the transactional factors involved in HSCT recovery would allow for the introduction of relevant psychological preventive interventions. First, such results suggest that people should be provided with positive personal resources and adaptive coping strategies before hospitalization. This type of care appears relevant in that it will help patients better face this challenging hospitalization period and the next steps of the allograft, and help to improve psychological recovery after HSCT [64–67].

5. Conclusions

There are psychological considerations specific to HSCT, especially with regard to protective isolation and its consequences [68]. Exploring the factors which contribute to better mental and physical health at each step of the process, and their relationships, is fundamental in order to identify which interventions can be implemented in an effective way. However, the role of psychological factors and interventions on the physiological markers of HSCT success or fail rates are under-explored.

Author Contributions: M.C. and M.D. developed the questionnaire and the general protocol, collected, and treated the data, and wrote the paper. I.B. contributed to the compliance with ethical standards through ethics considerations management during the protocol time. A.C. helped in the compliance with ethical standards and collected questionnaires in Estaing Hospital. M.-T.R. served as the coordinator for the collection of questionnaires in Nancy-Brabois Hospital. J.-O.B. served as the coordinator for the collection of questionnaires in Estaing Hospital. R.P.d.L.T. contributed to the protocol and the questionnaire development and served as coordinator-in-chief at Saint-Louis Hospital. All authors have read and agreed to the published version of the manuscript.

Funding: This research received no external funding.

Institutional Review Board Statement: The study was conducted in accordance with the Declaration of Helsinki, and approved by the Institutional Review Board (or Ethics Committee) of Ethical Committee Sud-Est III (protocol code IRB 2017-026 B).

Informed Consent Statement: Informed consent was obtained from all patients included in the study.

Data Availability Statement: Dataset of this study is available on: https://doi.org/10.6084/m9.figshare.23266445 (accessed on 30 May 2023).

Acknowledgments: We would like to thank all the nursing staff and the clinical investigators (Frédérique Thomas Lallement and Tiana Andriamasy) of Saint-Louis Hospital, Nancy-Brabois Hospital and Estaing Hospital. We also thank promotor AGRAH and, finally, all the patients of the EGMOS association.

Conflicts of Interest: The authors declare no conflict of interest.

References

1. El-Jawahri, A.R.; Bs, H.B.V.; Traeger, L.N.; Ba, J.N.F.; Keenan, T.; Gallagher, E.R.; Greer, J.A.; Pirl, W.F.; Jackson, V.A.; Spitzer, T.R.; et al. Quality of life and mood predict posttraumatic stress disorder after hematopoietic stem cell transplantation. *Cancer* **2016**, *122*, 806–812. [CrossRef]
2. Kisch, A.; Lenhoff, S.; Zdravkovic, S.; Bolmsjö, I. Factors associated with changes in quality of life in patients undergoing allogeneic haematopoietic stem cell transplantation. *Eur. J. Cancer Care* **2012**, *21*, 735–746. [CrossRef]
3. Sun, C.-L.; Kersey, J.H.; Francisco, L.; Armenian, S.H.; Baker, K.S.; Weisdorf, D.J.; Forman, S.J.; Bhatia, S. Burden of morbidity in 10+ year survivors of hematopoietic cell transplantation: Report from the bone marrow transplantation survivor study. *Biol. Blood Marrow Transpl.* **2013**, *19*, 1073–1080. [CrossRef] [PubMed]
4. Pillay, B.; Lee, S.J.; Katona, L.; De Bono, S.; Burney, S.; Avery, S. A prospective study of the relationship between sense of coherence, depression, anxiety, and quality of life of haematopoietic stem cell transplant patients over time. *Psycho-Oncology* **2014**, *24*, 220–227. [CrossRef] [PubMed]
5. Artherholt, S.B.; Hong, F.; Berry, D.L.; Fann, J.R. Risk factors for depression in patients undergoing hematopoietic cell transplantation. *Biol. Blood Marrow Transpl.* **2014**, *20*, 946–950. [CrossRef] [PubMed]
6. Hefner, J.; Kapp, M.; Drebinger, K.; Dannenmann, A.; Einsele, H.; Grigoleit, G.-U.; Faller, H.; Csef, H.; Mielke, S. High prevalence of distress in patients after allogeneic hematopoietic SCT: Fear of progression is associated with a younger age. *Bone Marrow Transpl.* **2014**, *49*, 581–584. [CrossRef] [PubMed]
7. Sarkar, S.; Scherwath, A.; Schirmer, L.; SchulzKindermann, F.; Neumann, K.; Kruse, M.; Dinkel, A.; Kunze, S.; Balck, F.; Kroger, N.; et al. Fear of recurrence and its impact on quality of life in patients with hematological cancers in the course of allogeneic hematopoietic SCT. *Bone Marrow Transpl.* **2014**, *49*, 1217–1222. [CrossRef] [PubMed]
8. Biagioli, V.; Piredda, M.; Mauroni, M.R.; Alvaro, R.; De Marinis, M.G. The lived experience of patients in protective isolation during their hospital stay for allogeneic haematopoietic stem cell transplantation. *Eur. J. Oncol. Nurs.* **2016**, *24*, 79–86. [CrossRef] [PubMed]
9. Biagioli, V.; Piredda, M.; Alvaro, R.; de Marinis, M. The experiences of protective isolation in patients undergoing bone marrow or haematopoietic stem cell transplantation: Systematic review and metasynthesis. *Eur. J. Cancer Care* **2016**, *26*, e12461. [CrossRef]

10. Biagioli, V.; Piredda, M.; Annibali, O.; Tirindelli, M.C.; Pignatelli, A.; Marchesi, F.; Mauroni, M.R.; Soave, S.; Del Giudice, E.; Ponticelli, E.; et al. Factors influencing the perception of protective isolation in patients undergoing haematopoietic stem cell transplantation: A multicentre prospective study. *Eur. J. Cancer Care* **2019**, *28*, e13148. [CrossRef]

11. Prieto, J.M.; Atala, J.; Blanch, J.; Carreras, E.; Rovira, M.; Cirera, E.; Espinal, A.; Gasto, C. Role of depression as a predictor of mortality among cancer patients after stem-cell transplantation. *J. Clin. Oncol.* **2005**, *23*, 6063–6071. [CrossRef] [PubMed]

12. Baliousis, M.; Rennoldson, M.; Dawson, D.; Mills, J.; das Nair, R. Perceptions of hematopoietic stem cell transplantation and coping predict emotional distress during the acute phase after transplantation. *Oncol. Nurs. Forum* **2017**, *44*, 96–107. [CrossRef] [PubMed]

13. Tecchio, C.; Bonetto, C.; Bertani, M.; Cristofalo, D.; Lasalvia, A.; Nichele, I.; Bonani, A.; Andreini, A.; Benedetti, F.; Ruggeri, M.; et al. Predictors of anxiety and depression in hematopoietic stem cell transplant patients during protective isolation. *Psycho-Oncology* **2013**, *22*, 1790–1797. [CrossRef]

14. Lazarus, R.F.; Folkman, S.C.S. *Stress, Appraisal and Coping*; Springer Publishing Company: New York, NY, USA, 1984.

15. Efficace, F.; Breccia, M.; Cottone, F.; Okumura, I.; Doro, M.; Riccardi, F.; Rosti, G.; Baccarani, M. Psychological well-being and social support in chronic myeloid leukemia patients receiving lifelong targeted therapies. *Support. Care Cancer* **2016**, *24*, 4887–4894. [CrossRef]

16. Paterson, C.; Jones, M.; Rattray, J.; Lauder, W. Exploring the relationship between coping, social support and health-related quality of life for prostate cancer survivors: A review of the literature. *Eur. J. Oncol. Nurs.* **2013**, *17*, 750–759. [CrossRef]

17. Bárez, M.; Blasco, T.; Fernández-Castro, J.; Viladrich, C. Perceived control and psychological distress in women with breast cancer: A longitudinal study. *J. Behav. Med.* **2008**, *32*, 187–196. [CrossRef]

18. Groarke, A.; Curtis, R.; Kerin, M. Global stress predicts both positive and negative emotional adjustment at diagnosis and post-surgery in women with breast cancer. *Psycho-Oncology* **2013**, *22*, 177–185. [CrossRef]

19. Li, Y.; Yang, Y.; Zhang, R.; Yao, K.; Liu, Z. The mediating role of mental adjustment in the relationship between perceived stress and depressive symptoms in hematological cancer patients: A cross-sectional study. *PLoS ONE* **2015**, *10*, e0142913. [CrossRef] [PubMed]

20. Hulbert-Williams, N.; Storey, L.; Wilson, K. Psychological interventions for patients with cancer: Psychological flexibility and the potential utility of Acceptance and Commitment Therapy. *Eur. J. Cancer Care* **2015**, *24*, 15–27. [CrossRef]

21. Moritz, S.; Jahns, A.K.; Schröder, J.; Berger, T.; Lincoln, T.M.; Klein, J.P.; Göritz, A.S. More adaptive versus less maladaptive coping: What is more predictive of symptom severity? Development of a new scale to investigate coping profiles across different psychopathological syndromes. *J. Affect. Disord.* **2015**, *191*, 300–307. [CrossRef]

22. Thompson, R.J.; Mata, J.; Jaeggi, S.M.; Buschkuehl, M.; Jonides, J.; Gotlib, I.H. Maladaptive coping, adaptive coping, and depressive symptoms: Variations across age and depressive state. *Behav. Res. Ther.* **2010**, *48*, 459–466. [CrossRef] [PubMed]

23. Hodges, K.; Winstanley, S. Effects of optimism, social support, fighting spirit, cancer worry and internal health locus of control on positive affect in cancer survivors: A path analysis. *Stress Health* **2012**, *28*, 408–415. [CrossRef] [PubMed]

24. Manne, S.; Myers-Virtue, S.; Kashy, D.; Ozga, M.; Kissane, D.; Heckman, C.; Rubin, S.C.; Rosenblum, N. Resilience, positive coping, and quality of life among women newly diagnosed with gynecological cancers. *Cancer Nurs.* **2015**, *38*, 375. [CrossRef]

25. Brunault, P.; Champagne, A.-L.; Huguet, G.; Suzanne, I.; Senon, J.-L.; Body, G.; Rusch, E.; Magnin, G.; Voyer, M.; Réveillère, C.; et al. Major depressive disorder, personality disorders, and coping strategies are independent risk factors for lower quality of life in non-metastatic breast cancer patients. *Psycho-Oncology* **2016**, *25*, 513–520. [CrossRef] [PubMed]

26. Allart, P.; Soubeyran, P.; Cousson-Gélie, F. Are psychosocial factors associated with quality of life in patients with haematological cancer? A critical review of the literature. *Psycho-Oncology* **2012**, *22*, 241–249. [CrossRef] [PubMed]

27. Montgomery, C.; Pocock, M.; Titley, K.; Lloyd, K. Predicting psychological distress in patients with leukaemia and lymphoma. *J. Psychosom. Res.* **2003**, *54*, 289–292. [CrossRef] [PubMed]

28. O'connor, M.; Guilfoyle, A.; Breen, L.; Mukhardt, F.; Fisher, C. Relationships between quality of life, spiritual well-being, and psychological adjustment styles for people living with leukaemia: An exploratory study. *Ment. Health Relig. Cult.* **2007**, *10*, 631–647. [CrossRef]

29. Barata, A.; Gonzalez, B.D.; Sutton, S.K.; Small, B.J.; Jacobsen, P.B.; Field, T.; Fernandez, H.; Jim, H.S. Coping strategies modify risk of depression associated with hematopoietic cell transplant symptomatology. *J. Health Psychol.* **2016**, *23*, 1028–1037. [CrossRef]

30. Fife, B.L.; Huster, G.A.; Cornetta, K.G.; Kennedy, V.N.; Akard, L.P.; Broun, E.R. Longitudinal study of adaptation to the stress of bone marrow transplantation. *J. Clin. Oncol.* **2000**, *18*, 1539–1549. [CrossRef]

31. Rodrigue, J.R.; Boggs, S.R.; Weiner, R.S.; Behen, J.M. Mood, coping style, and personality functioning among adult bone marrow transplant candidates. *Psychosomatics* **1993**, *34*, 159–165. [CrossRef]

32. Wells, K.J.; Booth-Jones, M.; Jacobsen, P.B. Do coping and social support predict depression and anxiety in patients undergoing hematopoietic stem cell transplantation? *J. Psychosoc. Oncol.* **2009**, *27*, 297–315. [CrossRef] [PubMed]

33. Jacobsen, P.B.; Sadler, I.J.; Booth-Jones, M.; Soety, E.; Weitzner, M.A.; Fields, K.K. Predictors of posttraumatic stress disorder symptomatology following bone marrow transplantation for cancer. *J. Consult. Clin. Psychol.* **2002**, *70*, 235. [CrossRef]

34. Cayrou, S.; Dickès, P.; Gauvain-Piquard, A.; Rogé, B. The mental adjustment to cancer (MAC) scale: French replication and assessment of positive and negative adjustment dimensions. *Psycho-Oncol. J. Psychol. Soc. Behav. Dimens. Cancer* **2002**, *12*, 8–23. [CrossRef] [PubMed]

35. Watson, M.; Greer, S.; Young, J.; Inayat, Q.; Burgess, C.; Robertson, B. Development of a questionnaire measure of adjustment to cancer: The MAC scale. *Psychol. Med.* **1988**, *18*, 203–209. [CrossRef] [PubMed]

36. Caron, J. L'Échelle de provisions sociales: Une validation québécoise. *Santé Ment. Au Québec* **1996**, *21*, 158–180. [CrossRef]
37. Cutrona, C.E.; Russell, D.W. The provisions of social relationships and adaptation to stress. *Adv. Pers. Relatsh.* **1987**, *1*, 37–67.
38. Cohen, S.; Kamarck, T.; Mermelstein, R. A global measure of perceived stress. *J. Health Soc. Behav.* **1983**, *24*, 385–396. [CrossRef]
39. Bellinghausen, L.; Collange, J.; Botella, M.; Emery, J.L.; Albert, É. Validation factorielle de l'échelle française de stress perçu en milieu professionnel. *Santé Publique* **2009**, *21*, 365–373. [CrossRef]
40. Jutras, S. L'IPAH, version canadienne-française de l'Échelle de Levenson mesurant le lieu de contrôle tridimensionnel. *Can. J. Behav. Sci./Rev. Can. Des Sci. Du Comport.* **1987**, *19*, 74. [CrossRef]
41. Levenson, H. Multidimensional locus of control in psychiatric patients. *J. Consult. Clin. Psychol.* **1973**, *41*, 397–404. [CrossRef]
42. Gandek, B.; Ware, J.E.; Aaronson, N.K.; Apolone, G.; Bjorner, J.B.; Brazier, J.E.; Bullinger, M.; Kaasa, S.; Leplege, A.; Prieto, L.; et al. Cross-validation of item selection and scoring for the sf-12 health survey in nine countries: Results from the iqola project. International quality of life assessment. *J. Clin. Epidemiol.* **1998**, *51*, 1171–1178. [CrossRef] [PubMed]
43. Zigmond, A.S.; Snaith, R.P. The hospital anxiety and depression scale. *Acta Psychiatr. Scand.* **1983**, *67*, 361–370. [CrossRef]
44. Dambrun, M.; Ricard, M.; Després, G.; Drelon, E.; Gibelin, E.; Gibelin, M.; Loubeyre, M.; Py, D.; Delpy, A.; Garibbo, C.; et al. Measuring happiness: From fluctuating happiness to authentic–durable happiness. *Front. Psychol.* **2012**, *3*, 16. [CrossRef] [PubMed]
45. Weathers, F.W.; Litz, B.T.; Herman, D.S.; Huska, J.A.; Keane, T.M. *The PTSD Checklist (PCL): Reliability, Validity, and Diagnostic Utility*; Annual Convention of the International Society for Traumatic Stress Studies: San Antonio, TX, USA, 1993; Volume 462.
46. Yao, S.-N.; Cottraux, J.; Note, I.; De Mey-Guillard, C.; Mollard, E.; Ventureyra, V. Evaluation of Post-traumatic Stress Disorder: Validation of a measure, the PCLS. *L'Encephale* **2003**, *29 Pt 1*, 232–238.
47. Connor, K.M.; Davidson, J.R.T. Development of a new resilience scale: The Connor-Davidson Resilience Scale (CD-RISC). *Depress. Anxiety* **2003**, *18*, 76–82. [CrossRef]
48. Guihard, G.; Morice-Ramat, A.; Deumier, L.; Goronflot, L.; Alliot-Licht, B.; Bouton-Kelly, L. Évaluer la résilience des étudiants en santé en France: Adaptation et mesure de l'invariance de l'échelle CD-RISC 10. *Mes. Eval. En Educ.* **2018**, *41*, 67–96. [CrossRef]
49. Krueger, J.I.; Heck, P.R. Putting the *P*-Value in its Place. *Am. Stat.* **2019**, *73*, 122–128. [CrossRef]
50. Zhou, E.S.; Penedo, F.J.; Bustillo, N.E.; Benedict, C.; Rasheed, M.; Lechner, S.; Soloway, M.; Kava, B.R.; Schneiderman, N.; Antoni, M.H. Longitudinal effects of social support and adaptive coping on the emotional well-being of survivors of localized prostate cancer. *J. Support. Oncol.* **2010**, *8*, 196–201. [CrossRef]
51. Kvillemo, P.; Bränström, R. Coping with breast cancer: A meta-analysis. *PLoS ONE* **2014**, *9*, e112733. [CrossRef]
52. Fang, C.Y.; Daly, M.B.; Miller, S.M.; Zerr, T.; Malick, J.; Engstrom, P. Coping with ovarian cancer risk: The moderating effects of perceived control on coping and adjustment. *Br. J. Health Psychol.* **2006**, *11*, 561–580. [CrossRef]
53. Braamse, A.M.J.; Gerrits, M.M.J.G.; van Meijel, B.; Visser, O.; van Oppen, P.; Boenink, A.D.; Cuijpers, P.; Huijgens, P.C.; Beekman, A.T.F.; Dekker, J. Predictors of health-related quality of life in patients treated with auto- and allo-SCT for hematological malignancies. *Bone Marrow Transpl.* **2012**, *47*, 757–769. [CrossRef] [PubMed]
54. Schumacher, A.; Sauerland, C.; Silling, G.; Berdel, W.E.; Stelljes, M. Resilience in patients after allogeneic stem cell transplantation. *Support. Care Cancer* **2014**, *22*, 487–493. [CrossRef] [PubMed]
55. Amonoo, H.L.; Brown, L.A.; Scheu, C.F.; Millstein, R.A.; Pirl, W.F.; Vitagliano, H.L.; Antin, J.H.; Huffman, J.C. Positive psychological experiences in allogeneic hematopoietic stem cell transplantation. *Psycho-Oncology* **2019**, *28*, 1633–1639. [CrossRef] [PubMed]
56. Nipp, R.D.; El-Jawahri, A.; Fishbein, J.N.; Eusebio, J.; Stagl, J.M.; Gallagher, E.R.; Park, E.R.; Jackson, V.A.; Pirl, W.F.; Greer, J.A.; et al. The relationship between coping strategies, quality of life, and mood in patients with incurable cancer. *Cancer* **2016**, *122*, 2110–2116. [CrossRef] [PubMed]
57. Hack, T.F.; Degner, L.F. Coping responses following breast cancer diagnosis predict psychological adjustment three years later. *Psycho-Oncol. J. Psychol. Soc. Behav. Dimens. Cancer* **2003**, *13*, 235–247. [CrossRef]
58. Cousson-Gélie, F. How do patients cope with cancer? The latest advances in cancer coping strategies and the cognitive models that underlie them. *Rev. Neuropsychol.* **2019**, *11*, 289–293.
59. Folkman, S. The case for positive emotions in the stress process. *Anxiety Stress. Coping* **2008**, *21*, 3–14. [CrossRef]
60. Kangas, M.; Henry, J.L.; Bryant, R.A. Correlates of acute stress disorder in cancer patients. *J. Trauma. Stress Off. Publ. Int. Soc. Trauma. Stress Stud.* **2007**, *20*, 325–334. [CrossRef]
61. Kim, S.-D.; Kim, H.-S. Effects of a relaxation breathing exercise on anxiety, depression, and leukocyte in hemopoietic stem cell transplantation patients. *Cancer Nurs.* **2005**, *28*, 79–83. [CrossRef]
62. Hoyt, M.A.; Stanton, A.L.; Irwin, M.R.; Thomas, K.S. Cancer-related masculine threat, emotional approach coping, and physical functioning following treatment for prostate cancer. *Health Psychol.* **2013**, *32*, 66–74. [CrossRef]
63. Karekla, M.; Panayiotou, G. Coping and experiential avoidance: Unique or overlapping constructs? *J. Behav. Ther. Exp. Psychiatry* **2011**, *42*, 163–170. [CrossRef]
64. Casellas-Grau, A.; Font, A.; Vives, J. Positive psychology interventions in breast cancer. A systematic review. *Psycho-Oncology* **2014**, *23*, 9–19. [CrossRef] [PubMed]
65. Feros, D.L.; Lane, L.; Ciarrochi, J.; Blackledge, J.T. Acceptance and Commitment Therapy (ACT) for improving the lives of cancer patients: A preliminary study. *Psycho-Oncology* **2013**, *22*, 459–464. [CrossRef] [PubMed]
66. Grossman, P.; Zwahlen, D.; Halter, J.P.; Passweg, J.R.; Steiner, C.; Kiss, A. A mindfulness-based program for improving quality of life among hematopoietic stem cell transplantation survivors: Feasibility and preliminary findings. *Support. Care Cancer* **2015**, *23*, 1105–1112. [CrossRef] [PubMed]

67. Ochoa, C.; Casellas-Grau, A.; Vives, J.; Font, A.; Borràs, J.-M. Positive psychotherapy for distressed cancer survivors: Posttraumatic growth facilitation reduces posttraumatic stress. *Int. J. Clin. Health Psychol.* **2016**, *17*, 28–37. [CrossRef]
68. Amonoo, H.L.; Massey, C.N.; Freedman, M.E.; El-Jawahri, A.; Vitagliano, H.L.; Pirl, W.F.; Huffman, J.C. Psychological considerations in hematopoietic stem cell transplantation. *Psychosomatics* **2019**, *60*, 331–342. [CrossRef]

 *healthcare*

Article

Dynamically Changing Mental Stress Parameters of First-Year Medical Students over the Three-Year Course of the COVID-19 Pandemic: A Repeated Cross-Sectional Study

Morris Gellisch [1,*], Martin Bablok [1], Gabriela Morosan-Puopolo [1], Thorsten Schäfer [2] and Beate Brand-Saberi [1]

[1] Department of Anatomy and Molecular Embryology, Institute of Anatomy, Medical Faculty, Ruhr University Bochum, 44801 Bochum, Germany; martin.bablok@rub.de (M.B.); gabriela.morosan-puopolo@rub.de (G.M.-P.); beate.brand-saberi@rub.de (B.B.-S.)

[2] Center for Medical Education, Ruhr-University Bochum, 44801 Bochum, Germany; thorsten.schaefer@ruhr-uni-bochum.de

* Correspondence: morris.gellisch@rub.de

Abstract: Numerous research results have already pointed towards the negative influence of increased mental stress on educational processes and motivational criteria. It has also been shown that the global public health crisis induced by COVID-19 was related to anxiety symptoms and elevated levels of distress. To holistically elucidate the dynamics of the pandemic-related mental stress of first-year medical students, the associated parameters of three different cohorts were measured at the beginning of the pandemic-related restrictions on university life in Germany (20/21), at the peak of the COVID-19-related restrictions (21/22) and during the easing of the restrictions in the winter term 22/23. In a repeated cross-sectional study design, the constructs of worries, tension, demands and joy were collected from first-year medical students ($n = 578$) using the Perceived Stress Questionnaire. The results demonstrate significantly increased values of the constructs worries ($p < 0.001$), tension ($p < 0.001$) and demands ($p < 0.001$) at the peak of the pandemic related restrictions compared to the previous and following year as well as significantly decreasing values of general joy of life during the observed period of 3 years (all p-values < 0.001). A confirmatory factor analysis was performed to verify the questionnaire's factor structure regarding the addressed target group during the pandemic (CFI: 0.908, RMSEA: 0.071, SRMR: 0.052). These data, collected over a period of three years, provide information regarding dynamically manifesting mental stress during the COVID-19 pandemic, and refer to new areas of responsibility for the faculties to adequately counteract future crisis situations.

Keywords: stress; mental health; education; COVID-19; risk management; vulnerable populations

Citation: Gellisch, M.; Bablok, M.; Morosan-Puopolo, G.; Schäfer, T.; Brand-Saberi, B. Dynamically Changing Mental Stress Parameters of First-Year Medical Students over the Three-Year Course of the COVID-19 Pandemic: A Repeated Cross-Sectional Study. *Healthcare* **2023**, *11*, 1558. https://doi.org/10.3390/healthcare11111558

Academic Editors: Carlos Laranjeira and Ana Querido

Received: 7 April 2023
Revised: 22 May 2023
Accepted: 23 May 2023
Published: 26 May 2023

1. Introduction

At the turn of the year 2019/2020, the WHO Country Office in China received information regarding increased incidents of a novel viral pneumonia in the city of Wuhan, which, in retrospect, was the first signal of a developing pandemic [1]. In addition to global financial losses and drastic restrictions on leisure activities, 6.9 million reported deaths from COVID-19 infections occurred by mid-2022 (Institute for Health Metrics and Evaluation) [2]. The long-COVID syndrome appeared as an additional burdensome phenomenon, which describes long-term consequences of the virus infection, which can manifest themselves neurologically, psychologically and physiologically [3–5]. Emerging economic difficulties, the unpredictable nature of the COVID-19 pandemic and its consequences increase the risk of developing mental health problems, such as symptoms of depression and anxiety [6].

These serious consequences of the COVID-19 pandemic as well as the necessary regulations to limit in-person contact also posed an enormous challenge for educational processes in the relevant institutions [7–10]. In addition to self-reported declines in mental health and motivational factors, more objective changes such as different cortisol concentrations and altered sympathetic and parasympathetic activation patterns during remote learning compared to face-to-face learning could be demonstrated among German first-year medical students [11,12]. Mental stress among students, diffuse anxiety and difficulties in maintaining concentration during digital courses are widely reported burdens during the ongoing COVID-19 pandemic [13–17].

The extent to which these declines in mental health affect students' learning processes can be discussed against the background of numerous research findings that have demonstrated a negative connection between depressive symptoms and academic performance, an effect that has been observed internationally and across different disciplines [18–21]. These findings could also be validated from the opposite direction by empirically demonstrating that students with higher mental health status show increased motivation in terms of academic achievement and learning success [22]. The underlying mechanisms rely on the detrimental effect of mental health symptoms on certain predictors of academic achievement, such as academic self-efficacy, persistence, and study skills [23]. These mechanisms are again of substantial importance to the particularly vulnerable group of medical students, since numerous studies have indicated that medical students have an increased risk of suffering from depressive symptoms compared to the general population [24–26]. These findings could be further substantiated by investigating a cohort of Portuguese medical students, who showed a significantly higher expression of anxiety symptoms compared to non-medical students [27]. It is imperative to consider these findings in the light of the theoretical concept that psychological capital such as hope, optimism and resilience are currently discussed as the main predictors of academic achievement as well as general well-being [28–31].

This insight becomes exceptionally critical in the context of crisis situations which—as in the case of the COVID-19 pandemic—reduce resilience factors due to social isolation, effectuating dramatic consequences for mental and physical health [32–37].

The emergence of the relevant construct stress follows the psychological concept described by Lazarus in 1966, which states that stress arises when the demands exceed the individual resources [38]. The effects of stress on cognitive processes must be considered in a differentiated manner, since a moderate arousal can exert positive effects on learning and memory processes, while persistent, chronic, and intense stress stimuli can cause both cognitive and health impairments [39–42].

While several studies have already examined subjectively perceived stress in medical students during the COVID-19 pandemic [43–45], there is a lack of repeated data collection comparing the perceived stress parameters at the beginning of the pandemic-related consequences on university teaching (20/21), at the peak of the pandemic-related consequences (21/22) and during the easing of the restrictions associated with COVID-19 (22/23), in order to shed light on the dynamic effects within the course of this crisis situation. This described comparison of three different cohorts of first-semester medical students during the different stages of the COVID-19 pandemic is the focus of the present study.

The basis of this study focuses on the question of the direct effects of the intensity of crisis situations on perceived stress of the particularly vulnerable population of first-year medical students. This study design is specifically aimed at illuminating study entry conditions of medical students in consideration of how the COVID-19 pandemic dynamics modulate stress-associated parameters. In addition, this project aims to validate the factor structure of the PSQ as a survey tool during an ongoing crisis situation.

2. Materials and Methods

2.1. Study Design

The interventions associated with COVID-19 to limit the risk of infection also had a considerably strong effect on the modalities of medical teaching in Germany. While the winter semester 20/21 at the Ruhr-University Bochum began—in parts—with regular face-to-face courses, the winter semester 21/22 started with a complete hybrid teaching design with a majority of distance teaching to consider the safety measures decided in

Germany to reduce the COVID-19 infection risk. After this peak of the restrictions on in-person contact, the measures were eased at the beginning of the winter semester 22/23 so that regular face-to-face classes could be resumed. To make use of a survey instrument that also includes a resilience factor (general joy of life) in addition to stress-associated scales, the Perceived Stress Questionnaire (PSQ) was consistently used in this study for data collection for three consecutive years during the ongoing COVID-19 pandemic [46,47]. Levenstein et al. 1993 initially developed the PSQ for clinical psychosomatic research with a focus on the prognostic ability regarding the development of stress-related disorders [46]. The questionnaire consists of three dimensions of the stress reaction and one dimension of a stressor, which were deliberately chosen generically, enabling clinical utilizations of the PSQ as well as assessments in healthy adults [47]. In principle, the PSQ was identified as a valid and comprehensive assessment tool for stress research [48,49], which has already been utilized for the evaluation of perceived stress levels of medical students [26]. The 20-item instrument, which is subdivided into the subscales worries, tension, demands and joy, was used to assess the influence of the intensity of the COVID-19 pandemic-related consequences on perceived life stress. Therefore, the above-mentioned questionnaire was made available for first-semester medical students at the beginning of the respective winter semesters 20/21, 21/22 and 22/23 for three consecutive years. The questionnaire was completed at the university in paper format.

To investigate the academic performance of the students, the results of the final exam Anatomy I were used, since this exam provides an appropriate comparison due to its standardized form. In addition to the years 2020, 2021 and 2022, the academic performance in 2019 was indicated as a reference in order to provide additional information regarding the academic performance before the outbreak of the COVID-19 pandemic. Here, it should be noted that comparability is only possible to a limited extent, as the final examinations in the first semester at the Faculty of Medicine of the Ruhr University Bochum in 2020 were conducted online with only limited examination supervision, which was accompanied by an unusually high pass rate. While a uniform online exam policy is recommended for medical teaching, in which one camera should record the screen of the respective student and another camera the room [50], these technical configurations are often limited, so that the exam supervision in the case described here was limited to one camera, showing the students from the front during the online exam.

However, the pass rates for the years 2021 and 2022 can be used as a reference since these were conducted—as usual—in the presence of regular supervision.

2.2. Inclusion Criteria

All participants had to be properly enrolled medical students at the Ruhr University Bochum at the time of data collection. The recruitment for the study described here was deliberately aimed at including all genders, students of all ages as well as students from an immigrant background.

2.3. Participants

Five hundred and seventy-eight properly enrolled first-semester medical students (177 males: mean age = 21.52 ± 3.34 years; 399 females: mean age = 20.33 ± 2.90 years; 2 diverse: mean age = 19.50 ± 0.50 years (mean $\pm$ SD)) were participants in this study (Table 1). The observed COVID-19 restrictions explain the lower sample size in the winter semesters 20/21 ($n = 126$) and 2021/2022 ($n = 116$) compared to the winter semester 22/23 ($n = 336$).

Recruitment, data collection as well as obtaining informed written consent took place at the Medical Faculty of the Ruhr-University Bochum, Germany. The study procedure was conducted in agreement with the Declaration of Helsinki and approved by the ethics committee of the Medical Faculty at the Ruhr University Bochum (20–7135) and the ethics committee of the Professional School of Education (EPSE-2022-005).

Table 1. Demographic characteristics.

	2020	2021	2022	Total
Number of total participants	126	116	336	578
Male, n (%)	34 (26.98)	36 (31.03)	107 (31.85)	177 (30.62)
Female, n (%)	92 (73.02)	80 (68.97)	227 (67.56)	399 (69.03)
Diverse *, n (%)	-	-	2 (0.60)	2 (0.35)
Age, mean (SD)	20.03 (2.60)	22.00 (3.18)	20.40 (3.10)	20.63 (3.09)

Note. SD = Standard Deviation, * here, the term diverse is used to refer to persons who do not conform to socially defined male or female gender norms.

2.4. Statistical Analysis

All statistical calculations were performed using R-statistical software. Factor descriptions were calculated, reporting the factor ratings, the mean value of each item of the factor, the standard deviation and the skewness (Table 2).

Table 2. Factor description 2020–2022.

Factors	Items (n)	2020				2021				2022			
		Rating	Mean	SD	Skewness	Rating	Mean	SD	Skewness	Rating	Mean	SD	Skewness
Worries	5	3.33	0.66	0.20	0.18	4.67	0.93	0.21	0.49	3.74	0.75	0.24	0.47
Tension	5	4.03	0.81	0.18	0.51	5.05	1.01	0.17	0.05	3.61	0.72	0.22	0.28
Joy	5	6.75	1.34	0.20	−1.38	5.93	1.19	0.18	−0.09	5.20	1.04	0.18	−0.04
Demands	5	5.05	1.01	0.19	0.15	6.11	1.22	0.18	−0.37	4.48	0.90	0.14	−0.18

Note. n = number of items per construct, Rating = overall factor ratings, Mean = mean value of each item of the factor, SD = Standard Deviation.

To validate the factor structure of the questionnaire against the background of this particularly challenging pandemic situation, a confirmatory factor analysis (CFA) was calculated, which was evaluated by determining the fit indices.

3. Results

The data collected in this study indicate a connection between the respective phase of the pandemic, including the dynamic intensity of the associated consequences and the severity of perceived life stress of first-year medical students. Analyses of variance (ANOVAs) were performed to examine the differences between the distinct points in time, always including the different years (20/21, 21/22, 22/23) as between subject factors (Table 3). Avoiding the error of multiple comparisons, Bonferroni–Holm-corrected p-values were reported. The 95% confidence intervals were calculated to ensure better insight into the nature of the data structure (Table 3).

Table 3. Analysis of variance.

Factors	ANOVA				Post Hoc Analyses		
					2020 × 2021	2020 × 2022	2021 × 2022
	df	F	p	η^2		p [95% CI—Lower and Upper Bound]	
Worries	575	11.13	<0.001	0.037	<0.001 [0.065, 0.203]	0.081 [−0.014, 0.097]	<0.001 [−0.150, −0.035]
Tension	575	21.55	<0.001	0.070	<0.001 [0.040, 0.163]	0.048 [−0.092, 0.001]	<0.001 [−0.195, −0.092]
Demands	575	45.32	<0.001	0.136	<0.001 [0.057, 0.154]	<0.001 [−0.096, −0.018]	<0.001 [−0.203, −0.123]
Joy	575	31.52	<0.001	0.099	<0.001 [−0.134, −0.021]	<0.001 [−0.198, −0.106]	<0.001 [−0.121, −0.026]

Note. ANOVA = Analysis of variance, df = degrees of freedom, F = F-value, p = p-value, η^2 = partial eta-squared, CI = Confidence intervals.

Here, we demonstrate a significant increase in the factors worries, tension and demands, as well as a significant and steady decrease in the construct of general joy of life in the winter semester 2021/2022, at the peak of the COVID-19-associated restrictions (Figure 1).

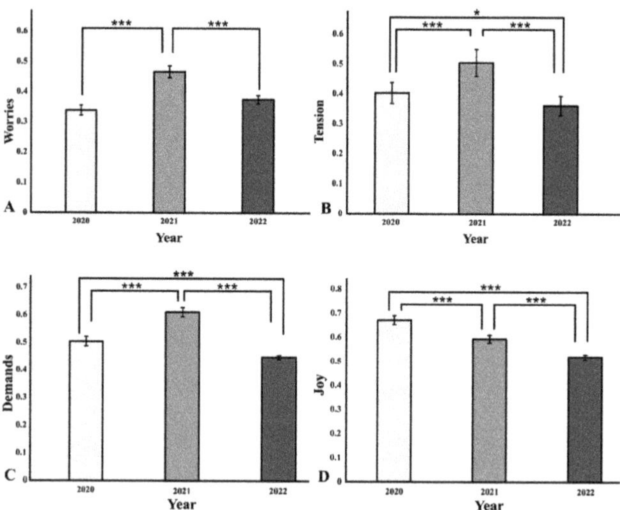

Figure 1. (**A–D**) Bar charts of perceived worries, tension, demands and joy among three different cohorts of first-year medical students over the period of three consecutive winter semesters (2020, 2021, 2022). Asterisks indicate: * = $p < 0.05$, *** = $p < 0.001$.

The examined factor worries differed significantly in the comparison of the three consecutive winter semesters (F(2, 575) = 11.13, $p < 0.001$, partial η^2 = 0.037). At the beginning of the winter semester 21/22, the factor of worries was significantly higher than 20/21 ($p < 0.001$) and 22/23 ($p < 0.001$), whereas no significant difference between the years 20/21 and 22/23 ($p = 0.08$) could be observed (Figure 1).

A significant difference for the factor of generally perceived tension was also found (F(2, 575) = 21.55, $p < 0.001$, partial η^2 = 0.070). The perceived tension at the beginning of the winter semester 21/22 was significantly higher than 20/21 ($p < 0.001$) and 22/23 ($p < 0.001$). Equally, the values for perceived tension in the winter semester 20/21 were significantly higher than at the beginning of the winter semester 22/23, albeit with a smaller effect size ($p = 0.048$) (Figure 1).

Similarly, significant differences were shown for the factor demands (F(2, 575) = 45.32, $p < 0.001$, partial η^2 = 0.136). At the beginning of the winter semester 21/22, the factor of demands was significantly higher than 20/21 ($p < 0.001$) and 22/23 ($p < 0.001$). It could also be demonstrated that the perceived demands in 20/21 were rated as significantly higher than in the winter semester 22/23 ($p < 0.001$) (Figure 1).

With regard to the factor joy, significant differences could also be identified over the course of the three consecutive years (F(2, 575) = 31.52, $p < 0.001$, partial η^2 = 0.099). While the factor perceived joy was most pronounced in the winter semester 20/21, it was already significantly reduced in the following year 21/22 ($p < 0.001$) and continued to decrease towards the winter semester 22/23 ($p < 0.001$) (Figure 1).

To verify the questionnaire's factor structure regarding the addressed target group of German medical students during this particularly challenging period and to investigate the correlations between the latent factors, a Confirmatory Factor Analysis was performed. The corresponding factor loadings were calculated and further analyzed (Table 4).

Table 4. Factor loadings.

Indicator	Estimate	p	95% Confidence Interval	
			Lower	Upper
w1	0.674	<0.001	0.601	0.746
w2	0.624	<0.001	0.558	0.690
w3	0.755	<0.001	0.686	0.824
w4	0.777	<0.001	0.707	0.848
w5	0.714	<0.001	0.639	0.788
t1r	0.632	<0.001	0.565	0.699
t2r	0.589	<0.001	0.521	0.657
t3	0.544	<0.001	0.475	0.613
t4	0.705	<0.001	0.639	0.772
t5	0.684	<0.001	0.607	0.761
j1	0.466	<0.001	0.402	0.531
j2	0.613	<0.001	0.544	0.682
j3	0.571	<0.001	0.499	0.644
j4	0.513	<0.001	0.454	0.572
j5	0.441	<0.001	0.362	0.519
d1	0.598	<0.001	0.530	0.666
d2	0.435	<0.001	0.372	0.498
d3	0.758	<0.001	0.686	0.829
d4r	0.523	<0.001	0.451	0.596
d5	0.659	<0.001	0.588	0.730

Note. p = p-value, w = worries, t = tension, j = joy, d = demands, r = reverse coded item.

The fit indices CFI (0.908), RMSEA (0.071) and SRMR (0.052) indicated a quite acceptable model fit, although the chi-square test was significant ($p < 0.001$). Strong positive correlations could be found between worries and tension (0.94), worries and demands (0.86), and tension and demands (0.876), whereas strong negative correlations could be observed between worries and joy (-0.72), tension and joy (-0.84), and joy and demands (-0.60) (Figure 2).

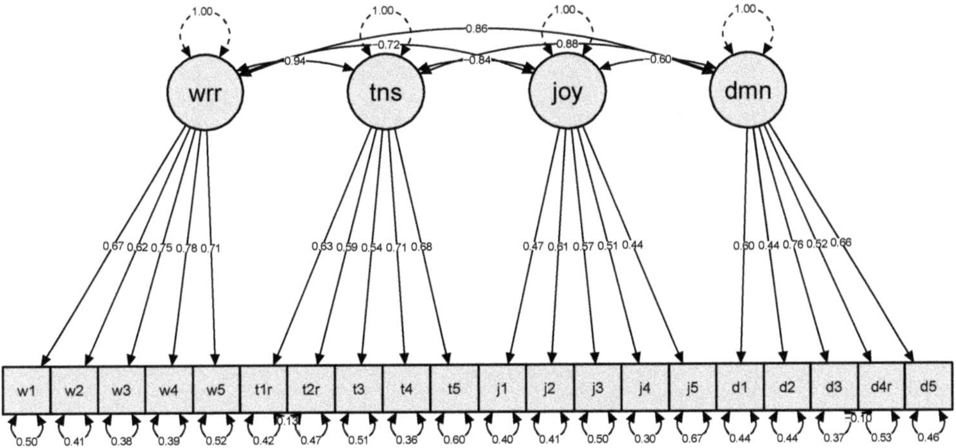

Figure 2. Confirmatory Factor Analysis. Abbreviations indicate: wrr: worries, tns: tension, joy = joy, dmn = demands.

The analysis of academic performance with regard to the reference year 2019 and the years of the ongoing pandemic 2020, 2021, 2022 revealed that students in 2020 had the highest pass rate. Except for the pass rate in 2020, the respective students have otherwise shown a slight downward trend in terms of performance since 2019, which, however, is not significant (Figure 3). The pass rate of the Anatomy I exam differed significantly when comparing the years 2019, 2020, 2021 and 2022 (F(3, 1533) = 8.31, $p < 0.001$, partial $\eta^2 = 0.016$). The pass rate in 2020 was significantly higher than in the previous year 2019 ($p = 0.009$) and significantly higher than in the following years 2021 ($p = 0.002$) and 2022 ($p < 0.001$). In 2019, the pass rate was slightly higher than in 2021 ($p = 0.651$) and 2022 ($p = 0.366$), but not significantly. In 2022, the pass rate was slightly reduced compared to 2021 ($p = 0.522$).

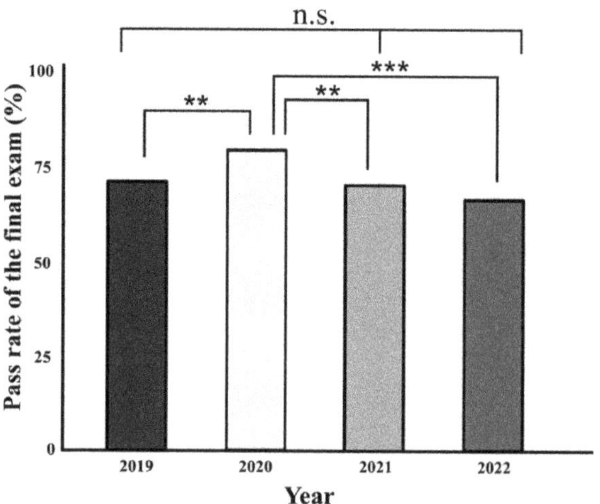

Figure 3. The figure indicates the pass rate of the Anatomy I exam of the first semester in medical studies over the years 2019, 2020, 2021 and 2022. Asterisks indicate: ** = $p < 0.01$, *** = $p < 0.001$. n.s.: not significant.

4. Discussion

This study points to the notion that there is a clear connection between the course of the COVID-19 pandemic and the associated variable intensity of the effects on social and university life and the expression of students' perceived life stress. A comprehensive corpus of scientific papers has already examined the effect of the COVID-19 pandemic on perceived stress levels of student populations across various disciplines [51–54]. While an analysis of 13 studies regarding the impact of the COVID-19 pandemic on the mental health status of medical students already pointed to increased levels of anxiety and stress [55], this study was able to identify the dynamics of perceived stress among first-year medical students caused in association with the ongoing pandemic.

Not only because of the described effects of experienced stress on cognitive processes and motivational factors [41,42,51,56,57], it is a relevant construct for the entire education sector. Additionally, it has already been empirically proven that academic stress is a relevant risk factor for mental health problems [52–54,56], which in turn can cause lower academic functioning and is considered a predictor of dropout among students in higher education [58–61].

Interestingly, in comparison to the obtained factors worries, tension and demands, the factor joy—which relates to general perceived joy of life—showed different dynamics in its expression over the given period of three years. In contrast to the factors worries, tension and demands, which increased at the peak of the COVID-19-related restrictions and then fell again, the factor joy recorded a continuous, significant decrease in the observed period of 3 years. This finding can also be discussed against the background of a more global data situation. Global survey data of more than 150 countries, analyzed and published by the Sustainable Development Solutions Network, revealed that in 2020, negative affect, as indicated by worry and sadness, increased by 8% compared to the preceding pre-pandemic years [62]. Against the backdrop of this globally collected data, this negative affect metric then dropped to 3% above baseline in the following year, which can be interpreted in terms of emerging resilience or habituation [62]. However, the data generated in this study, which show a continuous decline in general joy of life among first-year medical students, therefore underline the necessity of identifying local influencing factors for the evaluation of mental stress parameters in the respective environment. As a first interpretive approach regarding the lack of an increase in joy of life after the peak of the COVID-19-associated restrictions in the winter semester 21/22, the Russian invasion of Ukraine (24 February 2022) should be considered, since the serious consequences of this war of aggression caused and still cause great fear and uncertainty in the surrounding countries, including Germany [63]. The trend shown here regarding the steady decline in general joy of life among first-year medical students should be considered as an alarming signal, since joy of life is known to be a protective factor, strengthening resilience against mental illnesses [64,65]. This becomes particularly relevant in situations such as the COVID-19 pandemic, as designated protective factors such as perceived social support appear to be diminished [66–68].

Additionally, the data described here indicate that dynamics in academic performance during the pandemic may be difficult data to correlate, as the implications of COVID-19-associated changes in education also include alterations in assessment strategies. The sudden shift to online assessment is often discussed against the backdrop of safeguarding academic integrity due to the often-inadequate supervision of higher education exams [69–71]. In the data shown here, the Anatomy I exam was taken online only in 2020, which resulted in a significantly higher pass rate and should be discussed in light of the different conditions compared to the reference year 2019 and the subsequent years 2021 and 2022. Excluding the year 2020, however, a slight decrease in academic performance is evident, but this cannot be considered statistically significant. The fact that increased subjectively perceived stress does not necessarily have to be reflected in a sharp drop in academic performance is embedded in the context of the previous literature [16,17], which, conversely, cannot imply that a stable academic performance can indicate satisfactory external factors such as perceived stress.

In addition to the significance and social relevance of the data described, reference should be made to certain limitations contained in this study design. While the PSQ is a sufficiently validated and tested survey instrument for perceived life stress [47,49,72,73], here it could be shown that some factor loadings drop below 0.7; the factors tension and demands especially load quite inhomogeneously. These obtained findings encourage a discussion of the extent to which exceptional demanding situations such as the COVID-19 pandemic influence the factor structure of the PSQ.

Furthermore, although the data show a strong increase in the factors worries, tension and demands, no conclusions can be drawn about the individual attribution of certain stressors. This concern of individual data collection likewise limits the direct predictive power of subjectively perceived stress on academic performance, as this study focuses on the dynamic changes in the mean scores of the associated constructs. Follow-up studies should therefore break down specific subpopulations within vulnerable groups and follow individual parameter expressions to provide more holistic information on the impact of crisis situations on mental stress parameters along with academic achievement. For a more integrated comprehension, more objective, physiological markers of chronic stress could also be collected, such as immune markers, circulating glucocorticoids or catecholamines. Since the sample size in 2020 and 2021 is lower than in 2022, the potential for response bias as a possible source of error in standardized questionnaire-based surveys should be discussed. However, it has to be emphasized that at the beginning of each winter term, the medical students were randomly assigned to small groups by the Dean's Office of the Medical Faculty. Since, for infection prevention reasons, only a reduced number of randomly selected small groups were able to attend the seminars in presence in 2020 and 2021, correspondingly less data were generated within the survey period. While it would have been desirable to examine three similarly sized sample numbers, it should be noted that the acquisition of the three data sets was conducted under the same randomization and conditions.

The issue of perceived stress in young college entrants is a widely discussed and studied area in the scientific cosmos, reflecting the relevance of studying associated stressors and further predictors of general well-being. Our present results should contribute to this by illustrating the variable intensity of the influence of external stressors, such as the ongoing COVID-19 pandemic.

5. Conclusions

This repeated cross-sectional study was able to demonstrate that the factors worries, tension and demands of first-year medical students were significantly increased at the peak of the COVID-19-associated restrictions in the winter semester 21/22 compared to the previous (20/21) and the following (22/23) winter semesters. In addition, a continuous decline in general joy of life could be identified. Here, we describe the emerging dynamics of the influence of the COVID-19 pandemic on stress experienced by first-year medical students—a particularly vulnerable group regarding mental health parameters. Health hazards, political decisions and how the community deals with crisis situations are directly related to perception, behavior and ultimately to elementary mechanisms of a functioning society. These results should also be used to create awareness among the faculties in order to develop protective measures that take into account the influence of dynamic exogenous stressors on university life.

Author Contributions: Conceptualization, M.G., M.B., G.M.-P., T.S. and B.B.-S.; Methodology, M.G.; Software, M.G. and M.B.; Validation, M.G., M.B., G.M.-P., T.S. and B.B.-S.; Formal analysis, M.G., M.B. and B.B.-S.; Investigation, M.G.; Resources, M.G., G.M.-P., T.S.and B.B.-S.; Data curation, M.G., M.B., T.S.and B.B.-S.; Writing—original draft, M.G.; Writing—review & editing, M.G., M.B., G.M.-P., T.S.and B.B.-S.; Visualization, M.G.; Supervision, B.B.-S.; Project administration, M.G. and B.B.-S.; Funding acquisition, M.G. and B.B.-S. All authors have read and agreed to the published version of the manuscript.

Funding: Forschungsförderung an der Medizinischen Fakultät der Ruhr-Universität Bochum (Fo-RUM), Grant number: F1028–2021. OERContent.nrw project "Digital Histo NRW" by the Ministry of Culture and Science of the German State of North Rhine Westphalia (MKW NRW).

Institutional Review Board Statement: The study was conducted in accordance with the Declaration of Helsinki, and approved by the Ethics Committee of the Medical Faculty at the Ruhr University Bochum (20–7135) and the Ethics Committee of the Professional School of Education at the Ruhr University Bochum (EPSE-2022-005).

Informed Consent Statement: Informed consent was obtained from all subjects involved in the study.

Acknowledgments: The authors wish to thank the student participants for their contribution to the successful implementation of this study, especially during this particularly challenging time. Further, the authors would like to express their gratitude to Norma Hüllecremer, who assisted with data cleaning. The authors also thank Jan-Philipp Kersten, who assisted with data collection. We acknowledge support by the Open Access Publication Funds of the Ruhr-Universität Bochum.

Conflicts of Interest: The authors declare no conflict of interest.

References

1. Carvalho, T.; Krammer, F.; Iwasaki, A. The First 12 Months of COVID-19: A Timeline of Immunological Insights. *Nat. Rev. Immunol* **2021**, *21*, 245–256. [CrossRef]
2. Sachs, J.D.; Karim, S.S.A.; Aknin, L.; Allen, J.; Brosbøl, K.; Colombo, F.; Barron, G.C.; Espinosa, M.F.; Gaspar, V.; Gaviria, A.; et al. The Lancet Commission on Lessons for the Future from the COVID-19 Pandemic. *Lancet* **2022**, *400*, 1224–1280. [CrossRef] [PubMed]
3. Orrù, G.; Bertelloni, D.; Diolaiuti, F.; Mucci, F.; Di Giuseppe, M.; Biella, M.; Gemignani, A.; Ciacchini, R.; Conversano, C. Long-COVID Syndrome? A Study on the Persistence of Neurological, Psychological and Physiological Symptoms. *Healthcare* **2021**, *9*, 575. [CrossRef] [PubMed]

4. Ceban, F.; Ling, S.; Lui, L.M.W.; Lee, Y.; Gill, H.; Teopiz, K.M.; Rodrigues, N.B.; Subramaniapillai, M.; Di Vincenzo, J.D.; Cao, B.; et al. Fatigue and Cognitive Impairment in Post-COVID-19 Syndrome: A Systematic Review and Meta-Analysis. *Brain Behav. Immun.* **2022**, *101*, 93–135. [CrossRef]
5. Silva Andrade, B.; Siqueira, S.; de Assis Soares, W.R.; de Souza Rangel, F.; Santos, N.O.; dos Santos Freitas, A.; Ribeiro da Silveira, P.; Tiwari, S.; Alzahrani, K.J.; Góes-Neto, A.; et al. Long-COVID and Post-COVID Health Complications: An Up-to-Date Review on Clinical Conditions and Their Possible Molecular Mechanisms. *Viruses* **2021**, *13*, 700. [CrossRef]
6. Moreno, C.; Wykes, T.; Galderisi, S.; Nordentoft, M.; Crossley, N.; Jones, N.; Cannon, M.; Correll, C.U.; Byrne, L.; Carr, S.; et al. How Mental Health Care Should Change as a Consequence of the COVID-19 Pandemic. *Lancet Psychiatry* **2020**, *7*, 813–824. [CrossRef] [PubMed]
7. Cuschieri, S.; Calleja Agius, J. Spotlight on the Shift to Remote Anatomical Teaching During COVID-19 Pandemic: Perspectives and Experiences from the University of Malta. *Anat. Sci. Educ.* **2020**, *13*, 671–679. [CrossRef]
8. Loda, T.; Löffler, T.; Erschens, R.; Zipfel, S.; Herrmann-Werner, A. Medical Education in Times of COVID-19: German Students' Expectations—A Cross-Sectional Study. *PLoS ONE* **2020**, *15*, e0241660. [CrossRef]
9. Kaul, V.; Gallo de Moraes, A.; Khateeb, D.; Greenstein, Y.; Winter, G.; Chae, J.; Stewart, N.H.; Qadir, N.; Dangayach, N.S. Medical Education During the COVID-19 Pandemic. *Chest* **2021**, *159*, 1949–1960. [CrossRef]
10. Lorenzo-Lledó, A.; Lledó, A.; Gilabert-Cerdá, A.; Lorenzo, G. The Pedagogical Model of Hybrid Teaching: Difficulties of University Students in the Context of COVID-19. *Eur. J. Investig. Health Psychol. Educ.* **2021**, *11*, 1320–1332. [CrossRef]
11. Gellisch, M.; Wolf, O.T.; Minkley, N.; Kirchner, W.H.; Brüne, M.; Brand-Saberi, B. Decreased Sympathetic Cardiovascular Influences and Hormone-physiological Changes in Response to COVID-19-related Adaptations under Different Learning Environments. *Anat. Sci. Educ.* **2022**, *15*, 811–826. [CrossRef]
12. Gellisch, M.; Morosan-Puopolo, G.; Wolf, O.T.; Moser, D.A.; Zaehres, H.; Brand-Saberi, B. Interactive Teaching Enhances Students' Physiological Arousal during Online Learning. *Ann. Anat.—Anat. Anz.* **2023**, *247*, 152050. [CrossRef]
13. Srikhamjak, T.; Yanawuth, K.; Sucharittham, K.; Larprabang, C.; Wangsattabongkot, P.; Hauwadhanasuk, T.; Thawisuk, C.; Thichanpiang, P.; Kaunnil, A. Impact of the COVID-19 Pandemic on Mental Health and Lifestyle in Thai Occupational Therapy Students: A Mixed Method Study. *Eur. J. Investig. Health Psychol. Educ.* **2022**, *12*, 1682–1699. [CrossRef]
14. Hammoudi Halat, D.; Younes, S.; Safwan, J.; Akiki, Z.; Akel, M.; Rahal, M. Pharmacy Students' Mental Health and Resilience in COVID-19: An Assessment after One Year of Online Education. *Eur. J. Investig. Health Psychol. Educ.* **2022**, *12*, 1082–1107. [CrossRef]
15. Hendriksen, P.A.; Garssen, J.; Bijlsma, E.Y.; Engels, F.; Bruce, G.; Verster, J.C. COVID-19 Lockdown-Related Changes in Mood, Health and Academic Functioning. *Eur. J. Investig. Health Psychol. Educ.* **2021**, *11*, 1440–1461. [CrossRef]
16. Lemay, D.J.; Bazelais, P.; Doleck, T. Transition to Online Learning during the COVID-19 Pandemic. *Comput. Hum. Behav. Rep.* **2021**, *4*, 100130. [CrossRef]
17. Wilhelm, J.; Mattingly, S.; Gonzalez, V.H. Perceptions, Satisfactions, and Performance of Undergraduate Students during COVID-19 Emergency Remote Teaching. *Anat. Sci. Educ.* **2022**, *15*, 42–56. [CrossRef]
18. Hysenbegasi, A.; Hass, S.L.; Rowland, C.R. The Impact of Depression on the Academic Productivity of University Students. *J. Ment. Health Policy Econ.* **2005**, *8*, 145–151.
19. Andrews, B.; Wilding, J.M. The Relation of Depression and Anxiety to Life-Stress and Achievement in Students. *Br. J. Psychol.* **2004**, *95*, 509–521. [CrossRef]
20. Meilman, P.W.; Manley, C.; Gaylor, M.S.; Turco, J.H. Medical Withdrawals from College for Mental Health Reasons and Their Relation to Academic Performance. *J. Am. Coll. Health* **1992**, *40*, 217–223. [CrossRef]
21. Awadalla, S.; Davies, E.B.; Glazebrook, C. A Longitudinal Cohort Study to Explore the Relationship between Depression, Anxiety and Academic Performance among Emirati University Students. *BMC Psychiatry* **2020**, *20*, 448. [CrossRef]
22. Mahdavi, P.; Valibeygi, A.; Moradi, M.; Sadeghi, S. Relationship Between Achievement Motivation, Mental Health and Academic Success in University Students. *Community Health Equity Res. Policy* **2023**, *43*, 311–317. [CrossRef] [PubMed]
23. Credé, M.; Kuncel, N.R. Study Habits, Skills, and Attitudes: The Third Pillar Supporting Collegiate Academic Performance. *Perspect. Psychol. Sci.* **2008**, *3*, 425–453. [CrossRef] [PubMed]
24. Puthran, R.; Zhang, M.W.B.; Tam, W.W.; Ho, R.C. Prevalence of Depression amongst Medical Students: A Meta-Analysis. *Med. Educ.* **2016**, *50*, 456–468. [CrossRef] [PubMed]
25. Rotenstein, L.S.; Ramos, M.A.; Torre, M.; Segal, J.B.; Peluso, M.J.; Guille, C.; Sen, S.; Mata, D.A. Prevalence of Depression, Depressive Symptoms, and Suicidal Ideation Among Medical Students. *JAMA* **2016**, *316*, 2214. [CrossRef]
26. Heinen, I.; Bullinger, M.; Kocalevent, R.-D. Perceived Stress in First Year Medical Students—Associations with Personal Resources and Emotional Distress. *BMC Med. Educ.* **2017**, *17*, 4. [CrossRef]
27. Moreira de Sousa, J.; Moreira, C.A.; Telles-Correia, D. Anxiety, Depression and Academic Performance: A Study Amongst Portuguese Medical Students Versus Non-Medical Students. *Acta Med. Port.* **2018**, *31*, 454–462. [CrossRef]
28. Martínez, I.M.; Youssef-Morgan, C.M.; Chambel, M.J.; Marques-Pinto, A. Antecedents of Academic Performance of University Students: Academic Engagement and Psychological Capital Resources. *Educ. Psychol.* **2019**, *39*, 1047–1067. [CrossRef]
29. Donaldson, S.I.; Donaldson, S.I.; Chan, L.; Kang, K.W. Positive Psychological Capital (PsyCap) Meets Multitrait-Multimethod Analysis: Is PsyCap a Robust Predictor of Well-Being and Performance Controlling for Self-Report Bias? *Int. J. Appl. Posit. Psychol.* **2022**, *7*, 191–205. [CrossRef]

30. Donaldson, S.I.; Chan, L.B.; Villalobos, J.; Chen, C.L. The Generalizability of HERO across 15 Nations: Positive Psychological Capital (PsyCap) beyond the US and Other WEIRD Countries. *Int. J. Environ. Res. Public Health* **2020**, *17*, 9432. [CrossRef]

31. Chen, P.-L.; Lin, C.-H.; Lin, I.-H.; Lo, C.O. The Mediating Effects of Psychological Capital and Academic Self-Efficacy on Learning Outcomes of College Freshmen. *Psychol. Rep.* **2022**, 003329412210770. [CrossRef]

32. Holm-Hadulla, R.M.; Klimov, M.; Juche, T.; Möltner, A.; Herpertz, S.C. Well-Being and Mental Health of Students during the COVID-19 Pandemic. *Psychopathology* **2021**, *54*, 291–297. [CrossRef]

33. Elmer, T.; Mepham, K.; Stadtfeld, C. Students under Lockdown: Comparisons of Students' Social Networks and Mental Health before and during the COVID-19 Crisis in Switzerland. *PLoS ONE* **2020**, *15*, e0236337. [CrossRef]

34. Fu, W.; Yan, S.; Zong, Q.; Anderson-Luxford, D.; Song, X.; Lv, Z.; Lv, C. Mental Health of College Students during the COVID-19 Epidemic in China. *J. Affect. Disord.* **2021**, *280*, 7–10. [CrossRef]

35. Labrague, L.J.; De los Santos, J.A.A.; Falguera, C.C. Social and Emotional Loneliness among College Students during the COVID-19 Pandemic: The Predictive Role of Coping Behaviors, Social Support, and Personal Resilience. *Perspect. Psychiatr. Care* **2021**, *57*, 1578–1584. [CrossRef]

36. Tasso, A.F.; Hisli Sahin, N.; San Roman, G.J. COVID-19 Disruption on College Students: Academic and Socioemotional Implications. *Psychol. Trauma* **2021**, *13*, 9–15. [CrossRef]

37. Giovenco, D.; Shook-Sa, B.E.; Hutson, B.; Buchanan, L.; Fisher, E.B.; Pettifor, A. Social Isolation and Psychological Distress among Southern U.S. College Students in the Era of COVID-19. *PLoS ONE* **2022**, *17*, e0279485. [CrossRef]

38. Lazarus, R.S. *Psychological Stress and the Coping Process*, 1st ed.; McGraw-Hill Book Company: New York, NY, USA, 1966.

39. Shields, G.S.; Sazma, M.A.; McCullough, A.M.; Yonelinas, A.P. The Effects of Acute Stress on Episodic Memory: A Meta-Analysis and Integrative Review. *Psychol. Bull.* **2017**, *143*, 636–675. [CrossRef]

40. Sapolsky, R.M. Stress and the Brain: Individual Variability and the Inverted-U. *Nat. Neurosci.* **2015**, *18*, 1344–1346. [CrossRef] [PubMed]

41. Wolf, O.T. Stress and Memory Retrieval: Mechanisms and Consequences. *Curr. Opin. Behav. Sci.* **2017**, *14*, 40–46. [CrossRef]

42. Bierbrauer, A.; Fellner, M.-C.; Heinen, R.; Wolf, O.T.; Axmacher, N. The Memory Trace of a Stressful Episode. *Curr. Biol.* **2021**, *31*, 5204–5213.e8. [CrossRef] [PubMed]

43. Peng, P.; Hao, Y.; Liu, Y.; Chen, S.; Wang, Y.; Yang, Q.; Wang, X.; Li, M.; Wang, Y.; He, L.; et al. The Prevalence and Risk Factors of Mental Problems in Medical Students during COVID-19 Pandemic: A Systematic Review and Meta-Analysis. *J. Affect. Disord.* **2023**, *321*, 167–181. [CrossRef] [PubMed]

44. Jhajj, S.; Kaur, P.; Jhajj, P.; Ramadan, A.; Jain, P.; Upadhyay, S.; Jain, R. Impact of COVID-19 on Medical Students around the Globe. *J. Community Hosp. Intern. Med. Perspect* **2022**, *12*, 1. [CrossRef] [PubMed]

45. Liu, Z.; Liu, R.; Zhang, Y.; Zhang, R.; Liang, L.; Wang, Y.; Wei, Y.; Zhu, R.; Wang, F. Association between Perceived Stress and Depression among Medical Students during the Outbreak of COVID-19: The Mediating Role of Insomnia. *J. Affect. Disord.* **2021**, *292*, 89–94. [CrossRef]

46. Levenstein, S.; Prantera, C.; Varvo, V.; Scribano, M.L.; Berto, E.; Luzi, C.; Andreoli, A. Development of the Perceived Stress Questionnaire: A New Tool for Psychosomatic Research. *J. Psychosom. Res.* **1993**, *37*, 19–32. [CrossRef]

47. Fliege, H.; Rose, M.; Arck, P.; Walter, O.B.; Kocalevent, R.-D.; Weber, C.; Klapp, B.F. The Perceived Stress Questionnaire (PSQ) Reconsidered: Validation and Reference Values From Different Clinical and Healthy Adult Samples. *Psychosom. Med.* **2005**, *67*, 78–88. [CrossRef]

48. Østerås, B.; Sigmundsson, H.; Haga, M. Psychometric Properties of the Perceived Stress Questionnaire (PSQ) in 15–16 Years Old Norwegian Adolescents. *Front. Psychol.* **2018**, *9*. [CrossRef]

49. Kocalevent, R.-D.; Levenstein, S.; Fliege, H.; Schmid, G.; Hinz, A.; Brähler, E.; Klapp, B.F. Contribution to the Construct Validity of the Perceived Stress Questionnaire from a Population-Based Survey. *J. Psychosom. Res.* **2007**, *63*, 71–81. [CrossRef]

50. Bilen, E.; Matros, A. Online Cheating amid COVID-19. *J. Econ. Behav. Organ.* **2021**, *182*, 196–211. [CrossRef]

51. Gustems-Carnicer, J.; Calderón, C.; Calderón-Garrido, D. Stress, Coping Strategies and Academic Achievement in Teacher Education Students. *Eur. J. Teach. Educ.* **2019**, *42*, 375–390. [CrossRef]

52. Backovic, D.; Maksimovic, M.; Davidovic, D.; Ilic-Zivojinovic, J.; Stevanovic, D. Stress and Mental Health among Medical Students. *Srp. Arh. Celok. Lek.* **2013**, *141*, 780–784. [CrossRef]

53. Stansfeld, S.; Candy, B. Psychosocial Work Environment and Mental Health—a Meta-Analytic Review. *Scand. J. Work. Environ. Health* **2006**, *32*, 443–462. [CrossRef]

54. Cohen, S.; Janicki-Deverts, D.; Miller, G.E. Psychological Stress and Disease. *JAMA* **2007**, *298*, 1685. [CrossRef]

55. Mittal, R.; Su, L.; Jain, R. COVID-19 Mental Health Consequences on Medical Students Worldwide. *J. Community Hosp. Intern. Med. Perspect.* **2021**, *11*, 296–298. [CrossRef]

56. Pascoe, M.C.; Hetrick, S.E.; Parker, A.G. The Impact of Stress on Students in Secondary School and Higher Education. *Int. J. Adolesc. Youth* **2020**, *25*, 104–112. [CrossRef]

57. Lin, X.-J.; Zhang, C.-Y.; Yang, S.; Hsu, M.-L.; Cheng, H.; Chen, J.; Yu, H. Stress and Its Association with Academic Performance among Dental Undergraduate Students in Fujian, China: A Cross-Sectional Online Questionnaire Survey. *BMC Med. Educ.* **2020**, *20*, 181. [CrossRef]

58. Lie, S.A.; Tveito, T.H.; Reme, S.E.; Eriksen, H.R. IQ and Mental Health Are Vital Predictors of Work Drop out and Early Mortality. Multi-State Analyses of Norwegian Male Conscripts. *PLoS One* **2017**, *12*, e0180737. [CrossRef]

59. Jeffries, V.; Salzer, M.S. Mental Health Symptoms and Academic Achievement Factors. *J. Am. Coll. Health* **2022**, *70*, 2262–2265. [CrossRef]

60. Bruffaerts, R.; Mortier, P.; Kiekens, G.; Auerbach, R.P.; Cuijpers, P.; Demyttenaere, K.; Green, J.G.; Nock, M.K.; Kessler, R.C. Mental Health Problems in College Freshmen: Prevalence and Academic Functioning. *J. Affect. Disord.* **2018**, *225*, 97–103. [CrossRef]
61. Hjorth, C.F.; Bilgrav, L.; Frandsen, L.S.; Overgaard, C.; Torp-Pedersen, C.; Nielsen, B.; Bøggild, H. Mental Health and School Dropout across Educational Levels and Genders: A 4.8-Year Follow-up Study. *BMC Public. Health* **2016**, *16*, 976. [CrossRef]
62. Helliwell, J.F.; Huang, H.; Wang, S.; Norton, M. *World Happiness, Trust and Deaths under COVID-19*; UN Sustainable Development Solutions Network: New York, NY, USA, 2022.
63. Jawaid, A.; Gomolka, M.; Timmer, A. Neuroscience of Trauma and the Russian Invasion of Ukraine. *Nat. Hum. Behav.* **2022**, *6*, 748–749. [CrossRef] [PubMed]
64. Tugade, M.M.; Fredrickson, B.L. Resilient Individuals Use Positive Emotions to Bounce Back From Negative Emotional Experiences. *J. Pers. Soc. Psychol.* **2004**, *86*, 320–333. [CrossRef] [PubMed]
65. Bajaj, B.; Khoury, B.; Sengupta, S. Resilience and Stress as Mediators in the Relationship of Mindfulness and Happiness. *Front. Psychol.* **2022**, *13*. [CrossRef] [PubMed]
66. Kilic, R.; Nasello, J.A.; Melchior, V.; Triffaux, J.-M. Academic Burnout among Medical Students: Respective Importance of Risk and Protective Factors. *Public Health* **2021**, *198*, 187–195. [CrossRef] [PubMed]
67. Hutten, E.; Jongen, E.M.M.; Vos, A.E.C.C.; van den Hout, A.J.H.C.; van Lankveld, J.J.D.M. Loneliness and Mental Health: The Mediating Effect of Perceived Social Support. *Int. J. Environ. Res. Public Health* **2021**, *18*, 11963. [CrossRef] [PubMed]
68. Hefner, J.; Eisenberg, D. Social Support and Mental Health among College Students. *Am. J. Orthopsychiatry* **2009**, *79*, 491–499. [CrossRef]
69. Fuller, R.; Joynes, V.; Cooper, J.; Boursicot, K.; Roberts, T. Could COVID-19 Be Our 'There Is No Alternative' (TINA) Opportunity to Enhance Assessment? *Med. Teach.* **2020**, *42*, 781–786. [CrossRef]
70. Gamage, K.A.A.; de Silva, E.K.; Gunawardhana, N. Online Delivery and Assessment during COVID-19: Safeguarding Academic Integrity. *Educ. Sci.* **2020**, *10*, 301. [CrossRef]
71. Montenegro-Rueda, M.; Luque-de la Rosa, A.; Sarasola Sánchez-Serrano, J.L.; Fernández-Cerero, J. Assessment in Higher Education during the COVID-19 Pandemic: A Systematic Review. *Sustainability* **2021**, *13*, 10509. [CrossRef]
72. Moldt, J.-A.; Festl-Wietek, T.; Mamlouk, A.M.; Herrmann-Werner, A. Assessing Medical Students' Perceived Stress Levels by Comparing a Chatbot-Based Approach to the Perceived Stress Questionnaire (PSQ20) in a Mixed-Methods Study. *Digit. Health* **2022**, *8*, 205520762211390. [CrossRef]
73. Fliege, H.; Rose, M.; Arck, P.; Levenstein, S.; Klapp, B.F. Validierung Des "Perceived Stress Questionnaire" (PSQ) an Einer Deutschen Stichprobe. *Diagnostica* **2001**, *47*, 142–152. [CrossRef]

Article

Sex Differences in Mental Status and Coping Strategies among Adult Mexican Population: A Cross-Sectional Study

Aniel Jessica Leticia Brambila-Tapia [1,*], Fabiola Macías-Espinoza [2,*], Joel Omar González-Cantero [3], Reyna Jazmin Martínez-Arriaga [4], Yesica Arlae Reyes-Domínguez [5] and María Luisa Ramírez-García [5]

[1] Departamento de Psicología Básica, Centro Universitario de Ciencias de la Salud (CUCS), Universidad de Guadalajara, Guadalajara 44340, Jalisco, Mexico

[2] Departamento de Psicología Aplicada, Centro Universitario de Ciencias de la Salud (CUCS), Universidad de Guadalajara, Guadalajara 44340, Jalisco, Mexico

[3] Departamento de Ciencias del Comportamiento, Centro Universitario de los Valles (CUVALLES), Universidad de Guadalajara, Guadalajara 46600, Jalisco, Mexico

[4] Departamento de Salud Mental, Centro Universitario de Ciencias de la Salud (CUCS), Universidad de Guadalajara, Guadalajara 44340, Jalisco, Mexico

[5] Maestría en Psicología de la Salud, Centro Universitario de Ciencias de la Salud (CUCS), Universidad de Guadalajara, Guadalajara 44340, Jalisco, Mexico

* Correspondence: aniel.brambila@academicos.udg.mx (A.J.L.B.-T.); fabiola.macias@academicos.udg.mx (F.M.-E.)

Citation: Brambila-Tapia, A.J.L.; Macías-Espinoza, F.; González-Cantero, J.O.; Martínez-Arriaga, R.J.; Reyes-Domínguez, Y.A.; Ramírez-García, M.L. Sex Differences in Mental Status and Coping Strategies among Adult Mexican Population: A Cross-Sectional Study. *Healthcare* 2023, *11*, 514. https://doi.org/10.3390/healthcare11040514

Academic Editors: Ana Querido, Alyx Taylor and Carlos Laranjeira

Received: 28 December 2022
Revised: 24 January 2023
Accepted: 7 February 2023
Published: 9 February 2023

Abstract: We performed a cross-sectional study in order to determine the association between stress coping strategies and stress, depression, and anxiety, in which the Mexican population was invited to answer these variables by an electronic questionnaire. A total of 1283 people were included, of which 64.8% were women. Women presented higher levels of stress, depression, and anxiety than men; likewise, women showed a higher frequency of some maladaptive coping strategies (behavioral disengagement and denial) and lower levels of some adaptive ones (active coping and planning); additionally, maladaptive coping strategies were positively correlated with stress and depression in both sexes: self-blame, behavioral disengagement, denial, substance use, and self-distraction. Likewise, there were negative correlations between stress and depression and the adaptive strategies: planning, active coping, acceptance, and positive reframing. For women, religion presented negative correlations with stress, depression, and anxiety, and humor showed low positive correlations with stress, anxiety, and depression. In conclusion, most adaptive and maladaptive coping strategies are common in both sexes with the exception of religion, which seems to be adaptive in women and neutral in men, and humor, which seems to be adaptive in men and maladaptive in women. In addition, emotional and instrumental support seem to be neutral in both sexes.

Keywords: stress; coping strategies; anxiety; depression; sex

1. Introduction

Coping strategies are defined as an individual's attempts to use cognitive and behavioral strategies to manage and regulate pressures, demands, and emotions in response to stress [1]; which, in turn, is considered a "particular relationship between the person and the environment that is appraised by the person as taxing or exceeding his or her resources and endangering his or her well-being" [2].

Stress has been associated with a wide number of negative psychological variables including depression, anxiety, and somatization; for example, a positive moderate correlation between stress and somatization has been reported [3–5]. In addition, many reports have shown that all these negative psychological variables have been observed at higher levels in women than in men [4–6]. These differences have been explained by the presence of progesterone during the luteal phase in fertile women [7], and by the dysregulation of glucocorticoid receptors, which was observed in female rats [8]; however, it is possible that more molecular, cultural, and psychological explanations may be involved.

Additionally, stress has been associated with elevated levels of blood pressure, serum lipids, reactive C protein, and oxidative stress [9–12], which is why it has been associated with the appearance of chronic diseases.

Furthermore, it has been reported that men and women show different associations between some coping strategies and psychological and physiological variables [13–16]. In this sense, Blalock & Joiner [13] showed that cognitive avoidance, as a coping style, was associated with higher depression and anxiety in response to stressful life events in women when compared with men; in addition, this coping strategy was associated with lower levels of systolic blood pressure only in men [14]. Likewise, Kelly et al. [16] showed that self-blame was associated with more anxiety-trait in women and that positive restructuring was associated with less depression also in women in comparison with men. Additionally, Mazure & Maciejewski [15] showed that women are sensitive to a broader range of stressful life events than men and that in response to these, women are three times more prone to experience depression than men. This information supports the contention that women show higher responses to stress than men and that specific coping styles are more maladaptive in women than in men.

With respect to stress coping styles, there are a few instruments that measure them. The main ones are the Coping Strategies Inventory (CSI) [17] and the Brief-COPE [18]; these instruments share common coping strategies, but the Brief-COPE is the instrument that incorporates the measurement of more coping strategies than CSI, i.e., 14 strategies for the Brief-COPE vs. 8 strategies for the CSI. In this regard, Meyer [19] classified the stress coping strategies of the Brief-COPE instrument into two large dimensions: adaptive (including the use of emotional support, positive reframing, acceptance, spirituality/religion, humor, active coping, planning, and use of instrumental support) and maladaptive coping strategies (including denial, behavioral disengagement, self-blame, substance use, self-distraction, and venting). However, Cooper et al. reduced the Brief-Cope instrument to three dimensions: emotional-focused strategies (acceptance, emotional support, humor, positive reframing, and religion), problem-focused strategies (active coping, instrumental support, and planning), and dysfunctional-focused strategies (behavioral disengagement, denial, self-distraction, self-blame, substance use, and venting) [20].

However, none of these two main classifications has been proposed based on correlations with stress or any negative psychological variable; therefore, in this study, we have two main objectives: (1) to compare the different coping strategies by sex, in order to detect any potential difference that could explain the different responses to stress in each sex and their relation with the presence of stress, depression, and anxiety; and (2) to investigate the relationship of each coping strategy (of the Brief-COPE instrument) with the main psychological variables: stress, anxiety, and depression, according to sex, in order to determine the adaptive and maladaptive coping strategies in each sex, corroborated from an empirical approach.

This study has, therefore, two main hypotheses: (a) women present higher levels of maladaptive coping strategies and lower levels of adaptive strategies than men, and (b) the coping strategies considered as adaptive by Meyer show negative correlations with stress, depression, and anxiety; and those classified as maladaptive or dysfunctional strategies show positive correlations with stress, depression, and anxiety.

The study of these two important objectives, not previously performed, in a sample of the Mexican population would add important and useful information to the international literature on the topic, either for research or practice.

2. Subjects and Methods

This is a cross-sectional study performed in the period from September 2021 to March 2022. The target population was any adult (older than 18 years old) who resided in Mexico. The procedure consisted of the distribution of an electronic questionnaire that included sociodemographic variables and psychological instruments, which was sent via social networks (including Facebook, WhatsApp, e-mail, and the classroom application of Google). The reached population included university students, colleagues, and acquaintances of the research team in the first step and a more diverse population in the second step, given that the instrument was widely distributed with the snowball method. The study was conducted in accordance with the declaration of Helsinki and was approved by the ethics and research committee of the Health Sciences University Center (CUCS) of the University of Guadalajara (number: CI-06821). All the participants gave their consent to participate in the same questionnaire.

2.1. Sampling Strategy and Sample Size

The sampling approach was the snowball method, which corresponds to a non-random method, and the sample size was expected to be a minimum of 500 participants, in order to detect very low correlations (r < 0.2) as significant and add more potentially useful information to the international literature, considering that very low correlations can be useful in different research and practice contexts.

2.2. Variables Included

The socio-demographic data included age, sex, schooling, whether participants have a romantic partner, whether they have a job, and their socioeconomic level.

The psychological measures included stress, measured with the Cohen Perceived Stress Scale (CPSS) [21,22]; this scale consists of 14 questions with 5 answer options from "never" to "very frequently" (Cronbach´s alpha test: 0.855); depression was measured with the CES-D Scale (Cronbach's alpha test: 0.867) [23,24], which consists of 10 questions with 4 answer options from "none day" to "5–7 days in the week"; anxiety was measured with the GAD-7 Scale (Cronbach's alpha test: 0.923) [25,26]; this instrument consists of 7 questions with 4 answer options from "never" to "almost all the days". Coping strategies were measured with the Brief-COPE Scale; Cronbach´s alpha tests were above 0.6 for most subscales and above 0.5 for acceptance, self-distraction, behavioral disengagement, and denial. However, the subscale "venting" had a low Cronbach's alpha of 0.35, and therefore this subscale was not used for the analyses [18,27]. This instrument consisted of 28 questions with 4 answer options from "I never do it" to "I always do it", and in addition, this instrument presents 14 subscales (2 questions for each subscale).

We used the Spanish adaptations of scales because no Mexican adaptations were found; however, Spanish adaptations were considered understandable and adequate for the Mexican population and showed a desirable internal consistency in previous publications performed by the research group.

2.3. Statistical Analysis

In order to describe qualitative or categorical variables, we used frequencies and percentages and to describe quantitative variables we used means, and standard deviation; these were used instead of median and ranges (considering the non-parametric distribution of the data) because they better reflected the differences observed between sexes. In order to verify the distribution of the data, we used the Kolmogorov–Smirnov test. In order to compare categorical variables between sexes we used a chi-squared test. To compare the psychological variation between sexes, we used the Man–Whitney U test, considering the non-parametric distribution of these variables. To perform correlations between psychological variables, we used the Spearman correlation test. Finally, a multiple regression analysis, with the stepwise method for stress as a dependent variable, was performed for each sex, in order to determine the variables significantly correlated with stress after adjustment for confounders in both sexes. In this analysis, we excluded the variables anxiety and depression (in order to detect the coping strategies most associated with stress, excluding the variables most associated with stress: anxiety and depression).

All analyses were performed with the software SPSS v. 25, and a *p* value < 0.05 was considered significant.

3. Results

A total of 1283 participants were included, which represents an approximate 25.6% response rate (considering that the survey reached around 5000 persons); we eliminated the responses of participants that seemed to be incongruent in the data reported (i.e., those responses that included the same number in many questions, being even contradictory). No missing data were reported because all questions were marked as obligatory.

Of the included participants, 831 (64.8%) were women, and the sociodemographic data of participants are presented in Table 1. There were no differences in age, schooling, having a romantic partner, having a job, or socioeconomic level between the sexes.

Table 1. Sociodemographic variables in the studied population.

Variable	Men	Women	*p* Value
	N = 452	**N = 831**	
Age, mean ± SD	31.33 ± 11.60	31.47 ± 11.10	0.578
With romantic partner, n (%)	264 (58.40)	526 (63.30)	0.093
With job, n (%)	259 (57.30)	544 (65.50)	0.054
Educational level			
- Elementary school	1 (0.20)	1 (0.10)	
- High school	7 (1.50)	14 (1.70)	
- Preparatory	96 (21.20)	156 (18.80)	
- Bachelor's degree	249 (55.10)	459 (55.20)	0.328
- Technical career	22 (4.90)	38 (4.60)	
- Master's degree	46 (10.20)	119 (14.30)	
- Ph.D. degree	31 (6.90)	44 (5.30)	
Socioeconomic level			
- Very low	4 (0.90)	0 (0.00)	
- Low	68 (15.00)	130 (15.60)	0.07
- Medium	365 (80.80)	677 (81.50)	
- High	15 (3.30)	24 (2.90)	

SD: Standard deviation.

3.1. Comparison of Psychological Variables and Coping Strategies between Sexes

In Table 2 we show the means and standard deviations of coping strategies for each sex. We observed that the most frequent coping strategies in both sexes were acceptance, active coping, and planning while the least frequent were substance use, denial, and behavioral disengagement.

In the comparison of the psychological variables by sex, we observed that women had significantly higher levels of stress, anxiety, and depression. In relation to the coping strategies, women had higher levels of religion, behavioral disengagement, emotional support, and denial and lower levels of acceptance, humor, planning, and active coping, when compared with men (Table 2).

Table 2. Comparison of psychological variables and coping strategies between sexes.

Psychological Variable	Men (N = 452) Mean ± SD	Women (N = 831) Mean ± SD	*p* Value
Stress	2.68 ± 0.66	2.99 ± 0.64	<0.001
Depression	0.98 ± 0.74	1.30 ± 0.81	<0.001
Anxiety	0.99 ± 0.60	1.25 ± 0.65	<0.001
Coping strategies (Brief-COPE)			
Self-blame	1.28 ± 0.81	1.33 ± 0.84	0.415
Behavioral disengagement	0.46 ± 0.59	0.56 ± 0.61	0.001
Self-distraction	1.51 ± 0.81	1.58 ± 0.76	0.167
Denial	0.41 ± 0.62	0.53 ± 0.68	0.001
Substance use	0.30 ± 0.61	0.29 ± 0.64	0.232
Emotional support	1.11 ± 0.83	1.27 ± 0.82	<0.001
Instrumental support	1.21 ± 0.72	1.27 ± 0.73	0.317
Active coping	1.97 ± 0.70	1.88 ± 0.70	0.024
Planning	1.91 ± 0.76	1.75 ± 0.77	<0.001
Acceptance	1.99 ± 0.72	1.80 ± 0.68	<0.001
Positive reframing	1.56 ± 0.83	1.53 ± 0.75	0.586
Religion	0.81 ± 0.93	0.96 ± 0.91	<0.001
Humor	1.57 ± 0.94	1.31 ± 0.89	<0.001

The stress scale (CPSS) had a range of 1–5; the depression scale (CES-D) of 0–3; the anxiety scale (GAD-7) had a range of 0–3; and coping strategies (brief-COPE) had a range of 0–3.

3.2. Bivariate Correlations

In the correlations of the coping strategies with stress, anxiety, and depression by sex, we observed that, in both sexes, there were positive correlations between stress, depression, and anxiety and the following strategies: self-blame, behavioral disengagement, denial, substance use, and self-distraction. Likewise, in both sexes, there were negative correlations between stress, depression, and anxiety with planning, active coping, acceptance, and positive reframing. For women, religion also presented negative correlations with stress, depression, and anxiety that were not found in men. In addition, in men, a very low but significant negative correlation was found between humor and stress. However, in women, humor presented very low but positive significant correlations with stress, depression, and anxiety. In the case of emotional and instrumental support, they did not show significant correlations with stress or depression in any sex (Table 3).

Some sociodemographic variables also showed low negative but significant correlations with stress in both sexes, which are shown in Table 4.

3.3. Multivariate Analysis for Stress

In the multiple regression analysis for stress by sex, and adjusting for sociodemographic variables, we observed that self-blame was the variable most positively related to stress in both sexes. In the case of women, other positively correlated variables were detected: denial, behavioral disengagement, and self-distraction, while in men other positively correlated variables were denial, self-distraction, and substance use. The negatively associated variables with stress in women were active coping, acceptance, and positive reframing, while in men they were active coping, acceptance, and humor. In addition, some sociodemographic variables were also negatively associated with stress, including age and having a job in women and in men: schooling and socioeconomic level (Tables 5 and 6).

Table 3. Correlation between coping strategies and stress, depression, and anxiety by sex.

Coping Strategy	Men (N = 452)			Women (N = 831)		
	Stress	Depression	Anxiety	Stress	Depression	Anxiety
Self-blame	0.391 **	0.432 **	0.407 **	0.457 **	0.505 **	0.531 **
Behavioral disengagement	0.304 **	0.278 **	0.209 **	0.345 **	0.403 **	0.289 **
Self-distraction	0.238 **	0.302 **	0.264 **	0.143 **	0.222 **	0.217 **
Denial	0.227 **	0.235 **	0.170 **	0.328 **	0.402 **	0.336 **
Substance use	0.216 **	0.242 **	0.203 **	0.186 **	0.248 **	0.213 **
Emotional support	0.067	0.053	0.047	−0.023	0.006	0.061
Instrumental support	−0.002	0.003	0.052	0.048	0.030	0.114 **
Active coping	−0.327 **	−0.261 **	−0.146 **	−0.276 **	−0.210 **	−0.069 **
Planning	−0.318 **	−0.248 **	−0.140 **	−0.268 **	−0.219 **	−0.090 **
Acceptance	−0.210 **	−0.166 **	−0.133 **	−0.257 **	−0.217 **	−0.150 **
Positive reframing	−0.141 **	−0.148 **	−0.071	−0.217 **	−0.157 **	−0.092 **
Religion	−0.027	−0.026	0.040	−0.204 **	−0.209 **	−0.156 **
Humor	−0.097 *	−0.048	−0.015	0.081 *	0.111 **	0.120 **

* $p < 0.05$, ** $p < 0.01$. p value obtained with the Spearman correlation test.

Table 4. Bivariate correlations between sociodemographic variables and stress.

Variable	Men (N = 452)	Women (N = 831)
Age	−0.174 **	−0.302 **
Schooling	−0.141 **	−0.150 **
Having children	−0.198 **	−0.199 **
With romantic partner	−0.147 **	−0.163 **
With job	−0.100 *	−0.202 **
Socioeconomic level	−0.182 **	−0.114 **

* $p < 0.05$, ** $p < 0.01$. p value obtained with the Spearman correlation test.

Table 5. Multiple regression analysis for stress in women.

Variable	B	Beta Coefficient	p Value	Change in R^2
Constant	3.387	-	<0.001	-
Self-blame	0.256	0.335	<0.001	0.214
Active coping	−0.139	−0.152	<0.001	0.078
Age	−0.008	−0.136	<0.001	0.036
Denial	0.106	0.111	<0.001	0.024
Acceptance	−0.146	−0.154	<0.001	0.020
Have a job	−0.119	−0.088	0.003	0.009
Self-distraction	0.086	0.102	0.001	0.006
Positive reframing	−0.089	−0.104	0.002	0.007
Behavioral disengagement	0.092	0.088	0.005	0.006

Type of method for the regression model: Stepwise. R of the model = 0.633.

Table 6. Multiple regression analysis for stress in men.

Variable	B	Beta Coefficient	p Value	Change in R^2
Constant	3.387	-	<0.001	-
Self-blame	0.296	0.367	<0.001	0.176
Active coping	−0.184	−0.196	<0.001	0.114
Denial	0.109	0.104	0.009	0.024
Acceptance	−0.093	−0.102	0.024	0.019
Self-distraction	0.158	0.196	<0.001	0.030
Humor	−0.109	−0.157	<0.001	0.015
Substance use	0.133	0.125	0.001	0.014
Socioeconomic level	−0.172	−0.118	0.002	0.008
Schooling	−0.085	−0.099	0.045	0.005

Type of method for the regression model: Stepwise. R of de model = 0.637.

4. Discussion

As previously reported, higher levels of stress, depression, and anxiety were found in women when compared with men [3–6]. A factor possibly related to these differences is the higher prevalence of some maladaptive coping strategies (behavioral disengagement and denial) and the lower frequency of some adaptive ones (acceptance, active coping, and planning) in women than in men; these results corroborate our first hypothesis. However, religion showed a higher frequency in women, in whom it was negatively correlated with stress, depression, and anxiety, and humor was higher in men, in whom it was negatively related to stress, indicating that these two strategies (religion and humor) showed higher frequencies in the sex where they were more adaptive. Nevertheless, it is noteworthy that adaptive strategies were the most frequent coping strategies and maladaptive ones were the least frequent coping strategies in both sexes.

With respect to the analysis of correlations, we observed that in contrast to a previous report, which showed that avoidance coping was not associated with an increase in depression and anxiety in men, and another one showing that avoidance coping was actually associated with lower blood pressure in men [13,14], we found that avoidance coping (denial, self-distraction, and behavioral disengagement) showed positive correlations with stress, depression, and anxiety in both sexes. However, the most associated variable with stress, depression, and anxiety was self-blame. These correlations coincide with our previous report where the coping strategies associated with stress, depression, and anxiety were associated with somatization in both sexes, and some of them were also associated with a number of diseases, with the highest association for self-blame [3]. These results coincide with our second hypothesis, confirming that coping strategies classified as maladaptive by Meyer showed positive correlations with stress, depression, and anxiety, while most classified as adaptive ones showed negative correlations with these three variables. The only difference with respect to Meyer's classification is related to the coping strategies of religion and humor, which showed different correlations in each sex.

These associations are further corroborated in the multiple regression analysis performed by sex, where common maladaptive coping strategies were significantly associated with stress in both sexes (self-blame, denial, and self-distraction), with self-blame the most associated variable with stress. These results coincide with a meta-analysis performed among healthcare professionals, which showed that maladaptive coping strategies related to poor mental health outcomes were venting, denial, disengagement, self-blame, and substance use; they also found personal factors related to the use of maladaptive coping strategies, including being female, older than 50 years old, living alone, and having a history of personal trauma. In addition, environmental factors related to the use of these strategies were work stress, workload, and poor benefits [28]. This data also shows that environmental conditions are related to the use of adaptive or maladaptive coping strategies.

Based on these results, the adaptive strategies in both sexes would be active coping, planning, acceptance, and positive reframing; in addition, religion is also adaptive in women and humor in men. Maladaptive coping strategies in both sexes would be self-blame, behavioral disengagement, self-distraction, denial, and substance use; in addition, humor would be also maladaptive in women. Finally, neutral strategies would be emotional and instrumental support in both sexes and religion in men.

It is of interest that all sociodemographic variables showed significant negative correlations with stress in both sexes; this suggests that many other variables, mainly those related to constant conditions in life, including age, having a romantic partner, schooling, and having a job, also contribute to stress variability in both sexes.

The study has the following limitations: the sample was not randomly selected and is mainly young, so the representativeness of the Mexican population is diminished and restricted to the young and people with more education. In addition, the low response rate (around 26%) also diminished the representativeness of the targeted population. On the other hand, the cross-sectional design of the study does not permit us to demonstrate causality, so the presence of bilateral relationships between the studied variables is also plausible; in this sense, it is possible that increased stressful conditions and or negative psychological variables (mainly in women) could also increase maladaptive coping strategies and diminish adaptive ones, as previously mentioned. In the case of women, it is possible that biological factors, including hormones and hormone response variations [7,8], contribute to a higher frequency of stress, depression, and anxiety, which, in turn, could increase maladaptive coping strategies and diminish adaptive ones, producing a positive refeeding cycle. However, with respect to the theoretical limitations, we consider that the results reported in this study are as expected, according to the classification of the strategies performed by Meyer, showing negative correlations between adaptive strategies and stress, depression, and anxiety as well as positive correlations between maladaptive coping strategies and stress, depression, and anxiety. Nevertheless, in this report, emotional strategies (emotional and instrumental support) seemed to be neutral (neither adaptive, nor maladaptive) according to the correlations found, and religion and humor showed different correlations with stress in each sex.

In relation to objective number two, we also found a coincidence with the meta-analyses previously mentioned [28], which showed that the female sex was associated with the use of maladaptive coping strategies.

With reference to the implications for future research and potential intervention work, we consider that it is needed to perform experimental studies where subjects of both sexes are exposed to stressful situations in order to detect the coping strategies mainly used in each sex, along with the measurement of stress, depression, anxiety, and stress-coping strategies, before and after the experiment. These studies would be useful to corroborate the causal relationship between the variables analyzed in this study. Observational and longitudinal studies will also elucidate the causal relationship between stress-coping strategies and negative psychological variables in each sex. Finally, based on these results, it is important to mention that the implementation of intervention programs addressing an increase in emotional abilities, mainly emotional intelligence [29], that favor adaptive and diminish maladaptive coping, is needed and useful in order to diminish stress, depression, and anxiety. These programs should be administered at least from early education, as well as in labor and academic spaces with high-stress conditions.

5. Conclusions

In conclusion, women showed higher frequencies of stress, depression, and anxiety than men and higher levels of some maladaptive coping (behavioral disengagement and denial) strategies and lower frequencies of some adaptive ones (active coping and planning) in comparison with men. This could explain the higher frequency of these negative psychological variables in women. The coping strategies showed similar correlations with stress, depression, and anxiety in both sexes with the exception of humor, which was shown to be adaptive in men and maladaptive in women, and religion, which seemed to be adaptive only in women. Therefore, based on these correlations, we could determine common and specific adaptive, maladaptive, and neutral coping strategies according to sex.

In addition, the most associated coping strategy with stress in both sexes was self-blame, which indicates that intervention programs addressed to diminish this strategy could be effective in diminishing stress, depression, and anxiety; however, the diminishing of maladaptive coping strategies (mainly avoidance coping strategies) and the increasing of adaptive ones, mainly active-coping, are also essential. Further experimental and longitudinal studies will determine the causal associations between stress coping strategies and stress, anxiety, and depression in both sexes but mainly in women.

Author Contributions: Conceptualization, A.J.L.B.-T. and F.M.-E.; methodology, A.J.L.B.-T., F.M.-E., J.O.G.-C., R.J.M.-A., Y.A.R.-D. and M.L.R.-G., investigation, A.J.L.B.-T., F.M.-E., R.J.M.-A., J.O.G.-C., Y.A.R.-D. and M.L.R.-G., formal analysis, A.J.L.B.-T., writing-original draft preparation: A.J.L.B.-T., writing-review and editing, A.J.L.B.-T., F.M.-E., J.O.G.-C. and R.J.M.-A. All authors have read and agreed to the published version of the manuscript.

Funding: This research received no external funding.

Institutional Review Board Statement: The study was approved by the ethics and research committee of the Health Sciences University Center (CUCS) of the University of Guadalajara (number: CI-06821).

Informed Consent Statement: Informed consent was obtained from all subjects.

Data Availability Statement: Data that support the findings of the study are available upon a reasonable request.

Conflicts of Interest: The authors declare no conflict of interest.

References

1. McHugh, S.; Vallis, T.M. *Illness Behavior. A Multidisciplinary Model*; Plenum Press: New York, NY, USA, 1986; pp. 303–308.
2. Lazarus, R.S.; Folkman, S. *Stress, Appraisal and Coping*; Springer: New York, NY, USA, 1984; p. 19.
3. Brambila-Tapia, A.J.L.; Macías-Espinoza, F.; Reyes-Domínguez, Y.A.; Ramírez-García, M.L.; Miranda-Lavastida, A.J.; Ríos-González, B.E.; Saldaña-Cruz, A.M.; Esparza-Guerrero, Y.; Mora-Moreno, F.F.; Dávalos-Rodríguez, I.P. Association of Depression, Anxiety, Stress and Stress-Coping Strategies with Somatization and Number of Diseases According to Sex in the Mexican General Population. *Healthcare* **2022**, *10*, 1048. [CrossRef]
4. Brambila-Tapia, A.J.L.; Meda-Lara, R.M.; Palomera-Chávez, A.; De-Santos-Ávila, F.; Hernández-Rivas, M.I.; Bórquez-Hernández, P.; Juárez-Rodríguez, P. Association between personal, medical and positive psychological variables with somatization in university health sciences students. *Psychol. Health Med.* **2019**, *25*, 879–886. [CrossRef]
5. Schlarb, A.A.; Claßen, M.; Hellmann, S.M.; Vögele, C.; Gulewitsch, M.D. Sleep and somatic complaints in university students. *J. Pain Res.* **2017**, *18*, 1189–1199. [CrossRef]
6. Garaigordobil, M.; Pérez, J.I.; Mozaz, M. Self concept, self-esteem and psychopathological symptoms. *Psicothema* **2008**, *20*, 114–123.
7. Andréen, L.; Nyberg, S.; Turkmen, S.; van Wingen, G.; Fernández, G.; Bäckström, T. Sex steroid induced negative mood may be explained by the paradoxical effect mediated by GABAA modulators. *Psychoneuroendocrinology* **2009**, *34*, 1121–1132. [CrossRef]
8. Palumbo, M.C.; Dominguez, S.; Dong, H. Sex differences in hypothalamic–pituitary–adrenal axis regulation after chronic unpredictable stress. *Brain Behav.* **2020**, *10*, e01586. [CrossRef]
9. Wiernik, E.; Pannier, B.; Czernichow, S.; Nabi, H.; Hanon, O.; Simon, T.; Simon, J.-M.; Thomas, F.; Bean, K.; Consoli, S.M.; et al. Occupational Status Moderates the Association Between Current Perceived Stress and High Blood Pressure. *Hypertension* **2013**, *61*, 571–577. [CrossRef]
10. Kivimäki, M.; Pentti, J.; Ferrie, J.E.; Batty, G.; Nyberg, S.T.; Jokela, M.; Virtanen, M.; Alfredsson, L.; Dragano, N.; Fransson, E.I.; et al. Work stress and risk of death in men and women with and without cardiometabolic disease: A multicohort study. *Lancet Diabetes Endocrinol.* **2018**, *6*, 705–713. [CrossRef]
11. Ironson, G.; Banerjee, N.; Fitch, C.; Krause, N. Positive emotional well-being, health Behaviors, and inflammation measured by C-Reactive protein. *Soc. Sci. Med.* **2018**, *197*, 235–243. [CrossRef]
12. Eick, S.M.; Barrett, E.S.; Erve, T.J.V.; Nguyen, R.H.N.; Bush, N.R.; Milne, G.; Swan, S.H.; Ferguson, K.K. Association between prenatal psychological stress and oxidative stress during pregnancy. *Paediatr. Perinat. Epidemiol.* **2018**, *32*, 318–326. [CrossRef]

13. Blalock, J.A.; Joiner, J.T.E. Interaction of Cognitive Avoidance Coping and Stress in Predicting Depression/Anxiety. *Cogn. Ther. Res.* **2000**, *24*, 47–65. [CrossRef]
14. Martin, L.A.; Critelli, J.W.; Doster, J.A.; Powers, C.; Purdum, M.; Doster, M.R.; Lambert, P.L. Cardiovascular Risk: Gender Differences in Lifestyle Behaviors and Coping Strategies. *Int. J. Behav. Med.* **2013**, *20*, 97–105. [CrossRef] [PubMed]
15. Mazure, C.M.; Maciejewski, P.K. The interplay of stress, gender and cognitive style in depressive onset. *Arch. Women's Ment. Health* **2003**, *6*, 5–8. [CrossRef]
16. Kelly, M.M.; Tyrka, A.R.; Anderson, G.M.; Price, L.H.; Carpenter, L.L. Sex differences in emotional and physiological responses to the Trier Social Stress Test. *J. Behav. Ther. Exp. Psychiatry* **2008**, *39*, 87–98. [CrossRef]
17. Tobin, D.L.; Holroyd, K.A.; Reynolds, R.V.; Wigal, J.K. The hierarchical factor structure of the coping strategies inventory. *Cogn. Ther. Res.* **1989**, *13*, 343–361. [CrossRef]
18. Carver, C.S. You want to measure coping but your protocol' too long: Consider the brief cope. *Int. J. Behav. Med.* **1997**, *4*, 92–100. [CrossRef] [PubMed]
19. Meyer, B. Coping with Severe Mental Illness: Relations of the Brief COPE with Symptoms, Functioning, and Well-Being. *J. Psychopathol. Behav. Assess.* **2001**, *23*, 265–277. [CrossRef]
20. Cooper, C.; Katona, C.; Orrell, M.; Livingston, G. Coping strategies and anxiety in caregivers of people with Alzheimer's disease: The LASER-AD study. *J. Affect. Disord.* **2006**, *90*, 15–20. [CrossRef] [PubMed]
21. Cohen, S.; Kamarck, T.; Mermelstein, R. A global measure of perceived stress. *J. Health Soc. Behav.* **1983**, *24*, 385–396.
22. Remor, E. Psychometric properties of a European Spanish version of the perceived Stress Scale (PSS). *Span. J. Psychol.* **2006**, *9*, 86–93.
23. Radloff, L. The CES-D Scale: A self-report depression scale for research in the general population. *Appl. Psychol. Meas.* **1977**, *1*, 385–401. [CrossRef]
24. González-Forteza, C.; Jiménez-Tapia, J.A.; Ramos-Lira, L.; Wagner, F.A.A. Aplicación de la Escala de Depresión del Center of Epidemiological Studies en adolescentes de la Ciudad de México. *Salud Pública México* **2008**, *50*, 292–299. [CrossRef]
25. Spitzer, R.L.; Kroenke, K.; Williams, J.B.W.; Löwe, B. A Brief Measure for Assessing Generalized Anxiety Disorder. *Arch. Intern. Med.* **2006**, *166*, 1092. [CrossRef]
26. Garcia-Campayo, J.; Zamorano, E.; Ruiz, M.A.; Pardo, A.; Perez-Paramo, M.; Lopez-Gomez, V.; Freire, O.; Rejas, J. Cultural adaptation into Spanish of the generalized anxiety disorder-7 (GAD-7) scale as a screening tool. *Health Qual. Life Outcomes* **2010**, *8*, 8. [CrossRef]
27. Morán, C.; Landero, R.; González, M.T. COPE-28: Un análisis psicométrico de la versión en español del BriefCOPE. *Univ. Psychol.* **2010**, *9*, 543–552. [CrossRef]
28. Owen, C.P.; Djukic, M.; Whisenant, M.; Lobiondo-Wood, G. Factors of maladaptive coping in emergency healthcare professionals: A systematic review. *J. Nurs. Sch.* **2022**. [CrossRef]
29. Macías-Espinoza, F.; Brambila-Tapia, A.J.L.; Reyes-Domínguez, Y.A.; Ramírez-García, M.L. Association between Emotional Intelligence and Stress Coping Strategies According to Sex in Mexican General Population. *Int. J. Environ. Res. Public Health* **2022**, *19*, 7318. [CrossRef]

Article

Validating the German Short Basic Psychological Need Satisfaction and Frustration Scale in Individuals with Depression

Andreas Heissel [1,*], Alba Sanchez [2], Anou Pietrek [2], Theresa Bergau [2], Christiane Stielow [2], Michael A. Rapp [1] and Jolene Van der Kaap-Deeder [3]

[1] Social and Preventive Medicine, Department of Sports and Health Sciences, Intra-Faculty Unit "Cognitive Sciences", Faculty of Human Science, and Faculty of Health Sciences Brandenburg, Research Area Services Research and e-Health, University of Potsdam, 14469 Potsdam, Germany

[2] Social and Preventive Medicine, Department of Sports and Health Sciences, Faculty of Human Science, University of Potsdam, 14469 Potsdam, Germany

[3] Department of Psychology, Norwegian University of Science and Technology, 7034 Trondheim, Norway

* Correspondence: andreas.heissel@uni-potsdam.de; Tel.: +49-331-9774049

Abstract: Satisfaction and frustration of the needs for autonomy, competence, and relatedness, as assessed with the 24-item Basic Psychological Need Satisfaction and Frustration Scale (BPNSFS), have been found to be crucial indicators of individuals' psychological health. To increase the usability of this scale within a clinical and health services research context, we aimed to validate a German short version (12 items) of this scale in individuals with depression including the examination of the relations from need frustration and need satisfaction to ill-being and quality of life (QOL). This cross-sectional study involved 344 adults diagnosed with depression (M_{age} (SD) = 47.5 years (11.1); 71.8% females). Confirmatory factor analyses indicated that the short version of the BPNSFS was not only reliable, but also fitted a six-factor structure (i.e., satisfaction/frustration X type of need). Subsequent structural equation modeling showed that need frustration related positively to indicators of ill-being and negatively to QOL. Surprisingly, need satisfaction did not predict differences in ill-being or QOL. The short form of the BPNSFS represents a practical instrument to measure need satisfaction and frustration in people with depression. Further, the results support recent evidence on the importance of especially need frustration in the prediction of psychopathology.

Keywords: basic psychological need frustration; need satisfaction; mental health; ill-being; depression

Citation: Heissel, A.; Sanchez, A.; Pietrek, A.; Bergau, T.; Stielow, C.; Rapp, M.A.; Van der Kaap-Deeder, J. Validating the German Short Basic Psychological Need Satisfaction and Frustration Scale in Individuals with Depression. *Healthcare* **2023**, *11*, 412. https://doi.org/10.3390/healthcare11030412

Academic Editors: Carlos Laranjeira and Ana Querido

Received: 16 December 2022
Revised: 25 January 2023
Accepted: 27 January 2023
Published: 31 January 2023

1. Introduction

The recent body of work in the field of Self-Determination Theory (SDT, [1]) suggests that SDT can contribute to our understanding of the onset and course of psychopathological phenomena, such as depression. The Basic Psychological Needs Theory (BPNT), a central sub-theory within SDT, is especially relevant and states that there exist three innate, universal psychological needs that act as essential nutrients for autonomous motivation, optimal functioning, and well-being: autonomy (i.e., having a sense of volition and choice), competence (i.e., experiencing a sense of mastery), and relatedness (i.e., feeling connected to important others) [2].

Research within BPNT that initially focused on the benefits of need satisfaction for individuals' thriving demonstrated the association of need satisfaction with a variety of well-being indicators (e.g., vitality, self-esteem, and life satisfaction) (see [3] for an overview). An increasing focus of research has more recently investigated the dysfunctional side of human development, focusing on the concept of need frustration, which is referring to the active obstruction and undermining of the basic psychological needs for autonomy (i.e., feelings of pressure), competence (i.e., experiencing oneself as a failure), and relatedness (i.e., feeling socially excluded) [3,4]. The assumption that unmet personal needs lead to health costs and psychological distress is also present in other research areas such as the psychological pain theory ("psychache") ([5]; also see [6]). Conceptually, as well as empirically, the distinction between need satisfaction and need frustration is justified, as these constructs are not only negatively related to one another but also display different outcomes and antecedents [4,7]. Additionally, both experiences are asymmetrically related to one another, such that the absence of need satisfaction (e.g., perceiving limited choice) does not necessarily mean the presence of need frustration (e.g., feeling forced to engage in a certain activity), but the presence of need frustration always implies the absence of need satisfaction [4].

Previous research indeed yielded empirical support for the important distinction between need satisfaction and need frustration, with need satisfaction being more strongly related to indicators of optimal development (e.g., life satisfaction), whereas need frustration is imperative in the prediction of ill-being and even psychopathology [7]. To illustrate, Bartholomew et al. (2011) [8] initially confirmed the additional costs associated with need frustration within the sport context showing that experiences of need frustration, compared to low need satisfaction, are more strongly associated with ill-being (e.g., burn-out and depression). Results of further studies focusing on different age groups and diverse life domains (e.g., exercise, education, and work) pointed in a similar direction by demonstrating significant associations between need frustration and diverse indicators of ill-being such as depressive symptoms [9], poor sleep (e.g., [10]), and rigid and obsessive behavior [11]. In line with these studies, need frustration has also been found to account for the comorbidity between symptoms of psychopathology, attesting to the transdiagnostic role of the needs [9,12]. That is, besides an increasing number of studies showing the importance of need frustration in diverse symptoms of psychopathology, need frustration also accounts for why certain symptoms coincide and thereby represents a common mechanism in psychopathology.

The effects of basic psychological needs are assumed to be universal, regardless of individuals' gender, age, and culture. Indeed, previous research has shown the beneficial and detrimental effects of need satisfaction and need frustration, respectively, across cultures, ages, personalities, etc. [13–15]. However, it is currently unclear whether these effects also generalize to clinical samples, specifically to individuals with mental disorders such as depression. On a theoretical level, need frustration is expected to also play a crucial role in the prediction of clinical forms of depression [7]. That is, in SDT, it is assumed that permanent thwarting of the basic psychological needs hinders people's innate propensities and thus leads to a loss of motivation, even reaching a state of amotivation, reflecting discouragement and helplessness (i.e., core characteristics of depression). Such a lack of motivation is one possible mechanism that can then result in decreases in well-being [16]. Although several studies employing community samples have shown need frustration to relate to depressive symptoms (e.g., [9]), research in samples with mental disorders such as depression is generally lacking.

Currently, the most widely-used instrument to assess need satisfaction and need frustration is the Basic Psychological Need Satisfaction and Frustration Scale (BPNSFS; [13], which was originally developed by Chen and colleagues (2015) and is now available in 14 languages adapted for various contexts [17]. Numerous validation studies indicated evidence for a six-factor solution of this scale [4,9,13], referring to the satisfaction or frustration of one of the three basic psychological needs with each factor. Some studies also found support for higher-order models, thereby creating composite need scores, with the risk of overlooking need-specific effects [13,18].

To further validate SDT's universality hypothesis and to add to the emerging literature on the transdiagnostic role of basic psychological needs, this study aimed to 1. validate the BPNSFS as an assessment in the clinical context by examining the role of the basic needs in individuals with depression, and 2. develop a short version of the German BPNSFS. This is an important mission given the need for brief and valid assessments in both clinical and health service research contexts where time is of concern (e.g., lower psychological stress for clinical samples, lower dropout rates). It is conceivable that the questionnaire could be used to evaluate health treatments that target psychological health (e.g., psychotherapy or exercise therapy). Heissel et al. [19] use the questionnaire to explore rumination as a possible coping strategy to deal with experienced need frustration and a recent article suggests that the basic need questionnaire could also be practical in deriving specific treatment strategies if the respective profile is known [20]. Short versions of basic need instruments have proven to be valid and reliable instruments [21], thereby representing a good and economical alternative to the long versions. To do so, four different hypothetical models were tested: (1) a 3-factor model; (2) a 6-factor model [13]; (3) a 2-factor hierarchical model (6^2-factor); and (4) a 3-factor hierarchical model (6^3-factor) (Figure A2, Appendix B). It was hypothesized that the data would show a better fit for the 6-factor solution when compared to the 3-factor and the 6^3-factor hierarchical model and would show an acceptable fit for the 6^2-factor hierarchical model, indicating that the use of composite scores of need satisfaction or need frustration would be feasible. It was further hypothesized that the 6-factor model would show an acceptable to good fit for the subsamples with different severities of depressive symptoms and that measurement invariance could be established. Following this, the predictive validity of the 12-item version of the German BPNSFS was investigated and it was expected that need frustration especially relates to ill-being (depressive symptoms and anxiety) and need satisfaction mostly relates to physical and mental quality of life.

2. Materials and Methods

2.1. Participants and Procedure

Data from the baseline assessment of the project "STEP.De -Sports Therapy for Depression", assessing the implementation of sports therapy as an alternative treatment in depressed patients [22], were used. Patients were recruited through four local health insurance carriers in Berlin, Germany, between August 2018 and March 2021. Strict inclusion and exclusion criteria were followed for the clinical sample of mild to moderately severe depression diagnosed by psychotherapists using the Structural Clinical Interview I for Diagnostic and Statistical Manual of Mental Disorders 4 (DSM IV), Structured Clinical Interview for DSM (SCID) I, Axis 1, Section A, E, and I [23] (for full details, see [22]). The validation sample consisted of $N = 344$ patients (71.8% female) with mild to moderately severe depression (Table A1). The mean age was 47.5 years (range = 20–65, $SD = 11.1$). Figure A1 shows a population pyramid of the sample according to age and gender.

Most patients indicated that they were married or cohabiting (59.1%) or single (26.3%). With respect to the highest completed education level, 56.0% of the patients completed secondary school, and 34.9% had higher education. Finally, with regard to personal monthly net income, 57.5% of the patients earned between EUR 1000 and 2000 (middle income) and 32.1% earned more than EUR 2000 (high income) (Additional Information on measures in Appendix A.1). Regarding their employment status, 82.2% of the patients worked within the last three months. Finally, 97.0% of the participants spoke German as a first language.

The missing rate for the BPNSFS items was low (0 to 1.2%). The ethics committee of the Freie Universität Berlin (No.206/18) and the University of Potsdam (No.17/2018) approved the study registered under the trial registration number ISRCTN28972230. Informed written consent was obtained from all participants.

Power analyses were a priori calculated using the R package 'SampleSize4ClinicalTrials' [24] We report how we determined our sample size, all data exclusions (if any), all data inclusion/exclusion criteria, whether inclusion/exclusion criteria were established prior to data analysis (see [22]), all measures in the study, and all analyses including all tested models.

2.2. Measures

A short 12-item version of the previously validated German 24-item version of the BPNSFS [9] was used to assess basic psychological need satisfaction and frustration. Items were selected based on the best factor loadings found in the German validation of the scale [9], the intersection with the original English version and the Dutch version of the scale [13], and the clearest German wording by consensus within the researcher team. Four items, of which two measured satisfaction and two measured frustration (e.g., competence satisfaction "I am good at what I do", competence frustration "I feel disappointed with many of my performances"), all of them rated on a 5 point Likert scale with the range from 1 (completely disagree) to 5 (completely agree) assessed the three basic needs [4].

As indicators of subjective ill-being, depressive symptoms were measured with the German version [25] of the 21-item Beck Depression Inventory (BDI-II) [26] (e.g., "I feel sad most of the time"). Depression severity was classified according to Beck et al. [26]. The reliability in the present study had a Cronbach's $\alpha = 0.90$". Anxiety was measured with 5 items from the VDS90 (Verhaltensdiagnostik-System/behavioral diagnostic system) [27] (e.g., "I avoid anxiety-inducing situations as often as possible"). The reliability in the present sample had a Cronbach's $\alpha = 0.73$.

Health-related quality of life as an indicator of well-being was measured by the 12-item Short Form Survey (SF-12) [28] (e.g., "In general, would you say your health is: excellent, very good, good, fair, poor?"). For extended information please see Additional Information on measures in Appendix A.3.

2.3. Analytical Strategy

Analyses were performed using IBM SPSS (Version 27) and R (Version 1.2.5042). Four different hypothetical models were tested via confirmatory factor analysis (CFA): (1) a 3-factor model representing the needs for autonomy, competence, and relatedness (not differentiating between satisfaction and frustration); (2) a 6-factor model differentiating between autonomy satisfaction, competence satisfaction, relatedness satisfaction, autonomy frustration, competence frustration, and relatedness frustration [13]; (3) a 2-factor hierarchical model (6^2-factor) with six first-order (same as the previous model) and two second-order factors (need satisfaction and need frustration) included; (4) and a 3-factor hierarchical model (6^3-factor) including six first-order factors (same as the two previous models) and three second-order factors (autonomy, competence, and relatedness) (Figure A2, Appendix A.2). CFAs were conducted using the "lavaan" package in R [29].

To evaluate the 3-factor model and the 6^3-factor hierarchical model, frustration items were recoded to enable the creation of composite latent scores per need (Figure A2). Model comparison (CFA), measurement invariance (multigroup CFA), and predictive validity (structural equation modeling, SEM) were tested. For extended information about the analytical strategy please see Additional Information on analytical strategy in Appendix A.4.

3. Results

3.1. Descriptive Statistics and Correlations

As displayed in Table 1, correlational analyses showed that the three variables of need satisfaction were positively correlated with one another, while being negatively related to the three need frustration variables, which were equally positively correlated. In addition, need frustration correlated positively with ill-being and negatively with quality of life, whereas need satisfaction showed an opposite pattern of correlations. The scores for all items ranged from 1 to 5. Skewness values ranged between -0.85 and 1.17, thereby indicating minimal to moderate skewness. Kurtosis values ranged between -1.10 and 0.24, with some items showing a slightly flat distribution. Interitem correlations ranged from 0.08 to 0.55 in absolute value.

Table 1. Means, standard deviations, and correlations among the main study variables.

Measure	M	SD	1	2	3	4	5	6	7	8	9	10	11	12
1. Need satisfaction [a]	19.69	4.72	1											
2. Autonomy satisfaction	5.77	1.94	0.81 ***	1										
3. Competence satisfaction	6.19	2.09	0.80 ***	0.53 ***	1									
4. Relatedness satisfaction	7.71	1.92	0.73 ***	0.39 ***	0.37 ***	1								
5. Need frustration [a]	15.79	4.78	−0.51 ***	−0.34 ***	−0.46 ***	−0.39 ***	1							
6. Autonomy frustration	6.51	2.04	−0.29 ***	−0.23 ***	−0.26 ***	−0.20 ***	0.68 ***	1						
7. Competence frustration	5.21	2.25	−0.46 ***	−0.34 ***	−0.49 ***	−0.25 ***	0.81 ***	0.38 ***	1					
8. Relatedness frustration	4.11	2.10	−0.41 ***	−0.20 ***	−0.31 ***	−0.44 ***	0.74 ***	0.24 ***	0.43 ***	1				
9. Depressive symptoms	22.63	9.89	−0.53 ***	−0.41 ***	−0.48 ***	−0.33 ***	0.64 ***	0.44 ***	0.60 ***	0.41 ***	1			
10. Anxiety symptoms	0.92	0.74	−0.26 ***	−0.17 **	−0.29 ***	−0.15 **	0.28 ***	0.14 *	0.23 ***	0.23 ***	0.41 ***	1		
11. Physical quality of life	43.67	9.14	0.14 ***	0.03	0.19 **	0.14 *	−0.22 ***	−0.23 ***	−0.20 ***	−0.10	−0.37 ***	−0.26 ***	1	
12. Mental quality of life	31.52	9.16	0.36 **	0.28 ***	0.35 ***	0.17 **	−0.54 ***	−0.39 ***	−0.46 ***	−0.33 ***	−0.65 ***	−0.30 ***	0.01	1

Sample size ranged from $n = 305$ to $n = 344$ due to missing values in the variables. [a] Composite score over all three needs. * $p < 0.05$; ** $p < 0.01$; *** $p < 0.001$.

Next, the effects of sociodemographic variables were explored using a MANCOVA with gender, education level, and net income as between-subjects variables, age as a covariate, and the outcomes (depressive symptoms, anxiety symptoms, physical quality of life, and mental quality of life) as dependent variables. The results of the MANCOVA indicated that neither age (F (4269) = 2.37, ns), gender (F (4269) = 0.73, ns), education level (F (8538) = 1.57, ns), marital status (F (12,711.1) = 0.85, ns), or net income (F (8538) = 1.09, ns) yielded a significant multivariate main effect. Therefore, the sociodemographic variables were not included in the SEM analyses.

3.2. Confirmatory Factor Analyses

Chi-square tests were significant for all of the models, indicating that the models did not show a perfect model fit. As displayed in Table 2, the 3-factor model showed a poor fit, whereas the other models (especially the 6-factor model) showed an adequate fit. The 6-factor model showed a significant better fit than all other models, and this difference in fit was also found to be meaningful (ΔCFI > 0.01 for all model comparisons).

Table 2. Goodness-of-fit indices of the tested models (N = 344).

	χ^2	df	χ^2/df	CFI	TLI	SRMR	RMSEA (90% CI)	AIC	BIC	Model Comparison ΔSBS-χ^2 (Δdf)
6-factor model	68.869 **	39	1.77 [g]	0.963 [g]	0.937	0.039 [g]	0.049 [g] (0.029, 0.068)	12,536.838	12,570.926	
3-factor model	190.819 ***	51	3.74 [n]	0.821 [n]	0.768	0.068 [a]	0.095 [n] (0.081, 0.109)	12,653.019	12,679.086	112.42 (12) ***
6^2-factor hierarchical model	108.419 ***	47	2.31 [a]	0.920 [a]	0.888	0.051 [a]	0.066 [a] (0.050, 0.082)	12,569.265	12,598.005	35.22 (8) ***
6^3-factor hierarchical model	130.759 ***	45	2.91 [a]	0.894 [n]	0.845	0.060 [a]	0.078 [a] (0.062, 0.093)	12,591.771	12,621.849	63.575 (6) ***

df, Degrees of Freedom; CFI, Comparative Fit Index; TLI, Tucker–Lewis Index; SRMR, Standardized Root Mean Square Residual; RMSEA, Root Mean Square Error of Approximation; AIC, Akaike's Information Criterion; BIC, Bayesian Information Criterion; ΔSBSχ^2, Satorra–Bentler scaled chi-square difference. [g] Good value; [a] acceptable value; [n] unacceptable value. Models were compared with the 6-factor model. ** $p < 0.01$; *** $p < 0.001$.

Based on the superior fit of the 6-factor model, this model was adopted to continue the scale validation process. Table A2 shows the parameter estimates of the CFA presenting this model. Standardized factor loadings were above 0.50 as well as being significant at a $p < 0.001$ level, showing small robust standard errors (ranging from 0.04 to 0.08) [4]. By including residual item correlations between items 3 and 10, the assumption of conditional independence for further analyses was loosened. A similar wording of both items, which was indicated by modification indices (M.I. = 18.253), made the inclusion plausible.

3.3. Reliability, Convergent, and Discriminant Validity

The need satisfaction and frustration subscales showed an adequate internal consistency, with Cronbach's alpha at 0.75 and 0.69, respectively. The reliability of the 4-item autonomy (0.52) and relatedness (0.63) subscales was low, but adequate for competence (0.75). The coefficient omega (ω) [30] was calculated to evaluate the reliability of the three satisfaction subscales and the three frustration subscales in the 6-factor model, as it was considered a better indicator than Cronbach's alpha [31]. Cronbach's alpha was also calculated for comparison with other studies. Only the values of competence satisfaction and competence frustration exceeded the 0.70 threshold, showing adequate coefficient omega and Cronbach's alpha values (Table A3).

Convergent was determined by calculating Average Variance Extracted (AVE), which estimates the common variance between the indicators and their latent factors. Convergent validity was established when the values of AVE exceeded the cut-off of 0.50 [32], indicating that more than 50% of the variance of the construct is due to its indicators. The results showed that the values of AVE exceeded the cut-off value of 0.50 for competence satisfaction (0.55) and competence frustration (0.54), but not for the other four factors (Table A3).

The evaluation of discriminant validity was performed by using the Fornell and Larcker (1981) [33] method based on the comparison between the AVE of each factor with the shared variance between factors. Discriminant validity is supported if the AVE of each factor is higher than the squared correlations between the factor and all other factors in the model. Calculated squared correlations are shown in Table A4. According to the results, discriminant validity was demonstrated for competence frustration.

3.4. Measurement Invariance across Patients with Different Severities of Depression

Across patients with different severities of depression, the model and its measurement invariance were tested. Therefore, the sample was divided into two groups: minimal/mild and moderate/severe depressive symptoms [9,26]. The primary aim of the current study was to examine the unique contribution of psychological need frustration and need satisfaction in the prediction of adults' mental well-being and ill-being in a heterogeneous sample of adults ($N = 334$; $M_{age} = 43.33$, $SD = 32.26$; 53% females). Prior to this, validity evidence was provided for the German version of the Basic Psychological Need Satisfaction and Frustration Scale (BPNSFS) based on Self-Determination Theory (SDT). The results of the validation analyses found the German BPNSFS to be a valid and reliable measurement. Further, structural equation modeling (SEM) showed that both need satisfaction and frustration yielded unique and opposing associations with well-being. Specifically, the dimension of psychological need frustration predicted adults' ill-being. Future research should examine whether frustration of psychological needs is involved in the onset and maintenance of psychopathology (e.g., major depressive disorder). When calculating the 6-factor model for the two subsamples separately, fit indices remained good for the sample with minimal/mild depressive symptoms and acceptable to good for the sample with moderate/severe depressive symptoms (Table A5). Overall weak measurement invariance (imposing equality of factor loadings) across the two subsamples, could be established with the constrained model not differing significantly from the unconstrained model (ΔSBS-χ^2 (6) = 12.14, $p < 0.059$), and the SRMR and RMSEA indices for the constrained and unconstrained models not being substantially different (ΔCFI < -0.01, ΔSRMR < 0.03, ΔRMSEA < 0.015), with the exception of the CFI difference (although this value was only slightly higher than the recommended cutoff of -0.01 (ΔCFI $= -0.013$) [4]. Strong measurement invariance, which additionally requires equality constraints on the corresponding item intercepts, could not be established because the constrained model differed significantly from the unconstrained model (ΔSBS-χ^2 (6) $= 13.81$, $p < 0.032$), and the difference in CFI indices was found to be slightly larger than -0.01 (ΔCFI $= -0.015$).

3.5. Predictive Validity

The use of composite scores of need satisfaction and need frustration as predictors was substantiated by an acceptable fit to the data of the 6^2-factor hierarchical model [4]. Need frustration related positively to depressive symptoms and anxiety, while relating negatively to physical quality of life and mental quality of life. On the contrary differences in ill-being or quality of life were not predicted by need satisfaction. Residual correlations between e9 and e10 were included in the model after the inspection of the modification indices, and are justified as PCS-12 and MCS-12 are two subscales of the same questionnaire. The chi-square test for the model was significant (χ^2 (26) $= 72.88$, $p < 0.001$), indicating that the model did not show a perfect model fit. The model showed good SRMR (0.039), CFI values (0.954), and an acceptable RMSEA value (0.072). Figure A2 displays an overview of the SEM model. To assess whether the high standardized estimate between need frustration and depressive symptoms could be due to shared variance of need satisfaction and need frustration in the outcome, the existence of multicollinearity was investigated using the Variance Inflation Factor (VIF). The VIF was 1.30 for need satisfaction and need frustration and indicated that multicollinearity was not present to a critical degree.

4. Discussion

An increasing amount of research has now indicated that while need satisfaction is essential for individuals' well-being and striving, need frustration is imperative in explaining ill-being and even psychopathology including depressive symptoms [3]. There is a need for a shorter version of the BPNSFS to increase its applicability in the clinical context due to scanty research on the effects of need-based experiences in clinical samples [4].

In a validation of the 12-item BPNSFS, it was shown that both the scale is considered reliable and the reliability of the subscales' need satisfaction and need frustration is acceptable [4]. Focusing on the six factors (i.e., type of need X satisfaction/frustration), reliabilities were quite low (with the exception of competence satisfaction and frustration). This is understandable as each of these factors only contained two items. Further, CFA analyses confirmed our hypothesis that the 6-factor model and the 6^2-model (to a lesser extent) are the best-fitting models. Both models distinguish between the satisfaction and frustration dimensions of the three needs. Compared to the 3-factor model and the 6^3-model, the two models showed a better fit, as the former two did not distinguish between satisfaction and frustration of the needs explicitly. In the 24-item version of the German BPNSFS [7] and in previous validation studies in other languages [11], as well as theoretical assumptions [27], the same results could already be obtained. For example, the results of previously recently published Norwegian [4], Arabic [34], and Italian [35] BPNSFS validation studies, also indicated evidence for a 6-factor model. Compared to the 6-factor model, the fit of the 6^2-factor model was worse. Possibly, this expected result, based on the methodological and theoretical assumptions, may be due to the need-specific variance shared by the items of the different latent factors [7]. A worse fit for the hierarchical 6^2-factor model compared to the 6-factor model was also found by Heissel et al. (2018) [7].

In previous international studies, a worse fit for the hierarchical models compared to the reduced 6-factor model, was also shown (e.g., [11,29]). Due to its acceptable fit and the distinction between need satisfaction and need frustration as well as the distinction between different dimensions, the 6^2-factor model is nevertheless feasible [16].

Aiming to investigate whether the 6-factor model would be found in samples differing in the severity of depressive symptoms, we ran additional multigroup CFAs. Intercepts between groups were not equal; although, weak invariance was found overall. As would be expected, the group with moderate/severe depressive symptoms was found to experience less need satisfaction and more need frustration compared to the group with milder depressive symptoms.

Focusing on ill-being and quality of life as outcomes, the final aim of this study was to examine the predictive value of need satisfaction and need frustration as assessed by the brief version of the BPNSFS [4]. Need frustration (but not need satisfaction) was found to relate to depressive symptoms and anxiety (positively) and to physical and mental quality of life (negatively). These findings, therefore, indicate that need satisfaction and need frustration indeed represent distinct constructs and should be assessed separately. In line with previous work, need frustration, but not need satisfaction, was found to be essential in the prediction of ill-being [3]. Additionally, these results extend previous research by showing the importance of need-based experiences, and especially frustration experiences, in a clinical sample [12]. Surprisingly, and in contrast to the results obtained in previous studies [13,17], need satisfaction did not relate to quality of life once need frustration was accounted for. A possible explanation for this unexpected finding is related to the type of well-being indicators that were assessed in this study. That is, in contrast with previous research mainly focusing on the predictive value of need satisfaction in well-being indicators such as vitality, life satisfaction, or self-esteem (e.g., [13]), we focused on mental and physical quality of life. This quality of life was assessed with the Short Form Survey [28], which is quite strongly focused on limitations in daily life with only two items focusing on positive indicators of quality of life (i.e., feeling calm and peaceful, having a lot of energy). Given the predominant focus on maladaptive functioning within this questionnaire, it is understandable that especially need frustration was found to be predictive of quality of life. Another possible explanation for the absence of the predictive value of need satisfaction is related to the employed sample. In a clinical sample, it is possible that the extent of experienced need frustration is of stronger predictive value for indicators of both well-being and ill-being when compared to a more general sample. Further longitudinal research is needed to find out whether the absence of need frustration contributes additional valuable insights into human well-being beyond the presence of need satisfaction.

Limitations

A limitation of this study is the small sample size (in each subgroup), which may have caused a poorer estimation of the multigroup analysis (see [36]). Furthermore, due to the cross-sectional design, causality assumptions cannot be made, which in turn limits the results of the structural analysis.

The six factors of the proposed model are composed of two items each, not fulfilling the general requirement of three items to saturate a factor [32]. However, the use of the few best indicators has also been recommended, considering that one or two indicators are often sufficient, and encourages development of theoretically sophisticated models. Therefore, the two factors are retained in the 6-factor model [37].

5. Conclusions

The main findings showed an expected pattern of results within a sample of clinically depressed people, with high basic need frustration associated with greater self-reported symptomatology. Perhaps more important, the results of this study also yielded an economical and valid short form of the BPNSFS questionnaire that is useful for health science research. For the 12-item short version of the German BPNSFS, a six-factor solution differentiating between the three needs (autonomy, competence, and relatedness) in two dimensions (frustration, satisfaction) was feasible. The BPNSFS questionnaire in its short form turned out to be a beneficial instrument for specific health groups. In terms of predictive validity in a depressed sample, the recently added dimension of need frustration was found to be distinct from the satisfaction dimension and the only predictive value for all of the health-related outcome variables. Future research should investigate these associations incorporating different ways of using the scales (e.g., calculate balanced scores according to dimension). If these results can be verified in longitudinal studies, the assessment of basic need frustration alongside basic need satisfaction becomes crucial not only in the context of psychological ill-being. The great potential for future research can result in a deeper understanding of the onset and maintenance of ill-being and add substantial evidence in support of the underlying theory [4].

Author Contributions: Conceptualization, A.H. and M.A.R.; methodology, A.H., M.A.R., A.S., J.V.d.K.-D.; formal analysis, A.S.; writing—original draft preparation, A.H., T.B. and A.S.; writing—review and editing, A.H., A.S., A.P., T.B., C.S., M.A.R., J.V.d.K.-D. All authors have read and agreed to the published version of the manuscript.

Funding: The STEP.De study, from which this dataset stems was funded by the Innovationsausschuss des Gemeinsamen Bundesausschusses (G-BA; 01NVF17050; Innovation Fund of the Joint Federal Committee). Partly funding by an intramural junior research group grant to A.H. from the University of Potsdam, the Zentrale Forschungsförderung "UP—Innovative Ideen fördern" to A.H. and from the University of Potsdam Graduate School to A.H. Funded by the Deutsche Forschungsgemeinschaft (DFG, German Research Foundation)–Project Number 491466077.

Institutional Review Board Statement: The study protocol was approved by the ethics committee of the University of Potsdam (No. 17/2018) and the Freie Universität Berlin (No. 206/2018) and registered in the ISRCTN registry.

Informed Consent Statement: Informed written consent was obtained from all participants involved in the study.

Data Availability Statement: The datasets generated for this study are available upon reasonable request to the corresponding author.

Acknowledgments: We thank the STEP.De participants and team and all colleagues and partners involved that helped to execute this study.

Conflicts of Interest: The authors declare that the research was conducted in the absence of any commercial or financial relationships that could be construed as a potential conflict of interest. AH is the founder and CEO of the Centre for Emotional Health Germany GmbH supported by the Potsdam Transfer Centre from the University of Potsdam.

Appendix A. Supplementary Text

Appendix A.1. Additional Information of Patient Characteristics at Baseline

Education Level and Income

For education level, a variable with the three categories of low (lower secondary school), middle (secondary school diploma), and high education (university entrance qualification and university degree) levels was created. The income variable was categorized into low (EUR <1000), middle (EUR 1000–2000), and high (EUR >2000) personal monthly net income.

Appendix A.2. Additional Information on Hypothetical Models

Four different hypothetical models were tested via confirmatory factor analysis (CFA): (1) a 3-factor model representing the needs for autonomy, competence, and relatedness (not differentiating between satisfaction and frustration); (2) a 6-factor model differentiating between autonomy satisfaction, competence satisfaction, relatedness satisfaction, autonomy frustration, competence frustration, and relatedness frustration [13]; (3) a 2-factor hierarchical model (6^2-factor) including six first-order (same as the previous model) and two second-order factors (need satisfaction and need frustration); (4) and a 3-factor hierarchical model (6^3-factor) including six first-order factors (same as the two previous models) and three second-order factors (autonomy, competence, and relatedness) (Figure A2).

Appendix A.3. Additional Information on Measures

Ill-Being. With the German version [25] of the Beck Depression Inventory (BDI-II) depressive symptoms as one indicator of subjective ill-being could be measured [26]. The BDI-II is a 21-item self-report depression screening measure. Each item is rated on a 4-point Likert scale ranging from 0 to 3, with higher scores indicating higher levels of depressive symptoms. According to the BDI-II manual [26], a score of 0–13 indicates minimal depression, 14–19 mild depression, 20–28 moderate depression, and 29–63 severe depression. The reliability in the present sample had a Cronbach's $\alpha = 0.91$.

Anxiety as an indicator of subjective ill-being was measured with five items from the VDS90 (Verhaltensdiagnostik-System/behavioral diagnostic system) [27], a screening instrument designed to detect mental disorders in accordance with ICD-10 [38]. The scores range between 0 (*no mental health problems*) and 3 (*severe mental health problems*). The total anxiety score is the mean (average) of the five items. The reliability in the present study had a Cronbach's $\alpha = 0.74$.

Quality of life. Health-related quality of life as an indicator of well-being was measured by the 12-item Short Form Survey (SF-12) [28]. The SF-12 generates two main scores ranging from 1 to 100, the mental component summary (MCS-12) and the physical component summary (PCS-12), with higher scores indicating better quality of life.

Appendix A.4. Additional Information on Analytical Strategy

All models tested (3-factor model, 6-factor model, 6^2-factor hierarchical model, and 6^3-factor hierarchical model) used the robust maximum likelihood (robust ML) estimation method, with robust (Huber–White) standard errors to correct for the observed non-normality of the variables. A chi-square test and alternative fit indices were used to evaluate the models. To test the overall fit of the model, the chi-square test was performed. Goodness of fit was considered to provide a non-significant result at a 0.05 threshold [39]. The effects of sample size on chi-square were minimized by calculating a relative/normalized chi-square (ratio of chi-square test to degrees of freedom) [35]. A value <2 for the normed chi-square is considered a good model fit and a value <3 an acceptable model fit [40]. Due to a possible influence of model complexity or sample size on the fit indices, to evaluate approximate model fit, different types of fit indices have been used [36,37]. To check how well the proposed model fits the data, the following two absolute indices were used: the standardized root mean square residual (SRMR) and the robust root mean square error of approximation (RMSEA). For the SRMR, values <0.05 are considered good, and values <0.10 are acceptable [41]. For the RMSEA, values <0.05 reflect a good fit with the data, and <0.08 is an acceptable fit [42]. Two comparative fit indices that compare the chi-squared value to a baseline model were used: the robust Bentler Comparative Fit Index (CFI) and the robust Tucker–Lewis Index (TLI). Values >0.90 and >0.95 indicate an acceptable and good fit to the data, respectively [43].

Model comparison

The Satorra–Bentler scaled chi-square difference test (SBS-χ^2) [44] was used to better approximate chi-square under non-normality. In this chi-square difference test (SBS-χ^2), the usual normal value theory chi-square statistic was divided by a scaling correction. The difference in CFI (ΔCFI) [45] and two predictive fit indices, the Akaike Information Criterion (AIC) values [46] and the Bayesian Information Criterion (BIC) [45] were considered because the SBS-χ^2, as a chi-square fit test, is sensitive to sample size and violation of the normality assumption. This could result in potentially useful models being discarded [41]. To support this model, a reduction in CFI value of less than 0.01 should be considered when fitting the simplified model according to the guidelines of [45,47].

Models with smaller AIC and BIC values are preferred [41]. For the model that best fits the observed data, the parameter time points of the confirmatory factor analysis (CFA) are given.

Appendix A.4.1. Measurement Invariance

The measurement invariance of the 6-factor model in a series of multi-group CFAs with three levels of invariance (con-figural, weak, and strong invariance) [46] was tested to determine whether the 12-item BPNSFS has the same psychometric properties in patients with different levels of depressive symptom severity (minimal/mild vs. moderate/severe) according to the BDI-II [20]. Across all groups, configural invariance imposes the same factor structure. This contrasts with weak invariance, which imposes all factor loadings as the same across groups. Strong invariance additionally constrains the equality of intercepts. All model fits were tested using robust maximum likelihood (robust ML) and full information maximum likelihood (FIML) estimation. Model comparisons were processed using the Satorra–Bentler scaled chi-square difference test ($\Delta SBS\chi^2$) for two nested models [44], changes in fit indices, and AIC and BIC values. A change of -0.01 in CFI, supplemented by a change of 0.03 in SRMR or a change of 0.015 in RMSEA, would indicate non-invariance when testing weak invariance. When testing strong invariance, a change of ≥ -0.01 in CFI, supplemented by a change of ≥ 0.015 in SRMR or a change of ≥ 0.015 in RMSEA, would indicate non-invariance. Among the three indices, CFI is chosen as the main criterion [47].

Appendix A.4.2. Predictive Validity

To examine the unique predictive effects of basic psychological need satisfaction and frustration on ill-being and well-being, a structural equation model (SEM) was created with the "lavaan" package [29] and Ωnyx software [48]. Previous dimensionality analyses (CFA) indicated the use of the item parceling method [49]. As an indicator of the latent variables (need satisfaction and need frustration), sum scores for the six subscales (two items in each package) were used and calculated at the individual level. Unlike a more complex model that includes each item, the advantage of item parceling is that the ratio between the sample size and the number of free parameters is higher [49,50]. The latent variables were used as simultaneous predictors of ill-being (depressive symptoms measured with the BDI-II and anxiety measured with the VDS90) and well-being (physical and mental quality of life measured with the SF-12). Residual correlations between the item parcels of autonomy/competence/relatedness satisfaction and autonomy/competence/relatedness frustration were included in the model, due to the assumption that the parcels shared need-specific variance. The model fit was tested via the same method used in the CFA: robust ML and FIML estimation.

Appendix B. Supplementary Figures

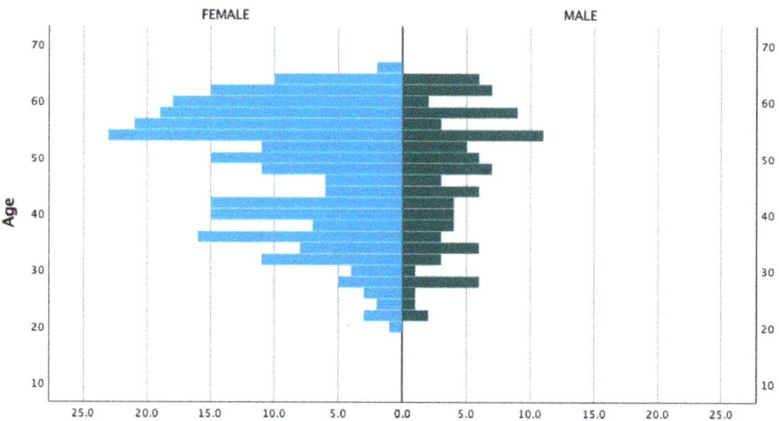

Figure A1. Population pyramid of the sample.

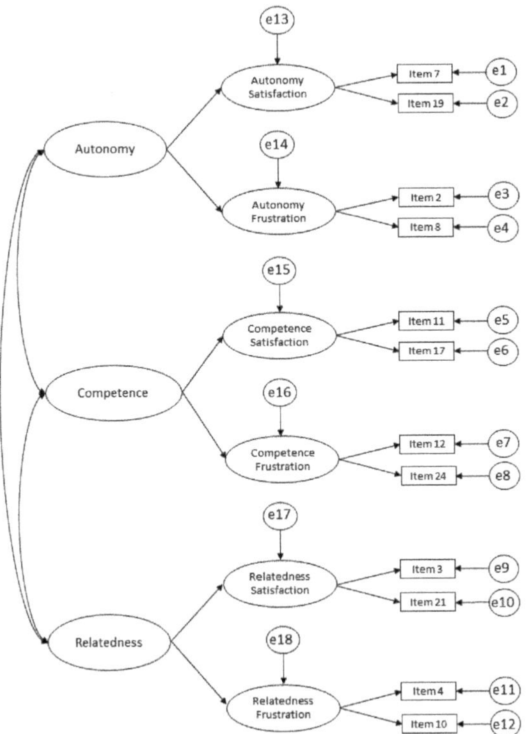

Figure A2. The 3-factor hierarchical model. *e*, error.

Appendix C. Supplementary Tables

Table A1. Characteristics of the sample (*N* = 344).

	n	No. (%)
Age (years), *M* (*SD*), range	339	47.5 (11.1), 20–65
Gender	341	
Female		245 (71.8)
Male		96 (28.2)
Marital status	338	
Single		89 (26.3)
Married/cohabiting		200 (59.1)
Separated/divorced		37 (10.9)
Widowed		12 (3.6)
Education level	332	
Lower secondary school		34 (9.0)
Secondary school		186 (56.0)
Higher education		116 (34.9)
Net income	327	
Low income		34 (10.4)
Middle income		188 (57.5)
High income		105 (32.1)
Worked within the last 3 months	338	
Yes		278 (82.2)
No		60 (17.8)
First language	328	
German		318 (97.0)
Other		10 (3.0)
Depressive symptoms (BDI-II)	344	
Minimal		63 (18.3)
Mild		74 (21.5)
Moderate		105 (30.5)
Severe		102 (29.7)

BDI-II, Beck Depression Inventory II.

Table A2. Parameter estimates of confirmatory factor analysis (CFA) for the 6-factor model (*N* = 344).

Items *	Satisfaction			Frustration			R^2
	Aut	Comp	Relate	Aut	Comp	Relate	
7. I feel that my decisions reflect what I really want.	0.55						0.30
19. I feel I have been doing what really interests me.	0.65						0.42
11. I feel capable at what I do.		0.68					0.46
17. I feel competent to achieve my goals.		0.80					0.63
3. I feel that the people I care about also care about me.			0.72				0.53
21. I experience a warm feeling with the people I spend time with.			0.50				0.25
2. Most of the things I do feel like "I have to."				0.54			0.29
8. I feel forced to do many things I wouldn't choose to do.				0.59			0.35
12. I feel disappointed with much of my performance.					0.71		0.51
24. I feel like a failure because of the mistakes I make.					0.76		0.57
4. I feel excluded from the group I want to belong to.						0.63	0.40
10. I feel that people who are important to me are cold and distant toward me.						0.61	0.37

Aut, Autonomy; Comp, Competence; Relate, Relatedness. * Numbers according to order of the original 24-item BPNSFS version.

Table A3. Coefficient omega and Average Variance Extracted (AVE) in the 6-factor model, and Cronbach's alpha (*N* = 344).

	Coefficient Omega	AVE	Cronbach's Alpha
Autonomy satisfaction	0.53	0.37	0.52
Competence satisfaction	0.71	0.55	0.7
Relatedness satisfaction	0.56	0.36	0.53
Autonomy frustration	0.48	0.32	0.48
Competence frustration	0.70	0.54	0.70
Relatedness frustration	0.56	0.42	0.55

Table A4. Correlation matrix between the latent factors of the 6-factor model.

	1	2	3	4	5	6
1. Autonomy satisfaction		0.76	0.53	0.21	0.3	0.08
2. Competence satisfaction	0.87		0.40	0.16	0.45	0.19
3. Relatedness satisfaction	0.73	0.63		0.14	0.14	0.30
4. Autonomy frustration	−0.46	−0.40	−0.37		0.40	0.19
5. Competence frustration	−0.56	−0.67	−0.37	0.63		0.44
6. Relatedness frustration	−0.28	−0.44	−0.55	0.44	0.66	

The scores in the left upper part of the table are squared correlations.

Table A5. Confirmatory factor analyses for subsamples with different severity of depressive symptoms and invariance testing.

	$SB\chi^2$	*df*	$SB\chi^2/df$	$\Delta SBS\chi^2$ (Δdf)	CFI	ΔCFI	SRMR	ΔSRMR	RMSEA	ΔRMSEA
Sample Overall (N = 344)	50.12	38	1.32 [g]		0.985 [g]		0.029 [g]		0.032 [g]	
Minimal/mild depressive symptoms (n = 137)	46.81	38	1.23 [g]		0.967 [g]		0.049 [g]		0.041 [g]	
Moderate/severe depressive symptoms (n = 207)	58.54 *	38	1.72 [g]		0.939 [a]		0.045 [g]		0.053 [a]	
Invariance level										
Configural	105.839 *	76	1.39		0.950		0.047		0.048	
Weak	118.128 **	82	1.44	12.29 (6)	0.937	−0.013	0.056	0.009	0.052	0.004
Strong	133.221 **	88	1.51	15.09 (6)	0.922	−0.015	0.063	0.007	0.056	0.004

$SB\chi^2$, Satorra–Bentler scaled chi-square; *df*, Degrees of Freedom; $\Delta SBS\chi^2$, Satorra–Bentler scaled chi-square difference; CFI, Comparative Fit Index; ΔCFI, Change in CFI when compared to the baseline model; SRMR, Standardized Root Mean Square Residual; ΔSRMR, Change in SRMR when compared to the baseline model; RMSEA, Root Mean Square Error of Approximation; ΔRMSEA, Change in RMSEA when compared to the baseline model. [g] Good value; [a] acceptable value; * $p < 0.05$; ** $p < 0.01$.

References

1. Ryan, R.M.; Deci, E.L. *Self-Determination Theory: Basic Psychological Needs in Motivation, Development, and Wellness*; Guilford Press: New York, NY, USA, 2017; ISBN 978-1-4625-2876-9.
2. Deci, E.L.; Ryan, R.M. The "What" and "Why" of Goal Pursuits: Human Needs and the Self-Determination of Behavior. *Psychol. Inq.* **2000**, *11*, 227–268. [CrossRef]
3. Vansteenkiste, M.; Ryan, R.M. On Psychological Growth and Vulnerability: Basic Psychological Need Satisfaction and Need Frustration as a Unifying Principle. *J. Psychother. Integr.* **2013**, *23*, 263–280. [CrossRef]
4. van der Kaap-Deeder, J.; Sanchez, A.; Johannessen, M.R.A.; Stenseng, F.; Saksvik-Lehouillier, I.; Heissel, A. The Validation of the Norwegian Basic Psychological Need Satisfaction and Frustration Scale: A Stratified Sampling Procedure. *Front. Psychol.* **2022**, *13*, 1032006. [CrossRef]
5. Namlı, Z.; Demirkol, M.E.; Tamam, L.; Karaytuğ, M.O.; Yeşiloğlu, C. Validity and Reliability Study of the Turkish Version of the Unbearable Psychache Scale. *Alpha Psychiatry* **2022**, *23*, 166. [CrossRef]
6. Shneidman, E.S. *Suicide as Psychache: A Clinical Approach to Self-Destructive Behavior*; Jason Aronson: Lanham, MD, USA, 1993; ISBN 0-87668-151-8.
7. Ryan, R.M.; Deci, E.L.; Vansteenkiste, M. Autonomy and Autonomy Disturbances in Self-Development and Psychopathology: Research on Motivation, Attachment, and Clinical Process. In *Developmental Psychopathology*; Cicchetti, D., Ed.; John Wiley & Sons, Inc.: Hoboken, NJ, USA, 2016; pp. 1–54. ISBN 978-1-119-12555-6.
8. Bartholomew, K.J.; Ntoumanis, N.; Thøgersen-Ntoumani, C. Self-Determination Theory and the Darker Side of Athletic Experience: The Role of Interpersonal Control and Need Thwarting. *Sport Exerc. Psychol. Rev.* **2011**, *7*, 23–27. [CrossRef]
9. Heissel, A.; Pietrek, A.; Flunger, B.; Fydrich, T.; Rapp, M.A.; Heinzel, S.; Vansteenkiste, M. The Validation of the German Basic Psychological Need Satisfaction and Frustration Scale in the Context of Mental Health. *Eur. J. Health Psychol.* **2018**, *25*, 119–132. [CrossRef]
10. Campbell, R.; Tobback, E.; Delesie, L.; Vogelaers, D.; Mariman, A.; Vansteenkiste, M. Basic Psychological Need Experiences, Fatigue, and Sleep in Individuals with Unexplained Chronic Fatigue. *Stress Health* **2017**, *33*, 645–655. [CrossRef] [PubMed]
11. Tóth-Király, I.; Bőthe, B.; Márki, A.N.; Rigó, A.; Orosz, G. Two Sides of the Same Coin: The Differentiating Role of Need Satisfaction and Frustration in Passion for Screen-based Activities. *Eur. J. Soc. Psychol.* **2019**, *49*, 1190–1205. [CrossRef]
12. Campbell, R.; Boone, L.; Vansteenkiste, M.; Soenens, B. Psychological Need Frustration as a Transdiagnostic Process in Associations of Self-Critical Perfectionism with Depressive Symptoms and Eating Pathology. *J. Clin. Psychol.* **2018**, *74*, 1775–1790. [CrossRef]
13. Chen, B.; Vansteenkiste, M.; Beyers, W.; Boone, L.; Deci, E.L.; Van der Kaap-Deeder, J.; Duriez, B.; Lens, W.; Matos, L.; Mouratidis, A.; et al. Basic Psychological Need Satisfaction, Need Frustration, and Need Strength across Four Cultures. *Motiv. Emot.* **2015**, *39*, 216–236. [CrossRef]
14. Nishimura, T.; Suzuki, T. Basic Psychological Need Satisfaction and Frustration in Japan: Controlling for the Big Five Personality Traits: Need Satisfaction and Frustration in Japan. *Jpn. Psychol. Res.* **2016**, *58*, 320–331. [CrossRef]
15. Rodríguez-Meirinhos, A.; Antolín-Suárez, L.; Brenning, K.; Vansteenkiste, M.; Oliva, A. A Bright and a Dark Path to Adolescents' Functioning: The Role of Need Satisfaction and Need Frustration across Gender, Age, and Socioeconomic Status. *J. Happiness Stud.* **2020**, *21*, 95–116. [CrossRef]
16. Haerens, L.; Aelterman, N.; Vansteenkiste, M.; Soenens, B.; Van Petegem, S. Do Perceived Autonomy-Supportive and Controlling Teaching Relate to Physical Education Students' Motivational Experiences through Unique Pathways? *Distinguishing between the Bright and Dark Side of Motivation. Psychol. Sport Exerc.* **2015**, *16*, 26–36. [CrossRef]

17. Liga, F.; Ingoglia, S.; Cuzzocrea, F.; Inguglia, C.; Costa, S.; Lo Coco, A.; Larcan, R. The Basic Psychological Need Satisfaction and Frustration Scale: Construct and Predictive Validity in the Italian Context. *J. Personal. Assess.* **2020**, *102*, 102–112. [CrossRef] [PubMed]

18. Frielink, N.; Schuengel, C.; Embregts, P.J.C.M. Psychometric Properties of the Basic Psychological Need Satisfaction and Frustration Scale—Intellectual Disability (BPNSFS-ID). *Eur. J. Psychol. Assess.* **2019**, *35*, 37–45. [CrossRef]

19. Heissel, A.; Pietrek, A.; Kangas, M.; der Kaap-Deeder, V.; Rapp, M.A. The Mediating Role of Rumination in the Relation between Basic Psychological Need Frustration and Depressive Symptoms. *J. Clin. Med.* **2023**, *12*, 395. [CrossRef]

20. Pietrek, A.; Kangas, M.; Kliegl, R.; Rapp, M.A.; Heinzel, S.; van der Kaap-Deeder, J.; Heissel, A. Basic Psychological Need Satisfaction and Frustration in Major Depressive Disorder. *Front. Psychiatry* **2022**, *13*, 962501. [CrossRef]

21. Eriksson, M.; Boman, E. Short Is Beautiful: Dimensionality and Measurement Invariance in Two Length of the Basic Psychological Need Satisfaction at Work Scale. *Front. Psychol.* **2018**, *9*, 965. [CrossRef]

22. Heissel, A.; Pietrek, A.; Schwefel, M.; Abula, K.; Wilbertz, G.; Heinzel, S.; Rapp, M. STEP.De Study—A Multicentre Cluster-Randomised Effectiveness Trial of Exercise Therapy for Patients with Depressive Symptoms in Healthcare Services: Study Protocol. *BMJ Open* **2020**, *10*, e036287. [CrossRef]

23. Wittchen, H.-U.; Zaudig, M.; Fydrich, T. Skid. In *Strukturiertes Klinisches Interview für DSM-IV*; Achse I Und II; Hogrefe: Göttingen, Germany, 1997.

24. Qi, H.; Zhu, F. Sample Size Calculation for the Comparison of Means or Proportions in Phase III Clinical Trials. Available online: https://cran.r-project.org/web/packages/SampleSize4ClinicalTrials/SampleSize4ClinicalTrials.pdf (accessed on 6 October 2022).

25. Hautzinger, M.; Keller, F.; Kühner, C. *BDI-II. Beck-Depressions-Inventar*; Pearson Assessment: Frankfurt, Germany, 2009; ISBN 3-8409-2349-2.

26. Beck, A.T.; Steer, R.A.; Brown, G.K. *BDI-II: Beck Depression Inventory*; Pearson: London, UK, 1996.

27. Sulz, S.K.D.; Grethe, C. Die VDS90-Symptomliste—Eine Alternative zur SCL90-R für Die Ambulante Psychotherapie-Praxis und das Interne Qualitätsmanagement? *Psychotherapie* **2005**, *10*, 38–48.

28. Ware, J.E.; Kosinski, M.; Keller, S.D. A 12-Item Short-Form Health Survey: Construction of Scales and Preliminary Tests of Reliability and Validity. *Med. Care* **1996**, *34*, 220–233. [CrossRef] [PubMed]

29. Rosseel, Y.; Jorgensen, T.D.; Rockwood, N.; Oberski, D.; Byrnes, J.; Vanbrabant, L.; Savalei, V.; Merkle, E.; Hallquist, M.; Rhemtulla, M.; et al. Lavaan: Latent Variable Analysis. 2020. Available online: https://research.tilburguniversity.edu/en/publications/lavaan-latent-variable-analysis (accessed on 6 October 2020).

30. Raykov, T. Estimation of Composite Reliability for Congeneric Measures. *Appl. Psychol. Meas.* **1997**, *21*, 173–184. [CrossRef]

31. Dunn, T.J.; Baguley, T.; Brunsden, V. From Alpha to Omega: A Practical Solution to the Pervasive Problem of Internal Consistency Estimation. *Br. J. Psychol.* **2014**, *105*, 399–412. [CrossRef] [PubMed]

32. Hair Jr, J.F.; Black, W.C.; Babin, B.J.; Anderson, R.E. *Multivariate Data Analysis*; Prentice Hall: London, UK, 2010; p. 785.

33. Fornell, C.; Larcker, D.F. Evaluating Structural Equation Models with Unobservable Variables and Measurement Error. *J. Mark. Res.* **1981**, *18*, 39. [CrossRef]

34. Zayed, K.N.; Omara, E.N.; Al-Rawahi, N.Y.; Al-Shamli, A.K.; Al-Atiyah, A.A.; Al-Haramleh, A.A.; Azab, M.S.; Al-Khasawneh, G.M.; Hassan, M.A. Psychometric Properties of the Arabic Version of the Basic Psychological Needs Satisfaction-Frustration Scale (BPNSFS). *BMC Psychol.* **2021**, *9*, 15. [CrossRef]

35. Costa, S.; Ingoglia, S.; Inguglia, C.; Liga, F.; Lo Coco, A.; Larcan, R. Psychometric Evaluation of the Basic Psychological Need Satisfaction and Frustration Scale (BPNSFS) in Italy. *Meas. Eval. Couns. Dev.* **2018**, *51*, 193–206. [CrossRef]

36. Hoyle, R.H.; Gottfredson, N.C. Sample Size Considerations in Prevention Research Applications of Multilevel Modeling and Structural Equation Modeling. *Prev. Sci.* **2015**, *16*, 987–996. [CrossRef]

37. Hayduk, L.A.; Littvay, L. Should Researchers Use Single Indicators, Best Indicators, or Multiple Indicators in Structural Equation Models? *BMC Med. Res. Methodol.* **2012**, *12*, 159. [CrossRef]

38. World Health Organization. *International Statistical Classification of Diseases and Related Health Problems: Instruction Manual*; World Health Organization: Geneva, Switzerland, 2004; Volume 2, ISBN 92-4-154653-0.

39. Barrett, P. Structural Equation Modelling: Adjudging Model Fit. *Personal. Individ. Differ.* **2007**, *42*, 815–824. [CrossRef]

40. Bollen, K.A. Structural Equation Models with Observed Variables. In *Structural Equations with Latent Variables*; John Wiley & Sons, Ltd.: Hoboken, NJ, USA, 2014; pp. 80–150. ISBN 978-1-118-61917-9.

41. Kline, R.B. *Principles and Practice of Structural Equation Modeling*, 4th ed.; Methodology in the Social Sciences; The Guilford Press: New York, NY, USA, 2015; ISBN 978-1-4625-2334-4.

42. Browne, M.W.; Cudeck, R. Alternative Ways of Assessing Model Fit. *Sociol. Methods Res.* **1992**, *21*, 230–258. [CrossRef]

43. Hu, L.; Bentler, P.M. Cutoff Criteria for Fit Indexes in Covariance Structure Analysis: Conventional Criteria versus New Alternatives. *Struct. Equ. Model. Multidiscip. J.* **1999**, *6*, 1–55. [CrossRef]

44. Satorra, A.; Bentler, P.M. A Scaled Difference Chi-Square Test Statistic for Moment Structure Analysis. *Psychometrika* **1999**, *66*, 507–514. [CrossRef]

45. Cheung, G.W.; Rensvold, R.B. Evaluating Goodness-of-Fit Indexes for Testing Measurement Invariance. *Struct. Equ. Model. Multidiscip. J.* **2002**, *9*, 233–255. [CrossRef]

46. Akaike, H. *Factor Analysis and AIC*; Springer: New York, NY, USA, 1987; Volume 16.

47. Chen, F.F. Sensitivity of Goodness of Fit Indexes to Lack of Measurement Invariance. *Struct. Equ. Model. Multidiscip. J.* **2007**, *14*, 464–504. [CrossRef]
48. von Oertzen, T.; Brandmaier, A.M.; Tsang, S. Structural Equation Modeling With Ωnyx. *Struct. Equ. Model. Multidiscip. J.* **2015**, *22*, 148–161. [CrossRef]
49. Bandalos, D.L. Finney Item Parceling Issues in Structural Equation Modeling. In *New Developments and Techniques in Structural Equation Modeling*; Marcoulides, G.A., Schumacker, R.E., Eds.; Lawrence Erlbaum Associates: Mahwah, NJ, USA, 2001; pp. 269–296. ISBN 978-0-8058-3593-9.
50. Bentler, P.M.; Chou, C.-P. Practical Issues in Structural Modeling. *Sociol. Methods Res.* **1987**, *16*, 78–117. [CrossRef]

Article

Childhood Adversities and Psychological Health of Adult Children of Parents with Mental Illness in Japan

Masako Kageyama [1,*], Taku Sakamoto [2], Ayuna Kobayashi [2], Akiko Hirama [2], Hiroyuki Tamura [2] and Keiko Yokoyama [3]

[1] Institute of Advanced Co-Creation Studies, Osaka University, 1-7 Yamadaoka, Suita 565-0871, Osaka, Japan
[2] KODOMO-PEER Tonoxbuilding, 3-5-1 Hirata, Ichikawa 272-0031, Chiba, Japan
[3] Department of Nursing, Faculty of Nursing, Yokohama Soei University, 1 Miho-cho, Yokohama 226-0015, Kanagawa, Japan
* Correspondence: kageyama@sahs.med.osaka-u.ac.jp; Tel./Fax: +81-6-6879-2553

Abstract: In this study, we seek to clarify whether the present-day experience of psychological distress among adults whose parents suffered from mental illness is related to their childhood experiences of abuse and neglect and their provision of emotional care for their parents during their school-age years. To this end, a web-based cross-sectional study was conducted. A total of 120 participants over the age of 20 who attended a self-help group responded (50% response rate); of these, 94 had a parent diagnosed with a mental illness, and these participants were included for data analysis purposes. Of the 94 respondents, 65 (69.2%) were highly distressed, as measured by a Kessler (K) 6 measure of ≥ 5. A logistic regression analysis revealed that the experience of providing emotional care for parents during school-age childhood was significantly related to high levels of distress in adulthood (OR = 3.48; 95% CI 1.21–9.96). For children of parents with mental illnesses, the effects of providing emotional care for parents during childhood may include long-term psychological distress. For this reason, mentally ill parents raising children need visiting community nurses or other professionals to provide emotional care on behalf of their children.

Keywords: adverse childhood experiences; young carers; emotional care; mental disorders

Citation: Kageyama, M.; Sakamoto, T.; Kobayashi, A.; Hirama, A.; Tamura, H.; Yokoyama, K. Childhood Adversities and Psychological Health of Adult Children of Parents with Mental Illness in Japan. *Healthcare* **2023**, *11*, 214. https://doi.org/10.3390/healthcare11020214

Academic Editors: Carlos Laranjeira and Ana Querido

Received: 15 December 2022
Revised: 7 January 2023
Accepted: 9 January 2023
Published: 10 January 2023

1. Introduction

Early life experiences make an important contribution to the health of individuals throughout their course of life. Adverse early life experiences such as child abuse and neglect have been a topic of political and academic interest since the 1960s [1]. Child abuse, neglect, and several other such experiences have been labeled as adverse childhood experiences (ACEs), which are defined as potentially traumatic events that occur in childhood (0–17 years). These include experiences of violence, abuse, or neglect; witnessing violence in the home; or having a family member attempt or die by suicide [2]. The cumulative effects of exposure to potentially traumatic events on the developing brain's stress response result in impairment in the development of brain structures and functions [3]. Decades of research have identified a robust, dose–response relationship between child abuse, neglect, and other forms of ACEs and an increased risk of physical and mental health problems across the lifespan [4–7].

One of the earliest and largest investigations into ACEs was the CDC-Kaiser ACE study [2], in which ACEs were categorized into three groups: abuse, neglect, and household challenges [2]. The category of household challenges included violence towards mothers, substance abuse in the household, mental illness in the household, parental separation or divorce, and incarcerated household members [5]. In the present study, among the ACEs, we focus on parental mental illness as a household challenge facing children, as well as abuse and neglect. In Japan, studies of ACEs have been scarce to date; however, there have been reports on the long-term effects of ACEs in adulthood in relation to self-harm ideation [8], inappropriate parenting by ACE-affected adults [9], and the onset of mental illness [10]. Regarding the association of ACEs with the development of mental illness in Japan, ACEs have been strongly associated with the development of mental illness in childhood but only weakly with the development of such illness in adulthood [10]. In Japan, the number of reports of child abuse and neglect and of patients with ACE-related mental illness continues to increase [11]. Because little is known of any specifically Japanese characteristics among the long-term effects of ACEs in adulthood, more research needs to be undertaken in this regard.

In this study, among the household challenges category of ACEs, we focus on parental mental illness. In western countries, a substantial amount of research has been conducted on children of parents with mental illness (COPMIs). One meta-analysis revealed an increased risk of developing a mental illness among the children of parents who suffer from severe mental illness [12]. Parental mental illness does not in itself guarantee psychosocial problems in their children. However, there is strong evidence that parents with mental illness are more likely to abuse, neglect, or maltreat their children [13]. In Japan, between one-third and one-half of child abuse cases where child protection is required involve parents with mental health problems [14]. Parental mental illness is known to be a high-risk factor for child abuse in Japan.

Nevertheless, the long-term effects on COPMIs that persist into adulthood have been little studied in western countries to date [15]. Some qualitative studies have reported long-term problems affecting COPMIs as adults, such as interpersonal problems [16], psychological distress [16], and difficulties in raising their own children [15]. Similarly, in Japan, there has been little research on long-term problems affecting COPMIs as adults, although one recent qualitative study reported long-term effects such as psychological distress, difficulties with emotions, an inability to trust people, and trauma persisting into adulthood [17].

In recent years, the experience of being a young carer has been identified in Japan as a particular ACE among COPMIs. In western countries generally, since the mid–1980s, there has been an increased awareness of the concept of young carers [18]. Young carers are defined as 'children and young persons under 18 who provide or intend to provide care, assistance, or support to another family member' [18]. Three meta-syntheses of qualitative studies describing children's subjective experiences revealed a relationship between children's experiences of performing housework and providing emotional care for their parents and feelings of distress among the children involved [16,19,20]. In a 2022 survey of Japanese high school students, high school students who cared for a family member with a disability or illness were significantly more distressed than those without such responsibilities [21].

This concept of young carers is especially relevant in the study of COPMIs. According to a national survey of young carers conducted in the United Kingdom, 29% of people who require care and 50% of mothers who require care have mental health problems [22]. Regarding young Japanese carers, the first national survey conducted in 2020 revealed that, among parents being cared for by their children, 14.3% of those cared for by middle- school-aged children (13–15 years) and 17.3% of those cared for by high-school-aged children (16–18 years) suffered from a mental illness [23]. Taking on such a responsibility can have physical, social, educational, and emotional impacts on the lives of children [24]. Among the many roles played by young carers, the provision of emotional care is more common when the person receiving the care has mental health problems [22], and this includes 'observing the care recipients' emotional state, providing supervision, or trying to cheer them up when they are depressed, etc.' [22]. Because changes in the behavior and personality of persons with mental illness can be distressing for close family members [25–29], the provision of emotional care can lead to psychological distress for carers, resulting in a lower quality of life [30]. For these reasons, in this study, we focus on the provision of emotional care by COPMIs and on any psychological distress as a result. We sought to identify those factors most likely to lead to psychological distress and possible means to alleviate their effects to improve the quality of life of carers.

This study aims to clarify whether the present-day experience of psychological distress among adult COPMIs is related to their experiences of being abused and neglected and of having provided emotional care to their parents during their childhood years, adjusted for factors potentially associated with dependent and independent variables, as follows: gender [31], age [32], presence or absence of siblings [32], personal experience of mental illness [33], physical condition [34], separation from parents in their childhood [6], having a parent without illness [22], the treatment status of their parents in their childhood [35,36], and the absence of a parental spouse in their childhood [7].

2. Methods

2.1. Research Design

The present study involves the analysis of data from the web-based, cross-sectional 'Survey of Children Raised by Parents with Mental Illness' [37], previously conducted by ourselves. This survey sought to obtain basic information to assess the support methods of COPMIs of elementary and high-school age. Respondents were asked to report their current life situation and to recall past childhood events.

2.2. Procedure

The survey was carried out among participants in the biggest self-help group in Japan (KODOMO-PEER) for the adult children of parents with mental illness. A web-based survey (URL supplied in the email) was emailed in October and November 2019. The REDCap (Research Electronic Data Capture) system was used for the web survey.

2.3. Study Samples

The inclusion criteria for this survey were as follows: individuals who had participated in KODOMO-PEER at least once in the past and who were at least 20 years old. The exclusion criterion covered all individuals less than 20 years old. Those eligible to participate in KODOMO-PEER are individuals whose parents have a mental illness that may or may not have been diagnosed by a psychiatrist.

Of the 240 people invited to participate in the survey, 120 responded (response rate 50%). In order to limit our analysis to the experiences of young carers (those who had to care for a parent when aged less than 18 years), two participants were excluded because one reported that their parent's mental illness began after they had graduated from high school and another did not supply an onset time. In addition, to limit the analysis sample to those whose parents had been diagnosed with a mental illness by a psychiatrist, we excluded nineteen participants whose parents were undiagnosed and five participants whose parents' diagnosis was unknown. In all, we included 94 respondents whose parents developed a diagnosed mental illness before the respondents themselves reached the age of 18.

2.4. Measures

2.4.1. Psychological Distress

We used the Kessler (K) 6, a short screening questionnaire consisting of six items that is used to screen for non-specific psychological distress [33]. The questions were designed to ask about psychological distress over the previous 30 days. They include 'Did you often feel nervous'? and 'How often did you feel hopeless'? as well as four other items. The degree of psychological distress is measured on a 5-point Likert scale with the following options: 0 (none of the time), 1 (a little of the time), 2 (some of the time), 3 (most of the time), and 4 (all of the time); possible scores, therefore, range from 0–24. The reliability and validity of the Japanese version of the K6 have previously been evaluated, and the two best cut-off points have been estimated as follows: 4/5, as an optimal lower threshold cut-point indicative of moderate mental distress; and 12/13, as a screening cut-off point indicative of severe mental illness [38]. For this study, we chose the 4/5 cut-off point because we did not screen for mental illness. The Cronbach's alpha was 0.92. The K6 score was used to categorize respondents into two groups: those with high (K6 $\geq$ 5) and low (K6 $\leq$ 4) levels of distress.

2.4.2. Independent Variables

We used three independent variables: exposure to aggressive acts, neglect, and emotional care.

Exposure to aggressive acts included experiences of child abuse. However, the survey did not ask about child abuse directly. Respondents were asked to consider the following two items with respect to their school-age life experience: 'There were constant fights between the adults at home' and 'There was an attack from my parents on me or my siblings'. These were operationally defined as 'exposure to aggressive acts'. If one or both items were selected, respondents were considered as having experienced exposure to aggressive acts.

In terms of neglect, those who chose one or both of the following items—'An adequate amount of food was not provided' or 'The laundry and cleaning were not well done'—were considered to have experienced 'neglect'.

Emotional care is the most common care role carried out by COPMIs [37]. Those who selected 'I provided emotional care, such as being close to my parents' as indicative of their childhood experience were deemed to fall into the 'emotional care' category.

Regarding the three variables, respondents were asked to recall and answer in terms of their childhood experience during their periods at elementary school, junior high school, and high school at the ages of 7–12, 13–15, and 16–18 years, respectively.

2.4.3. Control Variables

The control variables of adult COPMIs included gender (male/female), age (20–29/ 30–39/40–49/50 or older), having siblings (yes/no), suffering mental illness (suffered/not suffered), physical condition (poor/not poor), and separation from parents (experienced/not experienced). The variables related to the parents of respondents when the latter were of school age included whether one or both parents had a mental illness (one or both), whether the illness was treated (continued/discontinued or never treated), and the presence or absence of a spouse (presence/absence).

2.5. Data Analysis

First, the frequency distribution of variables was confirmed. Next, the background characteristics of high- and low-distress groups were compared using χ^2 tests. Finally, to examine the association between distress and the independent variables, a multiple logistic regression was performed. The high- and low-distress groups, as established by the K6 scores, were identified as the dependent variables. Exposure to aggressive acts, neglect, and emotional care were identified as the independent variables. All others were considered control variables after confirming variance inflation factors (less 2.0) due to possible multicollinearity among the variables. When multiple logistic regression was conducted, three independent variables and control variables were selected by the stepwise method. A post hoc logistic regression analysis was performed with a power of 0.81 using G*Power. All analyses, except for the power analysis, were performed in SAS Version 9.4 (SAS Institute Inc., Cary, NC, USA).

2.6. Ethical Approval

We explained in the text the purpose and method of the study, that participation in the study was voluntary, and that participants could withdraw at any time if they felt discomfort in recalling their past experiences while responding. Informed consent was given by means of this written explanation; explicit consent was then obtained using a checkbox. The anonymity and confidentiality of participants were maintained as researchers could not determine the email addresses and answers of individual respondents. The Research

Ethics Committee, Faculty of Medicine, Osaka University, approved the study protocol on 29 July 2019 (ID: 19152).

3. Results

3.1. Respondents' Demographics

As shown in Table 1, of the 94 respondents, 84.0% were female and 16.0% were male. In terms of age, those in their twenties formed the largest group, comprising 37.2% of the total. Of the remainder, 22.3% were in their thirties, 17.0% were in their forties, and 23.4% were aged fifty or more; 57.4% of respondents had siblings, and 42.6% did not. Concerning their present state of health, 24.5% stated that they suffered from mental illness, and 41.5% reported being in poor physical condition. Forty-one respondents (43.6%) had been separated from their parents for more than one month before graduating from high school; of these, twenty-eight (29.8%) reported that their parents had been admitted to psychiatric hospitals. None of the respondents reported that they had been under the child protection system as children.

Table 1. Demographic data of adult children from the high- and low-distress groups.

		All	Low-Distress Group (K6 $\leq$ 4)	High-Distress Group (K6 $\geq$ 5)		
		n = 94	n = 29	n = 65	x^2	p
		n (%)	n (%)	n (%)		
Adult children						
Gender	Male	15(16.0%)	4 (13.8%)	11 (16.9%)	0.147	1.000
	Female	79(84.0%)	25 (86.2%)	54 (83.1%)		
Age (years)	20–29	35 (37.2%)	6 (20.7%)	29 (44.6%)	9.190	0.027 *
	30–39	21 (22.3%)	7 (24.1%)	14 (21.5%)		
	40–49	16 (17.0%)	4 (13.8%)	12 (18.5%)		
	50 or older	22 (23.4%)	12 (41.4%)	10 (15.4%)		
Have siblings	Yes	54 (57.4%)	20(69.0%)	34 (52.3%)	2.276	0.131
	No	40 (42.6%)	9 (31.0%)	31 (47.7%)		
Mental illness	Suffered	23 (24.5%)	4 (13.8%)	19 (29.2%)	2.586	0.108
	Not suffered	71 (75.5%)	25 (86.2%)	46 (70.8%)		
Physical condition	Poor	39 (41.5%)	4 (13.8%)	35 (53.8%)	13.252	<0.001 *
	Not poor	55 (58.5%)	25 (86.2%)	30 (46.2%)		
Separation from parents	Experienced	41 (43.6%)	14 (48.3%)	27 (41.5%)	0.370	0.543
	Not experienced	53 (56.4%)	15 (51.7%)	38 (68.5%)		
Parents with mental illness						
Parents with mental illness	One	79 (84.0%)	25 (86.2%)	54 (83.1%)	0.146	1.000
	Both	15 (16.0%)	4 (13.8%)	11 (16.9%)		
Treatment	Continued,	49 (52.1%)	15 (51.7%)	34 (52.3%)	0.003	0.958
	Discontinued or never treated	45 (47.9%)	14 (48.3%)	31 (47.7%)		
Spousal presence	Presence	60 (63.8%)	17 (58.6%)	43 (66.2%)	0.493	0.483
	Absence	34 (36.2%)	12 (41.4%)	22 (33.8%)		

p-values were calculated for the differences between high- and low-distress groups using the χ^2 test. * $p < 0.05$.

With regard to which of their parents had been diagnosed with mental illness by a psychiatrist, 67.0% stated only their mothers and 17.0% stated only their fathers, while 16.0% said both parents had been diagnosed. The parents' diagnoses included schizophrenia (61.7%), depression (22.3%), anxiety disorder (12.8%), bipolar disorder (14.9%), developmental disorders (8.5%), personality disorders (9.6%), alcoholic addiction (5.3%), and others (14.9%) (duplicated). Seventy (74.5%) respondents estimated that the onset of their parents' illness was when they were younger than elementary-school age (6 years or younger), while eighteen (19.2%) stated that it began when they were at elementary school (age 7–12) and six (6.4%) recalled that it started when they were in junior high school (age 13–15). Concerning treatment for mental illness, 52.1% of respondents recalled that their affected parents were under continuous medical treatment, while 47.9% recalled either discontinued treatment or the absence of treatment. A total of 36.2% of respondents recalled the absence of a parental spouse due to divorce or other reasons. Control variables are shown in Table 1.

3.2. Psychological Distress

The average K6 score of the 94 respondents was 9.38 (SD 6.75). A total of 65 respondents (69.2%) reported K6 scores of over 5 (high-distress group), while 29 (30.9%) reported scores of 4 or less (low-distress group).

3.3. Independent Variables

As shown in Table 2, a total of 78.7% of respondents were exposed to aggressive acts when they were of school age, with 70.2% being exposed during their elementary school period, 71.3% when in junior high school, and 53.2% when in high school.

Table 2. Adverse experiences in childhood in the high- and low-distress groups.

		All	Low-Distress Group (K6 $\leq$ 4)	High-Distress Group (K6 $\geq$ 5)		
		n = 94	n = 29	n = 65		
		n (%)	n (%)	n (%)	x^2	p
Exposure to aggressive acts	Elementary	66 (70.2%)	16 (55.2%)	50 (76.9%)	4.536	0.033 *
	Junior high	67 (71.3%)	17 (58.6%)	50 (76.9%)	3.281	0.070
	High	50 (53.2%)	15 (51.7%)	35 (53.9%)	0.036	0.849
	Ever	74 (78.7%)	18 (62.1%)	56 (86.2%)	6.945	0.008 *
Neglect	Elementary	31 (33.0%)	7 (24.1%)	24 (36.9%)	1.483	0.223
	Junior high	35 (37.2%)	7 (24.1%)	28 (43.1%)	3.078	0.079
	High	25 (26.6%)	4 (13.8%)	21 (32.3%)	3.521	0.061
	Ever	46 (48.9%)	8 (27.6%)	30 (46.2%)	2.871	0.090
Emotional care	Elementary	56 (59.6%)	12 (41.4%)	44 (67.7%)	5.765	0.016 *
	Junior high	57 (60.6%)	11 (37.9%)	46 (70.8%)	9.060	0.003 *
	High	57 (60.6%)	13 (44.8%)	44 (67.7%)	4.392	0.036 *
	Ever	69 (73.4%)	16 (55.2%)	53 (81.5%)	7.141	0.008 *

p-values were calculated for the differences between high- and low-distress groups using the χ^2 test. * *p* < 0.05.

A total of 48.9% of respondents had experienced neglect, with 33.0% recalling neglect during their elementary schooling, 37.2% experiencing it while in junior high school, and 26.6% during high school.

Most respondents (73.4%) recalled providing emotional care for parents during their school-age childhood, with 59.6% providing such care during elementary school, 60.6% during junior high school, and 60.6% during high school.

3.4. Comparisons between the High-Distress and Low-Distress Groups

As shown in Table 1, compared to the low-distress group, respondents in the high-distress group were significantly younger (*p* = 0.027) and more likely to be in poor physical condition (*p* < 0.001). There were no other significant differences between respondent groups in terms of their gender, the presence or absence of siblings, their own experiences of mental illness, experiences of separation from parents, whether one or both parents had a mental illness, whether the parental illness was treated or not, or whether there was parental spousal absence.

As shown in Table 2, upon comparing independent variables between the high-distress and low-distress groups, we found that individuals in the high-distress group were significantly more likely than those in the low-distress group to have been exposed to aggressive acts while in elementary school ($p = 0.033$) and throughout their school-age childhood ($p = 0.008$). They were also more likely to have provided emotional care while in elementary school ($p = 0.016$), junior high school ($p = 0.003$), high school ($p = 0.036$), and throughout their school-age period ($p = 0.008$).

3.5. Independent Variables and Distress

As a result of multiple logistic regression with three independent variables and control variables selected via the stepwise method, we found significantly more respondents who had provided emotional care (OR = 3.48; 95% CI 1.21–9.96) among the independent variables and significantly more respondents in poor physical condition (OR = 7.15; 95% CI 2.17–23.58) among the control variables in the high-distress group (Table 3). The other independent variables, i.e., exposure to aggressive acts and neglect, were not significantly associated with psychological distress. Similarly, the other control variables, i.e., gender, age, having siblings, suffering mental illness as a COPMI, experiences of separation from parents, having one or both parents with mental illness, parental treatment or lack of treatment, and parental spousal presence, were not significantly associated with psychological distress.

Table 3. Risk factors for adult children with high distress (K6 $\geq$ 5).

		OR	95% CI	p
Provide emotional care	Never experienced	1.00	Reference	
	Ever experienced	3.48	1.21–9.96	0.020 *
Physical condition	Not poor	1.00	Reference	
	Poor	7.15	2.17–23.58	0.001 *

CI: confidence interval. * $p < 0.05$. Multiple logistic regression: The following variables were selected via stepwise method from exposure to aggressive acts, neglect, emotional care, gender, age, having siblings, suffering mental illness, physical condition, separation from parents, parents with mental illness (one or both), parental treatment, and the parental spousal presence.

4. Discussion

In this study, we sought to clarify whether any present-day psychological distress experienced by adult COPMIs is related to their experiences of being abused and neglected and of having provided emotional care for their parents during their school-age years. The main finding of this study is that the experience of providing emotional care to parents during childhood is significantly related to high levels of distress in adulthood.

4.1. Study Respondents' Demographics and Distress

The parents of most respondents (93.7%) experienced the onset of their mental illness before the respondent had graduated from elementary school. As a result, many respondents grew up experiencing their parents' illness. In terms of gender, the majority of the parents with mental illness were mothers (83.0%) and most of the respondents were daughters (84.0%). In Japan, primary carers for persons with mental illness are mostly women [39]. A Japanese national survey of junior high and high school students found that girls spend more time caring for other family members and bear a greater burden than boys [23]. In Japan, care roles are not only for adults but also for children, where a gender bias has arisen. It is thought that women who have more experience caring for their parents participate more in self-help groups to share their caregiving experiences.

In our survey, more than half of the respondents had parents who suffered from schizophrenia (61.7%), and 47.9% of the respondents reported that their parents' mental illness was untreated during the respondents' childhood. In general, only 21.9% of individuals with a mental illness seek treatment [40], and over three-quarters of children (78.5%) have parents who suffer from a mental illness but do not receive mental health care [41]; these findings are similar to those of the present study. The emotional care for untreated parents provided by COPMIs can be a heavy burden, and support for such caregivers is needed. However, none of our respondents reported being taken up by the child protection system, even when they had been neglected or exposed to mildly aggressive acts. In such cases, it can be difficult to identify children who need support because they are not being severely—and, thus, obviously—abused.

In the present study, the average K6 score among respondents was 9.38 (SD 6.75) and 65 (69.2%) respondents produced a K6 score of over 5 (high-distress group), which is higher than in the general population (3.6, SD 3.9) [42] or among carers for the elderly (4.29, SD 4.46) [43] or among general high school students (6.51, SD 5.87) [21]. In this study, 24.5% of respondents themselves suffered from a mental illness, which is higher than in the general population (7.6% in 12-month prevalence) [40], and there was no significant relationship between their diagnosis and their level of distress. One meta-analysis found that children of parents with severe mental illness had a 32% probability of developing severe mental illness by adulthood (age > 20) [12]. Regarding the association of parental mental illness with the development of mental illness in their children, in Japan, there appears to be a strong association with the development of such illness during childhood but not in adulthood [10]. A high prevalence of mental illnesses has also been reported among COPMIs under the age of 18, with a significantly higher risk of socioeconomic adversity as a background [44]. It may be that the parent's mental illness does not itself lead to the development of mental illness in their child. Rather, it is the deterioration of the domestic environment accompanying the onset of the parent's mental illness that negatively impacts the child's mental health.

4.2. Relationship between Aggressive Acts, Neglect, and Provision of Emotional Care and Levels of Distress

In our high-distress group, we found significantly more respondents who had provided emotional care. There was no significant relationship between childhood exposure to aggressive acts or neglect and present-day levels of distress. Studies on ACEs have identified significant relationships between child abuse or neglect and an increased risk of psychological health problems throughout the lifespan of affected children [6,45]. In our study, many respondents were exposed to aggressive acts (78.7%) and neglect (48.9%) from their elementary to high school periods. Because the cognitive dysfunction associated with many mental disorders impairs functionality, this may have impaired the parental performance of household chores [46], leading to neglect of their children. In Japan, physical violence toward family members occurs in 60% of schizophrenia sufferers [39]. Severe direct violence or neglect is categorized by the child protection system as abuse, but none of the respondents in this study were placed under child protection. Because they were considered insufficiently exposed to significant levels of violence or neglect, these experiences seemed to be unrelated to distress in the current study. However, these results do not mean that neglect or exposure to aggressive acts has no serious impact on children. The results merely suggest that for COPMIs, the long-term effects of emotional caregiving upon psychological distress are more significant compared with experiences of abuse or neglect.

In the current study, 73.4% of the respondents had the experience of providing emotional care to their parents. Mental illness requires families to become emotionally involved with the patients. Earlier qualitative studies have shown that children observe and react to the unpredictability of their parents' illness-related behavior [19,47]. As a result, children carefully monitor their parents for signs of change [48] and try to manage these behaviors to maintain a safe home environment [19]. The consequences of providing long-term emotional care may continue to affect children later in life. As adults, young carers may find it difficult to establish trust with others [49]. In their own childhoods, the normal roles of parent and child in care provision were reversed as they often had to provide care for their parents [20]. Therefore, they were never able to truly live as children. Such childhood experiences have been described in previous studies carried out in other countries and are often shared during KODOMO-PEER meetings in Japan.

4.3. Implications for Practice

The results of this study suggest that providing emotional care for parents with mental illness has long-term effects on the psychological health of the children involved. Many of our survey respondents grew up experiencing their parents' illness and providing emotional care for them. In Japan, although support for victims of child abuse is well established [11], support for young carers has only just begun. The findings reported here suggest a need to reduce the burden of emotional care in particular. Emotional care should not be provided by children but by visiting community nurses or other professionals on behalf of children. When health professionals support parents with childcare responsibilities, they also need to keep in mind that when children take on care roles, the burden typically falls most heavily on girls. However, whether COPMIs are caring for their parents or struggling in other ways with family matters, they are often reluctant to seek help because of stigma [50]. In Japan, more than 80% of COPMIs did not consult their schoolteachers about conditions at home [37]. When mentally ill parents are raising children, it is desirable for the treating medical facility to proactively assess the family's situation and link them to appropriate home-visiting services.

Finally, because many COPMIs continue to experience psychological distress in adulthood and need assistance, more support for such individuals is also needed.

4.4. Research Limitations and Future Research

The limitations of this study include the following: First, the sample may be unrepresentative of the general population because of its small size, the low response rate (50%), and its recruitment through a specific group. Nevertheless, this survey involved the largest sample size of adult children of parents with mental illness of any study carried out in Japan. In addition, respondents with high levels of distress may have been unable to respond due to poor health, and there may also have been bias among respondents. Respondents often find KODOMO-PEER through the internet, and this itself may reflect their awareness of concerns and difficulties originating in childhood and persisting into adulthood. Therefore, it is unclear if the respondents fairly and accurately represent all of the children of parents with mental illness. Secondly, we were unable to determine any economic circumstances in childhood that might have affected current psychological distress [6,44] because this study used variables from a previously conducted study. Thirdly, the memory of individual respondents may also serve as a source of bias. Our survey used a self-reporting questionnaire, and the accuracy of responses could not be guaranteed. Lastly, we did not explore any experiences of childhood prior to elementary school because this survey's main purpose was to focus on school-aged children.

Future research methods need to be more broadly based and not limited to self-help groups in order to reduce subject bias. In addition, prospective longitudinal surveys and interviews should be used to ensure accurate study information.

5. Conclusions

In this study, we sought to clarify whether the present-day experience of psychological distress among adult COPMIs is related to their experiences of being abused and neglected and of having provided emotional care for their parents during their childhood years, adjusted for factors potentially associated with dependent and independent variables. A logistic regression analysis revealed that the experience of providing emotional care for parents during childhood was significantly related to high levels of distress in adulthood. For children of parents with mental illnesses, the effects of providing emotional care for parents during childhood may include long-term psychological distress. The findings suggest a need to reduce the burden of emotional care provided by children.

Author Contributions: Conceptualization, M.K., T.S., A.K., A.H., H.T. and K.Y.; methodology, M.K.; software, M.K.; validation, M.K. and K.Y.; formal analysis, M.K. and K.Y.; investigation, M.K., T.S., A.K., A.H., H.T. and K.Y.; resources, M.K., T.S., A.K., A.H., H.T. and K.Y.; data curation, M.K.; writing—original draft preparation, M.K.; writing—review and editing, M.K.; visualization, M.K.; supervision, M.K. and K.Y.; project administration, M.K.; funding acquisition, M.K. All authors have read and agreed to the published version of the manuscript.

Funding: This research was funded by the JSPS KAKENHI Grant Number JP 19H03960. The APC was funded by the JSPS KAKENHI Grant Number JP 19H03960.

Institutional Review Board Statement: The study was conducted in accordance with the Declaration of Helsinki, and approved by the Institutional Research Ethics Committee of Faculty of Medicine, Osaka University (protocol code 19152; date of approval: 29 July 2019).

Informed Consent Statement: Informed consent was obtained from all participants involved in the study by using a checkbox before answering the questionnaire.

Data Availability Statement: The data of this study are not publicly available to protect the participants' privacy.

Acknowledgments: We would like to thank the research participants of KODOMO-PEER.

Conflicts of Interest: The authors have no conflict of interests to declare.

References

1. Connell-Carrick, K. A Critical Review of the Empirical Literature: Identifying Correlates of Child Neglect. *Child Adolesc. Soc. Work. J.* **2003**, *20*, 389–425. [CrossRef]
2. Center for Disease Control and Prevention Preventing Adverse Childhood Experiences (ACEs): Leveraging the Best Available Evidence. Atlanta, GA: National Center for Injury Prevention and Control, Centers for Disease Control and Prevention. 2019. Available online: https://www.cdc.gov/violenceprevention/pdf/preventingACEs.pdf (accessed on 14 December 2022).
3. Anda, R.F.; Felitti, V.J.; Bremner, J.D.; Walker, J.D.; Whitfield, C.; Perry, B.D.; Dube, S.R.; Giles, W.H. The Enduring Effects of Abuse and Related Adverse Experiences in Childhood: A Convergence of Evidence from Neurobiology and Epidemiology. *Eur. Arch. Psychiatry Clin. Neurosci.* **2006**, *256*, 174–186. [CrossRef] [PubMed]
4. Bellis, M.A.; Hughes, K.; Leckenby, N.; Hardcastle, K.A.; Perkins, C.; Lowey, H. Measuring Mortality and the Burden of Adult Disease Associated with Adverse Childhood Experiences in England: A National Survey. *J. Public Health United Kingd.* **2015**, *37*, 445–454. [CrossRef] [PubMed]
5. Felitti, V.J.; Anda, R.F.; Nordenberg, D.; Williamson, D.F.; Spitz, A.M.; Edwards, V.; Koss, M.P.; Marks, J.S. Relationship of Childhood Abuse and Household Dysfunction to Many of the Leading Causes of Death in Adults: The Adverse Childhood Experiences (ACE) Study. *Am. J. Prev. Med.* **1998**, *14*, 245–258. [CrossRef] [PubMed]
6. Hughes, K.; Bellis, M.A.; Hardcastle, K.A.; Sethi, D.; Butchart, A.; Mikton, C.; Jones, L.; Dunne, M.P. The Effect of Multiple Adverse Childhood Experiences on Health: A Systematic Review and Meta-Analysis. *Lancet Public Health* **2017**, *2*, e356–e366. [CrossRef]
7. Scully, C.; McLaughlin, J.; Fitzgerald, A. The Relationship between Adverse Childhood Experiences, Family Functioning, and Mental Health Problems among Children and Adolescents: A Systematic Review. *J. Fam. Ther.* **2020**, *42*, 291–316. [CrossRef]
8. Doi, S.; Fujiwara, T. Combined Effect of Adverse Childhood Experiences and Young Age on Self-Harm Ideation among Postpartum Women in Japan. *J. Affect. Disord.* **2019**, *253*, 410–418. [CrossRef]
9. Isumi, A.; Fujiwara, T. Association of Adverse Childhood Experiences with Shaking and Smothering Behaviors among Japanese Caregivers. *Child Abus. Negl.* **2016**, *57*, 12–20. [CrossRef]
10. Fujiwara, T.; Kawakami, N. Association of Childhood Adversities with the First Onset of Mental Disorders in Japan: Results from the World Mental Health Japan, 2002–2004. *J. Psychiatr. Res.* **2011**, *45*, 481–487. [CrossRef]
11. *Public Health of Japan 2021*; Japan Public Health Association: Shinjuku, Japan, 2021.
12. Rasic, D.; Hajek, T.; Alda, M.; Uher, R. Risk of Mental Illness in Offspring of Parents with Schizophrenia, Bipolar Disorder, and Major Depressive Disorder: A Meta-Analysis of Family High-Risk Studies. *Schizophr Bull* **2014**, *40*, 28–38. [CrossRef]
13. NSW Department of Community Services. *Parents with Mental Health Issues: Consequences for Children and Effectiveness of Interventions Designed to Assist Children and Their Families Literature Review*; NSW Department of Community Services: Ashfield, Australia, 2008.
14. Matsumiya, Y. Jido Gyautai to Oya No Mentaruherusu Mondai No Setten [A Review of Previous Research on the Relationship of Child Abuse and Parents with Mental Health Problems]. *J. Fac. Health Welf. Prefect. Univ. Hiroshima* **2012**, *12*, 103–115.
15. Patrick, P.M.; Reupert, A.E.; McLean, L.A. We Are More than Our Parents Mental Illness: Narratives from Adult Children. *Int. J. Env. Res. Public Health* **2019**, *16*, 839. [CrossRef] [PubMed]
16. Murphy, G.; Peters, K.; Jackson, D.; Wilkes, L. A Qualitative Meta-Synthesis of Adult Children of Parents with a Mental Illness. *J. Clin. Nurs.* **2011**, *20*, 3430–3442. [CrossRef] [PubMed]

17. Kosaka, H.; Kageyama, M. Seishin Shikkan No Oya o Motu Kodomo No Seijingo No Ikidurasa Ni Kansuru Situteki Kijututeki Kenkyu [A Qualitative Descriptive Study about the Living Difficulties of Adult Children of a Parent with a Mental Illness: Focused on Mother with Schizophrenia-Spectrum Disorder]. *J. Jpn. Acad. Nurs. Sci.* **2023**.
18. Becker, S.; Dearden, C.; Aldridge, J. Young Carers in the UK: Research, Policy and Practice. *Res. Policy Plan.* **2000**, *18*, 13–22.
19. Gladstone, B.M.; Boydell, K.M.; Seeman, M.V.; Mckeever, P.D. Children's Experiences of Parental Mental Illness: A Literature Review. *Early Interv Psychiatry* **2011**, *5*, 271–289. [CrossRef]
20. Yamamoto, R.; Keogh, B. Children's Experiences of Living with a Parent with Mental Illness: A Systematic Review of Qualitative Studies Using Thematic Analysis. *J. Psychiatr. Ment. Health Nurs.* **2018**, *25*, 131–141. [CrossRef]
21. Miyakawa, M.; Hamashima, Y.; Minami, T. Yangu Keara No Seisinteki Kutu: Saitama Kenritu Koko No Seito o Taisyo to Sita Situmonsi Chosa [Psychological Distress of Young Carers: A Questionnaire Survey among Saitama Prefectural High School Students]. *Nihon Koshu Eisei Zasshi* **2022**, *69*, 125–135.
22. Dearden, C.; Becker, S. *Young Carers in the UK: The 2004 Report*; Carers UK: London, UK, 2004.
23. Mitsubishi UFJ Research and Consulting. Yangukearano Jittainikansuru Chosahokokusyo. In *Research Report on the Actual Situation of Young Carers*; Mitsubishi UFJ Research and Consulting: Osaka, Japan, 2021.
24. Thomas, N.; Stainton, T.; Jackson, S.; Cheung, W.Y.; Doubtfire, S.; Webb, A. "Your Friends Don't Understand": Invisibility and Unmet Need: In the Lives of "Young Carers". *Child Fam. Soc. Work* **2003**, *8*, 35–46. [CrossRef]
25. Chaturvedi, S.K.; Hamza, A.; Sharma, M.P. Changes in Distressing Behavior Perceived by Family of Persons with Schizophrenia at Home—25 Years Later. *Indian J. Psychol. Med.* **2014**, *36*, 282–286. [CrossRef]
26. Vaddadi, K.; Soosai, E.; Gilleard, C.J.; Adlard, S. Mental Illness, Physical Abuse and Burden of Care on Relatives: A Study of Acute Psychiatric Admission Patients. *Acta Psychiatr. Scand.* **1997**, *95*, 313–317. [CrossRef] [PubMed]
27. Martens, L.; Addington, J. The Psychological Well-Being of Family Members of Individuals with Schizophrenia. *Soc. Psychiatry Psychiatr. Epidemiol.* **2001**, *36*, 128–133. [CrossRef] [PubMed]
28. Barrowclough, C.; Parle, M. Appraisal, Psychological Adjustment and Expressed Emotion in Relatives of Patients Suffering from Schizophrenia. *Br. J. Psychiatry* **1997**, *171*, 26–30. [CrossRef] [PubMed]
29. Lasebikan, V.O.; Ayinde, O.O. Family Burden in Caregivers of Schizophrenia Patients: Prevalence and Socio-Demographic Correlates. *Indian J. Psychol. Med.* **2013**, *35*, 60–66. [CrossRef]
30. Awad, A.G.; Voruganti, L.N.P. The Burden of Schizophrenia on Caregivers: A Review. *Pharmacoeconomics* **2008**, *26*, 149–162. [CrossRef]
31. Amirkhanyan, A.A.; Wolf, D.A. Parent Care and the Stress Process: Findings from Panel Data. *J. Gerontol. Ser. B Psychol. Sci. Soc. Sci.* **2006**, *61*, 248–255. [CrossRef]
32. Aldridge, J.; Becker, S. *Children Caring for Parents with Mental Illness:Perspectives of Young Carers, Parents and Professionals*; Policy Press: Bristol, UK, 2012.
33. Kessler, R.C.; Andrews, G.; Colpe, L.J.; Hiripi, E.; Mroczek, D.K.; Normand, S.L.T.; Walters, E.E.; Zaslavsky, A.M. Short Screening Scales to Monitor Population Prevalences and Trends in Non-Specific Psychological Distress. *Psychol. Med.* **2002**, *32*, 959–976. [CrossRef]
34. Breitborde, N.J.K.; López, S.R.; Chang, C.; Kopelowicz, A.; Zarate, R. Emotional Over-Involvement Can Be Deleterious for Caregivers' Health. *Soc. Psychiatry Psychiatr. Epidemiol.* **2009**, *44*, 716–723. [CrossRef]
35. Ohaeri, J.U. The Burden of Caregiving in Families with a Mental Illness: A Review of 2002. *Curr. Opin. Psychiatry* **2003**, *16*, 457–465. [CrossRef]
36. Friedman, S.H.; Mcewan, M.V. Treated Mental Illness and the Risk of Child Abuse Perpetration. *Psychiatr. Serv.* **2018**, *69*, 211–216. [CrossRef]
37. Kageyama, M.; Yokoyama, K.; Sakamoto, T.; Kobayashi, A.; Hirama, A. Previous Experiences of Japanese Children with Parents Who Have a Mental Illness, and Their Consultation Situation at School: A Survey of Grown-up Children (in Japanese). *Jpn. J. Public Health* **2021**, *68*, 131–143.
38. Prochaska, J.J.; Sung, H.Y.; Max, W.; Shi, Y.; Ong, M. Validity Study of the K6 Scale as a Measure of Moderate Mental Distress Based on Mental Health Treatment Need and Utilization. *Int. J. Methods Psychiatr. Res.* **2012**, *21*, 88–97. [CrossRef] [PubMed]
39. Kageyama, M.; Yokoyama, K.; Nagata, S.; Kita, S.; Nakamura, Y.; Kobayashi, S.; Solomon, P. Rate of Family Violence among Patients with Schizophrenia in Japan. *Asia Pac. J. Public Health* **2015**, *27*, 652–660. [CrossRef]
40. Ishikawa, H.; Kawakami, N.; Kessler, R.C. Lifetime and 12-Month Prevalence, Severity and Unmet Need for Treatment of Common Mental Disorders in Japan: Results from the Final Dataset of World Mental Health Japan Survey. *Epidemiol. Psychiatr. Sci.* **2016**, *25*, 217–229. [CrossRef] [PubMed]
41. Bassani, D.G.; Padoin, C.V.; Philipp, D.; Veldhuizen, S. Estimating the Number of Children Exposed to Parental Psychiatric Disorders through a National Health Survey. *Child Adolesc. Psychiatry Ment. Health* **2009**, *3*, 1–7. [CrossRef]
42. Sakurai, K.; Nishi, A.; Kondo, K.; Yanagida, K.; Kawakami, N. Screening Performance of K6/K10 and Other Screening Instruments for Mood and Anxiety Disorders in Japan. *Psychiatry Clin. Neurosci.* **2011**, *65*, 434–441. [CrossRef]
43. Oshio, T. The Association between Involvement in Family Caregiving and Mental Health among Middle-Aged Adults in Japan. *Soc. Sci. Med.* **2014**, *115*, 121–129. [CrossRef]

44. Pierce, M.; Abel, K.M.; Muwonge, J.; Wicks, S.; Nevriana, A.; Hope, H.; Dalman, C.; Kosidou, K. Prevalence of Parental Mental Illness and Association with Socioeconomic Adversity among Children in Sweden between 2006 and 2016: A Population-Based Cohort Study. *Lancet Public Health* **2020**, *5*, e583–e591. [CrossRef]
45. Gilbert, L.K.; Breiding, M.J.; Merrick, M.T.; Thompson, W.W.; Ford, D.C.; Dhingra, S.S.; Parks, S.E. Childhood Adversity and Adult Chronic Disease: An Update from Ten States and the District of Columbia, 2010. *Am. J. Prev. Med.* **2015**, *48*, 345–349. [CrossRef]
46. Trivedi, J. Cognitive Deficits in Psychiatric Disorders: Current Status. *Indian J. Psychiatry* **2006**, *48*, 10. [CrossRef]
47. Tanonaka, K. Seishin Shikkanno Oyaomotsu Kodomono Konnan [Difficulties Encountered by Children of a Parent with Mental Illness]. *Jpn. J. Public Health Nurs.* **2019**, *8*, 23–32. [CrossRef]
48. Simpson-Adkins, G.J.; Daiches, A. How Do Children Make Sense of Their Parent's Mental Health Difficulties: A Meta-Synthesis. *J. Child Fam. Stud.* **2018**, *27*, 2705–2716. [CrossRef] [PubMed]
49. O'Connell, K.L.C. What Can We Learn? Adult Outcomes in Children of Seriously Mentally Ill Mothers. *J. Child Adolesc. Psychiatr. Nurs.* **2008**, *21*, 89–104. [CrossRef] [PubMed]
50. Davies, G.; P. Deane, F.; Williams, V.; Giles, C. Barriers, Facilitators and Interventions to Support Help-Seeking amongst Young People Living in Families Impacted by Parental Mental Illness: A Systematized Review. *Early Interv Psychiatry* **2022**, *16*, 469–480. [CrossRef] [PubMed]

 *healthcare*

Article

The Effect of Emotional Labor on the Physical and Mental Health of Health Professionals: Emotional Exhaustion Has a Mediating Effect

Chien-Chih Chen [1], Yu-Li Lan [2,*], Shau-Lun Chiou [3,*] and Yi-Ching Lin [3]

[1] Department of Future Studies and LOHAS Industry, Fo Guang University, Yilan 262307, Taiwan
[2] Department of Health Administration, Tzu Chi University of Science and Technology, Hualien 970302, Taiwan
[3] Department of Rehabilitation, Taiwan Adventist Hospital, Taipei 10556, Taiwan
* Correspondence: yuhli@ems.tcust.edu.tw (Y.-L.L.); frank26480967@gmail.com (S.-L.C.)

Abstract: (1) Background: Workers who perform emotional labor for an extended period are prone to emotional exhaustion; in particular, when the work exceeds the range of one's emotional resources, it will produce job burnout. This study investigated the effects of emotional labor and emotional exhaustion on the physical and mental health of health professionals. (2) Methods: This study was cross-sectional and the sampling criteria were health professionals from August 2020 to July 2021, including rehabilitators, nutritionists, clinical psychologists, radiologists, respiratory therapists, pharmacists, medical examiners and audiologists. A questionnaire was used to collect data on participants' emotional labor, emotional exhaustion, physical health and mental health. A total of 120 valid questionnaires were obtained. (3) Results: Significant positive correlations were found between emotional labor and emotional exhaustion, physical and mental health and anxiety. A hierarchical regression analysis found that the effect of emotional labor on physical and mental health increased the predictive power to 59.7% through emotional exhaustion, and emotional exhaustion had a mediating effect on the relationship between emotional labor and physical and mental health. (4) Conclusions: This study provides a reference for managers of medical institutions to care for employees' work stress and physical and mental health, which will help institutions build a friendly and healthy workplace.

Keywords: emotional labor; emotional exhaustion; physical and mental health; medical professional technicians

Citation: Chen, C.-C.; Lan, Y.-L.; Chiou, S.-L.; Lin, Y.-C. The Effect of Emotional Labor on the Physical and Mental Health of Health Professionals: Emotional Exhaustion Has a Mediating Effect. *Healthcare* **2023**, *11*, 104. https://doi.org/10.3390/healthcare11010104

Academic Editors: Carlos Laranjeira and Ana Querido

Received: 21 October 2022
Revised: 22 December 2022
Accepted: 25 December 2022
Published: 29 December 2022

1. Introduction

As the population ages rapidly, the need for chronic disease care and the severity of disease increases. In addition to having a wealth of knowledge and experience in managing ever-changing medical technologies, medical staff also need to deal with hospital evaluations and maintain a good doctor–patient relationship. Their work performance behaviors can have a massive impact on patient outcomes and safety [1].

Healthcare work involves elevated levels of physical and psychosocial stress, resulting in low employee job satisfaction, burnout, turnover intentions and poor health [2–5]. In particular, during the COVID-19 pandemic, fear rose in medical staff and caused more health problems to surface, which not only had a huge impact on their psychological well-being, but also tested their resilience and ability to cope with stress. Therefore, the psychological and physical effects of COVID-19 on medical personnel have been the focus of attention [6].

The object of medical care services is the patient. During the extensive medical service process, it is necessary for health professionals to suppress their true feelings, so they are often required to maintain a good attitude and manage their emotions to provide quality healthcare services. Emotional labor is an individual's commitment to the management of emotions to create appropriate facial expressions and body movements in front of the public [7]. Emotional labor is divided into two levels of expression: surface and deep. Surface acting involves masking actual emotions such as using a fake smile to hide one's true feelings, whereas deep acting involves trying to feel and express desired emotions such as modifying one's feelings to suit the situation [7,8].

Empirical studies on emotional labor have found that the frequency of deep emotional performance of nursing staff correlates with higher emotional labor and a less ideal state of mental health [9]. Workers who perform emotional labor for an extended period are prone to emotional exhaustion [10,11]; in particular, when the work exceeds the range of one's emotional resources, it will produce job burnout [12].

If employees have negative emotions or cannot control their emotions rationally, and support and methods to eliminate emotional problems are not available, employees may experience job burnout and reduced job performance, which can result in emotional exhaustion [13]. When emotional exhaustion occurs, employees will feel that they are emotionally disconnected; their physical and emotional energy are exhausted, and they have negative emotions such as anxiety, tension, depression and irritability. Moreover, they will display discontentment toward work and lose their jobs. Lack of interest and enthusiasm [14,15], and then symptoms such as physical problems, lack of energy, deteriorating health and exhaustion follow [16]. Therefore, emotional exhaustion is one of the important indicators of physical and mental health [15].

Previous studies have demonstrated that emotional labor directly affects organizational outcomes and employee well-being [17–22]. However, few studies have investigated the relationship between emotional labor and physical and mental health. As for medical personnel, past research has focused on the emotional labor of nursing staff, and research on the effects of emotional labor and emotional exhaustion on physical and mental health of health professionals does not exist. Therefore, this study is expected to provide a reference for managers of medical institutions to care for employees' work stress and physical and mental health, which will help institutions build a friendly and healthy workplace.

2. Materials and Methods

2.1. Study Design and Participants

This study was cross-sectional and used purposive sampling. From August 2020 to July 2021, questionnaires were distributed to health professionals, including rehabilitators, nutritionists, clinical psychologists, radiologists, respiratory therapists, pharmacists, medical examiners and audiologists. The eligible sampling conditions included health professionals working in the hospital who were willing to participate in the study as indicated by their signed consent. The exclusion criteria included those employees who had submitted their resignation. A total of 120 valid questionnaires were obtained. In terms of gender, women and men accounted for 80% (96) and 20% (24), respectively. Ages 30–40 accounted for 41.7% (50), followed by 40–50 at 27.5% (33). More than half (56.7%, $n = 68$) of the participants were married, and 41.7% (50) were unmarried. Those without children were the majority, accounting for 54.2% (65), followed by those with two children, accounting for 26.7% (32). Most participants had a university degree, accounting for 67.5% (81), or had completed graduate school or more, accounting for 20.0% (24). In terms of occupation, rehabilitator accounted for 24.2% (29), followed by medical examiners, accounting for 20% (24) (Table 1).

Table 1. Descriptive statistics of participants ($n = 120$).

Variable	Item	n	(%)	Variable	Item	n	(%)
Gender	Male	24	20	Number of children	None	65	54.2
	Female	96	80		One	19	15.8

Table 1. *Cont.*

Variable	Item	n	(%)	Variable	Item	n	(%)
	20–30	15	12.5	Number of children	Two	32	26.7
	31–40	50	41.7		Three or more	4	3.3
Age	41–50	33	27.5		Rehabilitator	29	24.2
	51–60	17	14.2		Nutritionist	5	4.2
	61–70	5	4.2		Clinical psychologist	8	6.7
	College	15	12.5		Radiologist	20	16.7
Education	University	81	67.5	Professional	Respiratory therapist	8	6.7
	Master's degree	24	20	category	Pharmacist	23	19.2
	Unmarried	50	41.7		Medical examiner	24	20
Marital	Married	68	56.7		Audiologist	3	2.5
status	Divorce	1	0.8				
	Widowed	1	0.8				

2.2. Measures

A structured questionnaire was used as the research tool. Five experts and scholars were invited to evaluate the content validity of the questionnaire. They rated each item on the questionnaire with regard to "importance," "text clarity," and "appropriateness." The reliability was analyzed using Cronbach's alpha. The questionnaire included the following measures. Emotional labor items were rated using a five-point Likert scale [23], which included 7 items on surface acting and 4 items on deep acting. After factor analysis, two eigenvalues greater than 1 were extracted, the eigenvalues of a single factor were 3.42 and 4.12, respectively, and their sum explained 71.37% of the variance. The factor loading of each item is greater than 0.6, indicating that the item validity is good (Table 2). The reliability of deep effect and surface effect were 0.904 and 0.893, respectively, the Cronbach's alpha of emotional labor was 0.885, and the content validity was 0.881.

Table 2. Factor analysis results of emotional labor (n = 120).

Emotional Labor	Varimax Factor One Factor Loadings	Varimax Factor Two Factor Loadings
Surface acting		
I will feign appropriate emotion when serving patients.	0.800	
I will pretend to be in a good mood when serving patients.	0.854	
I treat patients like a play.	0.852	
When serving patients, I will feign job-appropriate emotions.	0.879	
In order to express the emotions required by the job, I will wear a mask to hide what I really feel inside.	0.842	
There is a gap between the emotions I express to the patient and what I feel inside.	0.621	
I would feign appropriate emotions to treat patients.	0.836	
Deep acting		
I try to feel the emotions that need to be expressed when serving patients.		0.809
I try to empathize with the emotions that have to be expressed in serving patients.		0.880
I try to feel the emotions that need to be expressed to the patient.		0.898
I try my best to serve patients with empathy.		0.843
Percentage of variation explained (%)	49.87	21.50
KMO (Kaister–Meyer–Olkin)		0.880
Bartlett's test of sphericity		$\chi^2 = 920.927$ $p < 0.001$

Emotional exhaustion was assessed using the Maslach Burnout Inventory (MBI-GS) [24] which contains five items that are responded to using a five-point Likert scale, with a good-enough reliability of 0.88. In this study, Cronbach's alpha was 0.919 and content validity was 0.933. In terms of physical and mental health, the China Health Questionnaire (CHQ-12) revised by Williams [25] in 1986 was used. On the basis of the 30 questions in the General Health Questionnaire (GHQ) [26], various questions about Chinese culture were added. The Cronbach coefficient of the scale was 0.83–0.92. The measure included 4 questions about physical health, 4 questions about anxiety, and 4 questions about depression and poor family relationships. The Cronbach's alpha of this study was 0.839 and content validity was 0.890.

2.3. Data Collection

After the study was approved by the Taiwan Adventist Hospital Institutional Review Board (108-E-21), the researcher explained the purpose of the study and relevant information to eligible participants. After the participant signed the written consent, they put their completed questionnaire and consent form into a secured box that was allocated for questionnaires.

2.4. Differences in Physical and Mental Health according to Demographic Characteristics

Differences in physical and mental health due to gender ($t = 0.358$, $p > 0.05$) and marital status ($t = -0.298$, $p > 0.05$) were evaluated using a t-test, and there were no significant differences. One-way analysis of variance was used to explore the differences in physical and mental health according to age ($F = 0.291$, $p > 0.05$), education ($F = 0.691$, $p > 0.05$), and profession ($F = 1.658$, $p > 0.05$). There were no statistically significant differences found among the groups.

3. Result

3.1. Descriptive Analysis for Each Scale

Table 3 shows the results for the average scores on the measures. The average score for emotional labor was 3.70 (1–5 points). The average score for surface acting was 3.42, and the average score for deep acting was 4.11. The average score for emotional exhaustion was 2.71 (1–5 points). Cutoff scores for the above were as follows: low group < 1.35, middle group 1.35–3.65, high group > 3.65. The average score for physical and mental health was 2.01 (1–4 points). Among them, the average scores were 1.72 for physical condition, 1.87 for anxiety, and 2.38 for depression and poor family relations. The average score for poor sleep was 2.13. Cutoff scores for the above were as follows: low group < 1.08, middle group 1.08–2.92, high group > 2.92.

Table 3. Descriptive analysis for each scale ($n = 120$).

Measurement Constructs/Items	Number of Items	Mean	Item Score Range
Emotional labor	11	3.70	1–5
Surface acting	7	3.42	1–5
Deep acting	4	4.11	1–5
Emotional exhaustion	5	2.71	1–5
Physical and mental health	12	2.01	1–4
Physical condition	4	1.72	1–4
Anxiety	3	1.87	1–4
Depression and poor family relations	4	2.38	1–4
Poor sleep	1	2.13	1–4

Therefore, the health professional's surface acting is moderate, deep acting is high, emotional exhaustion is moderate, poor physical condition is moderate, anxiety is moderate, depression and poor family relationships are moderate, and poor sleep is moderate.

3.2. Differences in Emotional Labor, Emotional Exhaustion, and Physical and Mental Health among Health Professionals

Kruskal–Wallis analysis was used to explore the differences in emotional labor (including surface acting and deep acting), emotional exhaustion, and physical and mental health in the different professions. The results show that surface acting ($p < 0.05$) and emotional exhaustion ($p < 0.05$) were significantly different among the health professionals categories (F = 3.491, $p < 0.05$).

The surface acting of the examiner and audiologist is the highest, and the surface acting of the clinical psychologist is the lowest. Emotional exhaustion was highest for examiner and pharmacist and lowest for radiologist and nutritionist.

3.3. The Relationship between Emotional Labor, Emotional Exhaustion, and Physical and Mental Health

Pearson correlation coefficients were used to analyze the relationships between the variables. Significant positive correlations were found between emotional labor and emotional exhaustion (r = 0.336, $p < 0.001$) and physical and mental health (r = 0.184, $p < 0.05$). The more surface acting (r = 0.918, $p < 0.001$) and deep acting (r = 0.659, $p < 0.001$), the more emotional labor. Emotional labor was also significantly positively correlated with anxiety (r = 0.225, $p < 0.05$), indicating that the greater the degree of anxiety, the more emotional labor. Emotional labor, physical condition, and depression and poor family relations did not reach a statistically significant difference, indicating that physical condition as well as depression and poor family relations have no correlation with emotional labor. Emotional exhaustion was significantly positively correlated with physical and mental health (r = 0.597, $p < 0.001$) and surface acting (r = 0.404, $p < 0.001$). This shows that the worse the physical and mental health and the more surface acting, the greater the emotional exhaustion. Deep acting and emotional exhaustion did not reach a statistically significant difference, indicating that there is no correlation between them. Emotional exhaustion was significantly positive with physical condition (r = 0.482, $p < 0.001$), anxiety (r = 0.596, $p < 0.001$), and depression and poor family relations (r = 0.294, $p < 0.001$). This shows that the worse the physical condition, the more anxiety and the more depression and poor family relations, the greater the degree of emotional exhaustion. Physical and mental health, surface acting and deep acting did not reach a statistically significant difference, indicating that there is no correlation between them. Physical and mental health and physical condition (r = 0.857, $p < 0.001$), anxiety (r = 0.878, $p < 0.001$), and depression and poor family relations (r = 0.600, $p < 0.001$) had a significantly positive correlation. This means that the worse the physical condition, the more anxiety, and the more depression and poor family relations, the worse the physical and mental health (Table 4).

Table 4. Correlations between emotional labor, emotional exhaustion, and physical and mental health.

Variable	Emotional Labor	Emotional Exhaustion	Physical and Mental Health
Emotional labor	1		
Emotional exhaustion	0.336 ***	1	
Physical and mental health	0.184 *	0.597 ***	1
Surface acting	0.918 ***	0.404 ***	0.135
Deep acting	0.659 ***	0.044	0.166
Physical condition	0.167	0.482 ***	0.857 ***
Anxiety	0.225 *	0.596 ***	0.878 ***
Depression and poor family relations	−0.070	0.294 ***	0.600 ***

*: $p < 0.05$; ***: $p < 0.001$.

3.4. The Mediating Effect of Emotional Exhaustion on the Relationship between Emotional Labor and Physical and Mental Health

A hierarchical regression analysis was used to explore emotional exhaustion as a mediator in the relation between emotional labor and physical and mental health. The results show that Model 1 (F = 4.138, $p < 0.05$), Model 2 (F = 14.996, $p < 0.001$) and Model 3 (F = 32.447, $p < 0.001$) were all statistically significant. The R^2 of emotional labor on physical and mental health was 18.4% ($\beta = 0.184$, $p < 0.05$), and the R^2 of emotional exhaustion on physical and mental health was 33.6% ($\beta = 0.336$, $p < 0.001$). The effect of emotional labor on physical and mental health increased the predictive power to 59.7% through emotional exhaustion, and emotional exhaustion had a mediating effect on the relationship between emotional labor and physical and mental health ($\beta = 0.603$, $p < 0.001$) (Table 5).

Table 5. The mediating effect of emotional exhaustion in the relation between emotional labor and physical and mental health.

Predictor	Physical and Mental Health		
	Model 1	Model 2	Model 3
Predictor			
Emotional labor	0.184 *		−0.019
Emotional exhaustion		0.336 ***	0.603 ***
R^2	0.184	0.336	0.597
$\triangle R^2$	0.184	0.152	0.261
F	4.138 *	14.996 ***	32.447 ***

*: $p < 0.05$; ***: $p < 0.001$.

4. Discussion

This study found that the higher the emotional labor of health professionals, the greater the emotional exhaustion. Similar to the discussion by scholars such as Hochschild [7], when employees are required to regulate their emotions at work to match the changes in the external environment, the appropriate external emotional performance may be inconsistent with the employee's true inner feelings; the external and internal emotions are alienated. This alienation cannot be maintained for a long time, which leads to stress overload and emotional exhaustion. Therefore, workers with high emotional labor are more likely to have emotional disorders and emotional exhaustion. Given the correlation between emotional labor and emotional exhaustion [27], the higher the emotional labor undertaken by an individual, the more there will be an increase in the frequency of emotional exhaustion [28].

The results of this study show that surface acting was positively correlated with emotional exhaustion, while deep acting had no significant relation to emotional exhaustion. The plausible reason is that surface acting only modifies the external emotional expression to achieve the purpose of emotional camouflage, which would make it easy to create conflict between inner and external emotions, resulting in emotional imbalance. Therefore, the higher the level of surface acting, the heavier the emotional labor required, which is akin to the view put forward by Grandey [29,30]. When an individual's inner emotions are repressed and desired emotions are expressed, the surface effects exhibited can deplete personal energy and affect employee well-being [30–32]. For example, nurses are more likely to experience emotional exhaustion when they engage in superficial performances [33]. Surface behavior is positively correlated with emotional exhaustion [31,34]. Conversely, deep performances tend to resolve the initial emotional dissonance, resulting in the same internal feelings and external performance [30,31]. Employees will experience positive emotional experiences to express positive emotions, and these experiences may provide relief of fatigue [35]. A prior study found that deep acting leads to better mood, better job performance and higher job satisfaction [29].

Long-term and constant emotional labor may damage an individual's physical and mental health [36,37]. This study found that the emotional labor of health professionals was significantly and positively correlated with anxiety and physical and mental health. Previous studies also found that surface acting emotional labor was positively associated with depression [38,39]. A study on nursing staff found that emotional labor can explain 21% of the variance in mental health status, showing that the greater the emotional labor of nursing staff, the less ideal was their mental health [9]. Additionally, a recent study found that episodic emotional labor was a strong predictor of depressive symptoms in nursing home health care workers two years later [40].

Past studies have discovered that in the process of performing medical care, nurses must suppress their emotions to provide professional services. Such emotional services can lead to work stress, deterioration of physical and mental health and emotional exhaustion [41], and directly affect the quality of patient care and the occurrence of medical negligence, which can even prompt them to leave the workplace [42]. Emotional labor can predict employee job satisfaction and emotional exhaustion. Nursing staff who use superficially disguised emotional labor are more likely to experience emotional exhaustion and have lower job satisfaction because they need to hide their true emotions and disguise unfelt emotions, increasing the degree of emotional dysregulation [43].

Moreover, the present study observed that surface acting emotional labor and emotional exhaustion were positively correlated, and the latter had a mediating effect on the relation between emotional labor and physical and mental health. Similar to previous studies, nurses are more likely to experience emotional exhaustion when they use surface acting emotional labor. As a way of regulating emotions, performances can help employees regain emotional resources and reduce emotional exhaustion [44]. In addition, Rogers et al. [38] found positive correlations between the surface acting emotional labor of doctors and work-related burnout and depression. There was a negative correlation between deep emotional labor and burnout, and work-related burnout mediated the relationship between surface emotional labor and depression.

The above literature suggests that the higher the level of emotional labor, the more serious the degree of emotional exhaustion, and the less ideal one's physical and mental health. Conversely, it also indicates the lower the emotional labor, the lower the degree of emotional exhaustion and the better one's physical and mental health. Therefore, it is inferred that there will be significant differences between emotional labor and emotional exhaustion and physical and mental health of different degrees.

This study adopted a cross-sectional data collection method to explore the emotional labor, emotional exhaustion, and physical and mental health of health professionals. It is suggested that future research conduct longitudinal studies to examine causal relationships and effects over time. In addition, the study included health professionals at one particular hospital, and there may be different results due to differences in the background variables of the research participants.

5. Conclusions

This study confirms that the emotional labor and emotional exhaustion of health professionals are quite serious. Emotional labor was significantly positively correlated with emotional exhaustion, physical and mental health, and anxiety. Emotional exhaustion was significantly and positively correlated with physical and mental health, physical condition, anxiety, and depression and poor family relationships. The predictive power of emotional labor on physical and mental health was improved through emotional exhaustion; emotional exhaustion had a mediating effect in the relation between emotional labor and physical and mental health.

We hope to inspire others to study how emotion management can help the health and well-being of healthcare professionals, thereby improving the quality of care healthcare professionals provide to patients. Therefore, developing emotional management skills necessary for health professionals to work effectively is essential to improve the quality of patient care and treatment outcomes, and to ensure that patient care is not compromised by the health professional's own emotional and health conditions. There is indeed a great need for channels to relieve and de-escalate their emotional labor to prevent emotional exhaustion from occurring. Related studies have found that social support through colleagues, supervisors and organizations in the workplace may reduce the negative effect of emotional labor [45,46]. Therefore, the implications of this study suggest that in addition to improving the working environment (including providing social support) to reduce the emotional labor of employees' surface acting behavior, hospital managers can provide supportive psychological counseling to reduce employees' emotional labor and emotional exhaustion. This in turn will improve the physical and mental health and well-being of employees.

Author Contributions: Conceptualization, C.-C.C., Y.-L.L. and S.-L.C.; methodology, C.-C.C. and Y.-L.L.; software, C.-C.C. and Y.-L.L.; validation, C.-C.C., Y.-L.L., Y.-C.L. and S.-L.C.; formal analysis, C.-C.C. and Y.-L.L.; investigation, Y.-C.L. and S.-L.C.; resources, Y.-C.L. and S.-L.C.; data curation, C.-C.C. and Y.-L.L.; writing—original draft, C.-C.C. and Y.-L.L.; writing—review and editing, C.-C.C. and Y.-L.L.; project administration, C.-C.C., Y.-L.L., Y.-C.L. and S.-L.C. All authors have read and agreed to the published version of the manuscript.

Funding: Taiwan Adventist Hospital provided funding for this research project.

Institutional Review Board Statement: The study was conducted according to the guidelines of the Declaration of Helsinki and approved by the Institutional Review Board of Taiwan Adventist Hospital (108-E-21).

Informed Consent Statement: Informed consent was obtained from all participants involved in the study.

Data Availability Statement: Data is available upon request from the corresponding author. Data are not publicly available due to privacy and ethical constraints.

Conflicts of Interest: The authors declare no conflict of interest.

References

1. Birkhauer, J.; Gaab, J.; Kossowsky, S.; Hasler, P.; Krummenacher, C.; Werner, C.; Gerger, H. Trust in the health care professional and health outcome: A meta-analysis. *PLoS ONE* **2017**, *12*, e0170988. [CrossRef] [PubMed]
2. Iliceto, P.; Pompili, M.; Spencer-Thomas, S.; Ferracuti, S.; Erbuto, D.; Lester, D.; Candilera, G.; Girardi, P. Occupational stress and psychopathology in health professionals: An explorative study with the multiple indicators multiple causes (MIMIC) model approach. *Stress* **2013**, *16*, 143–152. [CrossRef] [PubMed]
3. Mosadeghrad, A.M. Occupational stress and turnover intention: Implications for nursing management. *Int. J. Health Policy Manag.* **2013**, *1*, 169–176. [CrossRef]
4. Zhang, Y.; Punnett, L.; Gore, R. Relationships among employees' working conditions, mental health and intention to leave in nursing homes. *J. Appl. Gerontol.* **2014**, *33*, 6–23. [CrossRef] [PubMed]
5. Khamisa, N.; Oldenburg, B.; Peltzer, K.; Ilic, D. Work related stress, burnout, job satisfaction and general health of nurses. *Int. J. Environ. Res. Public Health* **2015**, *12*, 652–666. [CrossRef] [PubMed]
6. Sanchez-Gomez, M.; Giorgi, G.; Finstad, G.L.; Urbini, F.; Foti, G.; Mucci, N.; Zaffina, S.; León-Perez, J.M. COVID-19 pandemic as a traumatic event and Its associations with fear and mental health: A cognitive-activation approach. *Int. J. Environ. Res. Public Health* **2021**, *18*, 7422. [CrossRef]
7. Hochschild, A.R. *The Managed Heart: Commercialization of Human Feeling*; University of California Press: Berkeley, CA, USA, 1983.
8. Robbins, S.P.; Judge, T.A. *Organizational Behavior*; Prentice Hall: Boston, MA, USA, 2011.
9. Ching-Youn, L.; Pi-Ching, H.; Hui-Fang, S. The relationship between emotional labor and mental health among nurses in Catholic hospitals in Taiwan. *Taiwan J. Public Health* **2013**, *32*, 140–154.
10. Maslach, C. *Burnout, the Cost of Caring*; Prentice-Hall: Englewood Cliffs, NJ, USA, 1982.
11. Hsiu-Ching, T. The Study of the Relationship among Hospital Employees, Emotion Labor Load and Customer-Oriented Behavior. Master's Thesis, Institute of Human Resource Management, National Sun Yat-sen University, Kaohsiung, Taiwan, 2003.
12. Babakus, E.; Cravens, D.W.; Johnston, M.; Moncrief, W. The role of emotional exhaustion in sales force attitude and behavior relationships. *J. Acad. Mark. Sci.* **1999**, *27*, 58–70. [CrossRef]
13. Jackson, S.; Schwab, R.; Schuler, R. Toward an understanding of the burnout phenomenon. *J. Appl. Psychol.* **1986**, *71*, 630–640. [CrossRef]
14. Maslach, C.; Jackson, S.E. The measurement of experienced burnout. *J. Occup. Behav.* **1981**, *2*, 99–113. [CrossRef]
15. Wright, T.A.; Cropanzano, R. Emotional exhaustion as a predictor of job performance and voluntary turnover. *J. Appl. Psychol.* **1998**, *83*, 486–493. [CrossRef] [PubMed]
16. Kahill, S. Symptoms of professional burnout: A review of the empirical evidence. *Can. Psychol.* **1988**, *29*, 284–297. [CrossRef]
17. Kammeyer-Mueller, J.D.; Rubenstein, A.L.; Long, D.M.; Odio, M.A.; Buckman, B.R.; Zhang, Y.; Halvorsen-Ganepola, M.D. A meta-analytic structural model of dispositional affectivity and emotional labor. *Pers. Psychol.* **2013**, *66*, 47–90. [CrossRef]
18. Deng, H.; Walter, F.; Lam, C.K.; Zhao, H.H. Spillover effects of emotional labor in customer service encounters toward coworker harming: A resource depletion perspective. *Pers. Psychol.* **2017**, *70*, 469–502. [CrossRef]
19. Grandey, A.A.; Sayre, G.M. Emotional labor: Regulating emotions for a wage. *Curr. Psychol. Sci.* **2019**, *28*, 131–137. [CrossRef]
20. Hintsanen, M.; Puttonen, S.; Smith, K.; Törnroos, M.; Jokela, M.; Pulkki-Råback, L.; Hintsa, T.; Merjonen, P.; Dwyer, T.; Raitakari, O.T.; et al. Five-factor personality traits and sleep: Evidence from two population-based cohort studies. *Health Psychol.* **2014**, *33*, 1214–1223. [CrossRef]
21. Wagner, D.T.; Barnes, C.M.; Scott, B.A. Driving it home: How workplace emotional labor harms employee home life. *Pers. Psychol.* **2014**, *67*, 487–516. [CrossRef]

22. Diestel, S.; Rivkin, W.; Schmidt, K. Sleep quality and self-control capacity as protective resources in the daily emotional labor process: Results from two diary studies. *J. Appl. Psychol.* **2015**, *100*, 809–827. [CrossRef]
23. Diefendorff, J.M.; Croyle, M.H.; Gosserand, R.H. The dimensionality and antecedents of emotional labor strategies. *J. Vocat. Behav.* **2005**, *66*, 339–357. [CrossRef]
24. Schaufeli, W.; Leiter, M.; Maslach, C.; Jackson, S. Maslach Burnout Inventory-General Survey. In *The Maslach Burnout Inventory: Test Manual*; Maslach, C., Jackson, S.E., Leiter, M.P., Eds.; Consulting Psychologists Press: Palo Alto, CA, USA, 1996.
25. Cheng, T.A.; Williams, P. The design and development of a screening questionnaire (CHQ) for use in community studies of mental disorders in Taiwan. *Psychol. Med.* **1986**, *16*, 415–422. [CrossRef]
26. Goldberg, D.P. *The Detection of Psychiatric Illness by Questionnaire; A Technique for the Identification and Assessment of Non-Psychotic Psychiatric Illness*; Oxford University Press: Toronto, ON, Canada, 1972; p. 156.
27. Wu, T.Y.; Cheng, B.S. The relationships among frequency of encountering difficult customers, perceived service training utility, emotional labor, and emotional exhaustion-the viewpoint of conservation of resources theory. *J. Manag.* **2006**, *23*, 581–599.
28. Wen-Hua, C. The Comparison of Emotional Labor, Emotional Labor Performances Manners of First Line Service Personnel on Emotional Exhaustion and Mental Health. Master's Thesis, Department of Counselling and Guidance, National University of Tainan, Tainan, Taiwan, 2015.
29. Grandey, A.A. Emotion regulation in the workplace: A new way to conceptualize emotional labor. *J. Occup. Health Psychol.* **2000**, *5*, 95–110. [CrossRef] [PubMed]
30. Grandey, A.A. When "the show must go on": Surface acting and deep acting as determinants of emotional exhaustion and peer-rated service delivery. *Acad. Manag. J.* **2003**, *48*, 86–96. [CrossRef]
31. Brotheridge, C.M.; Grandey, A.A. Emotional labor and burnout: Comparing two perspectives of "people work". *J. Vocat. Behav.* **2002**, *60*, 17–39. [CrossRef]
32. Hülsheger, R.; Schewe, A.F. On the costs and benefits of emotional labor: A meta-analysis of three decades of research. *J. Occup. Health Psychol.* **2011**, *16*, 361–389. [CrossRef] [PubMed]
33. Halbesleben, J.R.B.; Buckley, M.R. Burnout in organizational life. *J. Manag.* **2004**, *30*, 859–879. [CrossRef]
34. Zammuner, V.L.; Galli, C. The relationship with patients: Emotional labor and its correlates in hospital employees. In *Emotions in Organizational Behavior*; Hartel, C.E., Zerbe, W.J., Ashkanasy, N.M., Eds.; Routledge: Mahwah, NJ, USA, 2005; pp. 251–285.
35. Schaubroeck, J.; Jones, J.R. Antecedents of workplace emotional labor dimensions and moderators of their effects on physical symptoms. *J. Organ. Behav.* **2000**, *21*, 163–183. [CrossRef]
36. Rafaeli, A.; Sutton, R.I. Expression of emotion as part of the work role. *Acad. Manag. Rev.* **1987**, *12*, 23–27. [CrossRef]
37. Bagdasarov, Z.; Connelly, S. Emotional labor among healthcare professionals: The effects are undeniable. *Narrat. Inq. Bioeth.* **2013**, *3*, 125–129. [CrossRef]
38. Rogers, M.E.; Creed, P.A.; Searle, J. Emotional labour, training stress, burnout, and depressive symptoms in junior doctors. *J. Vocat. Educ. Train.* **2014**, *66*, 232–248. [CrossRef]
39. Yoon, S.L.; Kim, J.H. Job-related stress, emotional labor, and depressive symptoms among Korean nurses. *J. Nurs. Scholarsh.* **2013**, *45*, 169–176. [CrossRef] [PubMed]
40. Suh, C.; Punnett, L. Surface-acting emotional labor predicts depressive symptoms among healthcare workers over a 2-year prospective study. *Int. Arch. Occup. Environ. Health* **2020**, *94*, 367–375. [CrossRef] [PubMed]
41. Mann, S.; Cowburn, J. Emotional labour and stress within mental health nursing. *J. Psychiatr. Ment. Health Nurs.* **2005**, *12*, 154–162. [CrossRef]
42. Chao-Kang, F.; Chung-Yi, L. Prevalence and factors associated with subjective fatigue among nurses. *Chin. J. Occup. Med.* **1998**, *5*, 129–138.
43. Laschinger, H.K.S.; Leiter, M.P. The impact of nursing work environments on patient safety outcomes: The mediating role of burnout/engagement. *J. Nurs. Adm.* **2006**, *36*, 259–267. [CrossRef]
44. Huei-Yin, C. A discussion on the factors influencing nurses' job satisfaction and emotional exhaustion. *J. Meiho Univ.* **2011**, *31*, 115–126.
45. Mikeska, J.; Harvey, E.J. The political CEO: An event study comparing consumer attributions of CEO behavior. *Soc. Sci. Q.* **2015**, *96*, 76–92. [CrossRef]
46. Flores, R.D.; Schachter, A. Who are the "illegals"? The social construction of illegality in the United States. *Am. Sociol. Rev.* **2018**, *83*, 839–868. [CrossRef]

 *healthcare*

Article

Integrating Mindfulness into the Subject of Physical Education—An Opportunity for the Development of Students' Mental Health

Roberto Delgado-Montoro [1], Alberto Ferriz-Valero [1], Olalla García-Taibo [2,*] and Salvador Baena-Morales [1,3,*]

1   Department of General and Specific Didactics, Faculty of Education, University of Alicante, 03690 Alicante, Spain
2   Department of Physical Education and Sport, Pontifical University of Comillas, CESAG-Mallorca, 07013 Palma de Mallorca, Spain
3   Faculty of Education, Valencian International University (VIU), 46002 Valencia, Spain
*   Correspondence: otaibo@cesag.org (O.G.-T.); salvador.baena@ua.es (S.B.-M.)

Citation: Delgado-Montoro, R.; Ferriz-Valero, A.; García-Taibo, O.; Baena-Morales, S. Integrating Mindfulness into the Subject of Physical Education—An Opportunity for the Development of Students' Mental Health. *Healthcare* 2022, 10, 2551. https://doi.org/10.3390/healthcare10122551

Academic Editors: Carlos Laranjeira and Ana Querido

Received: 20 October 2022
Accepted: 13 December 2022
Published: 16 December 2022

Publisher's Note: MDPI stays neutral with regard to jurisdictional claims in published maps and institutional affiliations.

Abstract: Stress, uncertainty, and the abuse of technologies are components that have a negative impact on the physical, social, and psychological health of young people. One of the aims of the Education for Sustainable Development (ESD) is to empower individuals to reflect on their actions, and mindfulness arises as one tool with an important potential to contribute on this matter. Therefore, the objective of this study was to assess the effects of mindfulness practices on the ability of students to focus their attention on external, internal or kinesthetic factors, awareness in acting, and acceptance. Consequently, a quasi-experimental study was developed to compare groups between the pre and post condition. The study participants were a total of 127 students (52 women) from 4th year of secondary school and 1st year of a achelor's degree (16.5 ± 1.5 years). The sample was assigned by academic convenience, with 54 students in the experimental group and 73 in the control group. The intervention was carried out for 4 weeks. During this period, the experimental group participated in mindfulness activities such as guided meditations at the end of the PE session or challenges that stimulated the student in daily actions. The control group continued with the planned programming in physical education class. These groups were subjected to the following test: (1) Mindfulness for School Scale (MSS) and (2) Child and Adolescent Mindfulness Measure (CAMM). To analyze the results, the normality of the sample was evaluated through the Mann–Whitney U test, resulting as non-parametric. The search for possible differences between the groups was carried out by using the Wilcoxon test. The statistics showed that the experimental group presented significant improvements ($p \leq 0.05$) in most of the measured parameters: external attention, kinesthetics attention, and mean of the CAMM. These results seem to show that the use of mindfulness could be an appropriate tool to be implemented in the school context in order to directly contribute to the mental health of high school students, and thus to an education for the sustainable development.

Keywords: health; physical education; mindfulness; attention; sustainability

1. Introduction

1.1. The Current Lifestyle and Its Problems for the Youngest

Nowadays, people's lifestyles are characterized by the presence of unhealthy behaviors, such as sedentary habits, due to the increased use of motorized transport and screens for work, education, and leisure time. The prevalence of physical inactivity and sedentary lifestyles has increased in recent years, standing at 28% of adults and 81% of adolescents globally [1]. Physical inactivity, in addition to being the fourth leading cause of premature mortality worldwide [2], can be one of the risk factors for the development of more than 35 chronic diseases and disorders [3], including obesity and overweight, which in turn rank as the fifth main cause of mortality worldwide. In addition to the consequences generated by physical inactivity at a physical level, it also has a negative impact at a psychological and social level. Moreover, the use of cell phones has not stopped growing in recent years [4], generating a social transformation and breaking with traditional interaction and socialization mechanisms [5]. This change has negatively impacted the social relations among young people, causing a deficit in communication skills; phobia; social isolation [6,7], deterioration of relationships with family members, teachers, partners, and among peers; worsening of academic and work performance [8]; and increasing depression and anxiety [9]. Many preadolescents show high levels of anxiety and/or behavioral problems that, without meeting certain clinical diagnostic criteria, imply restrictions on their daily lives [10]. This is closely related to technology addiction problems [11] and proof of this is that in the last decade the diagnoses of attention deficit hyperactivity disorder have increased by 30%, with nearly 250,000 Spanish minors taking psychostimulants to deal with this disorder [12]. Recently, a study that assessed the preoccupation and fears about the issues of the 21st century, such us climate changes, natural devastation, COVID-19, and war, showed that young people reported less psychological well-being than older adults, in terms of stress, anxiety, and depression [13]. Given this situation, the need to promote emotional and social competence and children's well-being is especially important during the transition from childhood to adolescence [14].

1.2. Education for Sustainable Development and Mindfulness

Faced with this current multidimensional problem, which affects us environmentally, physically, psychologically, and socially, the United Nations seeks to reverse this reality and contribute to the care of people in general, and young people in particular, through the design and establishment of the Sustainable Development Goals (SDGs), detailed in 169 specific targets. Therefore, the collaboration of institutions on the pursuit of these achievements is essential [15]. Among them, the educational context is a key factor to collaborate with the achievement of these European goals, especially in those ages in which education is compulsory. In this regard, UNESCO proposed the term Education for Sustainable Development (ESD), with the mission to empower students to make responsible decisions towards the achievement of a just society with economic and environmental integrity, both in the present and in future generations [16]. Education for Sustainable Development (ESD) aims to develop competencies that enable individuals to reflect on their actions, considering their current and future social, cultural, economic, and environmental impacts from a local and global perspective. Considering the deeply rooted origin of human cultures and behaviors, there is a need to implement new strategies that allow to change, in a profound way, some of their worldviews, values, and beliefs [17]. In this direction, mindfulness meets certain characteristics that allow us to contribute to the achievement of ESD. In recent years, this type of practice has been introduced in various sectors of society, particularly in education, showing several benefits, as will be approached below. Mindfulness is defined as a person's ability to focus attention on events, experiences, and states of the present moment, both external and internal [11]. Mindfulness refers to the state of bringing non-judgmental awareness to the present moment and implies two components: (1) self-regulation of attention and (2) orientating an individual to the present moment with curiosity, openness, and acceptance [18]. In other words, mindfulness represents a successful tool for promoting awareness, cultivating the ability to be attentive, and increasing subjective wellbeing. Therefore, the benefits generated in mindfulness facilitate emotional competence such as emotional intelligence, because of the ability to be aware of one's own emotions, the feelings of others, or to feel empathy, in line with the goals of ESD. Kabat-Zinn [19] rightly stated, "We do not know what specific knowledge our children will need ten or twenty years from now, because the world, and their work when they come to it, will be so different from ours. What we do know, is that they will need to know how to pay attention, how to focus and concentrate, how to listen and learn, and how to be in wise relationship with themselves—including their thoughts and emotions—and with others".

This change in the present scenario has attracted the attention and interest of psychologists, educationists, and researchers to the practice of mindfulness-based interventions with students, both with children and adolescents, or even with teachers, for enhancing their overall well-being. Most of the research conducted on mindfulness has focused on studying the impact of mindfulness practices on students. According to [20,21], different studies have shown that children and adolescents who practice mindfulness can experience improvements in three basic areas: in general well-being, in cognitive aspects, and in social and emotional skills. All of these aspects can be beneficial to them at some point in their lives when facing challenges and changes in their personal, biological, psychological, and psychological development. That is, mindfulness offers useful preparation beyond the classroom, as it provides people with important tools and strategies that could apply in a variety of settings throughout life [22].

Regarding the benefits and improvements in student well-being, different studies have shown that mindfulness helps to promote greater resilience to daily stress [22]. Children and adolescents are especially vulnerable to the harmful effects of toxic stress, which can cause attention problems, mood swings, emotional disturbances, sleep problems, and learning disabilities [23]. Therefore, reducing stress and anxiety levels has benefits both at a personal and academic level. In addition, this practice has been shown to be helpful in the treatment of depression in adolescence [24], as well as with symptoms of post-traumatic stress in children and adolescents [25]. In addition to reducing students' stress levels, it improves their attention and concentration, so that mindfulness could improve the learning process of students, thus raising their academic performance and, consequently, their results [22]. Regarding emotional intelligence, there is evidence that this type of practice improves the ability to manage emotions in children and adolescents, contributing to the development of emotional self-regulation skills [26] and fostering empathy and compassion with peers [27]. Precisely, emotional self-regulation and social interaction appears as new content within mental health in the last educational curriculum for the physical education subject in Spain (Real Decreto 217/2022).

1.3. Mindfulness in the Physical Education Curriculum

A review study on research that implemented mindfulness in PE classes concluded that these are scarce [28]. However, the analysis of the secondary physical education curriculum (Real Decreto 217/2022) suggests a close relationship between this subject and mindfulness. This newest approach for the physical education subject in Spain sets out a holistic perspective that highlights physical, psychological, social, and environmental aspects. Students will have to learn how to manage their emotions and social skills. In this regard, it will be necessary to integrate cognitive and motor skills, but also affective-motivational and interpersonal relationships. In terms of health contents, the "Active and healthy life" block addresses three components of health (physical, mental, and social); this means developing postural education, relaxation, and positive relationships among others, in order to achieve a healthy lifestyle.

In order to integrate healthy routines and responsible motor practice, developing the specific competence of "emotional self-regulation and social interaction" is needed. This means, "develop the processes aimed at regulating their emotional response to situations arising from the practice of physical activity and sport, and, on the other hand, developing social skills and promoting inclusive and constructive relationships. It is also emphasized the positive attitude to face challenges, regulating impulsivity, tolerating frustration and persevering in the face of difficulties. It involves the identification of emotions and feelings that are experienced within the motor practice, the positive expression of these and their proper management in order to constructively buffer the effects of unpleasant emotions and feelings that they generate, as well as to promote pleasant emotions. Likewise, in relation to one's own body, it involves the development of skills for the preservation and care of personal integrity". Regarding the collective level, the PE subject must specifically work on "putting into play social skills to face the interaction with the people with whom one converges in the motor practice. It involves dialoguing, debating, contrasting ideas and agreeing to resolve situations; expressing proposals, thoughts and emotions; listening actively, and acting assertively". In fact, one of the basic skills refers to reflection on negative attitudes towards physical activity derived from preconceived ideas, prejudices, stereotypes, or negative experiences. Having presented the characteristics and benefits of mindfulness and the particularities of the EF curriculum, there is no doubt that there is a close relationship between both and that meditation is an ideal tool to implement in the classroom. However, more studies on mindfulness MF in secondary education curriculum planning are needed to evaluate its benefits [29]. Thus, the objective of this study was to assess the effects of mindfulness practices on the ability of students to focus their attention on external, internal or kinesthetic factors, awareness in acting, and acceptance.

2. Materials and Methods

2.1. Design

The methodological design of the study was quasi-experimental with quantitative pre- and post-test measures. Two different groups were formed in the different courses of the 4th year of secondary education and 1st year of bachelor's degree. Only one of the two groups practiced activities focused on mindfulness in order to observe the differences obtained between the experimental group compared to the control group. The primary outcome measures were external, internal and kinesthetic attention, and mindfulness. Convenience sampling, a type of non-probabilistic or non-random sampling, was carried out. This type of sampling is valid when it includes members of the target population who meet certain practical criteria, such as easy accessibility, geographical proximity, availability at a given time, or willingness to participate in the study. This type of sampling is valid and common in quantitative studies, although the results obtained cannot be extrapolated to the overall study population [30]. In this research, the reference population would be adolescents. However, it should be noted that in a convenience sample, neither biases nor their probabilities are quantified [31]. Consequently, researchers can guarantee the extent to which participants will represent the population in terms of traits or research mechanism.

2.2. Participants

The total sample consisted of 127 students (52 females) from a high school in Alicante (Spain). The age range in the study was 16.5 ± 1.5. The experimental group (EG) was composed of 54 students of which 23 were girls and 31 were boys. The control group (CG) consisted of 73 individuals, 29 girls and 42 boys. In order to select the sample, inclusion criteria were assigned: no previous experience in meditation, being students at the school and agreeing to participate in the study. Exclusion criteria were: (1) having specific needs: e.g., autism or Down's syndrome, (2) students with learning difficulties in reading comprehension of the target language, and (3) students with physical incapacity to perform the practical classes

2.3. Intervention

The mindfulness training was carried out during 3 school weeks in the second and third trimester of the 2021/2022 school year. The pre-test evaluation was conducted 1 week before the start of the training and the post-test evaluation was conducted one week after the end of the intervention. The mindfulness program was implemented in a total of six consecutive PE sessions, two classes per week. In these sessions, a 10-min guided meditation was conducted by the person in charge of this study during the calm-down phase. This consisted of an exercise of relaxation and awareness, in which one had to pay attention to the breath and body, as well as to thoughts without judgment, while visualizing situations (spaces and sounds of nature) that lead to a state of presence and calm (see https://youtu.be/f44boj9hQRk (accessed on 5 May 2022)). The instructors who conducted the sessions had previous experience as facilitators and were trained in mindfulness. In addition, a series of challenges were also developed to keep them in touch with mindfulness during the Easter holidays, activities that stimulated the student into aware daily actions. These challenges consisted of: Day (1): Mindful eating; day (2): Mindful shower; day (3): Three conscious stretches while breathing with awareness; day (4): Mindful walk; day (5): 7-min guided meditation (focused on awareness of breathing, bodily sensations, thoughts, and feelings) (see https://www.youtube.com/watch?v=XXXaoKi7IY0 (accessed on 5 May 2022)). All students completed the intervention. Only one student did not attend one session due to sickness and then followed the rest of the sessions normally.

The collaboration and permission of the school and PE teachers was requested. All participants were informed of the aims of the study and signed an informed consent for the release of data for scientific use. The ethical aspects presented in the Declaration of Helsinki were respected in the design of the study. This research was approved by the ethics committee of the University of Alicante with code UA-2022-03-17. It was emphasized to the participants that the information would be totally confidential, treated statistically, and that neither the center nor the parents would be informed about the results of the questionnaires of each student. In order to carry out the study, a letter of application was sent indicating the outline of the project, with the objectives, relevance of the research, methodology, sample inclusion/exclusion criteria, risks and benefits, commitment to comply with data protection legislation, commitment to confidentiality and anonymity, information sheet, informed consent, and annexes with the tests or surveys to be taken or description of the tests to be carried out. The aforementioned scheme can be observed in Figure 1, which shows the number of participants and the processes they underwent.

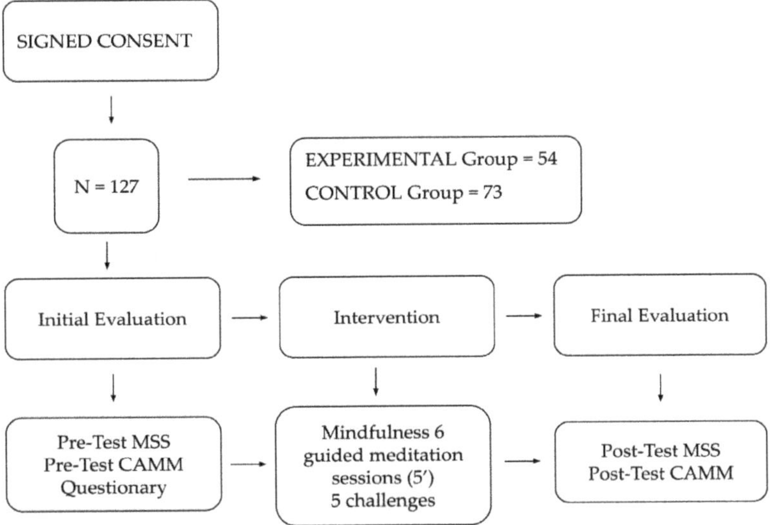

Figure 1. Intervention process.

2.4. Instruments

The MSS, which measures levels of mindfulness in the school environment, and the CAMM Questionnaire, which measures the level of mindfulness, were used as assessment instruments.

2.4.1. MSS (Mindfulness at School Scale)

The MSS [29] is a Likert scale whose score consists of an ordinal quantitative variable (1 = never to 5 = always). In this case, the 12 items of the questionnaire are written in a positive sense, so the result is the sum of the scale scores. The scale has an internal consistency of $\alpha = 0.84$ and a test-retest stability of 0.78. The MSS measures three factors which the authors describes and labels as follows:

- Kinesthetic attention: referring to the subject's ability to become aware of their movement and motor actions (items 5 and 12) ($\alpha = 0.74$).
- External attention: referring to the subject's ability to direct attentional resources towards events happening outside, attention to observation (items 1, 2, 3, 4, 6, 7, and 8). ($\alpha = 0.60$).
- Internal attention: referring to the subject's ability to direct attentional resources towards events occurring inwardly, attention to introspection (items 9, 10, and 11) ($\alpha = 0.66$).

2.4.2. CAMM (Child and Adolescent Mindfulness Measure)

The CAMM [32] is an assessment instrument composed of 10 items that measures acceptance and mindfulness with a Likert scale where the score consists of an ordinal quantitative variable (1 = never to 5 = always). In this case, the items of the questionnaire are written in a negative sense, so to obtain the final result the sum of the total score will be inverted. This questionnaire has mainly been used with children and adolescents aged 9–18 years. The test-retest reliability for the entire test according to the test authors is $\alpha = 0.88$. The factor structure, internal consistency, and construct validity of the individuals were tested. Although this instrument has a one-dimensional factor structure, there are measure items regarding two aspects of mindfulness: acting with awareness (e.g., "It's hard for me to pay attention to only one thing at a time") and accepting without judgment ("I get upset with myself for having certain thoughts") [33].

2.5. Statistical Analysis

According to [33], the statistical power of the sample size was calculated using the free software G*Power (See 3.1.9.6, University of Dusseldorf, Düsseldorf, Germany). The sample size, 54 participants and 73 participants per group, with an estimated medium effect size (0.5), and a significance of 95%, resulted in a power of 0.96. All continuous variables collected during the study were subjected to a normality test, specifically the Kolmogorov–Smirnov test. The data were also subjected to univariate statistical analysis for non-parametric samples, specifically the Mann–Whitney U and Wilcoxon test, to assess the differences between the experimental group (EG) and control group (CG) twice: pre- and post-intervention. The significance level was set at $p < 0.05$ in all cases.

A repeated-measures analysis of variance (ANOVA 2 × 2) mixed model was used when pre and post differences were identified, to give robustness to the analysis. The dependent variables were four: CAMM test, internal attention, external attention, and kinesthetic attention. Time (pre- and post-intervention) was the within-subject factor, whereas the group (control vs. experimental) was the between-subject factor. Levene test was used to check for homoscedasticity, the Mauchly test for sphericity, and the Box's test for the equivalence of covariance matrices. All the assumptions were correctly met in the dataset. The effect size of the ANOVA was calculated by the partial eta-squared (η^2_p). The statistical programs used for the statistical analysis were Statistics Product and Service Solutions (IBM® SPSS® Statistics Version 24.0.0.0) (International Business Machines Corp., Madrid, Spain) and Microsoft Excel® in its 2016 version (Microsoft Corp., Redmond, WD, USA).

3. Results

3.1. Descriptive Statistics and Baseline Differences

Table 1 shows the descriptive statistics for the CAMM and MSS tests for the control and experimental groups at pre- and post-test.

Table 1. Descriptive statistics.

| | | | Pre-Intervention Mindfulness at School Scale | | | | Post-Intervention Mindfulness at School Scale | | | |
			CAMM	EA [1]	IA [2]	KA [3]	CAMM	EA [1]	IA [2]	KA [3]
Control (n = 73)		Mean	1.86	3.24	3.79	3.37	1.88	3.18	3.69	3.32
		Standard error	0.07	0.07	0.10	0.10	0.07	0.07	0.09	0.10
		Median	1.90	3.14	3.66	3.50	1.90	3.14	3.66	3.50
		Standard deviation	0.62	0.67	0.85	0.87	0.60	0.62	0.78	0.86
		Range	2.90	3.14	3.00	3.50	3.20	2.71	3.00	3.50
Experimental (n = 54)		Mean	2.05	3.11	3.92	3.24	2.74	3.47	4.15	3.61
		Standard error	0.08	0.09	0.09	0.11	0.10	0.10	0.09	0.08
		Median	2.00	3.14	4.00	3.50	2.70	3.57	4.33	4.00
		Standard deviation	0.59	0.71	0.70	0.81	0.78	0.77	0.69	0.64
		Range	2.30	3.14	2.66	3.50	3.80	3.14	3.00	2.50

[1] External Attention, [2] Internal Attention, [3] Kinesthetic Attention.

At pre-test, both groups (EG vs. CG) presented similar starting values regarding the four research variables: external attention ($Z = -0.814$; $p = 0.416$), internal attention ($Z = -0.745$; $p = 0.456$), kinesthetic attention ($Z = -0.849$; $p = 0.396$) and CAMM test ($Z = -1.377$; $p = 0.168$).

3.2. Intra-Group Differences (Pre vs. Post-Test)

The Wilcoxon Test is shown in Table 2. This table will give information about the differences between the variables of each group (intra-group, pre vs. post-test). It can be observed that the EG shows significant differences in all study variables, except for internal attention that shows only a trend towards change. These variables show a positive change, i.e., to a higher score, which means that participants score higher in the post-test than in the pre-test. The CG shows a significant change in two variables (external and internal attention) and a trend towards change (Kinesthetic attention). To the contrary, in the CG, this significant change is negative, i.e., participants score smaller values in the post-test compared to pre-test.

Table 2. Wilcoxon test (intra-group differences, pre- vs. post-test).

| | | Experimental Group (EG) | | | | | Control Group (CG) | | | | |
		Positive	Negative	Ties	Z	Sig.	Positive	Negative	Ties	Z	Sig.
Mindfulness at School Scale	CAMM	43	9	2	4.602	<0.001	31	22	20	1.407	0.159
	EA [1]	36	15	3	2.056	0.040	8	21	44	−2.698	0.007
	I A [2]	28	14	12	1.755	0.079	2	13	58	−2.883	0.004
	KA [3]	30	15	9	2.766	0.006	2	8	63	−1.941	0.052

[1] External Attention, [2] Internal Attention, [3] Kinesthetic Attention.

3.3. Final Intergroup Differences

Post-test, both groups (EG vs. CG) presented different final values regarding the four research variables: external attention ($Z = -2.427$; $p = 0.015$), internal attention ($Z = -3.281$; $p = 0.001$), kinesthetic attention ($Z = -1.960$; $p = 0.050$), and CAMM test ($Z = -6.050$; $p = <0.001$). This means that the EG obtained higher values in all study variables.

3.4. Repeated-Measures Analysis of Variance Mixed Model (Two.Factor ANOVA)

Firstly, in regard to the CAMM, an interaction effect (Time × Treatment) was found ($F(1)$ 17.463, $p = < 0.001$; $\eta^2_p = 0.178$). That is, the mindfulness implementation in PE significantly increased CAMM items compared to the control group. Secondly, according to the internal attention, an interaction effect (Time × Treatment) was found ($F(1)$ 5.266, $p = < 0.022$; $\eta^2_p = 0.041$). That is, the mindfulness implementation in PE significantly increased internal attention compared to the control group. Finally, referring to the external and kinesthetic attention, no interaction effect (Time × Treatment) was found. That is, the mindfulness implementation in PE did not increase these variables significantly compared to the control group.

4. Discussion

The present study aimed to analyze how mindfulness influences the levels of external, internal, and kinesthetic attention; awareness in acting; and acceptance of adolescents. Considering the importance of the current moment in society as well as the curricular approach, which highlights the need to develop emotional self-regulation, mindfulness is a tool that could contribute to this purpose. However, few studies have developed these techniques in the PE classroom. This is the reason why the present study consisted of a mindfulness program integrated during the calm-down of the PE class as well as outside school hours, by daily mindfulness challenges to be implemented at home. After the end of the intervention, the data obtained showed very significant progress in the students in most of the variables analyzed, with a large number of positive ranges in the experimental group compared to the control group. On the other hand, no significant differences were observed between the groups in the case of external and kinesthetic attention.

The results from this study are aligned with other research, such us the study by [34], which showed how the practice of yoga (including mindfulness practice) in PE classes, based on different body postures, breathing exercises, relaxation, and meditation, contributed to numerous benefits at a physical, psychological, and social levels. In this sense, [35] also showed improvements in the levels of attention and state of calm after performing a mindfulness intervention with elementary school students. Moreover, there is the study by [36] that showed that introducing mindfulness in the school setting in Chile significantly decreased anxious and depressive symptomatology. In line with these results, although the present study did not include specific indicators of anxiety or depression, improving aspects such as introspection, acting with awareness, acceptance without judgment, and verbalization of your experiences, are deeply related to the emotional response. In a similar way, the benefits observed in this study may contribute to improved academic performance. In fact, [37] stated that including emotional training in the education curriculum is required since it is associated with higher academic results and a better transition to the next level of studies.

In order to increase the impact of the obtained results, increasing the study sample would be interesting. In fact, in the study by [38], they pointed out that enlarging the sample of mindfulness studies in education is necessary in order to verify the efficacy of this practice and generalize the obtained results. Furthermore, these authors mentioned that only 10% of the studies conducted with samples of university students had a control group; therefore, a positive aspect of the present article is that it compares the experimental condition with the control, which reinforces the strength of the results. On the other hand, although the study by [39], conducted with more than 400 high school students, concluded that there were no differences between men and women in overall relaxation habits, it would have been interesting to expand the detail by the comparison between genders. Moreover, analyzing the long-term impact of these interventions through longitudinal studies would be enriching, since the permanence of the achieved improvements over time has not been studied [40] and, therefore. it is not clear that a real change in lifestyle can be obtained. Nevertheless, it is a very encouraging outcome that short practices of only 10 min resulted in such good results. Moreover, the procedure presented facilitates the implementation of mindfulness in the PE classroom. After conducting this study, the response received by the participating students was very remarkable, since the vast majority showed absolute gratitude for having had the opportunity to carry out the proposed activities. Although some mindfulness activities can be easily implemented, in order to increase the success of the interventions, the training of teachers is very important [28]. In fact, after analyzing the PE curriculum, especially the latest update, teachers will require new tools according to the new content blocks to achieve the objectives, since the approach goes beyond the motor element. The PE teacher must update his or her training towards these new holistic trends in PE, which emphasizes emotional and social competencies. The PE curriculum demands student development and improving emotional self-regulation skills becomes essential. Among these new tools, mindfulness has been shown to be a successful and easily implementable option. Therefore, training teachers in mindfulness is required, first of all by applying it to themselves and their own lives, so that they can teach others in their educational context [41]. In line with the statement of these authors, the nature of these practices requires experimentation before applying it with students.

The subject of PE is not limited to the development of physical aspects, but also offers the possibility of working on psychological aspects. In fact, several studies showed that the practice of physical exercise reduces negative feelings and accentuates positive dimensions of psychological health [42]. In addition, PE provides the opportunity to work on interpersonal skills and decision making, aspects that will lead to an improvement of students' emotional intelligence [43]. Therefore, body practices in PE should enhance the stimulation of body sensations such as tension, relaxation, pleasure, fatigue, suffering, overload, etc., using methods that promote awareness of the body through internalization practices such as meditation, yoga, and taijiquan [44]. In this regard, as [45] states, developing self-knowledge is essential to train students capable of participating in a social transformation, which requires focusing education towards personal and social development. It is necessary to promote methodological procedures that develop the awareness of the participants, both students and teachers, that is, to integrate the cognitive, affective, moral, and spiritual dimensions [46]. To contribute in this regard, the PE subject could integrate activities towards reflection or meditation, which can be perfectly implemented into the calm-down part of the sessions. Accordingly, in this research, the mindfulness activities were included in the final section of the sessions, without affecting the corresponding main content.

Limitations and Future Proposals

As for the limitations and complications generated during the study and intervention, we can highlight certain aspects that were observed in the development of the activities. Firstly, the lack of mindfulness experience of the participants is a factor that could have adversely affected the experience if the students were not able to understand the questions in the questionnaires, since they are not used to this type of vision or activities. Therefore, making an introduction on this subject to both the control and experimental groups could have helped with this. Therefore, it is essential that mindfulness interventions are adapted to the different educational levels [28]. On the other hand, alternative tools could have been applied as those included by other studies such as [24], in which qualitative methods were used, consisted of collecting students' experiences through personal reflections in diaries, interviews, reflection groups, etc. In addition to these limitations, it must be acknowledged that this research involved convenience sampling. This sampling means that the results obtained cannot be extrapolated to the general population studied.

Besides, better control of the practice conditions would have been necessary, since during the experience of this intervention, we observed difficulties for students in focusing on the practice, because initially they were not able to maintain silence and concentrate on the activity, and there were situations of laughs and distractions among them. For this reason, it would have been essential to emphasize the familiarization with this type of activities to teach the importance and value of these practices, so that the students take them seriously. For this purpose, performing these tasks in a large space is relevant, guaranteeing distance among the participants, with dim lighting, preventing other people from observing the practice, aspects that guarantee certain intimacy. Besides, using appropriate sound equipment that facilitates audio monitoring helps the students to focus their attention on the activity instructions. In terms of strengths, the daily challenge proposal has been a very motivating component as it has made students aware of the multitude of ways in which mindfulness can be easily practiced and applied.

Finally, compared to other countries, research in Spain on implementing mindfulness in the PE class is practically missing, especially at the infant, primary, and secondary levels [28]. According to the education for sustainable development, further mindfulness interventions are necessary to contribute to understanding the benefits mental well-being can have on education, particularly to the subject of PE due to its curricular affinity.

5. Conclusions

In conclusion, the different issues addressed in this article related to the current lifestyle of the youngest are associated with negative psychological well-being, and the education for sustainable development aims to solve this situation by developing competencies of self-regulation. Mindfulness arises as a practice that consists of training awareness, which could enhance students' mental health. This study showed improvements in the ability of students to focus their attention on external, internal or kinesthetic factors, awareness in acting, and acceptance. These variables contribute to emotional self-regulation and improved mental health. Overall, this study justifies and supports the implementation of mindfulness practices in the physical education classes in line with the goals of the education for sustainable development. This is particularly relevant considering the topicality of the issue and the lack of related studies. Future research should focus on implementing actions that can contribute to the mental health of the students within the physical education curriculum, in line with the demands of the current education on sustainable development in facing the challenges of today's world.

Author Contributions: Conceptualization, R.D.-M., O.G.-T. and S.B.-M.; methodology, R.D.-M., S.B.-M. and A.F.-V.; software, A.F.-V.; formal analysis, R.D.-M. and A.F.-V.; investigation, R.D.-M. and S.B.-M.; resources, R.D.-M. and S.B.-M.; data curation, A.F.-V. and S.B.-M.; writing—original draft preparation, R.D.-M., O.G.-T. and S.B.-M.; writing—review and editing, S.B.-M. and O.G.-T.; visualization, R.D.-M. and S.B.-M.; supervision, R.D.-M. and S.B.-M.; project administration, S.B.-M. funding acquisition, S.B.-M. and A.F.-V. All authors have read and agreed to the published version of the manuscript.

Funding: This work was supported by the University of Alicante. Through the project: ODSEF PROJECT. Design and implementation of the Sustainable Development Goals for Physical Education (REDES ICE-2021-5489).

Institutional Review Board Statement: This research was approved by the ethics committee of the University of Alicante with code UA-2022-03-17.

Informed Consent Statement: Informed consent was obtained from all subjects involved in the study.

Data Availability Statement: Not applicable.

Conflicts of Interest: The authors declare no conflict of interest.

References

1. WHO. *Active: Paquete de Intervenciones Técnicas Para Acrecentar la Actividad Física*; World Health Organization: Geneva, Switzerland, 2019.
2. Ministry of Health. *Encuesta Nacional de Salud ENSE, España*; En SG Información Sanitaria: Madrid, Spain, 2015.
3. Mayo, X.; del Villar, F.; Jiménez, A. *Termómetro del Sedentarismo en España: Informe sobre la inactividad física y el sedentarismo en la población adulta española. En Observatorio de la Vida Activa y Saludable de la Fundación España Activa*; Centro de Estudios del Deporte; Universidad Rey Juan Carlos: Mostores, Spain; CSIC: Madrid, Spain, 2017.
4. Calderón Gómez, D. Approximation of the evolution of the digital divide among young people in Spain (2006–2015). *Rev. Española Sociol.* **2019**, *28*, 27–44.
5. Herrera-Mendoza, K.M.; Acuña Rodríguez, M.P.; Gil Vega, L. Motivación de jóvenes universitarios hacia el uso de teléfonos celulares/Motivation of youth students with use of cell pones. *Rev. Encuentros* **2017**, *15*, 91–105.
6. Enez Darcin, A.; Kose, S.; Noyan, C.O.; Nurmedov, S.; Yılmaz, O.; Dilbaz, N. Smartphone addiction and its relationship with social anxiety and loneliness. *Behav. Inf. Technol.* **2016**, *35*, 520–525. [CrossRef]
7. Primack, B.A.; Shensa, A.; Sidani, J.E.; Whaite, E.O.; Lin L yi Rosen, D.; Miller, E. Social media use and social isolation among young adults in the United States. *Am. J. Prev. Med.* **2017**, *53*, 1–8. [CrossRef] [PubMed]
8. Villavicencio, J.C.; Álava, V.P.C.; Loor, M.G.T. La ciberadicción en la conducta de los estudiantes. *Rev. Atlante Cuad. Educ. Desarro.* **2019**, *11763*, 1–13.

9.   Elhai, J.D.; Tiamiyu, M.; Weeks, J. Depression and social anxiety in relation to problematic smartphone use. *Internet Res.* **2018**, *28*, 315–332. [CrossRef]

10. Oland, A.A.; Shaw, D.S. Síntomas externalizantes e internalizantes puros versus concurrentes en los niños: El papel potencial de los hitos socio-evolutivos. *Rev. Psicol. Clin. Infant. Fam.* **2005**, *8*, 247–270.

11. Schonert-Reichl, K.A.; Roeser, R.W. Mindfulness in education: Introduction and overview of the handbook. In *Handbook of Mindfulness in Education: Integrating Theory and Research into Practice*; Schonert-Reichl, K.A., Roeser, R.W., Eds.; University of British Columbia: Vancouver, BC, Canada; Springer: Berlin/Heidelberg, Germany, 2016; pp. 3–16.

12. Redacción Médica. Los casos de TDAH en España han Aumentado un 30% en niños de 8 a 12 años. Redacción Médica. 2018. Available online: https://www.redaccionmedica.com/secciones/psiquiatria/los-casos-detdahenespana-han-aumentado-un-30-en-ninos-de-8-a-12-anos-6932 (accessed on 10 June 2022).

13. Barchielli, B.; Cricenti, C.; Gallè, F.; Sabella, E.A.; Liguori, F.; Da Molin, G.; Liguori, G.; Orsi, G.B.; Giannini, A.M.; Ferracuti, S.; et al. Climate Changes, Natural Resources Depletion, COVID-19 Pandemic, and Russian-Ukrainian War: What Is the Impact on Habits Change and Mental Health? *Int. J. Environ. Res. Public Health* **2022**, *19*, 11929. [CrossRef]

14. Hertzman, C.; Power, C. Un Enfoque de ciclo de vida para la salud y el desarrollo humano. In *Sociedades más Saludables: Del análisis a la acción*; Heymann, J., Hertzman, C., Barer, M.L., Evans, M.G., Eds.; Oxford University Press: Nueva York, NY, USA, 2006; pp. 83–106.

15. Sachs, J.D. From millennium development goals to sustainable development goals. *Lancet* **2012**, *379*, 2206–2211. [CrossRef]

16. UNESCO. Division for Inclusion, Peace and Sustainable Development, Education Sector. In *Education for Sustainable Development Goals: Learning Objectives*; United Nations Educational, Scientific and Cultural Organization: Paris, France, 2017.

17. Ives, C.D.; Freeth, R.; Fischer, J. Inside-out sustainability: The neglect of inner worlds. *Ambio* **2020**, *49*, 208–217. [CrossRef]

18. Hofmann, S.G.; Gómez, A.F. Mindfulness-based interventions for anxiety and depression. *Psychiatr. Clin.* **2017**, *40*, 739–749. [CrossRef] [PubMed]

19. Devi, S. Mindfulness in Education: The Need of the Hour. In *Handbook of Research on Clinical Applications of Meditation and Mindfulness-Based Interventions in Mental Health*; IGI Global: Hershey, PA, USA, 2022; pp. 337–356.

20. Mindful Schools. Research on Mindfulness. 2017. Available online: https://www.mindfulschools.org/about-mindfulness/research/#reference-12 (accessed on 2 August 2022).

21. Mindful Schools. Why Mindfulness is Needed in Education. 2017. Available online: https://www.mindfulschools.org/about-mindfulness/mindfulness-ineducation/ (accessed on 2 August 2022).

22. Broderick, P.C.; Metz, S.M. Working on the inside: Mindfulness for 72 adolescents. In *Handbook of Mindfulness in Education: Integrating Theory and Research into Practice*; Schonert-Reichl, K.A., Roeser, R.W., Eds.; Bennett Pierce Prevention Research Center, Pennsylvania State University: University Park, PA, USA; Springer: Berlin/Heidelberg, Germany, 2016; pp. 355–382.

23. Hawkins, K. *Mindful Teacher, Mindful School: Improving Wellbeing in Teaching and Learning*; SAGE: Riverside, CA, USA, 2017; ISBN 1526415585.

24. Raes, F.; Griffith, J.; Van der Gught, K.; Williams, J. School-Based Prevention and Reduction of Depression in Adolescents: A Cluster-Randomized Controlled Trial of a Mindfulness Group Program. *Mindfulness* **2015**, *5*, 477–486. [CrossRef]

25. Sibinga EM, S.; Webb, L.; Ghazarian, S.R.; Ellen, J.M. School-Based Mindfulness Instruction: An RCT. *Pediatrics* **2016**, *137*, e20152532. [CrossRef] [PubMed]

26. Lyons, K.E.; DeLange, J. Mindfulness matters in the classroom: The effects of mindfulness training on brain development and behavior in children and adolescents. In *Handbook of Mindfulness in Education: Integrating Theory and Research into Practice*; Schonert-Reichl, K.A., Roeser, R.W., Eds.; Springer: Berlin/Heidelberg, Germany, 2016.

27. Galla, B.M.; Kaiser-Greenland, S.; Black, D.S. Mindfulness training to promote selfregulation in youth: Effects of the Inner Kids program. In *Handbook of Mindfulness in Education: Integrating Theory and Research into Practice*; Schonert-Reichl, K.A., Roeser, R.W., Eds.; School of Education, University of Pittsburgh: Pittsburgh, PA, USA; Springer: Berlin/Heidelberg, Germany, 2016; pp. 295–311.

28. López Secanell, I.; Gené Morales, J. Revisión sistemática de la investigación sobre el uso del mindfulness en la educación física. *Cuad. Psicol. Deport.* **2021**, *21*, 83–98. [CrossRef]

29. León, B. Atención plena y rendimiento académico en estudiantes de enseñanza secundaria. *Eur. J. Educ. Psychol.* **2008**, *1*, 17–26. [CrossRef]

30. Etikan, I.; Musa, S.A.; Alkassim, R.S. Comparison of Convenience Sampling and Purposive Sampling. *Am. J. Theor. Appl. Stat.* **2016**, *5*, 1. [CrossRef]

31. Hatch, E.; Lazaraton, A. *The Research Manual: Design and Statistics for Applied Linguistics*; Newbury House Publishers: New York, NY, USA, 1991; Volume 28.

32. Greco, L.A.; Baer, R.A.; Smith, G.T. Assessing mindfulness in children and adolescents: Development and validation of the Child and Adolescent Mindfulness Measure (CAMM). *Psychol. Assess.* **2011**, *23*, 606. [CrossRef]

33. Kuby, A.K.; McLean, N.; Allen, K. Validation of the Child and Adolescent Mindfulness Measure (CAMM) with Non-Clinical Adolescents. *Mindfulness* **2015**, *6*, 1448–1455. [CrossRef]

34. Faul, F.; Erdfelder, E.; Buchner, A.; Lang, A.-G. Statistical Power Analyses Using G*Power 3.1: Tests for Correlation and Regression Analyses. *Behav. Res. Methods* **2009**, *41*, 1149–1160. [CrossRef]

35. Colón-Calvo, A.; Baena-Morales, S.; Ferriz-Valero, A.; García-Taibo, O. A new vision of mindfulness in physical education. Contributing to the social dimension of sustainable development. *J. Phys. Educ. Sport* **2022**, *22*, 2618–2626. [CrossRef]
36. Soriano Galeote, J.; Castellar Otín, C. *La Técnica de" Mindfulness" en Educación Primaria: Efecto de una Intervención Educativa*; TFG, University of Zaragoza: Zaragoza, Spain, 2017.
37. Langer, Á.I.; Schmidt, C.; Aguilar-Parra, J.M.; Cid, C.; Magni, A. Mindfulness y promoción de la salud mental en adolescentes: Efectos de una intervención en el contexto educativo. *Revista Med. Chile* **2017**, *145*, 476–482. [CrossRef] [PubMed]
38. Jordan, J.A.; McRorie, M.; Ewing, C. Gender differences in the role of emotional intelligence during the primary-secondary school transition. *Emot. Behav. Difficulties* **2010**, *15*, 37–47. [CrossRef]
39. Oblitas, L.A.; Soto, D.E.; Anicama, J.C.; Arana, A.A. Incidencia del mindfulness en el estrés académico en estudiantes universitarios: Un estudio controlado. *Terapia Psicol.* **2019**, *37*, 116–128. [CrossRef]
40. López-González, L.; Amutio, A.; Herrero-Fernández, D. The Relaxation-Mindfulness Competence of Secondary and High School students and its influence on classroom climate and academic performance. *Eur. J. Educ. Psychol.* **2018**, *11*, 5–17. [CrossRef]
41. Arévalo-Proaño, C.; Dávila, Y.; Álvarez-Cárdenas, F.; Peñaherrera-Vélez, M.J.; Vélez-Calvo, X. El mindfulness para mejorar procesos ejecutivos y cognitivos en niños con altas capacidades. Revista INFAD De Psicología. *Int. J. Dev. Educ. Psychol.* **2019**, *5*, 429–440. [CrossRef]
42. Secanell, I.L.; Lerma, M.B. "Mindfulness" y educación: Formación de los instructores de" mindfulness" en Educación Secundaria. *Didact. Rev. Investig. Didact. Específicas* **2019**, *6*, 126–143.
43. Estrada, P.R.; Vázquez, E.I.A.; Gáleas, Á.M.V.; Ortega, I.M.J.; Serrano, M.D.L.P.; Acosta, J.J.M. Beneficios psicológicos de la actividad física en el trabajo de un centro educativo. *Retos Nuevas Tend. Educ. Física Deporte Recreación* **2016**, *30*, 203–206.
44. Ruiz, G.; Lorenzo, L.; García, Á. El trabajo con la inteligencia emocional en las clases de educación física: Valoración de una experiencia piloto en educación primaria. *J. Sport Health Res.* **2013**, *5*, 203–210.
45. Aguiló, J. *Deportistas en el Prisma de lo Sagrado. Aportaciones Pedagógicas*; UOC: Barcelona, Spain, 2017.
46. Gallifa, J. Integral Thinking and its application to Integral Education. *J. Int. Educ. Pract.* **2019**, *2*, 15–27. [CrossRef]

Systematic Review

Hope Aspects of the Women's Experience after Confirmation of a High-Risk Pregnancy Condition: A Systematic Scoping Review

Mónica Antunes, Clara Roquette Viana and Zaida Charepe *

Center for Interdisciplinary Research in Health, Institute of Health Sciences, Universidade Católica Portuguesa, 1649-023 Lisbon, Portugal
* Correspondence: zaidacharepe@ucp.pt

Abstract: Background: Pregnancy is a period of transformation, hope, expectation, and worry for women and their families. A high-risk pregnancy refers to a pregnancy in which the mother and/or fetus are at greater-than-normal risk of complications, and it evokes a range of emotional and psychological experiences that largely depend on the care and support provided by health professionals. The purpose of this review is to summarize the existing literature on the lived experience of hope in women facing a high-risk pregnancy related to their own health and/or medical conditions related to the fetus. Methods: This review followed the Joanna Briggs Institute's methodology. No limits on a date were applied to the search. Identified titles and abstracts were screened to select original reports and were cross-checked for any overlap of cases. We included studies that emphasized the experience of hope of pregnant women dealing with a pregnancy complication. Main Results: According to the results of the present scoping review, we found two main dimensions: women experiencing a high-risk pregnancy themselves and prenatal diagnosis. In both cases, the women were in a dilemma between hope and hopelessness. Conclusion: The findings demonstrate that women facing high-risk pregnancies struggle with multiple fears and concerns about their own health and the fetus's health. Further research is needed to identify best practices for the care provided to the vulnerable populations.

Keywords: high-risk pregnancy; hope; life experience; mental health; pregnancy complications; prenatal diagnosis; review

Citation: Antunes, M.; Viana, C.R.; Charepe, Z. Hope Aspects of the Women's Experience after Confirmation of a High-Risk Pregnancy Condition: A Systematic Scoping Review. *Healthcare* **2022**, *10*, 2477. https://doi.org/10.3390/healthcare10122477

Academic Editor: Hooman Mirzakhani

Received: 29 October 2022
Accepted: 29 November 2022
Published: 8 December 2022

Publisher's Note: MDPI stays neutral with regard to jurisdictional claims in published maps and institutional affiliations.

1. Introduction

In 2017, approximately 810 women died from preventable causes related to pregnancy and childbirth [1]. Globally, more than 20 million women are at risk of high-risk pregnancies, which results in an estimated 830 deaths per day, more than 99% occurring in developing countries, and is more frequent among rural women and adolescents [2]. Nearly 22% of pregnant women develop a high-risk pregnancy [3]. As part of the sustainable development goals (SDG), countries have agreed on a new target to accelerate the decline of maternal mortality by 2030. The World Health Organization (WHO) considers high-risk pregnancies a major public health challenge, addressing the healthcare needs as a priority [1], and includes an ambitious target: decreasing the global maternal mortality rate to less than 70 per 100,000 births, with no country having a maternal mortality rate of more than twice that of the global average [1].

In most cases, the birth of a child is an experience filled with joy and happiness. However, when a mother is diagnosed with a medical condition or the child with a congenital anomaly, the parents' experience can take on a different meaning [4]. A high-risk pregnancy is defined as any pregnancy in which there is a medical factor, maternal or fetal, that may potentially adversely affect the outcome of the pregnancy [5]. The most common maternal complications are gestational diabetes, preeclampsia and eclampsia, depression, sexually transmitted diseases, preterm labor, and placenta previa [3], while the most frequent congenital anomalies are heart malformations, neural tube defects, and Down syndrome. An anomaly may be genetic, infectious, or environmental in origin. In most cases, however, the cause is unknown, which makes it more difficult for parents to understand and accept the situation [6]. In many countries, congenital anomalies are important causes of perinatal morbidity and mortality, which can lead to chronic disabilities which may have severe consequences on individuals, families, healthcare systems, and societies [7]. The sudden sense of grief, loss, and guilt coupled with a fear of the unknown future often produces a great deal of anguish for parents [5], especially in pregnant women.

Every pregnancy is unique, physiologically, and a natural episode in a woman's life, and all pregnant women experience physical, mental, and social changes in different manners [3]. The ways in which the changes inherent to this transition moment are integrated and experienced seem to be directly related to the woman's personality, marital status, family, and social support [8].

Under normal conditions, pregnancy is a natural transition for women and their families. When a pregnant woman is diagnosed as being high-risk, she may find it difficult to cope with this new reality, leading to psychological and emotional consequences such as fear, guilt, shock, grief, frustration, worry, loneliness, and isolation [9]. A high-risk pregnancy with complications is one of the risk factors causing pregnant women to experience psychosomatic problems such as anxiety, depression, and distress, and to suffer impairments in their health [10]. A qualitative study also showed that besides medical problems, women experience behavioral, affective, and emotional problems. Moreover, they are also at risk of sociocultural and financial strains that often lead to feelings such as uncertainty, concern, and insecurity [11].

Women's coping strategies during pregnancy demand multiple challenges in which hope and resilience play essential roles in managing stress and mental health. Hope is defined as the perceived ability to find pathways to desired goals and to motivate oneself to use those pathways [12]. As a focus of nursing practice, hope is defined as "(. . .) feelings of having possibilities, trust in others and in future, zest for life, expression of reasons and will to live, inner peace, optimism, associated with setting goals and mobilization of energy" [13] (p. 1). In a study published in 2019, the author defends that helping individuals and their family members to find meaning in suffering and striving to invoke a sense of constructive hope should be a fundamental aspect of health care [14]. Inspiring appropriate hope might be considered a concern to healthcare professionals whatever their specialty [14]. In this review, the lived experiences of women's hope comprises spheres and dimensions [14]. According to Dufault and Martocchio, there are six dimensions associated with the concept of hope [15]: affiliative (relationships with women and God that can be expressed by individuals who seek or are receptive to others' help in hope); affective (sensations and emotions that are part of the process of hope); cognitive and behavioral (interpret and judge in relation to hope and actions orientation towards the desired outcomes, respectively); contextual (life situations that surround, influence, and are a part of women's hope); and temporal (focuses upon hoping within the women's experience of time).

Therefore, the aim of this review was to assess the state of knowledge regarding the lived experience of hope among women facing high-risk pregnancies that may endanger the health of the mother and/or fetus. This openness to the possibility of working hope in a perspective other than cure, and the commitment of nurses in the practice of promoting hope as a duty of care and a standard of good clinical practice, has led to the need to investigate the concept and look for new ways to better-inspire hope in women who face pregnancies with complications in the context of their health and/or that endanger the health of the fetus.

2. Materials and Methods

The protocol and proposed systematic review were drawn and conducted in accordance with the JBI methodology for systematic reviews [16]: define search strategy, study selection, assessment of methodological quality, data extraction, and data synthesis. The protocol was registered prospectively with the Open Science Framework on 26 May 2022 (https://osf.io/u9ns8 (accessed on 12 September 2022)). Registration DOI: 10.17605/OSF.IO/U9NS8.

2.1. Review Questions

The following question guided this scoping review: What are the hope aspects related to the women's lived experience of a pregnancy that continues after confirmation of a high-risk pregnancy diagnosis? This question was divided into the following sub-questions: What type of evidence or study design exists in the area of women's life experience of pregnancy research in relation to hope aspects after confirmation of a high-risk pregnancy diagnosis?; What aspects of hope in the life experience of pregnant women have been addressed during health care associated with a high-risk pregnancy?; What are the gaps in the nursing research in relation to hope interventions in the context of care for women with high-risk pregnancies?

2.2. Inclusion Criteria

2.2.1. Participants

Both the exploratory and textual components of this review considered qualitative and quantitative studies that include women's lived experiences of pregnancy with a high-risk condition that may affect their own health or their baby's health, including the prenatal diagnosis of a congenital anomaly, regardless of race, nationality, level of education, or religious affiliation. The study could also include any patient with an age above 18, primigravida and multigravida, and in the second or third trimester of their pregnancy. This review excluded any studies focusing on healthcare professionals and pregnant women who were healthy and did not have any risk factors for high-risk pregnancies.

2.2.2. Concept

This review considered studies that explore woman's hopeful experiences when proceeding with a high-risk pregnancy diagnosis, which may have included maternal and/or fetal complications. The included studies may or may not have directly addressed hope in their proposals having, however, addressed hope-related experiences and related concepts in their findings or women's expectations. The aspects related to women's experiences of hope included psychological well-being; uncertainty; finding a sense of normality; finding a new meaning to life experiences; setting realistic goals and objectives; imagining possibilities and seeking alternative solutions; an ability to share their pregnancies with other women and/or health care professionals; and expressing positive personal transformation and identifying positive psychological factors. The aspects related to women's expectations will include difficulties in thinking about the child's future; condition-related expectations (positive or negative outlook); and concerns about acceptance by family and social networks.

2.2.3. Context

Women with high-risk pregnancies are usually referred to larger health centers for better treatment. In this review, women were assisted in high-risk maternal–fetal health consultations, which included the literature from any country or sociocultural setting.

2.2.4. Types of Sources

This review considered quantitative, qualitative, and mixed study designs for inclusion. Systematic reviews and meta-analyses were also considered in this review.

2.3. Search Strategy

The search strategy aimed to find studies published in the Portuguese, English, or Spanish languages, with no date limit. As recommended in all JBI types of reviews, a three-step search strategy was carried out [16].

An initial limited search took place on the EBSCO platform, where the databases CINAHL and MEDLINE (PubMed) were selected, as well as the Google Scholar platform. The search was performed using the keywords included in the PCC question. A second search with all identified keywords and index terms was undertaken across all of the included databases, using Boolean descriptors such as "OR" and "AND". A search with all identified keywords and index terms was used to develop a full search strategy for PubMed (see Appendix A). In the third stage, reference lists of sources selected from the full text and/or included in the review were examined. The databases searched include CINAHL Complete (from EBSCO); Pubmed; Nursing and Allied Health Collection (from EBSCO); PsycINFO; Mediclatina (from EBSCO); and Scopus.

2.4. Study/Source of Evidence Selection

Following the search, all identified citations were compiled and uploaded to Mendeley version 2.71.0 (Mendeley Ltd., Elsevier, Amsterdam, Netherlands), and duplicates were removed. After a pilot test, the titles and abstracts were screened by two or more independent reviewers for assessment against the inclusion criteria for the review. The resulting reference list was then uploaded to Rayyan (Qatar Computing Research Institute, Doha, Qatar). In the second phase of screening, the references of the selected studies were reviewed, and relevant studies were identified for a full review. The studies were classified into one of three categories: included, excluded, and uncertain. Later, the full texts of the "included" and "uncertain" studies were retrieved to identify potentially relevant studies, and their full citation details were imported into Rayyan.

The full texts of the selected citations were assessed in detail according to the inclusion criteria by two independent reviewers. Reasons for the exclusion of sources of evidence in full text that did not meet the inclusion criteria were recorded and reported in the scoping review. Any disagreements that arose between the reviewers at each stage of the selection process were resolved through discussion, or with an additional reviewer/s. The results of the search and the study-inclusion process are reported and presented in a preferred reporting items for systematic reviews and meta-analyses extension for scoping review (PRISMA-ScR) flow diagram [17].

2.5. Assessment of Methodological Quality

In describing the quality of the selected articles (n = 15), the studies were appraised by all of the authors. Divergent views regarding the critical appraisal were reviewed until a consensus was achieved. The Hawker et al. [16] assessment tool, with a four-grade scale (1 = very poor; 2 = poor; 3 = fair; 4 = good) was used. The total scores ranged between 9 and 36, and higher scores indicated a higher quality. An article's quality appraisal was centered on the following items: 1—abstract and title; 2—introduction and aims; 3—method and data; 4—sampling; 5—data analysis; 6—ethics and bias; 7—results; 8—transferability or generalizability; and 9—implications and usefulness.

2.6. Data Extraction

After the analyses of the title and abstract, duplicates and articles that did not correspond to the topic were excluded. Primary studies published in Portuguese, English, and Spanish were included, with no time limit. The search was conducted on 16 March 2022, with an update on 16 September 2022. A total of 479 articles were excluded, and another 4 full-text articles were excluded, two of them with an unsuitable concept and for the other two we did not receive a response from the authors to access the full text, leaving 15 for analysis (Figure 1). The titles and abstracts identified during the search were independently reviewed by the authors using the inclusion and exclusion criteria. The decision of whether to include or exclude studies was made by mutual agreement.

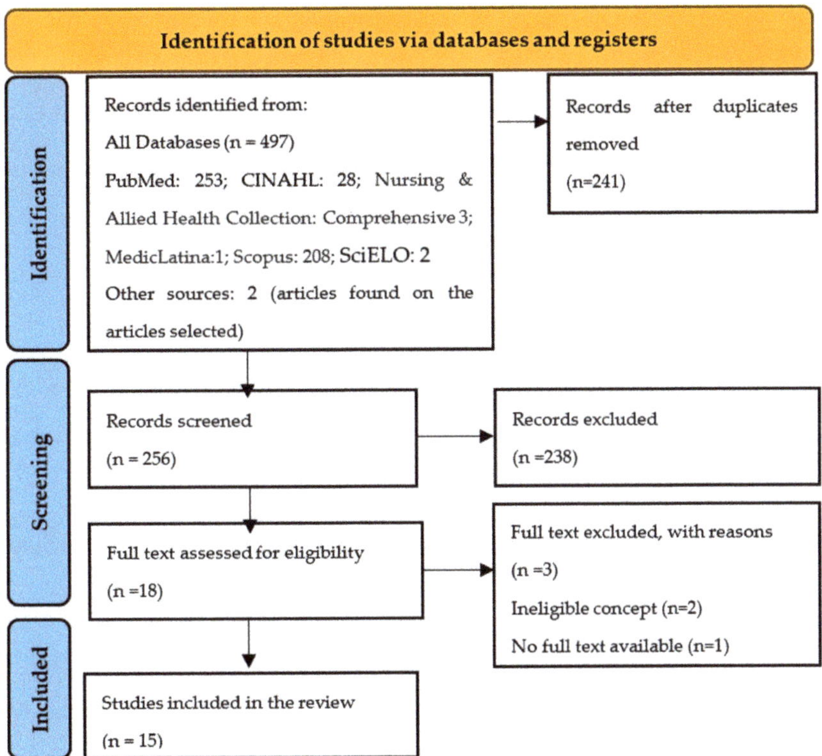

Figure 1. Flowchart of the selection and screening process of the systematic review articles according to the PRISMA method.

2.7. Data Synthesis

The 15 studies eligible for SR are described in Appendix B. The results are presented in narrative form. Considering the JBI guidelines [16], the synthesis of relevant data collected from each article was composed of the following elements: the identification of the article, hope and hopelessness experiences or expectations, aims, study design, study population/sample, context, population characteristics, typology, and main results.

3. Results

The generated demand resulted in 495 titles. After applying the inclusion/exclusion criteria and excluding duplicate studies, 15 studies were eligible. Of these, 14 were in the English language and 1 was in the Portuguese language (Brazil).

3.1. Characteristics of Sources of Evidence

The main characteristics of the fifteen articles were as follows: fifteen primary studies, four of which were conducted in Iran, two in the United States of America, one in Brazil, one in the United Kingdom, one in Belgium and the United Kingdom, one in Sweden, one in Australia, one in Paris, one in Africa, one in Malaysia, and one in Thailand. Only one of the studies found was quantitative (descriptive, prospective, and longitudinal); the rest of the studies were qualitative.

The selected studies were published between 2004 and 2021. The ages of the women ranged from 18 to 45 years. The sample sizes varied from nine to seventy-two women who used the services of specialized centers, hospitals, or clinics for the specific medical condition presented.

There are studies (see Appendix B) that have focused on women's experiences with maternal medical conditions [18–27], and other studies that have focused on the prenatal diagnosis of fetal pathology [28–32].

Few studies have addressed the aspects of hope during health care associated with high-risk pregnancies, and only one study was quantitative, while fourteen studies were qualitative.

The empirical studies described the lived experiences of women during pregnancy in the second and third trimesters with varied medical conditions.

Authors such as Behboodi-Moghadam, Khalajinia, Nasrabadi, Mohraz and Gharacheh (2016), and Sanders (2008) explored the experiences of women diagnosed with HIV at each stage of pregnancy [18,20,23]. This same medical condition was the focus of other studies that examined these experiences in the prenatal period in South Africa [25], Thailand [24], and Brazil [27]. On the other hand, authors such as Tong, Brown, Winkelmayer, and Craig and Jesudason described the beliefs, values, and experiences of pregnancy in Australian women with CKD to inform on pre-pregnancy counseling and pregnancy care [21].

The experiences and perceptions of women with high-risk pregnancies are focused on topics regarding health and care practice issues/needs [22,30]. Tayeh, Jouannic, Mansour, and Kesrouani and Attieh explored patients' perceptions of the prenatal diagnosis of fetal cardiac pathology and their reasons for deciding to continue the pregnancy despite being eligible for the medical termination of their pregnancy [30]. Norhayati, Hazlina, Hussain, Asrenee and Sulaiman examined women's experiences of near-misses and their perceptions of quality of care in a retrospective study [26].

A theoretical framework for the process of adaptation following a fetal anomaly diagnosis was provided based on the proposal in the study of Lalor, Begley and Galavan [28]. In this study, data was collected from Irish women's experiences of carrying a baby with a fetal abnormality to the end and beyond birth [28,29]. Integrated into the fetal abnormality, other authors describe the women's reactions to the discovery of fetal hydronephrosis in the context of uncertainty about prognosis [19] and the women's experiences during pregnancy with a child with a known, nonlethal congenital abnormality [32].

The concepts and strategies that women in the UK and Belgium use when considering maternal–fetal surgery as an option for the management of spina bifida in their fetus, and how this determines the acceptability of the intervention, were researched in [31].

3.2. Critical Appraisal within Sources of Evidence

The study's quality was high. Methods, ethics, and bias, as well as transferability, were the studies' main limitations, and the critical appraisals ranged from 32 to 36 (see Appendix C).

3.3. Results of Sources of Evidence

According to the results of the present systematic review, women who experience a high-risk pregnancy or a prenatal diagnosis of fetal abnormality found themselves in a dilemma between hope and hopelessness and, in some cases, a second dilemma between terminating or continuing the pregnancy.

To answer the question: "What are the hope aspects related to the women's life experience of a pregnancy that continues after confirmation of a high-risk pregnancy diagnosis?", we were able to find two subcategories within the main category of "Hope experiences in women with a high-risk pregnancy." We can distinguish hopelessness and hope experiences. In nine studies it was possible to analyze that the risk involved was related to the women themselves in the context of their medical conditions (see Table 1).

Table 1. Hopelessness and hope experiences in women with medical conditions.

Hopelessness Experiences	Hope Experiences
- Worries and concerns about the child, future pregnancy, relationships, and higher costs [18] - Stigma, stereotyping, discrimination, and judgment [20,23–25,27] - Emotional distress and ambivalence [23] - Negative self-image, loneliness, feelings of isolation, and blame [25] - Fears for their own life and the life of their baby, unable to become pregnant again, and postpartum complications [26] - Struggle, shock, anxiety, and depression [24] - Conscious of fragility and fear of genetic transmission [21] - Not being able to adjust to the complications, grieving for a long time, guilt, intolerance to pain, and irritability [22,26]	- Adaptation to challenges [18] - Belief that conditions will improve [18,27] - Positive about the future [18,25,28] - Children as a "divine gift", a chance to correct past mistakes, to be good, a loving mother [20,23] - Children as a meaning to their life and a motivator that helps to face challenges [24,27,28] - Pregnancy as construction of the female identity [27] - Medicines as hope, protecting their own life and protecting an unborn child [24,25] - Positive pregnancy experiences as a source of hope [21,24] - Religion, spirituality, faith in God [18,20,22,26,27] - Valuing life, gratitude, and focus on what is good [21]

The studies we found related well to the hope and hopelessness women experience in the context of prenatal diagnoses that could endanger the fetus's health (see Table 2).

Table 2. Hopelessness and hope experiences in women with prenatal diagnoses.

Hopelessness Experiences	Hope Experiences
- Disbelief, stress, and struggle [28] - Fear of developing a bond with a baby who may die [28] - Anxiety and fears about the unknown, worry, stress, and depression [19,28,29,32] - Feelings of grief, shock, anger, panic, distress, and guilt [29,32] - Loss of perfect baby and fear of potential complications of the treatment [31]	- Positive vision of the future [28,31] - Reconstruct the future and adjust earlier beliefs [28,31] - Strong feelings of responsibility about the treatment, seen as an opportunity that was given [31] - Determination to do anything that would improve their future child's health outcomes [19] - The sense of doing the right thing by continuing the pregnancy [29] - Hope for the best possible outcome and that everything would be all right at the end [29] - Positive attitude toward childbirth [29–31] - Hope to return to normal [30] - Other sources of hope, such as religious and spiritual beliefs, good sources of information, time to prepare, support from family and friends, staying busy at work, and empathizing with the baby [30]

Some studies have established a link between women's experiences of hope and the positive outcomes associated with this perception. The following results in women with medical conditions include the change from their experiences of uncertainty to new solutions and/or possibilities about a future uncertainty. These women focus their hope on the child and the privilege of experiencing pregnancy.

On the other hand, studies conducted with women with prenatal diagnoses are focused on outcomes that are associated with a positive future for the child and the realization that they made the right decision when they decided not to terminate the pregnancy.

The hope experiences related to women have more expression in the group of women who have the prenatal diagnosis without medical conditions in themselves. However, after a prenatal diagnosis of a fetal anomaly in the initial phase, some women hoped that they would hear that the diagnosis was a mistake [28].

Each person has different coping strategies when faced with stress [19]. Women should be clearly informed about the options and decisions they may need to make in cases of abnormal tests and prenatal screenings [29,32].

4. Discussion

In the study of high-risk pregnancies with different medical problems, hopelessness experiences are implicitly related to pregnancy worries, concerns about the child, future pregnancy, relationships and support with others, and higher costs [18].

When the risk was associated with the diagnosis of HIV during pregnancy, the attribute that emerged in all studies was fear of the cruelty of stigma, stereotyping, discrimination, and judgment [20,23–25,27].

Studies also showed the existence of concerns about the transmission of the virus to the baby and possible effects on their health [20]; emotional distress, ambivalence about pregnancy and motherhood [23], an association with negative self-image, loneliness, feelings of isolation, fears of loved ones, blame, and that they avoided any romantic relationships, and fears of being sick, going to the hospital, or dying [25]. In the same population, Ross et al. found that women perceived their lives as a struggle [24], such as with shock, fear, anxiety, and depression; with sharing one's struggling with others and that they struggle to care for their baby, especially after birth; and that they struggle through ups and downs. The consciousness of fragility, noxious self, denied motherhood, social jealousy, and fear of genetic transmission have been described in the study of women with chronic kidney disease [21].

Studies focused on maternal near-miss experiences showed many fears and concerns [22,26]: fear of being unable to become pregnant again, fear of raising their child without siblings, fear of carrying about their child without a mother, fear of remarriage of a spouse if the spouse wanted more children, fear of becoming pregnant again and experiencing postpartum complications, fear of not being able to adjust to the complications and grieving for a long time, and guilt, intolerance of pain, irritability, and postpartum depression.

In the optical of mental health, other emotions have an expression in this analysis, such as anxiety, discouragement, and numbness, fear for their own life and the life of their baby, feelings of death, and feelings of an incomplete self because they are a woman without a uterus and a baby [26].

In a study that looked at the experiences of women with various pathologies during pregnancy, hope was found to be related to the adaptation to challenges, the belief that conditions will improve, and being hopeful about the future [18].

Children were seen as a "divine gift", a chance to correct past mistakes, and to be good, loving mothers [20,23,27].

Religiousness resurfaced with the wish for a child to give meaning to their life's and to help with the construction of a female identity, because to be a mother was stronger than any problem [27]. Hope was also identified in protecting children from contracting the disease and from stigma in the case of HIV disease [23–25].

In general, the studies reported that a positive pregnancy experiences was an important source of support and hope for women [27]. By participating in spiritual practices, women believed that God would respond to their needs and take care of their children [20,25,27]. Spirituality and resorting to God and Imams are the most common attributes of hope found in the found studies [18,22,25–27].

Other forms of adaptations included the natural maternal disposition to focus on their child, which was a motivator to seek treatment and a source of strength to continue living [21,26].

When a prenatal diagnosis occurs, studies identified the following women's hopelessness attributes: the sense of disbelief, stress, doubting the struggle, shock, anger, and fear of developing a bond with the baby who may die [19,28,29,32]; anxiety and fears about the unknown [19]; guilt by the loss of the perfect baby [32].

Only one study analyzed the hopeless experience in the case of a woman who had maternal–fetal surgery [31]. Emotions such as uncertainty about their future child's quality of life, fear of potential complications of the surgery and the possibility of losing their unborn child, fear of not waking up after the operation, and post-traumatic stress and depression were related [31].

In the rebuilding phase after prenatal diagnosis, a positive vision of the future seems to develop (whatever that may be) as the woman processes her experiences in such a way as to reconstruct the future and adjust her earlier beliefs about pregnancy and the world in general [28]. For example, in the context of fetal surgery, women felt strong feelings of responsibility and determination to do anything to improve their future child's health outcomes [31].

The study conducted by Oscarsson et al. found that women's experiences of hope were based on going through a crisis and knowing that they were doing the right thing [19]. Irani et al. compared the emotional experiences of women that decided to continue or terminate the pregnancy after the prenatal diagnosis of fetal anomalies [29]. The results identified in this population include a dilemma between hope and worry maintained by a positive attitude toward childbirth to cope with the situation and/or a return to normality [29]. In another study that compared the experiences between continuing and terminating the pregnancy, the authors cited religious beliefs and convictions and the belief that the baby would survive after birth [30]. Other sources of hope were found when there were good sources of information, time to prepare, support from family and friends, spiritual beliefs, staying busy with work and other activities and empathizing with the baby [32].

5. Conclusions and Implications for Practice

This review demonstrated the meanings of the lived experiences of women with high-risk pregnancies due to maternal medical conditions or prenatal diagnoses.

The hope aspects related to the women's lived experiences of pregnancies that continued after the confirmation of a high-risk pregnancy diagnosis were addressed during health care and were analyzed with a personal woman's attributes.

The type of evidence or study design that exists in women's life experiences of pregnancy research in relation to hope aspects after the confirmation of a high-risk pregnancy diagnosis is mostly qualitative. In the studies analyzed, the qualitative design stood out in terms of methodology (14 of the 15 studies analyzed in this review). This aspect is related to the type of the main review question, as we intended to understand the experiences of hope and not measure them quantitatively.

In the studies that explored high-risk pregnancies, the main issues identified were women's fears for their own lives and the lives of their babies, the impossibility of getting pregnant again, postpartum complications, and an incomplete self. On the other hand, women facing a prenatal diagnosis of fetal anomalies faced the fear of developing an attachment to a baby who may die, the loss of the perfect baby, and the fear of possible complications of treatment. In both cases, women went through phases with feelings of emotional distress, ambivalence, disbelief, stress, struggle, anxiety, and depression.

Most women facing a high-risk pregnancy used religion, spirituality, and faith in God as coping mechanisms. Some others relied on good sources of information, time to prepare, support from family and friends, employment at work, and empathy with the baby. These women had a strong sense of responsibility for the treatment and saw it as an opportunity that they had been given. If they decided to continue with the pregnancy, they focused on the idea of doing the right thing. They hoped for the best possible outcome and that everything would be okay in the end.

In relation to hope interventions in the context of care for women with high-risk pregnancies, the gaps in the nursing research are about hope interventions. The results of different studies describe an implicit proposal to develop research about hope interventions in the context of care for women with high-risk pregnancies. However, we believe that women who are well informed about their situation and treatment are most likely to adapt and comply with treatment. In addition, recognizing the benefits of religious faith in situations of uncertainty is important and helps women to adjust to challenging situations.

It is extremely important to monitor women's emotional and psychological reactions after a prenatal anomaly diagnosis, not only throughout the pregnancy but also in the postnatal stages. Nurses and midwives have a privileged position in relation to women and can help them overcome difficult challenges in the present and future.

To establish a comparison between women's responses in the different conditions (maternal medical situations and prenatal diagnoses), it is highly important to increase our understanding of the impacts of these experiences in this population.

6. Limitations

The present scoping review was based mainly on qualitative studies conducted on a limited number of mothers with high-risk pregnancies, so the results cannot be generalized to similar populations.

Author Contributions: Contributions to conception and writing: M.A. and Z.C. Drafting the article or revising it critically for important intellectual content: M.A., Z.C. and C.R.V. All authors have read and agreed to the published version of the manuscript.

Funding: This research was funded by Universidade Católica Portuguesa—Center for Interdisciplinary Research in Health (CIIS). This work is financially supported by National Funds through FCT—Fundação para a Ciência e a Tecnologia, I.P., under the project UIDP/04279/2020.

Institutional Review Board Statement: Not applicable.

Informed Consent Statement: Not applicable.

Data Availability Statement: Not applicable.

Conflicts of Interest: The authors declare that there is no conflict of interest.

Appendix A

Scheme	Pubmed Search Equation	Results
#1	pregnant women OR maternal OR expectant mother	482,511
#2	high-risk pregnancy OR pregnancy complications OR obstetric complications OR medical condition OR obstetric health care	2,526,150
#3	hope OR hopelessness	89,314
#4	Qualitative Studies OR qualitative research OR phenomenological research OR Experiences OR Perceptions OR Attitudes OR Feelings OR meaning OR need	3,843,920
#5	1 AND 2 AND 3 AND 4	253

Appendix B

Author(s) Year Publication Country	Hope and Hopeless Experiences or/and Expectations	Aims	Study Design	Study Population/Sample	Context	Population Characteristics and Typology	Main Results
The lived experience of women with a high-risk pregnancy: A phenomenology investigation Badakhsh, Hastings-Tolsma, Firouzkohi, Amirshahi & Hashemi (2020) Iran [18]	Hopelessness experiences are implicit with pregnancy concerns, worries about the child, future pregnancy, relations and support with others and increase cost Hope is implicit on adaptation to the challenges, believing that the condition would improve and being hopeful about the future. Spirituality, resorting to the God and Imans.	To describe the lived experience of women during HRP	Phenomenological study	Pregnant women	Public health centre in a large urban city in south-east Iran	- Primigravida or multigravida - Second or third trimester of pregnancy - varied high-risk medical conditions - varied age and levels of education - able to speak Persian language	Four thematic categories were extracted: - Challenges of family in HRP - Challenges for future pregnancies - Challenge of anticipation for motherhood (involving others, increased costs, treatment, and life management) - Challenge of adaptation
Recasting Hope: A process of adaptation following fetal anomaly diagnosis Lalor, Begley & Galavan (2009) Ireland (United Kingdom) [28]	Hopelessness implicit on sense of incredulity, stress, the dually of the fight. Fear of developing a bond with the baby that could die. Initially some women hoped they would hear that the initial diagnosis was a mistake. Finally, in the rebuilding phase it would appear that the emergence of a positive vision of the future (whatever that might be), as the woman works through her experience in ways that permit her to reconstruct the future and adapt her previously held beliefs about pregnancy in particular and the world in general.	To provide a theorical framework of the process of adaptation following fetal anomaly diagnosis based on women's experiences of carrying a baby with a fetal abnormality up to the end and beyond the birth	Grounded theory study	Pregnant women carrying a baby with fetal abnormality	Fetal medicine unit of a major Dublin maternity hospital	All Irish women. Forty-one women, eleven primigravidae and 30 multigravidae. Thirthy-one women continued the pregnancy and ten travelled to UK to access termination of pregnancy services not available within the state. Forty women were married or partnered and although not all pregnancies were planned, all were wanted.	Recasting Hope, the process of adaptation following diagnosis is represented temporally as four phases: 'Assume Normal', 'Shock', 'Gaining Meaning' and 'Rebuilding'. Some mothers expressed a sense of incredulity when informed of the anomaly and the 'Assume Normal' phase provides an improved understanding as to why women remain unprepared for an adverse diagnosis. Transition to phase 2, 'Shock,' is characterised by receiving the diagnosis and makes explicit women's initial reactions. Once the diagnosis is confirmed, a process of 'Gaining Meaning' commences, whereby an attempt to make sense of this ostensibly negative event begins. 'Rebuilding', the final stage in the process, is concerned with the extent to which women recover from the loss and resolve the inconsistency between their experience and their previous expectations.
When fetal hydronephrosis is suspected antenatally—a qualitative study Oscarsson, Gottvall & Swahnberg (2015) Sweden [19]	Hopelessness experiences of anxiety and fears about the unknown, worry and stress. Experiences of hope are based on going through crisis by knowing that you are doing the right thing, told themselves that they would deal with it after delivery and that everything would be all right at the end	To explore women's reactions to the discovery of fetal hydronephrosis in the context of uncertainty regarding the prognosis	Qualitative study	Pregnant women	University clinic of Sweden	Ten women with antenatal diagnosis were invited to the interview 6-12 months after delivery. Mean age were 30.6 years. Six nulliparous and four parous.	The core category, 'Going through crisis by knowing that you are doing the right thing' illustrates the meaning of women's reactions and feelings. It illuminates the four categories: 'When the unexpected happens'—on the one hand, women had positive views that the suspicious malformation could be discovered; however, on the other hand, women questioned the screening. 'To live in suspense during pregnancy'—the suspicious malformation caused anxiety and was a stressful situation. 'Difficulties in understanding information'—the women thought they had limited knowledge and had difficulties in understanding the information. 'Suppress feelings and hope for the best'—the women tried to postpone the problem and thought they should deal with it after delivery

Author(s) Year Publication Country	Hope and Hopeless Experiences or/and Expectations	Aims	Study Design	Study Population/Sample	Context	Population Characteristics and Typology	Main Results
Pregnancy through the Lens of Iranian Women with HIV: A Qualitative Study Zahra Behboodi-Moghadam, Khalajinia, Nasrabadi, Mohraz & Gharacheh (2015) Iran [20]	Hopelessness experiences implicit on the concerns about transmitting the virus to the baby, effects on their health. Becoming a mother after a HIV diagnosis was a source of hope, value and esteem, children seen as a "divine gift", a chance to correct past mistakes. By participating in spiritual practices, women believed that God would respond to their needs, God would take care of their children.	To explore the experience of pregnancy among Iranian women with HIV	Qualitative study	Pregnant women	Counseling Center for Behavioral Diseases in Imam Khomeini Hospital in Tehran, Iran.	The study participants' age ranged from 22 to 39 years. Five of them had completed primary school and 7 had finished high school. Length of time since HIV diagnosis was 1 to 5 years. A total of 7 participants had children and 5 of them reported being pregnant for the first time. Nine of the women (75%) had been infected with HIV by their husbands and 3 of them (25%) through unknown route.	Four main themes emerged from the data: fear and hope, stigma and discrimination, marital life stability, and trust in God. Despite concerns about mother-to-child transmission of HIV, and uncertain life span, HIV-infected women tended to continue their pregnancy, and having children was viewed as a window of hope for them.
Perspectives on pregnancy in women with Chronic Kidney Disease (CKD): A Semistructured Interview Study Tong, Brown, Winkelmayer, Craig & Jesudason (2015) Australia [21]	Hopelessness showed by conscious of fragility, noxious self, denied motherhood, social jealousy, fear of genetic transmission. Hope implicit on the opportunity of getting pregnant, found that the baby didn't have the same disease, valuing life, gratitude in hindsight and focus on what is good.	To describe the beliefs, values, and experiences of pregnancy in women with CKD to inform prepregnancy counseling and pregnancy care.	Qualitative study	Pregnant women	Two renal units in Australia	41 women (95% response rate) aged 22 to 56 years with CKD stages 3 to 5 (n 5 5), receiving dialysis (n 5 5), or received a kidney transplant (n 5 31) from 2 renal units in Australia.	Six themes were identified: bodily failure (conscious of fragility, noxious self, critical timing, and suspended in limbo), devastating loss (denied motherhood, disempowered by medical catastrophizing, resolving grief, barriers to parenthood alternatives, and social jealousy), intransigent guilt (disappointing part- ners, fear of genetic transmission, respecting donor sacrifice, and medical judgment), rationalizing conse- quential risks (choosing survival, avoiding fetal harm, responding to family protectiveness, compromising health, decisional ownership, and unjustifiable gamble), strengthening resolve (hope and opportunity, medical assurance, resolute determination, and reticent hope), and reorientating focus (valuing life and gratitude in hindsight).
Lived experiences of women with maternal near miss: qualitative research Sabzevari, Yazdi,& Rad (2021) Iran [22]	Hopelessness implicit on fears and concerns. Unable to get pregnant again, fear of raising their child without siblings, concerns about their child without a mother, re-marriage of the spouse if the spouse wanted more children, fear of getting pregnant again and experiencing postpartum complications, failure to adjust to complications and prolonged mourning, feeling guilty, not tolerating pain, irritability, and postpartum depression. The only hope seemed to be believing in God's will and their survival and ability to support their children.	To understand experiences and perceptions of women with higher risk pregnancy relating to problems/needs of health and care practices.	Descriptive and qualitative	Pregnant women	The interviews were conducted at the office of the educational supervisor of Sabzevar Mobini Hospital or any other place that was conveni- ent to the mothers (e.g. their home).	The mothers were selected based on their ability to express their near-miss experience and willingness to be interviewed. The age range of the selected mothers was 19–36 years.	Five main categories were extracted, including fears and concerns, failure to accept and adapt, tolerating physical and psychological pain and hardships, death experience, and medical team mismanagement. Regret and fear of raising the child with siblings, fear of the re- marriage of the spouse, and fear of complications and costs were among the subcategories of fears and concerns. Lack of adaptation to the complications and prolonged mourning were the subcategories of failure to accept and adapt, and the subcategories of tolerating physical and psychological pain and hardships were a sense of guilt, tolerating physical pain, hopelessness, irritability, hatred toward the medical team, and postpartum depression. In addition, returning to normal life, and seeing/actually feeling death were the subcategories of the death experience. The subcategories of the medical team mismanagement included medical errors, lack of sup- port/negligence,

Author(s) Year Publication Country	Hope and Hopeless Experiences or/and Expectations	Aims	Study Design	Study Population/Sample	Context	Population Characteristics and Typology	Main Results
Emotional and cognitive experiences of pregnant women following prenatal diagnosis of fetal anomalies: A qualitative study in Iran Irani, Khadivzadeh, Nekah, Ebrahimipour & Tara (2019) Iran [29]	Hopelessness is defined by disbelief, distress, panic and shock during the time of diagnosis. When their pregnancy was terminated, women experienced perinatal loss such as guilt and shame during pregnancy termination, loss of their expected child, suffering and emotional distress process, and fear of recurrence in future pregnancies. Women that decided to continue their pregnancy had a dilemma between hope and worries. In general, women carrying babies tried to keep a positive attitude towards the birth, as a way to cope with the situation. They were hopeful about the best possible outcome or return of normality. Some women hope that the problem for the health of their unborn child in the case of abnormal findings will be resolved or is miner anomaly.	To explore the emotional and cognitive experiences of pregnant women following prenatal diagnosis of fetal anomalies in Mashhad, Iran.	Qualitative study	Pregnant women	Two tertiary referral centers for fetal anomaly at Mashhad University Hospitals, Omolbanin Hospital and Imam Reza Hospital in Mashhad, Iran.	The sample studied consisted of Persian speaking parents with prenatal diagnosis of fetal anomalies at the gestational week of 12–27. All the pregnant women with a suspected or definitive diagnosis of fetal anomaly as per the ultrasound or the combined test (NT, free β-hCG and PAPP-A) were eligible for participation.	Four categories and 10 subcategories emerged. Category one, grief reactions during the time of diagnosis, contained two subcategories: shocked and panicked, and distressed and disbelieved. Category two, perinatal loss through a pregnancy termination, contained four subcategories: guilt and shame during pregnancy termination, loss of their expected child, suffering and emotional distress process, and unmet needs by health professionals. Category three, fears of recurrence in future pregnancies, had two subcategories: worried about inadequate prenatal care in the future pregnancies and worried about abnormal fetus in next pregnancies. Finally, Category four, a dilemma between hope and worries contained two subcategories: hope for normality and worried about future.
Complexity of consenting for medical termination of pregnancy: Prospective and longitudinal study in Paris Tayeh, Jouannic, Mansour, Kesrouani &Attieh (2018) Paris [30]	Hope implicit in religious beliefs and convictions and believed that their child could exceptionally survive after birth.	To analyze the patients' perception of prenatal diagnosis of fetal cardiac pathology, and the reasons for choosing to continue with pregnancy despite being eligible to receive a medical termination of pregnancy.	Descriptive, prospective and longitudinal study	Pregnant women	Hôpital Necker—Enfants Malades in Paris, France	Eligible participants were pregnant women who decided to continue their pregnancy despite an unfavorable medical advice because they were carrying fetuses with incurable cardiac pathologies. Age between 23 and 44 years old.	Patient informed consent should be sought before any decision in neonatology, even if conflicting with the medical team's knowledge and the pregnant mother's benefits. Decisions to accept or decline pregnancy termination depend on the patients' psychological character, ideologies, convictions, and mistrust in the diagnosis/prognosis, or hope in the fetus survival.
Women's Voices: The Lived Experience of Pregnancy and Motherhood After Diagnosis With HIV Sanders (2008) New York (USA) [23]	Hopelessness characterized by periods of emotional distress, ambivalence in relation to pregnancy and mother- hood, and stigma. Hope in protecting children from contracting HIV and from HIV related stigma and hoped to rectify mistakes made with children born previously and to be a good, loving mother.	The study aimed to explore the meaning of pregnancy after diagnosis with HIV	Qualitative study	Pregnant women	Interviews conducted in two academic health centers in metropolitan New York.	Participants were a purposive sampling of 9 women, 34 to 53 years old, who had been diagnosed with HIV and were currently pregnant or who had become mothers postdiagnosis.	The result of the study included themes of extreme emotional distress after HIV diagnosis, feeling stigmatized, emotions related to the pregnancy and baby, experiences with health care providers, and as positive and supporting.

Author(s) Year Publication Country	Hope and Hopeless Experiences or/and Expectations	Aims	Study Design	Study Population/Sample	Context	Population Characteristics and Typology	Main Results
The lived experiences of rural women diagnosed with the human Immunodeficiency virus in the antenatal period Fords, Crowley & Merwe (2017) South Africa [25]	Hopelessness implicit in a negative self-image, loneliness, feelings of isolation, fear from loved ones, women experienced blame, fear, the cruelties of stigma, stereotyping and judging an avoided any romantic relationship. Fear of being ill, being in a hospital or dying. Women felt hope to live and see the future of their children, hope on their spiritual beliefs. The initiation of ART gave them hope as they were confident that the treatment would improve their health, extend their life, protect their unborn children and even cure HIV. The most important hope that they had for the future was that their unborn child would be HIV free	To explore the lived experiences of women diagnosed with HIV in the antenatal period in a rural area in the Eastern Cape province of South Africa.	Phenome-nological study	Pregnant women	Pregnant women residing in the Maluti local service area in the Eastern Cape who attended one the local clinics.	Ten women over the age of 18, diagnosed with HIV for the first time in the antenatal period of pregnancy.	Women diagnosed with HIV during pregnancy are ultimately concerned with the wellbeing of their unborn children, and this concern motivates their adherence to ART. Women's lived experiences are situated in their unique sociocultural context, and although some known challenges remain, counselling and support strategies need to be informed by exploring context-specific issues and involving the local community.
The Lived Experiences of HIV-Positive, Pregnant Women in Thailand Ross, Sawatphanit, Wilaiphan; Burke & Suwansujarid (2007) Thailand [24]	Hopelessness: Women perceived their lives as a struggle. Struggling alone and experiences of shock, fear, anxiety and depression; Sharing one's struggling, with fears of stigmatization and discrimination; struggling for the baby, most postpartum Thai mothers indicated that their babies were central to their determination in helping them to move on with their lives; Struggling through ups and downs Hope: Women found hope through the taking of antiretroviral medicines and, subsequently, showed a desire to fight the virus as long as possible for their children. For health care professionals, when a seropositive pregnant woman who has her baby as her hope is feeling "down," gently reminding her of her unborn baby could be a help in lifting her hope and spirit	The purpose of our study was to examine the lived experiences of 10 pregnant women in Thailand following their HIV diagnosis	Phenome-nological study	Pregnant women newly diagnose with HIV	Prenatal clinic at a government hospital in Thailand	All participants were Buddhist. They ranged in age from 18 to 29 years. Five were graduates of primary school, 3 had completed junior high school, and 2 had finished high school. Nine had a monthly low family income. Only one participant had a middle-class family. Eight were married or lived with a partner, and 2 were divorced.	All participants in these study decided not to end their pregnancies. This might be explained in the fact that all women were Buddhist, with Buddhists usually believing that terminating one's life or ending a pregnancy is a big sin. From a health care perspective, identifying helpful resources such as a peer/support group can be critical for a woman when she is ready to share her struggle with others. Peer support was found to be helpful for 7 women in our study and has been found to be effective across cultures and countries in reducing seropositive women's fears, depression, and attempts at suicide. In general, support from health care professionals to assist the HIV-positive mother's efforts to promote her baby's health will be of great value, regardless of the ultimate diagnosis of the baby's HIV status.

Author(s) Year Publication Country	Hope and Hopeless Experiences or/and Expectations	Aims	Study Design	Study Population/Sample	Context	Population Characteristics and Typology	Main Results
The experiences of women with maternal near miss and their perception of quality of care in Kelantan, Malaysia: a qualitative study Norhayati, Hazlina, Nik Hussain; Asrenee & Sulaiman (2017) Malaysia [26]	Hopelessness: several forms of negative emotions such as fear, anxiety, alarm, incomplete self, discouragement and numbness. Fear on their own lives and lives of their babies, fear for the prenatal outcomes and of undergoing surgery, fear of recurrence similar incidents and inability to conceive in the future, sense of death; incomplete self, because being a woman without a uterus and without a baby; sadness for not being able to have more children Hope: The women adapted to most of negative emotions and to difficult life events, like traumatic childbirth by anchoring their reasoning to religiosity and faith and appeared to have accepted the situation calmly. They responded to their situations positively, delegating the resolution to God and regarding what had happened as what God had planned for them. They were very grateful that God gave them a second chance. Other forms of adaptation included natural maternal disposition in which their attention were distracted and focused on their children and that was a motivator to seek treatment and as a source of strength to continue living. The competency of the healthcare providers in the form of adequate knowledge and skills in providing optimal care had gained trust from the women. Emotion and social support and improved relationship quality were associated with better mental health and well-being, reduced stress and protection from postpartum depression. In the current study, social support appeared to play a role in protecting the women from ill health. Despite their experiences, the women were relieved at having survived their acute, severe complications and looked forward to resuming their lives normally.	This study aimed to explore the experiences of women with maternal near miss and their perception of the quality of care in Kelantan, Malaysia	Qualitative phenomenological approach	All women screened for the presence of any vital organ dysfunction or failure based on the World Health Organization criteria for maternal near miss.	All women admitted to labour room, obstetrics and gynaecology wards and intensive care units in 2014	Thirty women who had experienced maternal near miss events were included in the analysis. All were Malays between the ages of 22 and 45. Almost all women (93.3%) had secondary and tertiary education and 63.3% were employed.	In appraising the maternal near miss events, the study found that the women viewed their experiences as frightening and that they experienced other negative emotions and a sense of imminent death. Their perceptions of the quality of their care were influenced by the competency and promptness in the provision of care, interpersonal communication, information-sharing and the quality of physical resources. These factors should be of concern to those seeking to improve services at healthcare facilities. The predisposition to seek healthcare was influenced by costs, self-attitude and beliefs.
The lived experience of pregnancy while carrying a child with a known, nonlethal congenital abnormality Hedrick (2004) USA [32]	Hopelessness is define by the loss of the perfect baby with feelings of grief, shock, anger and guilt. Hope defined with good sources of information, time to prepare, support from family and friends, spiritual beliefs and staying busy with work and other activities and empathizing with the baby	To gain an understanding of the experience of pregnancy while carrying a child with a known, nonlethal congenital abnormality	Phenomenological study	Pregnant women	Outpatient perinatal center at a large Midwestern hospital	Ages between 18 and 44 years. Gestacional age at the time of the diagnosis 17 to 26 weeks. Interviewed between 24 and 36 weeks. Fetal diagnosis of neural tube defect, cleft lip, congenital heart defect, renal anomaly, cystic malformation of the long and down syndrome	The pregnancy experience was of a paradoxical nature. Knowledge of fetal diagnosis with positive and negative consequences. Time is good but also the enemy; you grieve but you do not grieve; my baby's not perfect, but he's still mine

Author(s) Year Publication Country	Hope and Hopeless Experiences or/and Expectations	Aims	Study Design	Study Population/Sample	Context	Population Characteristics and Typology	Main Results
Feelings and expectations of pregnant women living with HIV: A phenomenological study Arcoverde, Conter, Silva & Santos (2015) Brazil [27]	Hopelessness: - Human suffering and anxiety resulting from stigma, prejudice and discrimination - fear of exposure, prejudice and discrimination produced by the stigma Hope: - Motherhood has such a strong meaning to these women that not even the possibility of transmitting the virus to the foetus can change their mind. The wish to be a mother is stronger than the problems faced throughout life. - Mothers with HIV focus their life on the uninfected child, which symbolizes the continuity and the hope to overcome their fears. The child can be the motivator that helps them to face the challenges imposed by the disease. - For them, positive pregnancy experiences were an important source of support and hope to carry on living and taking care of their own health. The participants accepted their pregnancy and asked for a healthy baby. Even if it is an unplanned pregnancy, the child becomes a motivating force, giving them reason to fight the disease. - A mother living with HIV will adhere more easily to treatment if she has been informed about the seriousness of the disease and about the possibilities of the child being infected. - Pregnant women make adjustments to deal with their seropositivity through religion, seeking in God's figure love, care, help, strength, forgiveness, and well-being. The support of religion brings hope conveying a feeling of comfort and has an impact on coping with HIV/AIDS - For these women, the unborn child is the motivation for them to rethink their life and to withstand the bad times. Religiousness resurfaced by the wish for a child gives meaning to their life and helps the construction of a female identity - Faith is a life support that helps these women to withstand the uncertainties of being pregnant and HIV positive. Their faith in God gave them the confidence to face the difficulties imposed by the disease and hope for better days for them and their children - Spirituality is a comfort that helps them to bear the pain of being HIV positive; it makes them believe in quality of life due to treatment and, in the hope of a miracle, that God may transform their lives completely	The objective of this research was to identify the feelings and expectations of such pregnant women about the disease and pregnancy	Phenomenological qualitative research based on Maurice Merleau-Ponty's philosophy of perception	Pregnant women with HIV diagnosis	Outpatient unit of the SAE of Foz do Iguaçu, in the state of Parana	Five pregnant women diagnosed with HIV monitored in SAE of Foz do Iguaçu, in the state of Paraná, participated in the study. They were married, aged between 20 and 35 years old, and had been diagnosed from one month to ten years before research began	The interviews conveyed the experiences of the women with HIV, their acceptance of the limitations imposed by the disease and showed how they dealt with the stigma surrounding HIV. Despite the prejudice, such pregnant women did not lose faith and hope. Pregnant women believe in the treatment and the possibility of their children being born healthy. The desire of motherhood increases their expectations about the care, which prevents complications from the infection. The study participants accepted the pregnancy, mainly because the desire to become a mother was stronger than anything else they could feel. The treatment was then accepted, as the only way to protect their children from a HIV infection. The researchers identified feelings of strength, will, and determination to overcome the problems which transcend the difficulties encountered throughout pregnancy. The fear of harming the child—symbol of perseverance, wishes and hopes—is faced and reignites their desire to carry on living in order to care for their children and to protect them. Communication and education for the construction of new concepts and ideas related to the phenomenon can indicate new ways to learn about change processes. Social movements, cultural changes, as well as social equality and inequality; action and intervention based on fair policies can also be one of the responses to the problem.

Author(s) Year Publication Country	Hope and Hopeless Experiences or/and Expectations	Aims	Study Design	Study Population/Sample	Context	Population Characteristics and Typology	Main Results
'We did everything we could'—a qualitative study exploring the acceptability of maternal-fetal surgery for spina bifida to parents Crombag, Sacco, Stocks, Vloo, Merwe, Gallagher, David, Marlow & Deprest (2021) Belgium and United Kingdom [31]	Hopelessness: - Uncertainty of impact of the surgery on their future child's quality of life. - Fear. Potential complications of the surgery - Fear of losing their unborn child and fear of not waking up after the operation - Post-traumatic stress and depression Hope: - Strong feelings of responsibility and determination to do anything to improve their future child's health outcomes - Only option to improve their child's outcome - an opportunity they were given - MFS provided hope for their child's future	To explore the concepts and strategies parents employ when considering maternal-fetal surgery (MFS) as an option for the management of spina bifida (SB) in their fetus, and how this determines the acceptability of the intervention	Qualitative study	Parents	Two MFS partner centres with specialist assessment (University Hospitals Leuven, Belgium; University College London Hospital, United Kingdom)	Parents whose fetuses with SB were eligible for MFS, Age above 18 years old.	MFS for SB remains highly acceptable from diagnosis until 3–6 months postnatally. For those opting for MFS, expectations seemed to be realistic yet were driven by hope and expectation of the best outcome. For parents opting for termination of pregnancy, the potential benefit of MFS seems to play a minimal role in their final decision

Appendix C

Article	Abstract and Title	Introduction and Aims	Method and Data	Sampling	Data Analysis	Ethics and Bias	Results	Transferability or Generalizability	Implications and Usefulness	TOTAL
Badakhsh, Hastings-Tolsma, Firouzkohi, Amirshahi & Hashemi (2020) [18]	4	4	4	4	4	4	4	4	4	36
Lalor, Begley & Galavan (2009) [28]	4	4	4	4	4	4	4	4	4	36
Oscarsson, Gottvall & Swahnberg (2015) [19]	4	4	4	4	4	4	4	4	4	36
Behboodi-Moghadam, Khalajinia, Nasrabadi, Mohraz & Gharacheh (2015) [20]	4	4	4	4	4	4	4	3	3	34
Tong, Brown, Winkelmayer, Craig & Jesudason (2015) [21]	4	4	4	4	4	4	4	4	4	36
Sabzevari, Yazdi, & Rad (2021) [22]	4	4	4	4	4	3	4	3	3	33
Irani, Khadivzadeh, Nekah, Ebrahimipour & Tara (2019) [29]	4	4	4	4	4	4	4	4	4	36
Tayeh, Jouannic, Mansour, Kesrouani &Attieh (2018) [30]	4	4	4	4	4	3	4	4	4	35
Sanders (2008) [23]	3	4	4	4	4	4	4	4	4	35
Fords, Crowley & Merwe (2017) [25]	4	4	4	4	4	4	4	4	4	36
Ross, Sawatphanit, Wilaiphan; Burke & Suwansujarid (2007) [24]	4	4	4	3	4	4	4	3	3	33
Norhayati, Hazlina, Nik Hussain; Asrenee & Sulaiman (2017) [26]	4	4	3	4	3	3	4	3	4	32
Hedrick (2004) [32]	4	4	3	4	3	3	4	4	3	32
Arcoverde, Conter, Silva & Santos (2015) [27]	4	4	4	4	3	4	3	3	3	32
Crombag, Sacco, Stocks, Vloo, Merwe, Gallagher, David, Marlow & Deprest (2021) [31]	4	4	4	4	4	4	4	4	4	36

References

1. WHO. Maternal Mortality. World Heal Organ. 2019. Available online: https://www.who.int/news-room/fact-sheets/detail/m (accessed on 24 May 2022).
2. Alkema, L.; Chou, D.; Hogan, D.; Zhang, S.; Moller, A.B.; Gemmill, A.; Fat, D.M.; Boerma, T.; Temmerman, M.; Mathers, C.; et al. National, regional and global levels and trend in MMR between 1990 and 2015 with scenario-based projections to 2030; a systematic analysis by the United Nations Maternal Mortality. *Lancet* **2016**, *387*, 462–474. [CrossRef] [PubMed]
3. James, D.K.; Steer, J.P. High-Risk Pregnancy: Management Options. 5th Edition. 2018, pp. 227–248. Available online: https://www.cambridge.org/us/academic/subjects/medicine/obstetrics-and-gynecology-reproductive-medicine/high-risk-pregnancy-management-options-5th-edition-1 (accessed on 6 June 2022).
4. Bonanno, L.; Bennett, M.; Pitt, A. The experience of parents of newborns diagnosed with a congenital anomaly at birth: A systematic review protocol. *JBI Database Syst. Rev. Implement Rep.* **2013**, *11*, 100–111. [CrossRef]
5. Leichtentritt, R.D.; Blumenthal, N.; Elyassi, A.; Rotmensch, S. High-Risk Pregnancy and Hospitalization: The Women's Voices. *Health Soc. Work.* **2005**, *30*, 39–47. [CrossRef] [PubMed]

6. WHO. Congenital Anomalies. 2017, pp. 1–2. Available online: https://www.who.int/news-room/fact-sheets/detail/c (accessed on 22 June 2022).
7. Atienza-Carrasco, J.; Linares-Abad, M.; Padilla-Ruiz, M.; Morales-Gil, I.M. Experiences and outcomes following diagnosis of congenital foetal anomaly and medical termination of pregnancy: A phenomenological study. *J. Clin. Nurs.* **2020**, *29*, 1220–1237. [CrossRef]
8. Sara Manuela Airosa da Silva Vinculação Materna durante e após a Gravidez: Ansiedade, Depressão, Stress e Suporte Social Universidade Fernando Pessoa Faculdade de Ciências Humanas e Sociais Mestrado Psicologia Clínica e da Saúde. Available online: https://bdigital.ufp.pt/bitstream/10284/3259/3/DM_16833.pdf (accessed on 20 September 2022).
9. Georgsson Öhman, S.; Saltvedt, S.; Grunewald, C.; Waldenström, U. Does fetal screening affect women's worries about the health of their baby? *Acta Obstet. Gynecol. Scand.* **2004**, *83*, 634–640. [CrossRef]
10. De Araújo, W.S.; Romero, W.G.; Zandonade, E.; Amorim, M.H.C. Efeitos do relaxamento sobre os níveis de depressão em mulheres com gravidez de alto risco: Ensaio clínico randomizado. *Rev. Lat. Am. Enferm.* **2016**, *24*. [CrossRef]
11. Oliveira, D.d.C.; Mandú, E.N.T. Women with high-risk pregnancy: Experiences and perceptions of needs and care. *Esc Anna Nery—Rev. Enferm.* **2015**, *19*, 93–101. [CrossRef]
12. Bailey, T.C.; Snyder, C.R. Satisfaction with life and hope: A look at age and marital status. *Psychol. Rec.* **2007**, *57*, 233–240. [CrossRef]
13. International Council of Nurses. CIPE®Versão 2019. Classificação Internacional Para a Prática de Enfermagem. Lisboa. Ordem dos Enfermeiros. 2019. Available online: https://www.icn.ch/what-we-do/projects/ehealth-icn (accessed on 6 June 2022).
14. Jones, A.C. Relevant Hope to Promote Therapeutic Change. Linkedin. 2019. Available online: https://www.linkedin.com/pulse/relevant-hope-promote-therapeutic-change-alun-charles-jones (accessed on 6 June 2022).
15. Martocchio, D.K.C. Hope: Its spheres and dimensions. *Nurs. Clin. N. Am.* **1985**, *20*, 382–389.
16. Institute, J.B. The Joanna Briggs Institute Reviewer's Manual 2020. Methodology for JBI Reviewers. 2020. Available online: https://jbi-global-wiki.refined.site/space/MANUAL/. (accessed on 20 March 2022).
17. Ricco, A.C.; Lillie, E.; Zarin, W.O.K. PRISMA extension for scoping reviews (PRISMA-ScR): Checklist and explanation. *Ann. Intern. Med.* **2018**, *169*, 467–473.
18. Badakhsh, M.; Hastings-Tolsma, M.; Firouzkohi, M.; Amirshahi, M.; Hashemi, Z.S. The lived experience of women with a high-risk pregnancy: A phenomenology investigation. *Midwifery* **2020**, *82*. [CrossRef]
19. Oscarsson, M.; Gottvall, T.; Swahnberg, K. When fetal hydronephrosis is suspected antenatally-a qualitative study. *BMC Pregnancy Childbirth* **2015**, *15*, 349. [CrossRef]
20. Behboodi-Moghadam, Z.; Khalajinia, Z.; Nasrabadi, A.R.N.; Mohraz, M.; Gharacheh, M. Pregnancy through the Lens of Iranian Women with H.I.V. *J. Int. Assoc. Provid. AIDS Care* **2016**, *15*, 148–152. [CrossRef]
21. Tong, A.; Brown, M.A.; Winkelmayer, W.C.; Craig, J.C.; Jesudason, S. Perspectives on pregnancy in women with CKD: A semistructured interview study. *Am. J. Kidney Dis.* **2015**, *66*, 951–961. [CrossRef]
22. Sabzevari, M.; Eftekhari Yazdi, M.; Rad, M. Lived experiences of women with maternal near miss: A qualitative research. *J. Matern. Neonatal Med.* **2021**, *35*, 7158–7165. [CrossRef]
23. Sanders, L.B. Women's Voices: The Lived Experience of Pregnancy and Motherhood After Diagnosis With HIV. *J. Assoc. Nurses AIDS Care* **2008**, *19*, 47–57. [CrossRef] [PubMed]
24. Ross, R.; Sawatphanit, W.; Draucker, C.B.; Suwansujarid, T. The lived experiences of HIV-positive, pregnant women in Thailand. *Health Care Women Int.* **2007**, *28*, 731–744. [CrossRef] [PubMed]
25. Fords, G.; Crowley, T.; Merwe, M. The Lived Experiences of Rural Women Diagnosed with the Human Immunodeficiency Virus in the Antenatal Period. 2017. Available online: https://www.ncbi.nlm.nih.gov/pmc/articles/PMC5639609/pdf/rsah-14-13794 30.pdf (accessed on 25 May 2022).
26. Norhayati, M.N.; Hazlina, N.H.N.; Asrenee, A.R.; Sulaiman, Z. The experiences of women with maternal near miss and their perception of quality of care in Kelantan, Malaysia: A qualitative study. *BMC Pregnancy Childbirth* **2017**, *17*, 189. [CrossRef]
27. Arcoverde, M.A.M.; Conter, R.S.; da Silva, R.M.M.; dos Santos, M.F. Feelings and Expectations of Pregnant Women Living With Hiv: A Phenomenological Study. *REME Rev. Min. Enferm.* **2015**, *19*, 561–566. [CrossRef]
28. Lalor, J.; Begley, C.M.; Galavan, E. Recasting Hope: A process of adaptation following fetal anomaly diagnosis. *Soc. Sci. Med.* **2009**, *68*, 462–472. [CrossRef]
29. Irani, M.; Khadivzadeh, T.; Nekah, S.M.A.; Ebrahimipour, H.; Tara, F. Emotional and cognitive experiences of pregnant women following prenatal diagnosis of fetal anomalies: A qualitative study in Iran. *Int. J. Community Based Nurs. Midwifery* **2019**, *7*, 22–31. [PubMed]
30. Tayeh, G.; Jouannic, J.M.; Mansour, F.; Kesrouani, A.; Attieh, E. Complexity of consenting for medical termination of pregnancy: Prospective and longitudinal study in Paris. *BMC Med. Ethics* **2018**, *19*, 33.
31. Crombag, N.; Sacco, A.; Stocks, B.; De Vloo, P.; Van Der Merwe, J.; Gallagher, K.; David, A.; Marlow, N.; Deprest, J. 'We did everything we could'—A qualitative study exploring the acceptability of maternal-fetal surgery for spina bifida to parents. *Prenat. Diagn.* **2021**, *41*, 910–921. [CrossRef] [PubMed]
32. Hedrick, J. The lived experience of pregnancy while carrying a child with a known, nonlethal congenital abnormality. *JOGNN J. Obstet. Gynecol. Neonatal Nurs.* **2005**, *34*, 732–740. [CrossRef]

Article

Development of Thai Sensory Patterns Assessment Tool for Children Aged 3–12 Years: Caregiver-Version

Revadee Sutthachai [1], Anuchart Kaunnil [1,*], Supaluck Phadsri [1], Ilada Pomngen [1], Mandy Stanley [2] and Tiam Srikhamjak [1,*]

[1] Department of Occupational Therapy, Faculty of Associated Medical Sciences, Chiang Mai University, Chiang Mai 50200, Thailand
[2] School of Medical and Health Sciences, Edith Cowan University, Joondalup, WA 6027, Australia
*   Correspondence: anuchart.kau@cmu.ac.th (A.K.); tiam.srikhamjak@cmu.ac.th (T.S.)

Abstract: Most existing tools for measuring sensory patterns of children have been developed in Western countries. These tools are complex and may not be culturally appropriate for other contexts that require specific knowledge in the clinical perspective. The aim of this study was to develop a simplified tool called the Thai Sensory Patterns Assessment (TSPA) tool for children. It is designed for children ages 3–12 years old to be completed by their caregiver. The process of creating the tool consisted of drafting a questionnaire and interpreting the result. Partial psychometrics were completed during item development, content validity of items was assessed by five expert ratings. Construct validity and internal consistency were assessed using data from 414 caregivers and intra-rater reliability was assessed with 40 caregivers. The two parts of the TSPA tool for children results, sensory preference, and sensory arousal, were designed to be presented as a sensory pattern in a radar chart/plot. The data analysis showed that both parts of the TSPA tool for children had acceptable psychometric properties with the retained 65 items. Only proprioceptive sensory arousal had a low Cronbach's α coefficient, suggesting more information sharing between caregivers and professionals is needed. This research is an initial study and must be continuously developed. Future development of this tool in technology platforms is recommended to support use within healthcare services.

Keywords: sensory processing patterns; sensory preference; sensory arousal; sensory assessment; validity; reliability; health promotion assessment; children

Citation: Sutthachai, R.; Kaunnil, A.; Phadsri, S.; Pomngen, I.; Stanley, M.; Srikhamjak, T. Development of Thai Sensory Patterns Assessment Tool for Children Aged 3–12 Years: Caregiver-Version. *Healthcare* **2022**, *10*, 1968. https://doi.org/ 10.3390/healthcare10101968

Academic Editors: Carlos Laranjeira and Ana Querido

Received: 22 August 2022
Accepted: 7 October 2022
Published: 8 October 2022

Publisher's Note: MDPI stays neutral with regard to jurisdictional claims in published maps and institutional affiliations.

1. Introduction

Sensory processing refers to the reception, modulation, integration, and organization of sensory stimuli as well as behavioral response to sensory input [1,2]. Sensory processing is the building blocks of perception, emotion, behavior, and development. Research studies in sensory processing investigate that individuals have different patterns of processing that can be observed behaviorally and neurophysiological [3,4]. Differential sensory patterns can be associated with human characteristics. Several previous studies indicated that sensory processing patterns may influence human behavior at all ages. The study in adulthood found that high and low external stimulation can be distinguished are related to higher fatigue frequencies in adulthood [5]. In addition, the study found that a calm atmosphere important strategy to control agitation, as well as sensory defensiveness in adults, which may be a tendency towards increased symptoms of anxiety and depression [6,7]. It is also studied in adolescents to demonstrate that adolescents' sensory avoidance may be related to pain experience, pain catastrophizing, and disability level [8]. Especially, children are one of the most important periods of human development, and sensory stimuli are most effective for them. Not only does sensory processing affect daily routine activities, social, cognitive, and sensorimotor development in children but also health and illness. For example, the study found that sensory processing patterns relate to conduct problems and inattentive and hyperactive behavior. It is also sensory processing patterns factors were significantly associated with the children's sleep patterns [9,10]. Furthermore, the study found that sensory processing patterns and children's traumatic experiences may specifically characterize individuals with affective disorders and prediction of their quality of life [11].

Currently, many sensory processing pattern assessments are used for the implementation of healthcare services. However, the most of previous existing tools are designed for healthcare professionals or therapists such as occupational therapists [12–14]. Furthermore, it required more specific knowledge in the clinical context or specific knowledge. Sensory processing issues are well known in the medical context but rarely in general people include caregivers of children. In Thailand, Tiam Srikhamjak and colleagues developed the Thai Sensory Patterns Assessment (TSPA) tool for assessing sensory processing patterns [15–18]. The tool was developed by modifying Dunn's sensory profile toward Thai adolescents and adults and developed continuously for more than 15 years. There are two parts of the TSPA tool for adolescents and adults, sensory preferences, and sensory arousal. Sensory preference is defined as the behavioral expression preferred in a particular sensory stimulus in daily life. Sensory arousal is defined as a behavioral expression that responds impulsively to a specific sensory stimulus in daily life. It is interesting that the TSPA tool is simple to use and can be used to link behavior responses to sensory stimuli in a variety of ways. The results are presented in picture form with each sensory domain specified and interpreted in simple layman's terms. Further, Thai research indicates that the TSPA tool for adolescents and adults can provide a feasible tool for identifying sensory preferences to match health promotion modalities appropriately. The previous study classified the level of the participants' sensory preferences concerning their cortisol levels by using Mindfulness-Based Flow Practice (MBFP) [19]. It is also a TSPA tool that is used to evaluate the process for planning healthcare services in clinical and community settings. As this tool is easy to use, it has been put to a variety of uses.

In Thailand, healthcare providers and health volunteers work with family caregivers. To enhance caregivers understanding of their children, our team is developing a simple tool. It enables family caregivers to improve the health and well-being of their children based on the results of an assessment tool. In Thailand, an assessment of sensory patterns was carried out, but the tool was intended for adults and adolescent. For children (caregiver version), there is no simple tool for sensory patterns assessment. To fill this gap, we developed the TSPA tool for children and use it by caregivers or healthcare professionals. In this study, it is an initial development and partial psychometric properties. For content and design, the resulting interpretation was further developed into platform applications for easier access and widely used in the future.

2. Materials and Methods

The developmental research was aimed to develop the TSPA tool for children and psychometric properties of content validity, construct validity, internal consistency, and intra–rater reliability. This research received ethical approval, from the Associated Medical Sciences Research Ethical Committees, Chiang Mai University, code AMSEC-64EX-037.

2.1. Procedure

2.1.1. Phase 1: Drafting a Questionnaire and Designing Result Interpreting for the TSPA Tool for Children

The purpose of this phase was to develop a simple questionnaire and create item to measure the behavioral response of children to sensory stimuli in everyday life and receive a report from the caregiver. It was based on the literature review about theoretical sensation, assessment of sensory processing patterns and the criteria of psychometric properties.

Previous study from the literature review found that the concept of existing tools had the purpose of measuring both atypical and typical people. For example, the tools by Jean Ayres [20,21] and Milleret et al. [22] aimed to measure sensory processing disorder, while those of Dunn and Brown [23,24] aimed to identify sensory processing patterns of typical persons. However, interpretation of sensory processing was limited to four patterns overall with these tools and they did not categorize each sensory domain. On the other hand, the TSPA tool for adolescents and adults developed more than 15 years by Srikhamjak et al. [15–18] aimed to measure typical persons by using simple sensory processing to interpret a variety of dimensions and categorize each sensory stimulus. aimed to measure typical persons by using simple sensory processing to interpret a variety of dimensions and categorize each sensory stimulus.

The TSPA tool for adolescents and adults has been continuously developed for simplicity in use since the original version developed in 2007 by Srikhamjak et al. In 2020, Pomngen et al. [16,17] reviewed the concept of the TSPA tool for adolescents and adults, finding that it is simple to use to cooperate with health providers and clients. In order to develop a new TSPA for children, the concept of the adult and adolescent versions of TSPA was adapted. Therefore, the first draft of a TSPA tool for children is divided into two parts (sensory preference and sensory arousal), each containing six sensory modalities. By requesting caregiver reports on the behavior of children in everyday life, we created items that are appropriate for measuring the children's response to sensory stimuli. Rating frequency criteria and scoring items are all included. In addition, we designed a simple and easy-to-use output interpretation for the TSPA tool for children that shows sensory patterns.

2.1.2. Phase 2: Examining Content Validity

We choose a content validity to examine was established using the index of item-objective congruence (IOC). This statistical procedure, developed by Rovinelli and Hambleton, 1997 [25], is best used in test development to assess content validity at the item development stage [26]. The content validity was performed by experts rating individual items and processes following guidelines and previous studies [27–30].

Determining qualification of five experts for content valid examine as follows (1) who specialize in sensory processing theory (2) who have experience using assessment instruments or measurements (3) experience in the field of research (4) have experience in the clinic of sensory processing in children at least five years.

We contact to expert in the field above mentioned and send the first draft of TSPA tool for children. In the IOC process, experts rated individual items based on whether they agree or disagree with the specific objectives. Individual experts rated using a 3-point scale as follow; 1 (a definite indicator of sensory preference and sensory arousal in each sensory modality), 0 (undecided), and −1 (not an indicator of sensory preference and sensory arousal in each sensory modality). In addition, giving suggestions for revised items.

IOC is calculated after expert ratings of individual items are completed. For content validity, IOC analysis took IOC values up to 0.80 to be considered acceptable [26] and removed if the item was the IOC value < 0.8. However, items with acceptable IOC values were revised if clarification was suggested by the expert.

After revising the TSPA tool for children of items that acceptable level was pilot testing in similarly caregivers. The pilot testing for checking is understandable and could be used by interviews and listening to feedback form caregivers between pilot tests. Therefore, the second draft of TSPA tool for children were used to collecting data for reliability and construct validity examine.

2.1.3. Phase 3: Examining of Reliability and Construct Validity
Internal Consistency and Construct Validity

Sample size: Based on the location and population, there were 44,192 caregivers. The sample size was estimated by the calculation formula of Yamane and determined reliability to 95% or to have an error of 0.05. Definition: n = number of sample size, N = number of populations, e = error by followed guidelines [31,32].

$$n = \frac{N}{1 + Ne^2}$$

A sample size of 396 was calculated. In factor analysis, a minimum sample size of at least 300 is recommended by Comrey and Lee [33]. Children aged 3–12 years are randomly sampled from schools in four districts in Chiang Mai and their caregivers at every class level to estimate the proportion of access samples.

Participant data were collected through multistage random sampling. In the first step, we randomly selected 30 schools from four districts in Chiang Mai, Thailand, and sent letters to the principals of those schools. We collected data from 16 schools and advertised our research in 16 schools. To participate in this study, parents or caregivers who have normal children aged 3–12 were invited. A previous study with access to participants and data collection [34] is the basis of this research.

Theses caregivers received the invitation letter includes information sheet explanation of the purpose and process of study and an information sheet. The purpose of this study was to collect data during the COVID-19 period and the social distancing rules that were in place at that time. The participant has the option of choosing response items on-site or online. Each caregiver filled out a consent form, provided demographic information, and completed the second draft of TSPA for children.

After caregivers have completed the TSPA tool for children in this phase, we collected the information in the first round of the internal consistency study. We selected the information according to the inclusion criteria for the data analysis. Construct validity was examined for the retention of items that were appropriate for factor analysis and removed items that did not meet criteria. The next step was to finalize the TSPA tool for children and to analyze the data by internal consistency.

Intra-Rater Reliability

After the data collection of the first round, we invite the caregiver to complete the TSPA tool for children twice (at 2 weeks intervals after the first round) to assess intra-rater reliability. Caregivers who participated were invited to enroll and the appointment date in the data collection. The caregivers completed the TSPA tool for children in the second round. Consequently, intra-rater reliability was analyzed based on first and second data from caregivers.

2.2. Data Analysis

All data analyses were performed using the IBM.SPSS. Statistics for Windows, Version 26.0.

Content validity was examined by experts who rated each item using a 3-point scale: 1 (a definite indicator of sensory preference and sensory arousal in each modality), 0 (undecided), and −1 (not an indicator of sensory preference and sensory arousal in each modality). The IOC value of ≥ 0.80 was considered an acceptable level and represented high content validity [27–30].

Construct validity was examined by factor analysis that indicated whether it was appropriate for each of the senses as individual subscales in the analysis. Factor structures of sensory preferences and sensory arousal in each sense were examined by principal components analysis with varimax rotation. The Kaiser criterion (Eigenvalues > 1) and proportion of total variance explained the criteria implemented for the number of factors to be extracted. Items that had poor factor loadings (<0.40) or were cross-loaded on two or more factors were removed [35–37].

Internal consistency reliability was established by using Cronbach's alpha coefficient, with criteria using the standard detailed by Arikanto, 1992: < 0.4 = poor, 0.41–0.70 = acceptable, 0.71–1.00 = excellent [38].

Inter-rater reliability was examined using the intraclass correlation coefficient (ICC), with interpretation using standards detailed by Cicchetti, 1999: < 0.40 poor, 0.40–0.59 acceptable, 0.60–0.74 good, 0.75–1.0, excellent [39].

3. Results

3.1. Phase 1: Drafting the TSPA for the Children Questionnaire

The first drafted TSPA tool for children was a caregiver observation-report questionnaire designed to evaluate the sensory processing patterns of normal children and divide into two parts: part 1—sensory preferences and part 2—sensory arousal. Each part comprised six sensory modalities in sight, sound, smell and taste, touch, vestibular, and proprioceptive. The first draft of the TSPA for children consisted of 99 items (part 1: 51 items and part 2: 48 items).

This tool determined caregivers to give information by choosing answers following the frequency (never, rarely, sometimes, often, always) of children's behavior response to sensory stimuli. The tool measured the frequency of response to sensory stimuli using the Likert scale from 1–5 and determined scoring for each item (1 = never, 2 = rarely, 3 = sometimes, 4 = often, 5 = always) in the sensory preference part. While sensory arousal part (1 = never, 2 = rarely, 3 = sometimes, 4 = often, 5 = always) in items with high arousal and (5 = never, 4 = rarely, 3 = sometimes, 2 = often, 1 = always) in items with low arousal.

For designed interpret results by determining score each item and then calculating percentages and present them for simple to use as a radar chart.

The radar chart presents sensory patterns in variety and individual. The radar chart/plot can show characteristics and properties that can present both lines of sensory preferences and arousal. A radar chart can show together a line/plot which is easy to use, and maybe a line shown in preference and arousal similar level or different level. Moreover, this pattern displays the integration of sensory preferences and arousal of each sense. For instance, in the Figure above, when the radar chart/plot showed low arousal and high preference that presents the tendency to the response of the seeking person. While however, if it displays high arousal and low preference that presents the tendency to the response of the sensitive person.

3.2. Phase 2: Examining Content Validity

The content validity of the tool was examined by five experts (a specialized pediatric occupational therapist, pediatric occupational therapy lecturer, pediatrician, family physician, and club president caregiver of autistic people). The index of the IOC was 0.8–1.00 The TSPA second draft had 93 accepted items consisting of part 1: 46 sensory preference items and part 2: 47 sensory arousal items. It also had 33 piloted caregivers who found the language clear and were able to collect data.

3.3. Phase 3: Reliability and Construct Validity

In following ethical research approval, the participants comprised 444 caregivers of normal children aged 3–12 years and who were interested in obtaining information. However, 30 caregivers did not meet the criteria and were excluded from the analysis because of caregiver incomplete response items and caregivers who have atypical children. Therefore, a total of 414 sample data of normal children were inclusion, consisted of age range, with 132 (31.9%) being 3–6 years old, and 282 (68.1%) 7–12 years. In this case, 196 (47.3%) were male and 218 (52.7%) females. Caregivers who complete items comprising 368 (88.9%) parents, 20 relatives (4.8 %), 2 nannies (0.5%) and 24 others (5.8 %). Data collected from analysis are shown as follows.

3.3.1. Internal Consistency

Cronbach's alpha coefficient was used to examine the internal consistency of the TSPA tool for children second draft, and the results are shown in Table 1. Cronbach's α coefficient showed that the TSPA was 0.92 (excellent) for overall items in part 1: sensory preference, and 0.80 (excellent) in part 2: sensory arousal.

Table 1. Internal consistency estimates of the TSPA before factor analysis ($n = 414$).

Part 1: Sensory Preference	No.of Items	α.			Part 2: Sensory Arousal	No.of Items	α.		
		3–6 Years	7–12 Years	Overall			3–6 Years	7–12 Years	Overall
Sight	7	0.81	0.77	0.79	Sight	7	0.69	0.57	0.61
Sound	8	0.73	0.77	0.78	Sound	7	0.61	0.55	0.57
Smell and Taste	8	0.76	0.74	0.74	Smell and Taste	8	0.74	0.71	0.72
Touch	8	0.69	0.65	0.68	Touch	10	0.77	0.72	0.73
Vestibular	7	0.83	0.78	0.81	Vestibular	7	0.52	0.53	0.53
Proprioceptive	8	0.73	0.73	0.73	Proprioceptive	8	0.22	0.20	0.20
Total	46	0.91	0.92	0.92	Total	47	0.81	0.79	0.80

3.3.2. Construct Validity

Factor analysis using the principal component was carried out for each subscale. The Kaiser-Mayer-Olkin (KMO) measure of sampling adequacy and Bartlett's Test of Sphericity were examined, and items were removed if their sampling adequacy was below 0.5. Factor extraction was based on examination of the Kaiser criterion of eigenvalues over 1.00. Additional consideration was given to the theoretical interpretation of the factor. Factors were not retained if they had fewer than three items. Items were deleted if they had factor loadings of less than 0.4, or were loaded on a factor that was not interpretable.

Part 1: Sensory preferences consisted of six modalities (sight, sound, smell and taste, touch, vestibular, and proprioceptive). By following the data analysis shown in Table 2, each modality was found to have a KMO range of 0.717–0.855, chi-square of 438.611–853.440 and significance < 0.001, which was appropriate for factor analysis. Sight comprised 6 items (factor loading rang of 0.666–0.804), with only one item removed. Sound comprised 6 items (factor loading rang of 0.527–0.758), with two removed. Smell and taste comprised 6 items, with two removed. Touch comprised 5 items (factor loading rang of 0.533–0.710), with two removed. Vestibular comprised 7 items (factor loading rang of 0.600–0.732), with none removed from the second daft. Proprioception comprised 5 items (factor loading rang of 0.604–0.719), with two removed. Therefore, part 1: sensory preference contained 35 items, and Table 2 presents the factor loading of items were retained in each sense.

Table 2. Factor loading of part 1: sensory preference.

Part 1: Sensory Preference							
Subscale	Items	Factor	KMO	Bartlett's Test		% of Variance	Eigenvalues
				Chi-Square	Sig.		
Sight	SPsight5	0.804	0.855	832.626	<0.001.	47.02	3.291
	SPsight2	0.803					
	SPsight1	0.732					
	SPsight4	0.728					
	SPsight6	0.697					
	SPsight7	0.666					
Sound	SPsound2	0.758	0.786	849.329	<0.001	40.22	3.218
	SPsound4	0.748					
	SPsound6	0.675					
	SPsound1	0.661					
	SPsound3	0.565					
	SPsound8	0.527					

Table 2. *Cont.*

Part 1: Sensory Preference							
Subscale	**Items**	**Factor**	**KMO**	**Bartlett's Test**		**% of Variance**	**Eigenvalues**
				Chi-Square	**Sig.**		
Smell and Taste	SPsmell&taste8	0.727	0.726	736.198	<0.001	36.403	2.912
	SPsmell&taste6	0.710					
	SPsmell&taste7	0.669					
	SPsmell&taste2	0.620					
	SPsmell&taste1	0.589					
	SPsmell&taste5	0.529					
Touch	SPtouch8	0.710	0.751	438.611	<0.001	31.928	2.554
	SPtouch2	0.669					
	SPtouch5	0.566					
	SPtouch6	0.520					
	SPtouch3	0.505					
Vestibular	SPvestibular6	0.732	0.810	853.440	<0.001	47.470	3.323
	SPvestibular2	0.724					
	SPvestibular3	0.716					
	SPvestibular4	0.695					
	SPvestibular5	0.689					
	SPvestibular7	0.658					
	SPvestibular1	0.600					
Proprioceptive	SPproprio1	0.719	0.717	700.833	<0.001	35.444	2.836
	SPproprio5	0.660					
	SPproprio3	0.636					
	SPproprio2	0.631					
	SPproprio6	0.604					

Part 2: Sensory arousal consisted of six modalities (sight, sound, smell and taste, touch, vestibular, and proprioception). By following the data analysis shown in Table 3, each modality yielded a KMO each modality yielded a KMO range of 0.660–0.822, chi-square of 313.129–682.784 and significance <0.001, which was appropriate for factor analysis. Sight comprised 5 items (factor loading rang of 0.569–0.679), with two items removed. Sound comprised 5 items (factor loading rang of 0.619–0.767), with two removed. Smell and taste comprised 6 items (factor loading rang of 0.503–0.741), with two removed. Touch comprised 6 items (factor loading rang of 0.449–0.732), with four removed. Vestibular comprised 4 items (factor loading rang of 0.569–0.791), with three removed. Proprioception comprised 4 items (factor loading rang of 0.433–0.692), with three removed. Therefore, part 2: sensory arousal contained 30 items, and Table 3 presents the factor loading of items were retained in each sense. By following the factor analysis, the TSPA questionnaire were contained total 65 items, comprise of 35 items of sensory preferences part and 30 items of sensory arousal part. Next, internal consistency for final and intra-rater reliability was examined.

Table 3. Factor loading of sensory arousal in each modality.

Part 2: Sensory Arousal							
Subscale	**Items**	**Factor**	**KMO**	**Bartlett's Test**		**% of Variance**	**Eigenvalues**
				Chi-Square	**Sig.**		
Sight	SAsight2	0.679	0.660	374.184	<0.001	31.308	2.192
	SAsight6	0.675					
	SAsight5	0.631					
	SAsight1	0.630					
	SAsight7	0.569					

Table 3. *Cont.*

				Part 2: Sensory Arousal			
				Bartlett's Test		**% of Variance**	**Eigenvalues**
Subscale	**Items**	**Factor**	**KMO**	**Chi-Square**	**Sig.**		
Sound	SAsound5	0.767	0.758	435.991	<0.001	35.502	2.485
	SAsound7	0.705					
	SAsound3	0.699					
	SAsound6	0.696					
	SAsound4	0.619					
Smell and Taste	SAsmell&taste6	0.741	0.770	682.784	<0.001	35.355	2.828
	SAsmell&taste4	0.695					
	SAsmell&taste7	0.678					
	SAsmell&taste3	0.677					
	SAsmell&taste5	0.644					
	SAsmell&taste8	0.503					
Touch	SAtouch6	0.732	0.822	659.627	<0.001	31.173	3.117
	SAtouch5	0.692					
	SAtouch8	0.672					
	SAtouch4	0.653					
	SAtouch7	0.537					
	SAtouch9	0.449					
Vestibular	SAves4	0.791	0.738	405.672	<0.001	34.115	2.388
	SAves5	0.757					
	SAvestibular1	0.666					
	SAvestibular7	0.569					
Proprioceptive	SAproprio2	0.692	0.677	313.129	<0.001	27.060	2.165
	SAproprio1	0.611					
	SAproprio5	0.520					
	SAproprio7	0.433					

3.3.3. Internal Consistency of Items in the Final Factor Analysis

After the items were removed from the factor analysis in the TSPA questionnaire, the consistency of response to those of part 1 and 2 were examined again. The results showed Cronbach's α. coefficient of 0.92 (excellent) and 0.81 (excellent) in part 1 and 2, respectively. Internal consistency of Cronbach's α. coefficient is shown in Table 4.

Table 4. Internal consistency after items were removed from the factor analysis.

Part 1: Sensory Preference	No. of Item	α.			Part 2: Sensory Arousal	No. of Item	α.		
		3–6 Years	7–12 Years	Overall			3–6 Years	7–12 Years	Overall
Sight	6	0.81	0.82	0.83	Sight	5	0.71	0.62	0.65
Sound	6	0.72	0.75	0.76	Sound	5	0.76	0.72	0.74
Smell and Taste	6	0.77	0.73	0.74	Smell and Taste	6	0.76	0.73	0.74
Touch	5	0.61	0.59	0.62	Touch	6	0.76	0.70	0.72
Vestibular	7	0.78	0.77	0.81	Vestibular	4	0.68	0.70	0.69
Proprioceptive	5	0.70	0.70	0.64	Proprioceptive	4	0.42	0.50	0.47
Total	35	0.90	0.91	0.92	Total	30	0.83	0.80	0.81

3.3.4. Intra-Rater Reliability

Intra-rater reliability was examined in the data of 40 normal children aged 3–12 years. The caregivers of these 40 children completed the TSPA tool for children twice to examine consistency between the ratings provided by the same rater. The intraclass correlation coefficient for part 1: sensory preference with 35 items was 0.74 (good) and part 2: sensory preference with 30 items was 0.79 (excellent). Intra-rater reliability was examined in the data of 40 children aged 3–12 years, as shown in Table 5.

Table 5. Intra-rater reliability and intraclass correlation coefficient of the TSPA.

Part 1: Sensory Preference	No. of Items	ICC	Part 2: Sensory Arousal	No. of Items	ICC
Sight	6	0.79	Sight	5	0.75
Sound	6	0.64	Sound	5	0.77
Smell and Taste	6	0.55	Smell and Taste	6	0.63
Touch	5	0.65	Touch	6	0.56
Vestibular	7	0.66	Vestibular	4	0.78
Proprioceptive	5	0.67	Proprioceptive	4	0.64
Total	35	0.74	Total	30	0.79

4. Discussion

This study aimed to develop and examine psychometric properties of the TSPA tool: caregiver-version for children aged 3–12 years. The results showed that both parts of the tool had acceptable content validity, construct validity, internal consistency, intra-rater reliability. Only proprioceptive sense in part 2: sensory arousal had a lowest score for Cronbach's α coefficient (Table 4), as found in previous studies [15–18], in which proprioceptive sense in sensory arousal part had the lower than others. This might be due to the fact that this sensory modality differed from other senses. Firstly, there are two types of proprioceptive: conscious and non-conscious. In non-conscious proprioceptive, impulses that arise from this type of sensation are delayed to the cerebellum rather than to the cerebral cortex [40,41]. This naturally causes proprioceptive sense to be rarely noticed. Secondary, in the clinical aspect, the behavior of children who actively seek proprioceptive sense such as hitting, pushing and rough play are often associated with displays of aggression [42]. This might lead to the perspective of caregivers being subjective and bias. These conditions and previous study demonstrate no single measure of proprioceptive due to the complexity [43]. Therefore, to decrease bias and more strong information may considered proprioceptive sense integrate with others assessment. It is important that the caregiver understands the purpose of this tool and how the results are used. It should be explained that sensory patterns are reflections of who we are, and not a pathology that needs to be remedied. Once understood, sensory patterns of the child open the door to an enriched life, which leads to the authors' suggestion of cooperation between caregiver and pediatric professions in sharing more data.

However, although further verification of the TSPA tool for children is needed, this new tool showed much potential and differing points from previous tools. Firstly, the radar chart was used to present interpretation of the sensory processing pattern that provides a brief picture for easily understanding the individual. A line that intersects the web can easily be perceived and interpreted as the tendency between high and low scores. Secondly, by integrating the level of preferences and arousal in each sense, information for deeply understanding the behavior of children can be achieved in a variety of dimensions. For example, Figure 1. shows a radar chart of sensory processing patterns, which reports a particular preference in sight and sound. While result integrating preferences and arousal in each sense which report sight has high preference and low arousal than the others. This is similar to Dunn's sensory seeking pattern, but that sought only sight sense and not senses overall, which is different to the TSPA and previous tools. The TSPA tool can interpret a variety of sensory processing patterns, while Dunn's pattern is limited to four. Once family caregivers and pediatric professionals including healthcare professionals have insight into the variety of sensory patterns, the child is offered a way to prefer what it needs to do in growing up in its own way.

This TSPA tool for children enhances caregiver's understanding of their children. Firstly, we knew the level of sensory preference and sensory arousal to describe the behavior response and meaningful activity of their children. Which result interpretation by radar chart help caregiver understand how much preference and arousal each sensory modality of individual children. Information on sensory patterns can help the caregiver's understanding of sensory needs with impact on children's satisfying experiences led to the quality of life and relationship in the family. Importantly, understand the identity of each child and their unique set of talents. It is important to recognize acceptance in children and not force them, but rather support them in choosing their way. In-home and community contexts, caregivers can use results to promote children's health. To build intrinsic motivation to participate in daily routines in self-care, eating, and sleeping, we need to pay attention to sensory preferences. When children demonstrate healthy behavior, they can be rewarded based on sensory preference. Sensory arousal is used in sensory-based environments to create safe and comfortable areas for children. There is a relationship between high sensory arousal and stress, blood pressure, and mental health [44]. Similarly, sleep problems are frequently associated with children who are easily over-aroused and are tactile sensitive. There is a link between sensory hypersensitivity and lower sleeping quality among children [45–47].

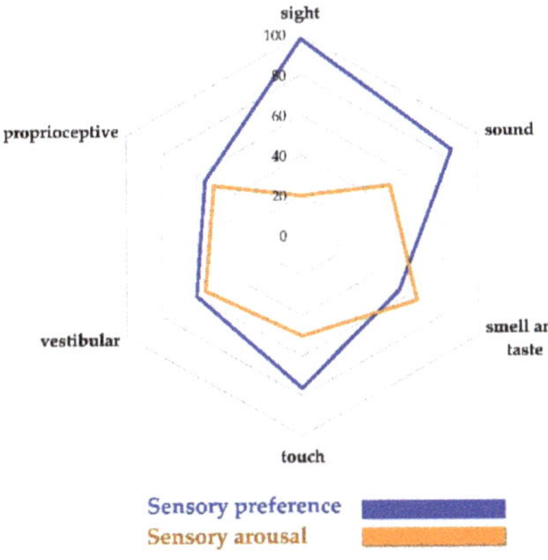

Figure 1. Radar chart of TSPA interpretation.

By integrating preferences and arousal into the radar chart, healthcare professionals can deeply understand the interpretation of the results. Radar charts revealed many patterns, including sensory preferences outside sensory arousal, sensory arousal outside preference, and sensory preferences balanced with arousal. It may be helpful for healthcare professionals to obtain an understanding of the sensory needs of individual children and the services that are provided by the client's center. To increase the effectiveness of health programs, the sensory preferences of children should be taken into account. It is possible to obstruct effective sensory arousal while providing an appropriate amount of stimulation. Furthermore, the result can be applied to describe children not only from a neuroscience perspective but from an educational and social perspective as well. According to a previous study [48], sensory processing patterns are strongly related to persons with introversion and extroversion. In this tool, caregivers would respond to items, and the results were calculated as a percentage and presented as a radar chart. Nevertheless, in this paper-based study, only items from the TSPA tool for children were examined. It is the first development of a TSPA tool for children that is developed on the platform application for the next steps. Future applications can evaluate both caregivers and healthcare professionals by the program can automatically result and strategies ready to use.

Finally, it will be feasible to use the TSPA tool for children on the telehealth in the future, as the process of collecting data continued during the COVID-19 pandemic. The hybrid approach was used for collecting data, both onsite and over distances via online. From the hybrid approach, the TSPA tool was found to be available online, and its benefits are convenience for the caregiver, saving time and costs, and safety in a crisis. Similar study previously predicted that the hybrid approach will become the norm in the future, with telehealth being used to support families from a distance [49]. Implication for future may be developing this tool on a technological platform, from which telehealth supports healthcare services for children in the future.

However, this is first to studies TSPA tool for children in Thai context and studies some psychometric property. For more strong potential of this tools and benefit for widely others children is need to continuously developed. Thus, further research should be study criteria cut score, convergent and discriminant validity and if possible, may be examine for cross cultural children in others context.

5. Conclusions

In summary, the development and examination of the TSPA tool for children were satisfactory. This assessment has been content validity and construct validity and internal consistency and intra-rater reliability for measuring sensory processing patterns of normal children aged 3–12 years. To pediatric profession and family caregivers can cooperate use this tool to understand and promote for their children. It is recommended that users should consider this point (proprioceptive sense) when using TSPA for children, as well as other information from the child's medical history. In future research, the proprioceptive of sensory arousal could be revised and increased examined, along with predictive validity, discriminative validity, and cut score criteria. Implication, to appropriate with world changing and increase healthcare services accessed the researcher may be to develop this TSPA on technology platform in the future.

Author Contributions: Conceptualization, R.S. and T.S.; methodology, R.S., A.K. and T.S.; software, R.S., I.P. and T.S.; validation, R.S., A.K., I.P., S.P. and T.S.; formal analysis, R.S., A.K. and T.S.; investigation, R.S., A.K., I.P., S.P. and T.S.; resources, R.S., I.P. and T.S.; data curation, R.S., A.K., I.P. and T.S.; writing—original draft preparation, R.S., A.K., I.P., S.P. and T.S.; writing—review and editing, R.S., A.K., I.P., S.P., M.S. and T.S.; visualization, R.S., A.K. and T.S.; supervision, A.K. and T.S.; project administration, R.S., A.K. and T.S.; funding acquisition, T.S. All authors have read and agreed to the published version of the manuscript.

Funding: This research was supported by Faculty of Associated Medical Sciences, Chiang Mai University [AMS-2021].

Institutional Review Board Statement: The research was ethical approval, from the Associated Medical Sciences Research Ethical Committee, Chiang Mai University, code AMSEC-64EX-037.

Informed Consent Statement: Inform consent was obtained from all participants involved in the study.

Data Availability Statement: The data in this study are available on request from the corresponding author.

Acknowledgments: We want to thank all caregivers of children provide information for this study, as well as Chiang Mai kindergarten and primary school for support our data collection.

Conflicts of Interest: The authors declare no conflict of interest.

References

1. Dunn, W. The impact of sensory processing abilities on the daily lives of young children and their families: A conceptual model. *Infants Young Child.* **1997**, *9*, 23–35. [CrossRef]
2. Kandel, E.R.; Schwartz, J.H.; Jessell, T.M. *Principles of Neural Science*, 4th ed.; McGrow-Hill: New York, NY, USA, 2000.
3. Brown, C.; Tollefson, N.; Dunn, W.; Cromwell, R.; Filion, D. The Adult Sensory Profile: Measuring patterns of sensory processing. *Am. J. Occup. Ther.* **2001**, *55*, 75–82. [CrossRef] [PubMed]
4. Miller, E.; Miller-Kuhaneck, H. The relationship among sensory preferences, play preferences, motivation, and mastery in guiding children's play: A review of the literature, part 2. *Am. Occup. Ther. Assoc. Sens. Integr. Spec. Interest Sect. Q.* **2006**, *29*, 1–4.
5. Rijk, A.E.; Schreurs, K.M.; Bensing, J.M. Complaints of fatigue: Related to too much as well as too little external stimulation. *J. Behav. Med.* **1999**, *22*, 549–573. [CrossRef]
6. Ragneskog, H.; Kihlgren, M. Music and other strategies to improve the care of agitated patients with dementia: Interviews with experienced staff. *Scand. J. Caring Sci.* **1997**, *11*, 176–182. [CrossRef]
7. Kinnealey, M.; Fuiek, M. The relationship between sensory defensiveness, anxiety, depression and perception of pain in adults. *Occup. Ther. Int.* **1999**, *6*, 195–206. [CrossRef]

8.   Genizi, J.; Halevy, A.; Schertz, M.; Osman, K.; Assaf, N.; Segal, I.; Srugo, I.; Kessel, A.; Engel-Yeger, B. Sensory processing patterns affect headache severity among adolescents with migraine. *J. Headache Pain.* **2020**, *21*, 48. [CrossRef]

9.   Nesayan, A.; Gandomani, R.A.; Movallali, G.; Dunn, W. The relationship between sensory processing patterns and behavioral patterns in children. *J. Occup. Ther. Sch. Early Interv.* **2018**, *11*, 124–132. [CrossRef]

10.   Foitzik, K.; Brown, T. Relationship between sensory processing and sleep in typically developing children. *Am. J. Occup. Ther* **2020**, *72*, 1–9. [CrossRef]

11.   Serafini, G.; Gonda, X.; Pompili, M.; Rihmer, Z.; Amore, M.; Engel-Yeger, B. The relationship between sensory processing patterns, alexithymia, traumatic childhood experiences, and quality of life among patients with unipolar and bipolar disorders. *Child Abus. Negl.* **2016**, *62*, 39–50. [CrossRef]

12.   Shahbazi, M.; Mirzakhani, N. Assessment of sensory processing characteristics in children between 0 and 14 years of age: A systematic review. *Iran. J. Child Neurol.* **2021**, *15*, 29–46.

13.   Wan Yunus, F.; Liu, K.P.; Bissett, M.; Penkala, S. Sensory-based intervention for children with behavioral problems: A sys-tematic review. *J. Autism Dev. Disord.* **2015**, *45*, 3565–3579. [CrossRef]

14.   Schoen, S.A.; Miller, L.J.; Brett-Green, B.A.; Nielsen, D.M. Physiological and behavioral differences in sensory processing: A comparison of children with autism spectrum disorder and sensory modulation disorder. *Front. Integr. Neurosci.* **2009**, *3*, 29. [CrossRef]

15.   Srikhamjak, T.; Sawlom, S.; Munkhetvit, P.; Apikimonkon, H. *Thai Sensory Profile Assessment Tool*; Chiang Mai University: Chiang Mai, Thailand, 2007. (In Thai)

16.   Pomngen, I. Development of the Sensory Patterns Assessment. Master's Thesis, Chiang Mai University, Chiang Mai, Thailand, 24 February 2020.

17.   Pomngen, I.; Srikhamjak, T.; Putthinoi, S. Development of Thai's Sensory Patterns Assessment Tool for Adolescents and Adults. *J. Health Sci. Res.* **2020**, *14*, 76–85.

18.   Srikamjak, T.; Saolarm, S.; Munkhetvit, P.; Apikonekorn, H. The sensory profile assessment tool: TSPA. In Proceedings of the Opening Word Optimizing Occupational Therapy, The 5th Asia Pacific Occupational Therapy Congress, Chiang Mai, Thailand, 19–24 November 2011; Volume A0193, p. 183.

19.   Kanchanawong, T.; Prasoetsang, T.; Limvongvatana, F.; Ooraikul, L.; Kanupan, S.; Srikamjak, T.; Gomutbutra, P. The Feasi-bility of the Thai Sensory Profile Assessment Tool (TSPA) for Classifying the participants for Mind-Body intervention. *medRxiv* **2022**, 1–15. [CrossRef]

20.   Ayres, A.J. *Sensory Integration and the Child*; Western Psychological Services: Los Angeles, CA, USA, 1970.

21.   Alterio, C.J. *Clinical Oriented Theory for Occupational Therapy*; Wolters Kluwer: New York, NY, USA, 2019.

22.   Miller, L.J.; Anzalone, M.E.; Lane, S.J.; Cermak, S.A.; Osten, E.T. Concept evolution in sensory integration: A proposed nosology for diagnosis. *Am. J. Occup. Ther.* **2007**, *61*, 135–139. [CrossRef] [PubMed]

23.   Dunn, W. *Sensory Profile: User's Manual*; Psychological Corporation: San Antonio, TX, USA, 1999.

24.   Brown, C.; Dunn, W. *The Adolescent/Adult Sensory Profile: User's Manual*; Psychological Corporation: San Antonio, TX, USA, 2002.

25.   Rovinelli, R.J.; Hambleton, R.K. On the use of content specialists in the assessment of criterion-referenced test item validity. *ERIC* **1997**, *2*, 49–60.

26.   Thorn, D.W.; Deitz, J.C. Examining content validity through the use of content experts. *OTJR Occup. Particip. Health* **1989**, *9*, 334–346. [CrossRef]

27.   Turner, R.C.; Carlson, L. Indexes of item-objective congruence for multidimensional items. *Int. J. Test.* **2003**, *3*, 163–171. [CrossRef]

28.   Turner, R.C.; Mulvenon, S.W.; Thomas, S.P.; Balkin, R.S. Computing indices of item congruence for test development validity assessments. In Proceedings of the 27th Annual SAS Users' Group International Conference, Miami, FL, USA, 14–17 April 2002.

29.   Blanche, E.I.; Parham, D.; Chang, M.; Mallinson, T. Development of an adult sensory processing scale (ASPS). *Am. J. Occup. Ther.* **2014**, *68*, 531–538. [CrossRef]

30.   Ismail, F.K.M.; Zubairi, A.M.B. Item Objective Congruence Analysis for Multidimensional Items: Content Validation of a Reading Test in Sri Lankan University. *Engl. Lang. Teach.* **2022**, *15*, 106–117. [CrossRef]

31.   Yamane, T. *Statistics: An Introductory Analysis*, 2nd ed.; Harper and Row: New York, NY, USA, 1967.

32.   Uakarn, C.; Chaokromthong, K.; Sintao, N. Sample Size Estimation using Yamane and Cochran and Krejcie and Morgan and Green Formulas and Cohen Statistical Power Analysis by G* Power and Comparisions. *APHEIT Int. J.* **2021**, *10*, 76–86.

33.   Comrey, A.L.; Lee, H.B. *A First Course in Factor Analysis*; Psychology Press: London, UK, 2013.

34.   Lai, C.Y.; Yung, T.W.; Gomez, I.N.; Siu, A.M. Psychometric properties of sensory processing and self-regulation checklist (SPSRC). *Occup. Ther. Int.* **2019**, *2019*, 8796042. [CrossRef] [PubMed]

35.   Gorsuch, R.L. *Factor Analysis*, 2nd ed.; Erlbaum: Hillsdale, NJ, USA, 1983.

36.   Shrestha, N. Factor analysis as a tool for survey analysis. *Am. J. Appl. Math. Stat.* **2021**, *9*, 4–11. [CrossRef]

37.   Levine, T.R. Confirmatory factor analysis and scale validation in communication research. *Commun. Res. Rep.* **2005**, *22*, 335–338. [CrossRef]

38.   Arikunto, S. *Dasar-Dasar Evaluasi Pendidikan*, 3rd ed.; Bumi Aksara: Jakarta, Indonesia, 2021.

39.   Cicchetti, D.V.; Sparrow, S.A. Developing criteria for establishing interrater reliability of specific items: Applications to as sessment of adaptive behavior. *Am. J. Ment. Defic.* **1981**, *86*, 127–137.

40. Siegel, A.; Sapru, H.N. *Essential Neuroscience: Holistic Approach*, 3rd ed.; Wolters Kluwer Health/Lippincott Williams & Wilkins: Philadelphia, PA, USA, 2015; pp. 249–253.

41. Proske, U.; Gandevia, S.C. The proprioceptive senses: Their roles in signaling body shape, body position and movement, and muscle force. *Physiol. Rev.* **2012**, *92*, 1651–1697. [CrossRef]

42. Lopez, M.; Swinth, Y. A group proprioceptive program's effect on physical aggression in children. *J. Occup. Ther. Sch. Early Interv.* **2008**, *1*, 147–166. [CrossRef]

43. Hillier, S.; Immink, M.; Thewlis, D. Assessing proprioception: A systematic review of possibilities. *Neurorehabilit. Neural Repair* **2015**, *29*, 933–949. [CrossRef]

44. Elbert, T.; Dworkin, B.R.; Rau, H.; Pauli, P.; Birbaumer, N.; Droste, C.; Brunia, C.H.M. Sensory effects of baroreceptor activation and perceived stress together predict long-term blood pressure elevations. *Int. J. Behav. Med.* **1994**, *1*, 215–228. [CrossRef]

45. Reynolds, S.; Lane, S.J.; Thacker, L. Sensory processing, physiological stress, and sleep behaviors in children with and without autism spectrum disorders. *OTJR Occup. Particip. Health* **2012**, *32*, 246–257. [CrossRef]

46. Shani-Adir, A.; Rozenman, D.; Kessel, A.; Engel-Yeger, B. The relationship between sensory hypersensitivity and sleep quality of children with atopic dermatitis. *Pediatr. Dermatol.* **2009**, *26*, 143–149. [CrossRef] [PubMed]

47. Shochat, T.; Tzischinsky, O.; Engel-Yeger, B. Sensory hypersensitivity as a contributing factor in the relation between sleep and behavioral disorders in normal children. *Behav. Sleep Med.* **2009**, *7*, 53–62. [CrossRef] [PubMed]

48. Greven, C.U.; Lionetti, F.; Booth, C.; Aron, E.N.; Fox, E.; Schendan, H.E.; Pluess, M.; Bruinng, H.; Acevedo, B.; Bijttebier, P.; et al. Sensory processing sensitivity in the context of environmental sensitivity: A critical review and development of research agenda. *Neurosci. Biobehav. Rev.* **2019**, *98*, 287–305. [CrossRef] [PubMed]

49. Camden, C.; Silva, M. Pediatric teleheath: Opportunities created by the COVID-19 and suggestions to sustain its use to support families of children with disabilities. *Phys. Occup. Ther. Pediatr.* **2021**, *41*, 1–17. [CrossRef]

Article

Vulnerability through the Eyes of People Attended by a Portuguese Community-Based Association: A Thematic Analysis

Carlos Laranjeira [1,2,3,*], Inês Piaça [1], Henrique Vinagre [1], Ana Rita Vaz [1], Sofia Ferreira [1], Lisete Cordeiro [4] and Ana Querido [1,2,5]

[1] School of Health Sciences, Polytechnic of Leiria, Campus 2, Morro do Lena, Alto do Vieiro, Apartado 4137, 2411-901 Leiria, Portugal

[2] Centre for Innovative Care and Health Technology (ciTechCare), Polytechnic of Leiria, Campus 5, Rua de Santo André—66–68, 2410-541 Leiria, Portugal

[3] Research in Education and Community Intervention (RECI I&D), Piaget Institute, 3515-776 Viseu, Portugal

[4] InPulsar (Associação para o Desenvolvimento Comunitário), Rua José Gonçalves LT 55—LJ 3 PISO-1, 2410-121 Leiria, Portugal

[5] Center for Health Technology and Services Research (CINTESIS), NursID, University of Porto, 4200-450 Porto, Portugal

* Correspondence: carlos.laranjeira@ipleiria.pt

Abstract: Vulnerability is associated with the individual's social and biological conditions, but also the conditions of their enveloping environment and society, leading to terms such as vulnerable populations or risk groups. This study aimed to give a voice to people with experiences of vulnerability and explore their perspectives, using a descriptive qualitative design. Purportedly vulnerable adults were recruited and interviewed with semi-structured questions on vulnerability. Data were organized, using WebQDA software, and submitted to thematic content analysis, as proposed by Clark and Braun, which generated a thematic tree. The study included six men and six women with a mean age of 43.8 [SD = 14.17] years old. Thematic analysis generated three themes: (1) Conceptions about vulnerability, (2) Barriers imposed by vulnerability, and (3) Strategies for dealing with vulnerability. The results highlight that vulnerability is a highly dynamic process of openness to circumstances that influence individual outcomes. However, there is a lack of conceptual clarity. Although being vulnerable is perceived as something negative, we need to transform the social mindset, because vulnerability also has the potential to change priorities in life for the better.

Keywords: vulnerability; qualitative study; social risk; community intervention

Citation: Laranjeira, C.; Piaça, I.; Vinagre, H.; Vaz, A.R.; Ferreira, S.; Cordeiro, L.; Querido, A. Vulnerability through the Eyes of People Attended by a Portuguese Community-Based Association: A Thematic Analysis. *Healthcare* **2022**, *10*, 1819. https://doi.org/10.3390/ healthcare10101819

Academic Editor: Daniele Giansanti

Received: 13 August 2022
Accepted: 19 September 2022
Published: 21 September 2022

Publisher's Note: MDPI stays neutral with regard to jurisdictional claims in published maps and institutional affiliations.

1. Introduction

Vulnerability is a key concept to health and social welfare and is commonly applied in academic research and policy formulation. Indeed, vulnerable people are "individuals or groups who require support with social, health, or economic problems" [1] (p. 2). Although there is no precise definition of vulnerability or vulnerable persons, the literature suggests common indicators of vulnerability and/or vulnerable groups [2]. Typically, it refers to a lack of physical, psychological, and social well-being, and hence the risk of lagging behind or becoming socially isolated from society [3,4]. Other indicators commonly stressed in the literature include: "(a) accumulation of problems or limitations (multi-complex problems); (b) feelings of powerlessness and distrust; (c) disrupted communication; (d) limited or no access to resources; (e) marginality; (f) imbalance in burden and capacity; (g) dependency situation; and (h) low self-esteem" [2] (p. 2). The ongoing COVID-19 pandemic has raised the vulnerability of many people, with several health implications. The disease and its repercussions—including dietary deficits, physical and mental health issues (potentially decreasing awareness of distinct hazards), and more limited access to healthcare services—were factors exacerbated by the pandemic [3,5].

According to Herring [6], there are two main conceptualizations of vulnerability: the dominant perspective in the literature classifies some persons or groups as vulnerable, whereas the alternative perspective emphasizes universal human vulnerability [6,7]. The latter view is uncommon in the literature, appearing largely in a psychological or even therapeutical context [8], but is highly promising, as stressed by Martha Fineman. According to her, universal vulnerability bears with it the impending or ever-present risk of damage, pain, and tragedy [1,9]. However, Fineman does not dispute that certain people or groups are more exposed to vulnerability than others [1]. Vulnerability is both universal and specific. As a multifaceted experience, it can take a variety of forms, accumulate and be passed down through generations. Fineman also used vulnerability as an interpersonal notion, referring to relational interactions, particularly those between people and society [9]. Fineman [9] concentrates on the social processes that generate vulnerability, as well as the duty of the state and its institutions/systems to decrease their risks and effects. Indeed, vulnerability analysis must consider individual circumstances, social processes, society, and its institutions [1,10]. Thus, exploring the vulnerability experiences offers a richer and wider perspective of vulnerability.

1.1. Theoretical Framework

Vulnerability is a key concept in health and social care disparities [11]. Based on this assumption, Flaskerud and Winslow [12] developed a conceptual model that characterises vulnerability as a complex interaction between susceptibility, risk, resource availability and health status. From a social justice stance, the authors propose that the interplay between these concepts may guide research and support evidence-based policy. With their knowledge and experience, institutions and vulnerable groups can help develop care strategies to mitigate the effects of vulnerability [13].

Based on the ecological systems theory model created by Bronfenbrenner [14], addressing vulnerability necessitates identifying its causes and drivers, as well as clarifying its expanding concept. In this sense, several factors influencing vulnerability in an individual's family, community and society at large cannot be taken out of this context [14,15]. As the risk of being impacted by exogenous events, vulnerability can be influenced by economic variables (either linked to welfare or development), socio-political factors, or environmental variables. These sources of vulnerability correlate with dimensions commonly mentioned in the sustainable development agenda. Vulnerability appears as the polar opposite of sustainability in these three domains [16]; it is a danger to sustainability.

1.2. Research Problem

Portugal has been indicated as one of the most vulnerable countries in the European Union [17]. In 2019, 9.8% of the Portuguese population was in persistent poverty, with 6% in poverty over the previous four years; unemployment or family disruption were the most common causes [18]. Thus, groups that are considered vulnerable in both economic and social terms [19]. Moreover, living in extreme poverty contributes to the perpetuation of risky behaviours, resulting in a greater risk of social and health problems.

The "exclusion of vulnerable groups from codesign processes may result in a failure to challenge dominant constructions of health and social care that may unintentionally reinforce oppression and existing inequities" [20] (p. 285). Growing evidence shows that research co-production may do more than just help put research findings into practice; it can also boost the possibility that public services are appropriately adapted to the requirements of the communities they serve [20,21]. In contrast, evidence shows conceptual confusion surrounding the meaning and consequences of the concept of vulnerability [22], requiring additional research among people with first-hand experiences of being vulnerable.

To the best of our knowledge, there is a dearth of qualitative research on vulnerability from the viewpoint of the vulnerable. To address this gap, using a participatory bottom-up approach, this study aims to explore the perceptions and experiences of vulnerability from the perspective of vulnerable people and identify strategies they used to reduce vulnerability.

2. Materials and Methods

2.1. Study Design

This study used a qualitative descriptive design. Through a naturalistic inquiry, this study aimed to comprehend the specifics of vulnerability in the natural context and based on the perspective of those involved [2,23]. The COnsolidated criteria for REporting Qualitative research (COREQ) checklist were followed [24].

2.2. Study Setting, Participants and Recruitment

Gathering data with people in a vulnerable state can be challenging [25]. Access to the target population often needs cooperation with local social services. Therefore, our study scenario was a non-governmental organization (InPulsar-Association for Community Development) located in the region of Leiria (Portugal), whose mission is to contribute to the social and economic inclusion of vulnerable populations. Their intervention process is based on the participatory action-research methodology: a network intervention based on observation, reflection, planning and evaluation of the various actors. The aim is to achieve the emancipation and autonomy of beneficiaries and their inclusion in the local community. Training and empowerment are also encouraged.

This organization was selected because it provided the research team with access to the vulnerable population. Potential participants were approached and informed about the study by one institutional facilitator (L.C.), with previous knowledge and experience working with vulnerable people.

The sample size was based on purposeful sampling because we used a qualitative exploratory research technique. The following criteria were applied in the selection process: (1) participants who fulfilled the current criteria and/or characteristics of "vulnerable people"; (2) adults (at least 18 years old); (3) understand the Portuguese language and have reflective capacity; and (4) participants with a personal sense of vulnerability or the experience of being vulnerable.

Gender and age distribution diversity were also considered. Eligible responders undergoing medical or psychological therapy were excluded from the study, to avoid any detrimental influence on the person's treatment. People with cognitive disabilities or weak communication abilities were also excluded.

2.3. Data Collection

The study was conducted in April 2022 and employed a semi-structured interview guide to perform the in-depth interviews. The authors developed the interview guide based on the previous literature [2]. Interviews began with getting to know the participant and then exploring the meaning of vulnerability for the individual, the forms it assumes, how it is viewed, and when in life it was experienced. Interviews ended with discussing insights into potential measures to lessen perceived vulnerability, in order to grasp the perceived vulnerability on a personal and interactional level.

All interviews took place in a private room of InPulsar where distractions were minimal. Each interview was digitally recorded and lasted approximately 30 min (varying between 20 and 50 min). No repeat interviews were carried out.

2.4. Data Analysis

Transcripts of interview recordings were imported into WebQDA (Universidade de Aveiro, Aveiro, Portugal), a qualitative software program, for thematic content analysis according to the Braun and Clarke [26] guidelines. First, data familiarization was accomplished by reading all transcripts. Second, pertinent information was grouped into understandable codes. Third, codes were categorized into probable vulnerability-related topics. Fourth, thematic validity was confirmed by reviewing all codes and the complete data set. Fifth, topics were defined and named, producing a final thematic tree. Finally, a literature review was used to help write the report [27].

Coding involved two (co)researchers: the lead researcher (C.L.) coded all transcripts, which were then independently co-coded by the co-researcher with a scientific background and competence in data processing (A.Q.). When differences or disagreements arose during co-coding, the coders reflected until consensus was obtained. To reach a final decision, a third researcher (L.C.) was consulted. Selected transcript quotes from Portuguese were translated, italicized, and inserted into the following report.

2.5. Study Rigour

The interviews were planned and performed by the research team, and all interviewers (final year nursing students) were supervised by leading co-researchers (C.L. and A.Q.), who have an insider's viewpoint and practice knowledge of vulnerability. To guarantee reliability and validity, the interviewers underwent interview training and reflexive sessions, including discussion about the interview guide, and, during the data collection period, participated in regular feedback sessions based on recorded interviews.

Findings were peer-reviewed by co-authors (investigator triangulation) and resulted in the revision of theme and sub-themes names to rightfully capture the essence of the findings. We also kept an audit trail, which included field notes, coded transcripts, and comments and revisions from group coding meetings, following the best practices in qualitative research [28].

2.6. Ethics

The study was conducted after approval by the Ethics Committee of the Polytechnic of Leiria (CE/IPLEIRIA/02/2022). All participants provided written informed consent, and all sociodemographic data, interview recordings, transcriptions, and files including IDs were saved and password-protected, accessible only to members of the study team. Respondents received no compensation for participating in the study.

3. Results

3.1. Sample Description

The 12 respondents varied in age from 24 to 67 years old (mean = 49; SD = 14.17), with six males and six women (Table 1). The manifestations of vulnerability reported by participants included being homeless, being a migrant, having an infectious disease (e.g., HIV and hepatitis), being drug dependent, living socioeconomic difficulties (unemployment), and experiencing a process of loss and grief. Three respondents worked, while the remainder received social benefits. Three participants attributed the cause of their homelessness and their inability to find suitable accommodation to a lack of family support, substance abuse, and other mental health concerns.

Some respondents reported having a mental or physical health problem, or both. Depression and anxiety were the most often reported mental health disorders. Participants recognized the significance that substance abuse had in their mental health, but also highlighted how drugs and alcohol provided short-term respite. Use of alcohol and drugs serves to block traumatic experiences or deal with painful childhood memories, usually including abuse, loss, and bereavement. Physical health difficulties such as hepatitis C, physical impairments, and HIV were quite common. As greater emphasis was given to their mental health, housing, and drug abuse, these were rarely regarded as a priority.

Table 1. Participants' demographics.

Participant	Age (Years)	Sex	Educational Level
P1	43	Male	7th year (3rd cycle of basic education)
P2	56	Female	3rd year (1st cycle of basic education)
P3	24	Female	9th year (3rd cycle of basic education)
P4	41	Female	7th year (3rd cycle of basic education)
P5	25	Female	12th grade (secondary school)
P6	49	Male	Attended basic school but never finished
P7	49	Female	2nd year (1st cycle of basic education)
P8	65	Male	4th year (1st cycle of basic education)
P9	34	Male	8th year (3rd cycle of basic education)
P10	42	Male	7th year (3rd cycle of basic education)
P11	67	Male	9th year (3rd cycle of basic education)
P12	31	Female	9th year (3rd cycle of basic education)

3.2. Qualitative Findings

The data from our interviews can be summarized in terms of three major themes: (1) Conceptions about vulnerability, (2) Barriers imposed by vulnerability, and (3) Strategies for dealing with vulnerability. Three subthemes were identified within the first theme: ontology condition that spreads, being alone "without network" and being exposed to external pressure (others). In the second theme, there were also three subthemes: discrimination/stigma, difficulties in social reintegration, and "my condition is difficult". Lastly, in the third theme, we found four subthemes: the ability to ask for help/seek support, motivation, and commitment to behavioural change, not exposing others to the same risks, and ignoring the disapproving look of others. All themes and subthemes are represented in Figure 1 and are considered crucial in determining participants' understanding of vulnerability. Of course, some elements of participants' comprehension are related to more than one theme. However, this ought to be seen as a good interpretation of perceptions and attitudes generally, which are never made up of distinct ideas but rather relate to one another in various ways. Each theme is discussed below, and meaningful quotes were included to support the findings.

3.2.1. Conceptions about Vulnerability

An individual's notion, belief, or sense of anything known, experienced, or imagined is referred to as a conception. There are several meanings attributed to vulnerability, derived from each participant's life context. The first theme describes vulnerability as an ontological condition that spreads, that is, the person sees vulnerability as something intrinsic to their life and that runs through their existence:

> P1: *"It's part of a period of life that I lived, when I was living on the street, that's when I felt this effect the most (. . .) Yes, it's a snowball, and it affects our whole life (. . .) I lived on the street because the money I received from the unemployment fund was not enough to pay for a room and now due to drug consumption I continue to live on the street and spend all my money on drugs".*

> P10: *"During my childhood my father beat me because yes . . . so I felt vulnerable, because I couldn't do anything, I thought why was he doing that to me, why? And he felt vulnerable because he couldn't do anything. Now he still feels vulnerable because I can't stop consuming".*

From the analysis of the discourses, it is evident that vulnerability is not a unique and isolated phenomenon, but rather a condition that can accompany the subject throughout life, interfering with their functioning.

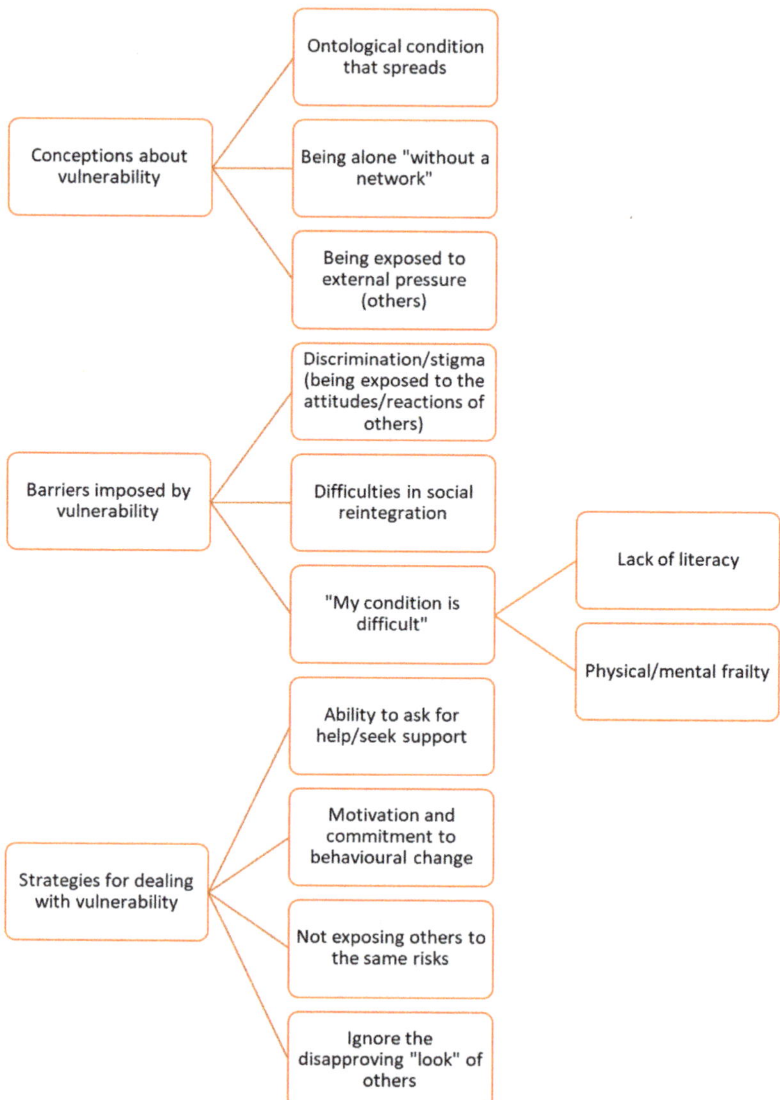

Figure 1. Thematic map of themes and subthemes.

Another frequent theme in this dimension was being alone "without a network". Four participants addressed how loneliness interferes with the way they perceive and experience vulnerability:

> P5: *"The truth is that I need a friend to be able to talk, so I don't get to this point of keeping everything, and then when it explodes it's like this (. . .), after my father died, after 6 months my brother left home and my mother worked at night (. . .) I was always alone, so I've been used to being alone for a long time (. . .) It's been very difficult, I don't know what to do with my life, what I can do myself . . . "*

P7: "I have no support from anyone, not even the family (. . .) anyone! I only have the support of my son, nothing else, I stopped talking to them, because I am of a nature that they do not admit. I did so much for them, but they don't recognize . . . "

P9: "I prefer to be alone in my corner, because I have no one to help me"

P10: "Alone (. . .) I don't have anyone, my father died, my wife died, I mean I don't have anyone else"

Lack of support and loneliness are conditions that promote situations of vulnerability, since participants do not have anyone to share their feelings and concerns and feel isolated. Participants mentioned failures in the quality of the relationship networks, then there are less relationships than desired or retribution is not proportional.

Finally, the theme of being exposed to external pressure (others) arose associated with the adoption of risky behaviours, beyond their personal will. In this sense, P4 accepted what was asked, suggested or instilled, as a way of pleasing others.

P4: "Yes, it's not just for me, it's because of my environment, I think people are like that, a little weak and they always go after each other; and it is easy, for example, to find someone who has never smoked anything; and they say then don't you want to try it? Or they see me smoking and ask to try it. If they didn't call me I wouldn't go. I never have money, but when I do, I think straight away, of course I do (. . .) Dealers call me every day, write "there are big scenes in the neighbourhood, you get 50€ for two", "you take 50€ and you get a bonus", scenes like that . . . "

Vulnerability resulting from social pressure seems to arise from an imbalance between the socio-emotional system and the cognitive control system. If on the one hand there is a need for acceptance, on the other hand, there is a lack of personal and social skills (e.g., assertiveness) to say "no" to psychoactive substances.

3.2.2. Barriers Imposed by Vulnerability

Some participants mentioned barriers associated with their vulnerability, referring to the importance of identifying vulnerability early in the hope of avoiding its perpetuation. The main topic addressed in this dimension was discrimination/stigma, i.e., the exposure of participants to stereotypes and negative attitudes of others, namely their own family and society in general.

P4: "Yes, I think that people, at least my family, think I'm a drug addict and look at me as a drug addict, and that's why I don't get a job, because I'm not all hot to go to work".

P6:" Oops, when I was on the street, they looked at me differently".

P7: "Yes, I do, the recriminatory look of people".

P8: "They look from the side (. . .) some people, not all".

P9: "There were people who discriminated against me, others who tried to help me and I didn't want to; sometimes not accepting help makes things more difficult!"

P10: "I say I have leishmaniasis and people look at me in a different way! And they ask, can you catch Filipe? If you get that scared, how much more if I said I have HIV, it's not . . . "

P9 underlined that, despite some prejudice, other people were available to help, but refusing that help perpetuated isolation, a characteristic of the most vulnerable. Some participants also pointed out that stigma affected childhood and adolescence, as well as the search for employment in adulthood.

P3:" I've been criticized many times and they said you're like this because you want to, and you don't have a job because you don't want to. But it's not quite like that, I was made fun of at school for not wearing designer clothes, or for being chubby, or because my mother doesn't have a profession that is said to be worthy. My mother, always did everything to not miss us with anything . . . "

P2: *"I have missing teeth and when I look for a job, there are people who look at me and say "we just want younger people". My image doesn't help me, I know, but what can I do . . . "*

Akin to discrimination/stigma, participants also referred difficulty in social reintegration.

P1: *"At the moment, what I feel is difficulty in integrating myself into society again, it seems that I am dependent on everything (. . .) I even wanted to attend professional training, but they said that it would be difficult to enter".*

P7: *"At my age, it's so hard to find a job! I've been looking for an alternative for months and nothing . . . "*

Social reintegration assumes the character of reconstructing losses, and its objective is to enable people to fully exercise their right to citizenship. However, difficulties in social reintegration are attributed to deficient living conditions, namely in the search for employment and/or education/training.

The third theme—my condition is difficult—is subdivided into two sub-themes. The first sub-theme refers to the lack of literacy, which portrays how a lack of education became an obstacle in the lives of the participants. In this regard, P7 mentioned being sad for not having the necessary resources to secure a job.

P7: *"(. . .) I have no studies. I would like to know, even more, read and write, and study at a school. I need studies and I don't have them. I feel sad, I feel that instead of going up, I go down, because with a lack of studies everything is more difficult. Even for cleaning tasks, I need a driver's license, I've been to about three or four interviews and they all ask for a driver's license and they don't accept me because of that, but it's not just me, it's with several people. Do you need a license to clean windows? For God's sake, if they wanted to help a person in need, they wouldn't, they wouldn't".*

In the case of participant P2, the lack of studies made it impossible for him to answer certain questions during the interview:

Interviewer: *"(. . .) Can you tell me what you mean by vulnerability and human fragility? Do you know?"; P2: "No, because I have no studies".*

Poor literacy appeared to be an important barrier that exacerbates vulnerability and limits access to available social resources. As a second sub-theme, physical/mental frailty emerged, with greater emphasis on mental frailty. There were several participants who focused on stress, depression, anxiety, and other psychopathological manifestations.

P6: *"Wow, that's not very easy. When I found out I had this [HIV] it wasn't easy, you see? To accept that I had this wasn't very easy, it didn't really fit in my head. When I fell in the hospital, I was told what this was, now it took years and years to get it into my head. It was not easy. And even now I get anxious and stressed".*

P7: *"My condition is difficult (. . .) my depression, there are things that I cannot deal with, they have already seen, that I am under enormous pressure. I went to treatments, I've been hospitalized a few times, they [professionals] know everything, but this is not easy. When a person is discouraged, sad, a person is forced to smoke a cigarette to kill the stress".*

P9: *"I also had that . . . persecution. I also have these thoughts, that . . . " Interviewer: Do you think people were following you?" P9: "Yes, yes".*

Physical and mental fragility were apparently a constant, with repercussions on people's functioning, justifying the need for monitoring, supervision, and sometimes medical treatment.

3.2.3. Strategies for Dealing with Vulnerability

Vulnerability identifies the extent and intensity of individual susceptibility to risk, while resilience is the ability to withstand and overcome risk. In this sense, participants needed to adopt strategies, such as seeking support from those who could help them. The first theme portrays the ability to ask for help/seek support, where some participants underlined its importance in dealing with vulnerability. They highlighted the importance of the professionals providing their services at InPulsar.

P2: "I turn to people who are not my family, and these are the people who give me the most support. I came to ask for help because I wanted to change".

P3:" I think I'm having it now with InPulsar than I had before. Previously, it was not easy, I was already unemployed and was denied help many times, and now, as far as I can see, I still have no reason to complain".

P5: "They [InPulsar] always have their doors open to help; They have professionals in the social, health and legal areas. And they even give us psychological support".

P7: "Yes, they have helped me, nobody points the finger at me, everyone helps and supports me in here, they never point the finger at me".

P10: "They are really tireless". I'm so glad I came here and ask for help in time.

The request for help proved to be crucial as a strategy to adjust to vulnerability, namely through community institutions that help them and promote their autonomy and inclusion. This support can be carried out not only through material and financial support, but above all through emotional support.

Some participants underlined the importance of motivation and commitment to behavioural change.

P1: "When I was living on the street, I had many moments when I thought that this was not life and that I had to change something. After all, living on the street is not a system".

P6: "I am waiting to take a course, Carpentry (...). I still have time, since I didn't learn at 18, I'm learning now at 50".

P7: "I have to be a strong woman. After all, I have a son, it's the most important thing in my life ... It's not with a social support check (189 euros) that I'm going to take my life forward. I need to find a job to give you a better life!"

P10: "I won't give up, I won't! I take methadone, it's one of the things I want too, it's a goal, to end methadone".

Another theme identified by the participants refers to not exposing others to the same risks. HIV infection alters a person's awareness of the results of their behaviour towards sexuality. This perception transforms the sexual act and the intimate interpersonal relationship into a challenge. Some participants showed some resentment for having contracted HIV, believing that they were infected because of the insincerity of their sexual partner, whom they blame. However, they expressed concern about protecting others, despite sometimes hiding that they are HIV positive:

P1:" Yes. I got HIV from someone who had it, but at the time he couldn't tell me he had it and we had sex and it happened ... he never told me he had HIV, I was very angry. But since then, I've never exchanged syringes with anyone, I've never had sex without a condom, always respecting others".

P10: "(...) I've had several girls, since I have HIV it was always with a "condom" and I haven't found the exact girl to say like that (...) Look, I have HIV".

Any kind of risk should be avoided, even more so when other individuals are concerned. However, issues such as sexually transmitted diseases are embarrassing, because they involve intimacy, are sometimes difficult to address, which can increase the individual's vulnerability.

Finally, six participants mentioned ignoring the disapproving "look" of others as a coping mechanism focused on avoidance.

P1: "I don't know. Because I lived in Belgium for 12 years and I know what it is and for me I can ignore it".

P2: "Walking around with my face uncovered and I like to pass by and they don't point at me (. . .) they're ready to look at me, but I don't care. Move on".

P3: "I never waste time on what others think about me".

P5: "I don't know, I don't notice it! Sometimes it's my boyfriend who says (. . .) but I don't care".

P6: "Now I don't know, but before it was more difficult".

P11: "When they discriminated against me, it was without me knowing, it was behind the scenes. I recognize that sometimes pretending not to see yourself may not be enough. Then we have to give it a go!"

The "Ignore" strategy seems to help minimize suffering. However, P11 warned that prolonged maintenance of avoidance may be insufficient to deal with prejudice, and problem-focused strategies may be necessary, as these are apparently more effective.

4. Discussion

This qualitative study involved an in-depth exploration of the perceptions and experiences regarding vulnerability issues. A thematic analysis of the responses and their commonalities suggested several themes and sub-themes, which allow us to portray vulnerability as a concept that is both individual and universal. There is a high symbolic load associated with vulnerability, regardless of gender, age, socioeconomic stratum or cultural origin [29]. Vulnerability "is a product of situations people are living in, the so-called *vulnerable situations*. This, in turn, requires a deep and nuanced identification and analysis of elements of vulnerability that might intersect and result in differentiated degrees of vulnerability" [30] (p. 6).

According to Carmo and Guizardi [31], the concept of vulnerability is a constitutive condition of the human being, which corroborates the view, also present in our study, that vulnerability is an ontological condition that spreads. Conditions for vulnerability depend on several factors, such as biological factors, life experiences, and the environment of each individual [32,33].

Human beings cannot survive without the help of others; therefore their existence is marked by moments of greater or lesser vulnerability [34]. Our findings agree with the available evidence showing that vulnerability is typically associated with situations of social disadvantage that lead to poverty; marginalization, feelings of powerlessness/distrust, and limited or no access to resources [2,35,36]. Furthermore, lower levels of education predict higher levels of vulnerability [37]. Thus, vulnerability is caused by the lack of means and capacity of populations to protect themselves.

Our results clearly indicate that vulnerability is composed of an intrinsic dimension (that is, non-modifiable) and an extrinsic component, and therefore subject to fluctuations conditioned by the environment [38]. Both components were addressed by the participants, through the themes: an ontological condition that spreads, being alone (without a network), and being exposed to external pressure (others). UNESCO [39] stresses that the human condition, by itself, implies vulnerability, with every human being exposed to susceptibilities, whether physical or mental. Therefore, we need recognize that, at some point in life, we may not have the ability or the means to protect ourselves, as there is always the possibility of being confronted with pathologies, deficiencies, and damages from the environment or other human beings that can lead to death.

Another theme seen in the literature is life disruption caused by a lack of support and societal pressure [40], which refers to a variety of possibly overlapping factors, such as those originating from a 'broken' family due to substantial socioeconomic issues or other crises [41].

Regarding the barriers imposed by vulnerability, our findings show that discrimination and stigma are prevalent phenomena, experienced by some participants across life. Such prejudice made participants feel insecure and unable to affect their own future; instead, decisions were made about them and for them by individuals who didn't comprehend their circumstances. This sense of powerlessness, evident in their perception that they have no option, was previously recorded in a Gypsy travelling community [42], where it leads to self-segregation and contributes to a general sense of mistrust. This lack of trust fosters a sense of helplessness over one's own destiny and leads to vulnerability.

Poor physical and mental health, harmful risk-taking behaviours (e.g., drug misuse, HIV, and sexually transmitted infection) [41] and public health consequences [3,43] are frequently identified, in the literature, as barriers to and negative consequences of vulnerability. Our findings demonstrate that lack of literacy and individual frailty increased vulnerability to stressors and limited social reintegration [37]. These barriers include concerns about fitting in and becoming a well-adjusted member of society. According to Guignon [44], one can only feel authentic in a world that values unique abilities, embraces variety, provides equal opportunity, and values criticism and controversial views; a world that ensures there are no restrictions to free expression. Although some individuals feel stigmatized by society, they do not retaliate in the same way, displaying a compassionate concern for others [45].

Other elements mentioned during the interviews may be seen as a promise to overcome vulnerability and stigmatization. The ability to ask for and seek help allows the subject to overcome their difficulties, or else, provide them with tools to be able to fight their weaknesses and the threats to their balance [34]. On the other hand, the motivation/commitment to change encourages the individual to create new behaviours or stop those that are harmful to their condition [46].

From a transactional perspective, Lazarus and Folkman [47] characterize the coping construct as the individual's cognitive and behavioural efforts to deal with internal or external challenges, evaluated as exceeding their resources. They conceptualize the categories of coping along the three dimensions of control or confrontation, avoidance, and task-oriented emotion. Control or confrontation strategies consist of proactive cognitive actions and reassessments, highlighting the themes "ability to ask for help/seek support", "motivation/commitment to change", and "not exposing others to the same risks", where participants reveal a clear concern with changing their behaviour, through commitment and concern for others. Avoidance strategies, based on actions and cognitions that suggest an escape from the problem, were also identified, the main example being "ignoring the disapproving 'look' of others". Finally, emotion-focused strategies were not identified by the participants, although some manifestations of stress, anxiety and depression were highlighted as interfering with their functioning and emotional regulation.

4.1. Study Strengths and Limitations

This study provides a timely overview of the challenges faced by vulnerable people, as seen from their perspective. This might help academics, who want to involve vulnerable people in the community and make informed suggestions on providing public services. Furthermore, by keeping a reflective research journal and peer-reviewing the developed themes and sub-themes, the research was transparent and rigorous throughout the stages of data collection and analysis.

Despite its strengths, the current study has several limitations. First, the obtained sample size was smaller than expected. Overall, participants revealed poor knowledge of vulnerability factors, due to a lack of motivation or literacy, resulting in communication issues throughout the interview process. This limitation is consistent with earlier research on contributors' unwillingness to give their opinions owing to stigma and trauma, or lack of trust in the researchers' intentions [48,49]. However, the small sample size was offset by the richness of our participants' descriptions of their experiences. Second, interviews could potentially elicit unpleasant emotions and influence contributors' health and well-being.

Psychological support was offered when some emotional overburdening arose during interviews. Third, there was no data source triangulation to obtain various viewpoints (including researchers, service providers and service users). Fourth, and to protect contributors' privacy, we did not collect individual socio-economic information. Fifth, only those who used support services were 'heard', which may imply that the least engaged individuals were excluded. Sixth, the study did not expressly focus on health vulnerability per se, therefore no specific health questions were addressed; however, participant beliefs and experiences of vulnerability clearly influenced their health, so this area deserves further investigation. Finally, this study outlined one way to investigate the problem of vulnerability in the Portuguese context. It made no mention of the structural reasons of vulnerability, such as limited income support, or the plethora of other systemic problems that have contributed to rising vulnerability levels. This might limit the generalizability of our findings to other contexts. Further study should take these structural and systemic aspects into account. Research may help national and local governments and organizations set specialized standards, hire workers with the required skills to address specific requirements, and inform people in suitable ways and according to the differences among groups presently depicted as vulnerable [29]. Socioecological models [14,15] provide useful frameworks to consider levels and interconnections in research. Mixed-method studies are also suggested as they can: (a) measure indicators of physical, mental, and social vulnerability; and, additionally, (b) understand the lived experience of those involved, in order to create multidisciplinary intervention programs that meet their real needs.

In addition, further research is needed to investigate practitioners' understandings and perceptions of vulnerability, to better grasp what needs to be done to address health disparities encountered by vulnerable populations.

4.2. Implications for Practice

This study enables us to reflect on the dimensions of vulnerability. Understanding the social determinants of vulnerability is necessary to achieve satisfactory care for human groups. Only by considering a population's context and the sociocultural factors contributing to vulnerability, can we promote actions that meet and respect the collective and individual needs. Therefore, this study help can broaden our thinking and awaken our consciousness of the organizing values of health care as a social practice. This entails political–cultural–social participation to assist the health care of vulnerable human groups, as well as the adoption of attitudes and behaviours that establish and enhance health care and maintenance activities from all perspectives. Another relevant implication is the need for skills training opportunities for healthcare professionals and social workers to support them in engaging with vulnerable groups [25].

In recent decades, scholars in the Western world have paid close attention to the idea of vulnerability. Health and social care workers must explicitly apply critical views to the concept of vulnerability in order to build a more comprehensive understanding of vulnerability. By doing so, educators will be better prepared to teach about vulnerability, question the concept's structural components and anticipate collectivist methods to address problems such as vulnerability in community contexts [50]. Torralba i Roselló [51] mentioned that the lack of a pedagogy of vulnerability has serious effects on care processes in a broad sense. Practitioners must recognize and grasp these broader societal forces, as well as ensure that services are culturally sensitive. Practitioners should also be encouraged to share information on best practices, thus contributing toward effective interventions to improve health and social outcomes [52].

5. Conclusions

Thematic analysis of the collected data yielded three key themes: conceptions about vulnerability; barriers imposed by vulnerability; and strategies for dealing with vulnerability. Our findings revealed that vulnerability is a very dynamic process of openness to conditions that impact individual outcomes. However, there is a conceptual gap: being vulnerable is perceived as something negative, but vulnerability also has the potential to change priorities in life for the better. This research considered the larger themes of vulnerability, helping to transform the link between how individuals might use life's catastrophic and adverse occurrences as possibilities for positive growth, thus contributing to change our mindset concerning vulnerability. We must recognize that a person's strength may sometimes be expressed via their vulnerability, as this allows them to reorganize and be a part of a more integrative and inclusive society.

Author Contributions: Conceptualization, C.L. and A.Q.; Methodology, C.L. and A.Q.; Software, C.L.; Validation, C.L.; Formal analysis, C.L. and A.Q.; Investigation, A.Q., C.L., L.C., I.P., H.V., A.R.V. and S.F.; Resources, C.L. and A.Q.; Data curation, C.L. and A.Q.; Writing—original draft preparation, C.L.; Writing—review and editing, C.L., A.Q., L.C., I.P., H.V., A.R.V. and S.F.; Visualization, C.L., A.Q., I.P., H.V., A.R.V. and S.F.; Supervision, C.L.; Project administration, C.L.; Funding acquisition, A.Q. and C.L. All authors have read and agreed to the published version of the manuscript.

Funding: This work is funded by national funds through FCT—Fundação para a Ciência e Tecnologia, I.P. (UIDB/05704/2020 and UIDP/05704/2020) and under the Scientific Employment Stimulus-Institutional Call—[CEECINST/00051/2018].

Institutional Review Board Statement: The study was conducted in accordance with the Declaration of Helsinki and approved by the Institutional Review Board of IPLeiria (protocol approval no. CE/IPLEIRIA/02/2022).

Informed Consent Statement: Informed consent was obtained from all subjects involved in the study.

Data Availability Statement: The data are available upon reasonable request.

Acknowledgments: We acknowledge all the volunteers who participated in the interviews to make this study possible.

Conflicts of Interest: The authors declare no conflict of interest. The funders had no role in the design of the study; in the collection, analyses, or interpretation of data; in the writing of the manuscript; or in the decision to publish the results.

References

1. Virokannas, E.; Liuski, S.; Kuronen, M. The contested concept of vulnerability–A literature review. *Eur. J. Soc. Work* **2018**, *23*, 327–339. [CrossRef]
2. Numans, W.; Regenmortel, T.V.; Schalk, R.; Boog, J. Vulnerable persons in society: An insider's perspective. *Int. J. Qual. Stud. Health Well-Being* **2021**, *16*, 1863598. [CrossRef] [PubMed]
3. Mezzina, R.; Gopikumar, V.; Jenkins, J.; Saraceno, B.; Sashidharan, S.P. Social Vulnerability and Mental Health Inequalities in the "Syndemic": Call for Action. *Front. Psychiatry* **2022**, *13*, 894370. [CrossRef]
4. Gordon, B.G. Vulnerability in Research: Basic Ethical Concepts and General Approach to Review. *Ochsner J.* **2020**, *20*, 34–38. [CrossRef] [PubMed]
5. Kar, S.K.; Arafat, S.M.; Marthoenis, M.; Kabir, R. Homeless mentally ill people and COVID-19 pandemic: The two-way sword for LMICs. *Asian J. Psychiatry* **2020**, *51*, 102067. [CrossRef]
6. Herring, J. *Vulnerable Adults and the Law*; Oxford University Press: Oxford, UK, 2016.
7. Fineman, M.A. Equality, autonomy, and the vulnerable subject in Law and politics. In *Vulnerability: Reflections on a New Ethical Foundation for Law and Politics*; Fineman, M.A., Grear, A., Eds.; Ashgate: Surrey, UK, 2013; pp. 13–28.
8. Bielby, P. Not 'us' and 'them': Towards a normative legal theory of mental health vulnerability. *Int. J. Law Context* **2019**, *15*, 51–67. [CrossRef]
9. Fineman, M. The vulnerable subject and the responsive state. *Emory Law J.* **2010**, *60*, 251–275. Available online: https://scholarlycommons.law.emory.edu/elj/vol60/iss2/1 (accessed on 12 August 2022).
10. Fineman, M. Vulnerability and Social Justice. *Valpso. Univ. Law Rev.* **2019**, *53*, 341–370. [CrossRef]
11. Grabovschi, C.; Loignon, C.; Fortin, M. Mapping the concept of vulnerability related to health care disparities: A scoping review. *BMC Health Serv. Res.* **2013**, *13*, 94. [CrossRef]
12. Flaskerud, J.; Winslow, B.J. Conceptualizing vulnerable population's health-related research. *Nurs. Res.* **1998**, *47*, 69–78. [CrossRef]
13. Macedo, J.; Santos, A.; Costa, L.; Lima, A.; Lima, J.; Vasconcelos, B. Vulnerability and its dimensions: Reflections on Nursing care for human groups. *Rev. Enferm. UERJ* **2020**, *28*, e39222. [CrossRef]
14. Bronfenbrenner, U.; Evans, G. Developmental science in the 21st century: Emerging questions, theoretical models, research designs and empirical findings. *Soc. Dev.* **2000**, *9*, 115–125. [CrossRef]

15. Hamwey, M.; Allen, L.; Hay, M.; Varpio, L. Bronfenbrenner's Bioecological Model of Human Development: Applications for Health Professions Education. *Acad. Med.* **2019**, *10*, 1621. [CrossRef] [PubMed]
16. Jozaei, J.; Chuang, W.; Allen, C.; Garmestani, A. Social vulnerability, social-ecological resilience and coastal governance. *Glob. Sustain.* **2022**, *5*, E12. [CrossRef]
17. Horta, A.; Gouveia, J.P.; Schmidt, L.; Sousa, J.C.; Simões, S. Energy poverty in Portugal: Combining vulnerability mapping with household interviews. *Energy Build.* **2019**, *203*, 109423. [CrossRef]
18. Peralta, S.; Carvalho, B.; Esteves, M. Portugal, Balanço Social 2021. Um Retrato do País e dos Efeitos da Pandemia. Lisboa: Nova SBE Economics for Policy Knowledge Center, Fundação "la Caixa", BPI, 2021. Available online: https://www.novasbe.unl.pt/Portals/0/Files/Reports/SEI%202021/Relat%F3rio%20Balan%E7o%20Social_Janeiro%202022.pdf (accessed on 12 August 2022).
19. Peroni, P.; Timmer, A. Vulnerable groups: The promise of an emerging concept in European Human Rights Convention law. *Int. J. Const. Law* **2013**, *11*, 1056–1085. [CrossRef]
20. Mulvale, G.; Moll, S.; Miatello, A.; Robert, G.; Larkin, M.; Palmer, V.; Powell, A.; Gable, C.; Girling, M. Codesigning health and other public services with vulnerable and disadvantaged populations: Insights from an international collaboration. *Health Expect.* **2019**, *22*, 284–297. [CrossRef]
21. Parveen, S.; Barker, S.; Kaur, R.; Kerry, F.; Mitchell, W.; Happs, A.; Fry, G.; Morrison, V.; Fortinsky, R.; Oyebode, J. Involving minority ethnic communities and diverse experts by experience in dementia research: The caregiving HOPE study. *Dementia* **2018**, *17*, 990–1000. [CrossRef]
22. Bracken-Roche, D.; Bell, E.; Macdonald, M.E.; Racine, E. The Concept of 'Vulnerability' in Research Ethics: An In-Depth Analysis of Policies and Guidelines. *Health Res. Policy Syst.* **2017**, *15*, 8. [CrossRef]
23. Sutton, J.; Austin, Z. Qualitative research: Data collection, analysis, and management. *Can. J. Hosp. Pharm.* **2015**, *68*, 226–231. [CrossRef]
24. Tong, A.; Sainsbury, P.; Craig, J. Consolidated criteria for reporting qualitative research (COREQ): A 32-item checklist for interviews and focus groups. *Int. J. Qual. Health Care* **2007**, *19*, 349–357. [CrossRef] [PubMed]
25. Amann, J.; Sleigh, J. Too Vulnerable to Involve? Challenges of Engaging Vulnerable Groups in the Co-production of Public Services through Research. *Int. J. Public Adm.* **2021**, *44*, 715–727. [CrossRef]
26. Braun, V.; Clarke, V. Using thematic analysis in psychology. *Qual. Res. Psychol.* **2006**, *3*, 77–101. [CrossRef]
27. Braun, V.; Clarke, V. *Successful Qualitative Research: A Practical Guide for Beginners*; Sage: London, UK, 2013.
28. Padgett, D.K. *Qualitative Methods in Social Work Research*, 3rd ed.; Sage Publications Inc.: Los Angeles, CA, USA, 2017.
29. Ten Have, H.; Gordijn, B. Vulnerability in light of the COVID-19 crisis. *Med. Health Care Philos.* **2021**, *24*, 153–154. [CrossRef]
30. Kuran, C.; Morsut, C.; Kruke, B.; Krüger, M.; Segnestam, L.; Orru, K.; Nævestad, T.O.; Airola, M.; Keränen, J.; Gabel, F.; et al. Torpan Vulnerability and vulnerable groups from an intersectionality perspective. *Int. J. Disaster Risk Reduct.* **2020**, *50*, 101826. [CrossRef]
31. Carmo, M.; Guizardi, F. The concept of vulnerability and its meanings for public policies in health and social welfare. *Cad. Saúde Pública* **2018**, *34*, e00101417. [CrossRef]
32. Borges, C.; Schneider, D. Vulnerabilidade, família e o uso de drogas: Uma revisão integrativa de literatura. *Psic. Ver.* **2021**, *30*, 9–34. [CrossRef]
33. Calpe-López, C.; Martínez-Caballero, M.A.; García-Pardo, M.P.; Aguilar, M.A. Resilience to the effects of social stress on vulnerability to developing drug addiction. *World J. Psychiatry* **2022**, *12*, 24–58. [CrossRef]
34. Almeida, C.M. A Representação da Vulnerabilidade Humana Como Motor Para a Recuperação do Paradigma do Cuidar em Saúde; The Representation of Human Vulnerability as an Engine for the Recovery of the Health Care Paradigm. Doctoral Thesis, Universidade Católica Portuguesa (UCP), Porto, Portugal, 2014.
35. Bühlmann, F. Trajectories of Vulnerability: A Sequence-Analytical Approach. In *Social Dynamics in Swiss Society. Life Course Research and Social Policies*; Tillmann, R., Voorpostel, M., Farago, P., Eds.; Springer: Cham, Switzerland, 2018; Volume 9, pp. 129–144.
36. Paugam, S. *Les Formes Élémentaires de la Pauvreté*; PUF: Paris, France, 2005.
37. Obasuyi, F.; Rasiah, R.; Chenayah, S. Identification of Measurement Variables for Understanding Vulnerability to Education Inequality in Developing Countries: A Conceptual Article. *SAGE Open* **2020**, *10*, 2158244020919495. [CrossRef]
38. Weisser-Lohmann, E. Ethical aspects of vulnerability in research. *Poiesis Prax.* **2012**, *9*, 157–162. [CrossRef]
39. United Nations Educational, Scientific and Cultural Organization [UNESCO]. *The Principle of Respect for Human Vulnerability and Personal Integrity*; Report of the International Bioethics Committee of UNESCO; UNESCO: Paris, France, 2013.
40. Venter, I.; Wyk, N. Experiences of vulnerability due to loss of support by aged parents of emigrated children: A hermeneutic literature review. *Community Work Fam.* **2019**, *22*, 255–266. [CrossRef]
41. Dorsen, C. Vulnerability in homeless adolescents: Concept analysis. *J. Adv. Nurs.* **2010**, *66*, 2819–2827. [CrossRef]
42. Heaslip, V.; Hean, S.; Parker, J. Lived experience of vulnerability from a Gypsy Roma Traveller perspective. *J. Clin. Nurs.* **2016**, *25*, 1987–1998. [CrossRef] [PubMed]
43. Lajoie, C.; Fortin, J.; Racine, E. Enriching our understanding of vulnerability through the experiences and perspectives of individuals living with mental illness. *Account Res.* **2019**, *26*, 439–459. [CrossRef]
44. Guignon, C. *On Being Authentic*; Routledge: London, UK, 2004.

45. Pfaff, K.; Krohn, H.; Crawley, J.; Howard, M.; Zadeh, P.M.; Varacalli, F.; Ravi, P.; Sattler, D. The little things are big: Evaluation of a compassionate community approach for promoting the health of vulnerable persons. *BMC Public Health* **2021**, *21*, 2253. [CrossRef] [PubMed]
46. Hardcastle, S.J.; Hancox, J.; Hattar, A.; Maxwell-Smith, C.; Thøgersen-Ntoumani, C.; Hagger, M.S. Motivating the unmotivated: How can health behavior be changed in those unwilling to change? *Front. Psychol.* **2015**, *6*, 835. [CrossRef] [PubMed]
47. Biggs, A.; Brough, P.; Drummond, S. Lazarus and Folkman's psychological stress and coping theory. In *The Handbook of Stress and Health: A Guide to Research and Practice*; Cooper, C.L., Quick, J.C., Eds.; Wiley Blackwell: West Sussex, UK, 2017; pp. 351–364.
48. Condon, L.; Bedford, H.; Ireland, L.; Kerr, S.; Mytton, J.; Richardson, Z.; Jackson, C. Engaging gypsy, Roma, and traveller communities in research: Maximizing opportunities and overcoming challenges. *Qual. Health Res.* **2019**, *29*, 1324–1333. [CrossRef]
49. Dovey-Pearce, G.; Walker, S.; Fairgrieve, S.; Parker, M.; Rapley, T. The burden of proof: The process of involving young people in research. *Health Expect.* **2019**, *22*, 465–474. [CrossRef]
50. Tomm-Bonde, L. The Naïve nurse: Revisiting vulnerability for nursing. *BMC Nurs.* **2012**, *11*, 5. [CrossRef]
51. Torralba i Roselló, F. *Ética del Cuidar*; Editorial MAPFRE: Madrid, Spain, 2002.
52. Mc Conalogue, D.; Maunder, N.; Areington, A.; Martin, K.; Clarke, V.; Scott, S. Homeless people and health: A qualitative enquiry into their practices and perceptions. *J. Public Health* **2021**, *43*, 287–294. [CrossRef]

 *healthcare*

Article

Psychological Well-Being Increment as Post-Traumatic Growth in Women with Breast Cancer: A Controlled Comparison Design Using Propensity Score Matching

Ren-Hau Li [1,2,*,†], Hsiu-Ling Peng [1,2], Ming-Hsin Yeh [3] and Jiunnhorng Lou [4,*,†]

[1] Department of Psychology, Chung Shan Medical University, Taichung 40201, Taiwan; phl@csmu.edu.tw
[2] Clinical Psychological Room, Chung Shan Medical University Hospital, Taichung 40201, Taiwan
[3] Department of Surgery, Chung Shan Medical University Hospital, Taichung, Taiwan, Institute of Medicine, School of Medicine, Chung Shan Medical University, Taichung 40201, Taiwan; cshy1629@csmu.edu.tw
[4] Department of Nursing, Hsin Sheng College of Medical Care and Management, Taoyuan 325004, Taiwan
*   Correspondence: davidrhlee@yahoo.com.tw (R.-H.L.); stujhl@gmail.com (J.L.)
†   These authors contributed equally to this work.

Abstract: The aim of this study was to confirm post-traumatic growth with respect to the psychological well-being of women with breast cancer compared to women without disease. Propensity score was used to match the two groups according to age, religious beliefs, education level, monthly income, and marital status. A psychological well-being scale with six factors was used, including positive relations with others (PR), autonomy (AU), environmental mastery (EM), personal growth (PG), purpose in life (PL), and self-acceptance (SA). A total 178 women with vs. 178 women without breast cancer were compared by matching with propensity scores, using factorial invariance tests to reduce measurement errors. The results showed that women with breast cancer had significantly higher psychological well-being for all the six factors ($\Delta\chi^2 = 37.37$, $p < 0.001$) and higher variability in terms of PR, AU, and PL than women without breast cancer ($\Delta\chi^2 = 45.94$, $p < 0.001$). Furthermore, women with breast cancer exhibited a significantly higher association between PG and PL and a significantly lower association between PG and EM than women without breast cancer ($\Delta\chi^2 = 44.49$, $p < 0.001$). This implies that psychological well-being could assess broader and more subtle post-traumatic growth in women with breast cancer and that growth was more associated with internal life value than with external environmental control.

Keywords: breast cancer; propensity score; factorial invariance tests; psychological well-being; post-traumatic growth; latent variables

Citation: Li, R.-H.; Peng, H.-L.; Yeh, M.-H.; Lou, J. Psychological Well-Being Increment as Post-Traumatic Growth in Women with Breast Cancer: A Controlled Comparison Design Using Propensity Score Matching. *Healthcare* **2022**, *10*, 1388. https://doi.org/10.3390/healthcare10081388

Academic Editors: Carlos Laranjeira and Ana Querido

Received: 22 June 2022
Accepted: 23 July 2022
Published: 25 July 2022

Publisher's Note: MDPI stays neutral with regard to jurisdictional claims in published maps and institutional affiliations.

1. Introduction

Post-traumatic growth, a positive phenomenon that usually occurs after stressful events, is a type of transformational change that occurs as a positive response to challenges to one's core beliefs following genuinely traumatic events [1]. However, stressful events may also result in severely negative outcomes in persons with cognitive, emotional, or behavioural distress or injuries, called traumas, such as receiving a diagnosis of breast cancer. Trauma can be briefly defined as 'a life-altering event that is seismic enough to impact one's assumptive world, prioritising one's subjective response to significantly challenging events' [2]. Individuals who have undergone trauma may exhibit positive and/or negative responses. Some negative responses may lead to psychological disease, known as post-traumatic stress disorder, as per the Diagnostic and Statistical Manual of Mental Disorders: Fifth Edition (DSM-V). In contrast, some positive responses or benefits may result in growth, including strengthening of individuals, families, and communities; discovery of previously unrecognised abilities and talents; and tightening of relationships and solidarity [3]. Post-traumatic growth is also viewed as a positive psychological change experienced due to struggles with highly challenging life circumstances and is usually negatively associated with depression, anxiety, and stress [4,5], which are also associated with poor breast cancer prognosis [6].

Breast cancer, the most frequent malignancy in women worldwide, has a curability rate of 70–80% in patients with early-stage, non-metastatic disease. However, advanced breast cancer with distant organ metastases is considered incurable [7]. Breast cancer is usually viewed as a trauma [8] associated with anxiety, depression, suicide, and neurocognitive and sexual dysfunctions, according to a systematic review [9]. Furthermore, it leads to psychological and physiological distress, such as stress, demoralisation, and sleeping disturbances [10]. In 2020, breast cancer surpassed lung cancer as the most commonly diagnosed cancer, with an estimated 2.3 million new cases (11.7%) worldwide. Furthermore, it is the leading cause of cancer deaths among women [11]. In Taiwan, breast cancer is the most rapidly growing cancer and the leading cause of cancer deaths among women, with over 10,000 women diagnosed and more than 2000 women dying from breast cancer annually [12]. Although breast cancer is usually prevalent in middle-aged and older women, its prevalence has been increasing among youth in Taiwan [13,14].

Positive psychology, which encourages facing diseases with positive attitudes, has been advocated in the last 20 years [15], and researchers have started to consider the positive benefits of traumas, such as quality of life, post-traumatic growth, social support, and well-being [16,17]. Approximately 30–70% of participants in various studies reported benefits from trauma [18], and 97% of women with breast cancer experienced post-traumatic growth [19]. However, most studies on positive psychology-related variables of breast cancer did not confirm the post-traumatic growth phenomena via a control group design or even by matching participants via confounding or sociodemographic variables, much less by comparisons among latent variables to avoid measurement errors.

Only one study compared post-traumatic growth in women with and without breast cancer using the matching method [20]. A study including 774 women with breast cancer and 666 randomly sampled women without breast cancer adjusted for age and education in a regression analysis found that although the group with breast cancer had significantly higher post-traumatic growth scores in two of the five subscales, there were no significant differences in terms of the total post-traumatic growth scale [21]. Cordova et al. compared 70 breast cancer survivors to 70 age- and education-matched healthy women. Their findings indicated that the group with breast cancer had higher post-traumatic growth scores; however, there were no significant differences in depression and psychological well-being [20]. The lack of a significant difference in depression conflicted with previous systematic reviews and meta-analyses [9,22]. In addition, it is worth noting that no significant difference was observed in psychological well-being was. Because only one study used a matching method [20], we found it necessary to implement a comparative design with advanced methods.

In the present research, in to reach achieve more equivalent groups in random assignment design, we used propensity score to match two groups of women with/without breast cancer based on five sociodemographic variables—age, presence or absence of religious belief, educational level, monthly income level, and marital status—which are often mentioned in the literature [9,20,23,24]. The advantage of the propensity score method is that it reduces selection bias in observational studies while simultaneously considering many confounding variables in matching [25]. In addition, we operationally defined the difference between the two groups on the psychological well-being scale as post-traumatic growth and discussed their differences. Comparisons were executed for latent variables via a factorial invariance test through confirmatory factor analysis in structural equation modelling [26]. The primary aim of the study was to use a control-group setting with advanced statistical control and techniques to confirm the existence of post-traumatic growth in women with breast cancer. The secondary aim was to use the Psychological Well-Being Scale to identify post-traumatic growth and its related phenomena. We assumed that the psychological well-being increments (differences between groups) would lead to the discovery of some extent of post-traumatic growth for women with breast cancer.

2. Materials and Methods

2.1. Procedures

A cross-sectional survey was mainly conducted in Taiwan from August 2013 to September 2015. However, a few participants with breast cancer were recruited from December 2020 to April 2021 before the outbreak of COVID-19 in Taiwan. The questionnaire included questions about age, religious beliefs, education level, monthly income, marital status, and a psychological well-being scale. This study was approved by the Institutional Review Board of Chung Shan Medical University Hospital (CSMUH Nos. CS-13203 and CS1-20158). We adopted a convenience sampling method to recruit women with or without breast cancer from the central area of Taiwan. The participants were informed of the research contents, their rights, and privacy protection and were administered the scale by trained researchers and assistants after they agreed to participate in the study. Collected data were checked for the purpose of matching similar sociodemographic variables to guarantee post-traumatic growth mainly from events associated with breast cancer. Propensity score matching was executed using SPSS statistical software version 23 to find qualified participants for later analysis and factorial invariance tests.

2.2. Participants

A total of 351 women with breast cancer agreed to participate. We also recruited 226 women without breast cancer, who were included in the control group. Because the two sample groups had different distributions of sociodemographic variables, we matched participants according to their age, religious beliefs, educational level, monthly income, and marital status using propensity scores. We set a caliper of 0.05 in terms of propensity scores to select 178 women with and 178 women without breast cancer. The former group had a mean age of 57.56 years and standard deviation of 6.70, with an age range of 39.0 to 76.0. The latter had a mean age of 57.78 years and standard deviation of 6.61, with an age range of 42.7 to 77.4. Among 178 women with breast cancer, 88.2% had undergone surgery, 65.7% had received chemotherapy, 58.4% had received radiation therapy, and 41.6% had received hormone therapy. In addition, 10.2% had received at least one of the above-mentioned treatments, and 83.1% had received at least two of the above-mentioned treatments; however, 6.7% had not received any of the above-mentioned treatments.

2.3. Instrument

We adopted a brief 18-item psychological well-being scale from Li's version based on Ryff's efforts, which was reported to have validity and reliability in middle-aged and older samples [27]. The scale included six factors/subscales: positive relations with others (PR), autonomy (AU), environmental mastery (EM), personal growth (PG), purpose in life (PL), and self-acceptance (SA). Each subscale has three items, and each item was rated on a 6-point Likert scale. The higher the scale score, the higher the level of psychological well-being. The scale had Cronbach's reliability coefficient alpha of 0.92 and 0.82–0.90 for each subscale in the present study. Confirmatory factor analysis of the scale showed that chi-square (χ^2) = 393.97, degree of freedom (df) = 120, $p < 0.001$, χ^2/df = 3.28, comparative fit index (CFI) = 0.98, non-normed fit index (NNFI) = 0.97, goodness-of-fit index (GFI) = 0.89, standardized root mean squared residual (SRMR) = 0.049, and root mean error of approximation (RMSEA) = 0.080, which indicated an acceptable model fit. The factor loadings ranged from 0.72 to 0.93, with a mean factor loading of 0.82.

2.4. Statistical Analysis

The propensity score method is a statistical technique used to reduce selection bias in observational study samples to equate the distribution of confounding covariates, such as age, religious beliefs, educational level, marital status, and monthly income level, between the two groups. Based on logistic regression, the propensity score represents the probability that each participant was assigned to a particular group, given a set of confounding observed covariates [25]. The matching processes involved pairing participants in the two groups based on similar propensity scores (probability difference within 0.05), which resulted in two groups with similar distributions of confounding covariates, as shown in the Results section. The skewness coefficients of the total score on the psychological well-being scale and its six subscale scores ranged from -0.268 to 0.237, and the kurtosis coefficients ranged from -0.586 to 0.661, i.e., close to zero, satisfying normality in the sample of the 356 matched participants.

Next, factorial invariance tests were used to confirm that differences in the observed scores between groups reflected true differences in psychological constructs and that they were free from the interference of measurement errors [26]. The tests were used to verify that the measurement models for groups of women with and without breast cancer had full or partial metric and scalar invariance, which was necessary to guarantee the next feasible comparisons between the parameter estimates in the structural models of the two groups. The testing processes were executed using chi-squared difference ($\Delta\chi^2$) tests between the nested models. The metric invariance and scalar invariance tests were used to verify the extent of equality for factor loadings and intercepts in the process of confirmatory factor analysis, respectively. Metric invariance ensured that the unit of the scale was consistent, and scalar invariance ensured that the origin of scale was consistent between the two groups. Whereas the two groups had the same unit and origin of the scale, their comparisons in each factor of psychological well-being were meaningful. Partial invariances in the measurement models were also capable of guaranteeing that comparisons in the structural model between groups were feasible [28,29]. We used Lisrel 8.8 statistical software to conduct factorial invariance tests.

3. Results

3.1. Sociodemographic Information after Matching

A t-test showed no significant difference in age ($t = -0.315$, $p = 0.753$). The other sociodemographic variables of the matched samples showed consistent distributions between the two groups, according to chi-squared tests, as shown in Table 1. The non-significant results for the sociodemographic variables indicated that matching through propensity scores was successful.

3.2. Factorial Invariance Tests

Table 2 shows the steps of the factorial invariance tests involving measurement and structural models. In the measurement model, the configural model invariance test checked that the two groups (with and without breast cancer) had the same number of factors with acceptable model-fit indices, such as RMSEA = 0.092 (<0.10), CFI = 0.969 (>0.090), and SRMR = 0.058 and 0.057 (<0.08). The full metric invariance test confirmed the same factor loadings between the two groups, with a non-significant chi-squared difference ($\Delta\chi^2$) of 17.90. Although full scalar invariance was not satisfied due to a $\Delta\chi^2$ of 32.24 with $df = 12$, reaching a significance level of $p < 0.001$, partial scalar invariance was satisfied, with a non-significant $\Delta\chi^2$ of 15.83 and $df = 10$, and only two intercepts were freely estimated. The next test, which involved invariance of measurement errors, also indicated that partial invariance of error variances was satisfied ($\Delta\chi^2 = 18.89$, $df = 12$); however, it was not important for parameter comparisons in the structure model [28]. Therefore, the next invariance tests, which involved factor covariances, variances, and means in the structural model, were based on model D instead of model F to avoid excessive distortion in the next invariance tests [30].

Table 1. Sociodemographic information of the participants (n = 356).

Sociodemographic Variable	n (%)		Chi-Square (χ^2)
	Breast Cancer	Normal	
Religious belief			0.11
No	19 (10.7%)	21 (11.8%)	
Yes	159 (89.3%)	157 (88.2%)	
Educational level			2.79
Illiterate	5 (2.8%)	9 (5.1%)	
Elementary school	31 (17.4%)	34 (19.1%)	
Junior high school	26 (14.6%)	19 (10.7%)	
Senior high school or university	110 (61.8%)	112 (62.9%)	
Graduate degree or above	6 (3.4%)	4 (2.2%)	
Monthly income			7.56
Less than TWD 20 thousand	97 (54.5%)	78 (43.8%)	
TWD 20–50 thousand	56 (31.5%)	61 (34.3%)	
TWD 50–80 thousand	14 (7.9%)	29 (16.3%)	
TWD 80 thousand	11 (6.2%)	10 (5.6%)	
Marriage			2.85
Unmarried/single	9 (5.1%)	14 (7.9%)	
Married/cohabited	142 (79.8%)	130 (73.0%)	
Divorced/separated	15 (8.4%)	16 (9.0%)	
Widowed	12 (6.7%)	18 (10.1%)	

Table 2. Tests of factorial invariance of the psychological well-being scale between women with and without breast cancer (n = 356).

Models	Compared Model	χ^2 (df)	RMSEA	CFI	SRMR	$\Delta\chi^2$ (Δdf)
A. Configural invariance		600.62 (240)	0.092	0.969	0.058/0.057	
B. Full metric invariance	A	618.52 (252)	0.091	0.968	0.065/0.062	17.90 (12)
C. Full scalar invariance	B	650.76 (264)	0.091	0.966	0.065/0.058	32.24 *** (12)
D. Partial scalar invariance	B	634.35 (262)	0.090	0.967	0.065/0.059	15.83 (10)
E. Full invariance of error variances	D	698.49 (280)	0.092	0.963	0.067/0.063	64.14 *** (18)
F. Partial invariance of error variances	D	653.24 (274)	0.088	0.972	0.066/0.062	18.89 (12)
G. Full invariance of factor variances	D	680.29 (268)	0.093	0.964	0.144/0.211	45.94 *** (6)
H. Partial invariance of factor variances	D	641.67 (265)	0.090	0.967	0.108/0.107	7.32 (3)
I. Full invariance of factor covariances	D	678.84 (277)	0.091	0.964	0.122/0.094	44.49 *** (15)
J. Partial invariance of factor covariances	D	653.42 (275)	0.088	0.966	0.112/0.079	19.07 (13)
K. Full invariance of latent means	G	717.66 (274)	0.096	0.961	0.137/0.193	37.37 *** (6)

*** $p < 0.001$.

Because the unit and origin of scale were the same between the two groups for metric invariance and scalar invariance, partial invariance tests of factor covariances, variances, and means provided information regarding the true differences in terms of psychological well-being between women with and without breast cancer. As shown in Table 3, we found that the factor means (latent means) were differed across six factors between the two groups ($\Delta\chi^2 = 37.37$, $df = 6$, $p < 0.001$). The group of women with breast cancer consistently had significantly higher means than the group of women without breast cancer, such as higher means for positive relations with others (PR: 3.38 vs. 3.13), autonomy (AU: 3.18 vs. 3.04), environmental mastery (EM: 3.78 vs. 3.43), personal growth (PG: 4.00 vs. 3.64), purpose in life (PL: 3.74 vs. 3.53), and self-acceptance (SA: 3.46 vs. 3.25).

Table 3. Invariant and noninvariant factor loadings, intercepts, error variances, and mean differences between women with and without breast cancer.

Factors	Items	Factor Loadings	Intercepts	Error Variances	Latent Mean
PR	PR1	0.81	−0.45	0.33	
	PR2	0.77	0.13	0.39	
	PR3	0.79	0.39	0.39	3.38/3.13
AU	AU4	0.73	0.48	0.58/0.36	
	AU5	0.83	−0.32	0.31	
	AU6	0.79	−0.14	0.45/0.29	3.18/3.04
EM	EM7	0.84	−0.22/−0.31	0.28	
	EM8	0.89	0.01	0.22	
	EM9	0.81	0.21/0.31	0.33	3.78/3.43
PG	PG10	0.76	0.28	0.41	
	PG11	0.92	−0.15	0.11/0.20	
	PG12	0.91	−0.15	0.17	4.00/3.64
PL	PL13	0.85	−0.37/−0.24	0.27	
	PL14	0.91	−0.38	0.26/0.13	
	PL15	0.75	0.80/0.66	0.44	3.74/3.53
SA	SA16	0.72	0.22	0.62/0.36	
	SA17	0.91	−0.77	0.15	
	SA18	0.76	0.57	0.53/0.34	3.46/3.25

Note: All estimates are presented in a completely standardised common metric solution. For factor loadings, intercepts, and error variances, non-invariant estimates are presented as a pattern of with/without breast cancer, and invariant estimates are presented as a single value. PR, positive relations with others; AU, autonomy; EM, environmental mastery; PG, personal growth; PL, purpose in life; SA, self-acceptance.

Table 4 shows the covariances and variances of the six factors for the two groups under a completely standardised common metric solution. Women with breast cancer had significantly larger variances in three of the six factors ($\Delta\chi^2 = 45.94$, $df = 6$, $p < 0.001$)—positive relations with others (1.14 vs. 0.86), autonomy (1.38 vs. 0.62), and purpose in life (1.31 vs. 0.69)—than women without breast cancer. In addition, women with breast cancer had a significantly larger correlation coefficient ($\Delta\chi^2 = 44.49$, $df = 15$, $p < 0.001$) between personal growth and purpose in life (0.86 vs. 0.61) and a lower correlation coefficient between personal growth and environmental mastery (0.48 vs. 0.68) than women without breast cancer.

Table 4. Interfactor correlation of the six factors of PWB between women with and without breast cancer ($n = 356$).

	PR	AU	EM	PG	PL	SA
PR	1.14/0.86					
AU	0.51	1.38/0.62				
EM	0.64	0.60	1.00			
PG	0.66	0.54	0.48/0.68	1.00		
PL	0.53	0.46	0.50	0.86/0.61	1.31/0.69	
SA	0.66	0.57	0.66	0.66	0.68	1.00

Note: All estimates are presented in a completely standardised common metric solution; hence, some values are higher than 1.00. Non-invariant estimates between groups are presented as a pattern of with/without breast cancer; invariant estimates are presented as a single value. PR, positive relations with others; AU, autonomy; EM, environmental mastery; PG, personal growth; PL, purpose in life; SA, self-acceptance.

4. Discussion

In the present study, using the propensity score method, we found that the matched samples had similar sociodemographic distributions between women with and without breast cancer. The research results support the hypothesis that women with breast cancer have higher levels of psychological well-being than women without breast cancer across six factors.

Benefits or positive changes that arise from trauma can be viewed as adversarial growth, stress-related growth, benefit finding, and positive psychological changes [3,31]. These gains in the aftermath of suffering are known as post-traumatic growth and were developed to form a post-traumatic growth inventory [32]. This inventory includes 21 items across five factors: new possibilities, relating to others, personal strength, spiritual change, and appreciation of life. In terms of psychological well-being, it is also a type of eudaimonism approach, such as the post-traumatic growth inventory [33], as opposed to hedonism-like well-being, i.e., it originates from growth as a result of a struggle with suffering, not simply happiness resulting from a sensory feeling. Therefore, psychological well-being generally refers to broader growth experiences of persons facing problems or challenges, from light, trivial problems to serious, stressful events—not just trauma. We believe that using the psychological well-being scale could help to evaluate broader overall growth in women with breast cancer. Furthermore, we believe that the post-traumatic growth inventory measures deep awareness of changes involving trauma.

Our findings that all six factors differed significantly between the two groups are roughly consistent with those reported by Brix et al. [21], who found statistically significant post-traumatic growth in two aspects—appreciation of life and relating to others— but not in new possibilities, personal strength, spiritual change, or overall post-traumatic growth. Their aspect 'appreciation of life' was similar to our factor 'purpose in life', and their aspect 'relating to others' was similar to our factor 'positive relations with others.' Therefore, Brix et al. did not characterise post-traumatic growth in the other three aspects or according to an overall scale. As they indicated, 'severity of breast cancer played a role in the development of post-traumatic growth', not just having breast cancer or not. Moreover, given advancements in medical technology, breast cancer is no longer considered a serious or incurable disease. Currently, it is a relatively mild cancer, with a 91% 5-year survival rate after diagnosis, making it the 5th most survivable of 23 cancers [34]. Therefore, the severity of breast cancer could be interpreted as a subjective feeling and not merely a diagnosis with respect to the stages of cancer or the corresponding treatments. In the present study, broadly speaking, the various types of growth among women with breast cancer were significant compared to those experienced by women without breast cancer in terms of psychological well-being. This could imply that although some women with breast cancer did not experience post-traumatic growth, they actually grew in terms of psychological well-being. Therefore, we viewed the difference (increment) of psychological well-being between the two groups as a kind of post-traumatic growth.

Cordova et al. [20] reported significantly higher levels of post-traumatic growth in relation to others, appreciation of life, and spiritual change, which was similar to the results reported by Brix et al. However, they did not find significant differences in terms of psychological well-being, unlike our results. Cordova et al. used only three of the six subscales of the psychological well-being scale, namely personal growth, purpose in life, and self-acceptance, and reported no significant difference between the groups. Moreover, they recruited participants who had undergone surgery, chemotherapy, and radiotherapy, which ensured the development of post-traumatic growth [35]. However, we found that three of our six factors had significantly higher variability in women with breast cancer. This may imply that the sample in the study by Cordova et al. was too homogenous and not sufficiently large, which led to low variance and insignificance. This inference was directly supported by comparison of the sizes of standard deviations with those reported by Brix et al. [21] and indirectly supported by the conflicts with a systematic review by

Carreira et al. [9] regarding the significance of depression. More evidence is required to confirm this hypothesis.

We also found that the correlation coefficient between personal growth and purpose in life was larger in women with breast cancer than in women without breast cancer. Furthermore, the correlation coefficient between personal growth and environmental mastery was smaller in women with breast cancer than in women without breast cancer. As for the connotations of environmental mastery and purpose in life [36], environmental mastery meant that people had good control of living situations, daily life, and finances, whereas purpose in life meant making positive plans and struggling to fulfil them. However, the former was more subject to many external factors, and the latter was more focused on self-determination. This result may imply that women with breast cancer were able to pursue their internal values to a greater extent than they were able to control the outer environment in association with their personal growth. That is, even if women with breast cancer had significantly higher scores for all six factors, their post-traumatic growth may have been more associated with the achievement of an internal value rather than built-in outer environmental controls. This result is roughly consistent with empirical research on patient-perceived changes in the system of values after cancer diagnosis involving terminal and instrumental values [37].

Other cancer-related background variables, including post-traumatic growth, such as cancer stage, time since diagnosis or operation, and treatment type [4,20,21], could not be considered in the matching, because the control group had no such variables. However, these variables may have affected the research results. They should be controlled experimentally or statistically and should not be simply described or listed as characteristics of the participants. Furthermore, we did not consider matching some related psychological variables, such as personality traits, coping style, and social support. Although the authors of many studies have found that such variables were associated with post-traumatic growth [38–41], we suggest that personality traits are much more worthy of consideration in future research, as they are essentially innate dispositions and are not easily changed in response to stressful events or trauma.

5. Conclusions

To confirm post-traumatic growth in women with breast cancer in an observational study, we used propensity scores to match sociodemographic variables between women with and without breast cancer. In addition, we used a factorial invariance test to reduce measurement errors and found a difference in psychological well-being between the two groups in terms of latent variables. Women with breast cancer had significantly higher levels of psychological well-being than women without breast cancer, including positive relations with others, autonomy, environmental mastery, personal growth, purpose in life, and self-acceptance. These findings confirm women with breast cancer experienced post-traumatic growth to some extent. Women with breast cancer exhibited significantly larger variability than women without breast cancer among the three factors of psychological well-being of positive relations with others, autonomy, and purpose in life. This may imply that psychological well-being can assess broader and more subtle post-traumatic growth. Furthermore, women with breast cancer had a significantly larger associations between personal growth and purpose in life and a significantly lower association between personal growth and environmental mastery than did women without breast cancer. This may imply that post-traumatic growth is more associated with internal life value than with external environmental control.

The main limitation of the present research is that conducting a study like ours using an experimental design with randomization would be unethical; hence, a longitudinal panel study with a control group design is preferable. In addition, indicators of severity of breast cancer, such as time since diagnosis, stage, and treatment type, may have influenced the results of our study; therefore, they should be considered subjective measures of severity as a psychological variable to be controlled in future studies.

Author Contributions: Conceptualization, R.-H.L.; methodology, R.-H.L.; software, R.-H.L.; validation, R.-H.L.; formal analysis, R.-H.L.; investigation, R.-H.L. and J.L.; resources, H.-L.P., M.-H.Y. and J.L.; data curation, R.-H.L. and H.-L.P.; writing—original draft preparation, R.-H.L.; writing—review and editing, J.L.; visualization, R.-H.L.; supervision, R.-H.L. and J.L.; project administration, R.-H.L. and M.-H.Y.; funding acquisition, R.-H.L. and J.L. All authors have read and agreed to the published version of the manuscript.

Funding: This research was funded by the Ministry of Science and Technology (MOST 109-2410-H-040-006).

Institutional Review Board Statement: The study was conducted in accordance with the Declaration of Helsinki and was approved by Chung Shan Medical University Hospital (CSMUH No: CS-13203 and CS1-20158).

Informed Consent Statement: Informed consent was obtained from all the subjects involved in the study. Written informed consent was obtained from the patients for the publication of this paper.

Data Availability Statement: The data presented in this study are available upon request, owing to privacy restrictions.

Acknowledgments: We appreciate the contribution of the women who participated in this study.

Conflicts of Interest: The authors declare no conflict of interest. The funders had no role in the study design; collection, analyses, or interpretation of data; writing of the manuscript; or decision to publish the results.

References

1. Tedeschi, R.G.; Shakespeare-Finch, J.; Taku, K.; Calhoun, L.G. *Posttraumatic Growth: Theory, Research, and Applications*; Routledge: New York, NY, USA, 2018.
2. Kadri, A.; Gracey, F.; Leddy, A. What factors are associated with posttraumatic growth in older adults? A systematic review. *Clin. Gerontol.* **2022**, 1–18. [CrossRef] [PubMed]
3. Berger, R. *Stress, Trauma, and Posttraumatic Growth*; Routledge: New York, NY, USA, 2015. [CrossRef]
4. Casellas-Grau, A.; Ochoa, C.; Ruini, C. Psychological and clinical correlates of posttraumatic growth in cancer: A systematic and critical review. *Psychooncology* **2017**, *26*, 2007–2018. [CrossRef] [PubMed]
5. Thakur, M.; Sharma, R.; Mishra, A.K.; Singh, K. Posttraumatic growth and psychological distress among female breast cancer survivors in India: A cross-sectional study. *Indian J. Med. Paediatr. Oncol.* **2022**, *43*, 165–170. [CrossRef]
6. Wang, X.; Wang, N.; Zhong, L.; Wang, S.; Zheng, Y.; Yang, B.; Zhang, J.; Lin, Y.; Wang, Z. Prognostic value of depression and anxiety on breast cancer recurrence and mortality: A systematic review and meta-analysis of 282,203 patients. *Mol. Psychiatry* **2020**, *25*, 3186–3197. [CrossRef]
7. Harbeck, N.; Penault-Llorca, F.; Cortes, J.; Gnant, M.; Houssami, N.; Poortmans, P.; Ruddy, K.; Tsang, J.; Cardoso, F. Breast cancer. *Nat. Rev. Dis. Primers* **2019**, *5*, 66. [CrossRef]
8. Cordova, M.J.; Giese-Davis, J.; Golant, M.; Kronenwetter, C.; Vickie, C.; Spiegel, D. Breast cancer as trauma: Posttraumatic stress and posttraumatic growth. *J. Clin. Psychol. Med. Settings* **2007**, *14*, 308–319. [CrossRef]
9. Carreira, H.; Williams, R.; Müller, M.; Harewood, R.; Stanway, S.; Bhaskaran, K. Associations between breast cancer survivorship and adverse mental health outcomes: A systematic review. *J. Natl. Cancer Inst.* **2018**, *110*, 1311–1327. [CrossRef]
10. Peng, H.L.; Hsueh, H.W.; Chang, Y.H.; Li, R.H. The mediation and suppression effect of demoralization among breast cancer patients after primary therapy: A structural equation model. *J. Nurs. Res.* **2021**, *29*, e144. [CrossRef]
11. Sung, H.; Ferlay, J.; Siegel, R.L.; Laversanne, M.; Soerjomataram, I.; Jemal, A.; Bray, F. Global cancer statistics 2020: GLOBOCAN estimates of incidence and mortality worldwide for 36 cancers in 185 countries. *CA Cancer J. Clin.* **2021**, *71*, 209–249. [CrossRef]
12. Health Promotion Administration, Ministry of Health and Welfare. Breast Cancer Prevention. 2022. Available online: https://www.hpa.gov.tw/Pages/Detail.aspx?nodeid=205&pid=1124 (accessed on 25 January 2022).
13. Formosa Cancer Foundation. Breast Cancer. 2022. Available online: https://www.canceraway.org.tw/pagelist.php?keyid=62 (accessed on 25 May 2022).
14. Chien, L.H.; Tseng, T.J.; Chen, C.H.; Jiang, H.F.; Tsai, F.Y.; Liu, T.W.; Hsiung, C.A.; Chang, I.S. Comparison of annual percentage change in breast cancer incidence rate between Taiwan and the United States-A smoothed Lexis diagram approach. *Cancer Med.* **2017**, *6*, 1762–1775. [CrossRef]
15. Seligman, M.E.P. *Authentic Happiness: Using the New Positive Psychology to Realize Your Potential for Lasting Fulfillment*; Atria Books: New York, NY, USA, 2002.
16. McDonough, M.H.; Sabiston, C.M.; Wrosch, C. Predicting changes in posttraumatic growth and subjective well-being among breast cancer survivors: The role of social support and stress. *Psychooncology* **2014**, *23*, 114–120. [CrossRef] [PubMed]
17. Schroevers, M.J.; Helgeson, V.S.; Sanderman, R.; Ranchor, A.V. Type of social support matters for prediction of posttraumatic growth among cancer survivors. *Psychooncology* **2010**, *19*, 46–53. [CrossRef] [PubMed]

18. Weiss, T.; Berger, R. (Eds.) *Posttraumatic Growth and Culturally Competent Practice: Lessons Learned from Around the Globe*; Wiley: Hoboken, NJ, USA, 2010.
19. Zsigmond, O.; Vargay, A.; Józsa, E.; Bányai, É. Factors contributing to post-traumatic growth following breast cancer: Results from a randomized longitudinal clinical trial containing psychological interventions. *Dev. Health Sci.* **2019**, *2*, 29–35. [CrossRef]
20. Cordova, M.J.; Cunningham, L.L.; Carlson, C.R.; Andrykowski, M.A. Posttraumatic growth following breast cancer: A controlled comparison study. *Health Psychol.* **2001**, *20*, 176–185. [CrossRef]
21. Brix, S.A.; Bidstrup, P.E.; Christensen, J.; Rottmann, N.; Olsen, A.; Tjønneland, A.; Johansen, C.; Dalton, S.O. Post-traumatic growth among elderly women with breast cancer compared to breast cancer-free women. *Acta Oncol.* **2013**, *52*, 345–354. [CrossRef]
22. Pilevarzadeh, M.; Amirshahi, M.; Afsargharehbagh, R.; Rafiemanesh, H.; Hashemi, S.; Balouchi, A. Global prevalence of depression among breast cancer patients: A systematic review and meta-analysis. *Breast. Cancer Res. Treat.* **2019**, *176*, 519–533. [CrossRef]
23. Bellizzi, K.M.; Blank, T.O. Predicting posttraumatic growth in breast cancer survivors. *Health Psychol.* **2006**, *25*, 47–56. [CrossRef]
24. Shaw, A.; Joseph, S.; Linley, P.A. Religion, spirituality, and posttraumatic growth: A systematic review. *Ment. Health Relig. Cult.* **2005**, *8*, 1–11. [CrossRef]
25. Rosenbaum, P.R. *Design of Observational Studies*, 2nd ed.; Sage: Thousand Oaks, CA, USA, 2020.
26. Hair, J.F.; Black, W.C.; Babin, B.J.; Anderson, R.E. *Multivariate Data Analysis*, 8th ed.; Cengage: Boston, MA, USA, 2018.
27. Li, R.H. Reliability and validity of a shorter Chinese version for Ryff's psychological well-being scale. *Health Educ. J.* **2014**, *73*, 446–452. [CrossRef]
28. Little, T.D. *Longitudinal Structural Equation Modelling*; The Guilford Press: New York, NY, USA, 2013.
29. Li, R.H.; Kao, C.M.; Wu, Y.Y. Gender differences in psychological well-being: Tests of factorial invariance. *Qual. Life Res.* **2015**, *24*, 2577–2581. [CrossRef]
30. Brown, T.A. *Confirmatory Factor Analysis for Applied Research*, 2nd ed.; The Guilford Press: New York, NY, USA, 2015.
31. Helgeson, V.S.; Reynolds, K.A.; Tomich, P.L. A meta-analytic review of benefit finding and growth. *J Consult Clin Psychol* **2006**, *74*, 797–816. [CrossRef] [PubMed]
32. Tedeschi, R.G.; Calhoun, L.G. The posttraumatic growth inventory: Measuring the positive legacy of trauma. *J Trauma Stress* **1996**, *9*, 455–471. [CrossRef] [PubMed]
33. Tedeschi, R.G.; Calhoun, L.G. TARGET ARTICLE: Posttraumatic Growth: Conceptual Foundations and Empirical Evidence. *Psychol. Inq.* **2004**, *15*, 1–18. [CrossRef]
34. Byrnes, H. Cancers with the Highest and Lowest Survival Rates. 2021. Available online: https://247wallst.com/special-report/2021/02/04/cancers-with-the-highest-and-lowest-survival-rates/ (accessed on 18 July 2022).
35. Parikh, D.; Ieso, P.D.; Garvey, G.; Thachil, T.; Ramamoorthi, R.; Penniment, M.; Jayaraj, R. Post-traumatic stress disorder and post-traumatic growth in breast cancer patients—A Systematic Review. *Asian Pact. J. Cancer Prev.* **2015**, *16*, 641–646. [CrossRef]
36. Ryff, C.D. Happiness is everything, or is it? Explorations on the meaning of psychological well-being. *J. Personal. Soc. Psychol.* **1989**, *57*, 1069–1081. [CrossRef]
37. Greszta, E.; Siemińska, M.J. Patient-perceived changes in the system of values after cancer diagnosis. *J. Clin. Psychol. Med. Settings* **2011**, *18*, 55–64. [CrossRef]
38. Kolokotroni, P.; Anagnostopoulos, F.; Tsikkinis, A. Psychosocial factors related to posttraumatic growth in breast cancer survivors: A review. *Women Health* **2014**, *54*, 569–592. [CrossRef]
39. Koutrouli, N.; Anagnostopoulos, F.; Potamianos, G. Posttraumatic stress disorder and posttraumatic growth in breast cancer patients: A systematic review. *Women Health* **2012**, *52*, 503–516. [CrossRef]
40. Prati, G.; Pietrantoni, L. Optimism, social support, and coping strategies as factors contributing to posttraumatic growth: A meta-analysis. *J. Loss Trauma* **2009**, *14*, 364–388. [CrossRef]
41. Urcuyo, K.R.; Boyers, A.E.; Carver, C.S.; Antoni, M.H. Finding benefit in breast cancer: Relations with personality, coping, and concurrent well-being. *Psychol. Health* **2005**, *20*, 175–192. [CrossRef]

Article

The Malay Literacy of Suicide Scale: A Rasch Model Validation and Its Correlation with Mental Health Literacy among Malaysian Parents, Caregivers and Teachers

Picholas Kian Ann Phoa [1], Asrenee Ab Razak [1,*], Hue San Kuay [1], Anis Kausar Ghazali [2], Azriani Ab Rahman [3], Maruzairi Husain [1], Raishan Shafini Bakar [1] and Firdaus Abdul Gani [4]

[1] Department of Psychiatry, School of Medical Sciences, Universiti Sains Malaysia Health Campus, Kubang Kerian 16150, Kelantan, Malaysia; picholas@student.usm.my (P.K.A.P.); huesankuay@usm.my (H.S.K.); drzairi@usm.my (M.H.); raishanshafini@usm.my (R.S.B.)

[2] Biostatistics and Research Methodology Unit, School of Medical Sciences, Universiti Sains Malaysia Health Campus, Kubang Kerian 16150, Kelantan, Malaysia; anisyo@usm.my

[3] Department of Community Medicine, School of Medical Sciences, Universiti Sains Malaysia Health Campus, Kubang Kerian 16150, Kelantan, Malaysia; azriani@usm.my

[4] Department of Psychiatry and Mental Health, Sultan Haji Ahmad Shah Hospital, Temerloh 25000, Pahang, Malaysia; drfirdaus.gani@moh.gov.my

* Correspondence: asrenee@usm.my

Abstract: The 27-item Literacy of Suicide Scale (LOSS) is a test designed to measure the respondent's suicide knowledge. The purpose of this study is to examine the psychometric properties of the Malay-translated version of the LOSS (M-LOSS) and its association to sociodemographic factors and mental health literacy. The 27-item LOSS was forward–backward translated into Malay, and the content and face validities were assessed. The version was distributed to 750 respondents across West Malaysia. Rasch model analysis was then conducted to assess the scale's psychometric properties. The validated M-LOSS and the Malay version of the Mental Health Knowledge Schedule (MAKS-M) were then distributed to 867 respondents to evaluate their level of suicide literacy, mental health literacy, and their correlation. Upon Rasch analysis, 26 items were retained. The scale was found to be unidimensional, with generally satisfying separation and reliability indexes. Sex, socio-economic status, and experience in mental health were found to significantly impact the mean score for mental health literacy. This study also found a significant mean difference for suicide literacy across school types. Furthermore, while this study observed a weak but significant negative correlation between age and suicide literacy, no correlation was found between mental health and suicide literacy.

Keywords: psychometric analysis; Rasch model analysis; suicide literacy; mental health literacy; adolescent mental health

Citation: Phoa, P.K.A.; Razak, A.A.; Kuay, H.S.; Ghazali, A.K.; Rahman, A.A.; Husain, M.; Bakar, R.S.; Gani, F.A. The Malay Literacy of Suicide Scale: A Rasch Model Validation and Its Correlation with Mental Health Literacy among Malaysian Parents, Caregivers and Teachers. *Healthcare* **2022**, *10*, 1304. https://doi.org/10.3390/healthcare10071304

Academic Editors: Carlos Laranjeira and Ana Querido

Received: 10 June 2022
Accepted: 8 July 2022
Published: 14 July 2022

Publisher's Note: MDPI stays neutral with regard to jurisdictional claims in published maps and institutional affiliations.

1. Introduction

The COVID-19 pandemic forced the closure of all educational institutions, from pre-primary to tertiary institutions, in 200 countries. This closure halted face-to-face teaching and learning and affected 1.58 billion students worldwide. It is deemed the largest educational interruption in history [1]. While the prospect of learning from the comfort of our own home initially sounded appealing, the negative impacts of the biodisaster on mental health slowly surfaced, including those on children and adolescents. The social isolation, lack of certainty, fear of infection, and increased risk of household violence could lead to increased anxiety, depression, stress, and adjustment disorders, which could exacerbate existing mental health problems [2]. Compounding the issue, mental health issues such as depression and anxiety are associated with a higher risk of suicidal behaviors [3]. Since the outbreak, a recent study in the US found an increase in positive screening for suicide risk among adolescents [4]. A similar increasing trend in the number of suicide cases was also observed in Malaysia—from 609 cases in 2019 to 631 cases in 2020—while within the first quarter of 2021, a total of 336 suicide cases were reported, which is a matter of serious concern [5].

Mayne et al. also highlighted the suspension of mental health screening activities and in-person visits during the pandemic, possibly aimed at reducing clinic visits to lower the risk of COVID-19 infection, causing a decreasing trend on depression screening at primary care visits [4]. In this situation, adult stakeholders, i.e., parents, caregivers, and teachers, play an important role in identifying mental health issues and providing initial psychological first aid for adolescents in mental distress. Parents and caregivers were the second-most preferred source of mental health assistance among adolescents after their peers due to the presence of trust and interpersonal relationships [6,7]. Although teachers were not the preferred source of help, teachers were more likely to encounter teens suffering from mental illnesses, since they were required to meet and communicate with children and adolescents for most of their working hours. A qualitative study among parents and caregivers of adolescents suffering from mental illnesses stated that most of their children's mental health problems were initially detected by school personnel [8].

Of late, mental health services are pressured to enhance suicide risk assessment to lower the suicide rate. As a result, suicide is mostly seen as a psychiatric illness [9]. In this light, while virtually all mental illnesses are associated with an increased risk of suicide, suicide does not always indicate an underlying mental illness. A review article stated that up to 66.7% of suicide cases were absent of Axis 1 disorders, 37.1% did not have underlying personality disorders and Axis 1 disorders, and 37% of suicide cases did not have underlying Axis 1 disorders with mild conditions [10]. In high-income nations, psychological autopsy revealed that up to 10% of those who died by suicide did not have an underlying mental illness. The rates were higher in several Asian countries [11]. Additionally, several studies found that adverse childhood experiences may lead to suicide attempts among the youth, mediated by maladaptive personality development (e.g., aggression, impulsivity, learning difficulties, substance abuse, and chronic pain) [12,13]. While mental disorders and suicide are the most explored areas in suicidology, we have yet to come across studies that prove the link between suicide literacy and mental health literacy. This information might be useful in shaping future mental health and suicide awareness campaigns, ensuring that both issues receive equal attention.

Improving mental health and suicide literacy is an important first step in understanding the issues, reducing stigma, and enhancing confidence in providing help [14]. In Malaysia, studies on suicide literacy were rather limited to medical- and healthcare-related professionals and students [15–17]. To the best of our knowledge, there have been no studies on suicide literacy among parents, caregivers, and teachers. Thus, there are no available baseline data for these population subsets. To assess the level of knowledge on suicide, the Literacy of Suicide Scale (LOSS) was developed [18]. The scale is dichotomous and consists of 27 items encompassing the signs and symptoms, risk factors, causes, nature, treatment, and prevention of suicide. The Mental Health Knowledge Schedule (MAKS), on the other hand, is a two-part instrument consisting of 12 items that were designed to evaluate several dimensions of mental health literacy. This study aims to translate LOSS into Malay and validate it among Malaysian parents, caregivers, and teachers of secondary school adolescents. This study also aimed to compare the mean scores of mental health literacy and suicide literacy levels among the subgroups and investigate the correlations between sociodemographic factors, mental health literacy, and suicide literacy among the population. The study hoped to assess suicide literacy among the Malay-speaking population to quantify their level of understanding on suicide and evaluate the effectiveness of suicide awareness interventions. Investigation into the relationship between sociodemographic factors and the level of mental health and suicide literacies also allowed the conceptualization of factors that affect the maintenance of mental health and informs the policy and strategy development for mental health and suicide awareness interventions to encourage positive help-seeking behavior and confidence in providing appropriate mental assistance.

2. Materials and Methods

2.1. Study Design and Subjects

This study employed a cross-sectional study design. The targeted population comprised parents, caregivers, and teachers of adolescents in secondary school. The source population is the parent–teacher association of secondary schools in West Malaysia. The schools were selected via multistage stratified random sampling. First, government secondary schools in West Malaysia were stratified based on the type of schools (i.e., national secondary schools "SMK", fully residential schools "SBP", and religious secondary schools "SMA") and the locality of school (i.e., urban and rural). States in West Malaysia were then divided into four geographical zones (i.e., northern, southern, eastern, and central zones). For each stratum, simple random sampling was used to select one state from each zone and one district from each state. Finally, one school was randomly chosen from each district. A total of 24 schools were involved in this study. A link to the questionnaire and a poster with the QR code were circulated among the target sample via a key informant from each school's parent–teacher association for recruitment. The target sample comprised parents or caregivers of students in these schools and teachers who were still actively teaching in the selected schools. The inclusion criteria included Malaysian, aged 18 and above, and literate in Malay. Individuals who could not read in Malay were excluded from this study. The data were collected between March and September of 2021.To conduct a Rasch model analysis, a minimum sample size of 250 respondents, or about 25 respondents per response category, is necessary to achieve stable item calibration [19]. For this cross-sectional study, single proportion formula was adopted with 80% statistical power and at a significance level of 5%. We expected a dropout rate of 20%, and a design effect of 2 was applied, yielding an effective sample size of minimum 730 respondents.

2.2. Instrument

LOSS is a 27-item scale constructed to assess the knowledge on suicide among the respondents based on four categories: (a) cause and nature (CN), (b) risk factors (RF), (c) signs and symptoms (SS), and (d) treatment and prevention (TP) of suicide. Since LOSS is a dichotomous scale, respondents were required to choose "true" or "false" according to their understanding of suicide. Respondents were given 1 score for each correct answer, whereas 0 scores were given for incorrect answers. The total score was obtained by summing the scores of items answered correctly. The score ranged between 0 and 27; the higher the score, the higher the suicide literacy. During the development, the LOSS was validated via the item response theory approach. Each item was constructed based on the latent trait rather than factor analysis and internal consistency. In this light, several past studies that utilized LOSS have reported an acceptable Cronbach's alpha value of 0.71 [18,20].

MAKS is an instrument designed to assess general mental health knowledge. The schedule comprises 12 items in two parts; Part A evaluates knowledge of mental health stigma regarding help-seeking, recognition, support, employment, treatment, and recovery. Part B evaluates knowledge on mental health diagnoses. Items in MAKS are scored based on an ordinal scale between 1 to 5, and the response "Don't know" is represented by 3 on the scale, similar to "Neutral". The sum of scores ranged between 12 to 60, and a higher score indicates better mental health literacy. The original version of MAKS was found to have moderate to substantial internal reliability for Part A (alpha = 0.65) and test–retest reliability of 0.71 [21]; however, since the instrument purposely included multidimensional items, the low internal reliability was expected and permissible for testing various types of stigma knowledge. The current study adopted the Malay version of MAKS (MAKS-M), which was translated and adapted for Malaysian secondary school teachers and yielded an internal consistency of 0.62 [22]. The internal consistency for Part A and Part B of the scale were 0.54 and 0.71, respectively. Overall, the Cronbach's alpha of MAKS-M within this study's population was 0.68, which is consistent with the studies mentioned above.

2.3. Procedure

2.3.1. Translation and Validation of the LOSS

The Malay version of the LOSS (M-LOSS) was translated using the forward–backward translation method. The English version of LOSS was translated into Malay by two researchers. Subsequently, after discussion and reconciliation, the two independent forward translations were merged into one interim translation. The backward translation from Malay to English was performed by an independent translator who had not read the original English version of the scale. The backward translation of the scale was then compared to the original version. Alterations were made accordingly until a consensus Malay version was produced. This harmonization was then distributed among ten Malaysian adult subjects whose first language is Malay to conduct a cognitive debriefing. Several pieces of feedback were obtained, and amendments were made accordingly. The researchers then proofread the final version of the Malay-translated questionnaire to correct minor errors before the validation process.

The M-LOSS was then distributed to six experts—two psychiatrists, one community medicine specialist, one family medicine specialist, one psychologist, and one biostatistician—for content validation. The panel of experts rates each item based on the level of relevance based on a Likert scale ranging from 1 (i.e., not relevant) to 4 (i.e., highly relevant). The ratings were entered into Microsoft® Excel for Mac V.16.55 (Microsoft, Redmond, WA, USA) and processed with the content validity index (CVI) calculation. Items with the content validity index (I-CVI) value of 0.83 and above were retained [23–25].

The face validity of the M-LOSS was assessed by ten individuals—six parents and four teachers. All individuals are Malaysian citizens and were chosen via purposive sampling by the researchers. This step was performed via Google Form® (Google, Mountain View, CA, USA) survey. The individuals were required to rate the items based on clarity and comprehensibility. The ratings were based on a 4-point Likert scale from 1 (i.e., not clear and understandable) to 4 (i.e., very clear and understandable). The responses were gathered and calculated using Microsoft® Excel for Mac V.16.55 (Microsoft, Albuquerque, NM, USA) to obtain the face validity index (FVI). Items with an item level face validity index (I-FVI) of 0.83 or above were retained [26,27].

The questionnaire was then pilot tested among 139 parents and teachers using online Google Form® (Google, Mountain View, CA, USA). Through snowball sampling, the participants were selected among parents and teachers of secondary school students within the Kelantan state. The pilot test aimed to obtain feedback on administrative procedures, such as the instrument's timing and structure before data collection.

The 27-item M-LOSS was circulated among parent–teacher associations of selected schools via one key informant. The schools were sampled via multistage stratified random sampling as described previously. Brief information regarding the study was provided at the beginning of the form. Consequently, informed consent was obtained by clicking the option "agree" to enable them to proceed to subsequent sections. Once the validation was completed, the validated M-LOSS, the MAKS-M, and sociodemographics were circulated among the schools via a similar method as mentioned previously. All completed responses were automatically saved into the Google Form® database (See Figure 1: Study flowchart).

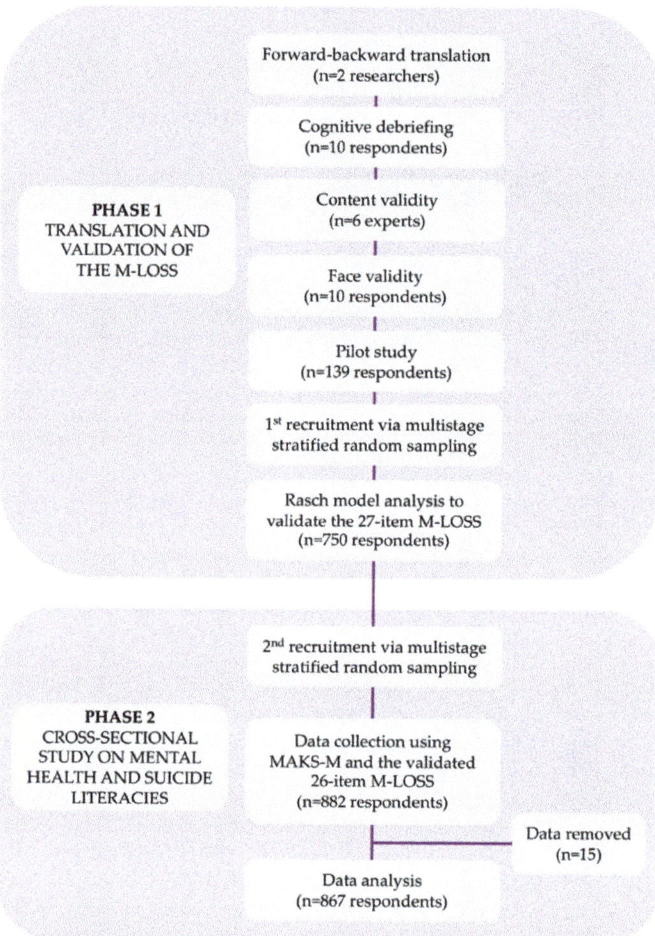

Figure 1. Study flowchart.

2.3.2. Statistical Analysis

The psychometric properties were evaluated via WINSTEPS® 3.68.2 (Linacre, USA) to analyze the construct validity through Rasch model analysis. Several important aspects of Rasch's analysis are outlined below.

Item polarity or point measure correlation (PTMEA Corr.) was performed to observe the correlations between the items within the scale for initial detection of the instrument's construct validity. A positive correlation indicates the items are pointing in the same direction, and good PTMEA Corr. should be 0.20 and above [28].

This study used inlier-pattern-sensitive fit statistics (INFIT) and outlier-sensitive fit statistics (OUTFIT) for Chi-squared-based tests of the model's fit for the item fit analysis. INFIT detects unexpected responses to items based on their ability level, whereas OUTFIT considers any discrepancies between observed and expected responses, disregarding the differences between the item difficulty and personal ability measures [29]. Item fit analysis is reported using mean squared (MNSQ) and z-standardized (ZSTD) scores, for which MNSQ is the Chi-squared calculation of fit statistics, and the ZSTD measures the probability of the MNSQ calculation occurring by chance. A productive item that fits into the model will have INFIT and OUTFIT MNSQ values within the range of 0.5 to 1.5. Meanwhile, if the ZSTD value exceeds 2.0, the item is considered erratic. On the other hand, if the MNSQ value is within the optimal range, the ZSTD value can be omitted, since the value is dependent on the MNSQ [30].

Principal component analysis of residual (PCAR) was used to test the dimensionality of the instrument, since unidimensionality is one of the assumptions in Rasch model analysis. Unidimensionality is met if the amount of variance explained by measures is more than 20%, the unexplained variance of the eigenvalue for the first contrast is less than 3, and the unexplained variance accounted for by the first contrast is less than 5% [31].

Item and person separation indexes were examined to separate the items, based on difficulty, and the respondents into several groups. The scales were expected to distinguish at least two groups of people or items. Therefore, the acceptable minimum value for the separation index is 1.50 [32]. This study also assessed the item and person reliability indexes. Item reliability index indicates if identical items' placements can be replicated if given to another sample with a similar skill level. In the meantime, the person reliability index indicates that a person's ranking could be replicated if he/she were given another set of items that evaluate similar constructs [33,34]. The reliability index ranged between 0 to 1. In this study, the acceptable value for the reliability index was set at 0.70 or above [32].

The Wright map visualizes the distribution of both the person measures and the location of the items. On the left of the map, person abilities are arranged from the highest to the lowest, whereas on the right, item difficulties are arranged from the hardest to the easiest from top to bottom.

Following the Rasch analysis, the descriptive data of the respondents' sociodemographics were obtained using IBM® SPSS® Statistics 27.0 (IBM, Armonk, NY, USA). Normality was assumed in this study based on the central limit theorem, whereby as the sample size grows higher (at least more than 30 samples), the distribution of sample means approaches a normal distribution [34]. This assumption was further verified by the value of skewness and kurtosis for each variable. Sample characteristics were tabulated, and the mean scores of MAKS-M, M-LOSS, and each subgroup were compared using the independent t-test and one-way ANOVA after data cleanup. The Pearson correlation analysis examined the correlation between age, suicide literacy, and mental health literacy. Results with $p < 0.05$ are considered statistically significant.

2.4. Ethical Consideration

The university's ethical committee has approved this study's proposal and instrument. Furthermore, permission was obtained from the author to adopt the questionnaire in this study. Information sheets outlining the study purpose, procedure, eligibility criteria, possible risks and benefits, confidentiality, and researchers' contact information were made available to the participants. Potential participants were informed that their participation was voluntary and that the data collected would be anonymous. After the full disclosure of the study, online informed consent was sought from individuals interested in participating in the study. Respondents who expressed their interest were redirected to the following sections in the Google Form.

3. Results

3.1. Psychometric Properties of M-LOSS

3.1.1. Translation

The 27-item M-LOSS was translated using the forward–backward translation method as described above. After the questionnaire was translated, ten Malaysians who are Malay native speakers were involved in the cognitive debriefing session. One subject commented that the word "psikotik" (psychotic) in item 11 is a medical term and should be replaced with a simpler word. However, due to the lack of appropriate synonyms, examples of psychotic symptoms such as hallucination and delusion (i.e., berhalusinasi, delusi) were added instead. Other than that, only several minor changes involving grammatical errors were made.

3.1.2. Content Validity

A total of 17 out of 27 items obtained universal agreement (UA), with an I-CVI value of 1.00. Items 6, 7, 9, 14, 18, 19, 20, 21, 22, and 23 obtained an I-CVI value of 0.83. Five out of six experts considered these items "relevant" or "very relevant". In general, all items were retained, as they obtained I-CVI values of 0.83 and above, with an overall average CVI of 0.94. The results were compiled and further discussed amongst the researchers. No changes were made in this phase (see Supplementary Materials Table S1).

3.1.3. Face Validity

Ten subjects were involved in the face validation study. A total of 20 out of 27 items achieved UA with an I-FVI value of 1.00. Items 3, 5, 6, 7, 8, 14, and 24 were rated "clear and comprehensive" or "very clear and comprehensive" by nine out of ten raters, resulting in an I-FVI score of 0.90. Hence, all items were retained, and no amendments were necessary (see Supplementary Materials Table S2).

3.1.4. Pilot Test ($n = 139$)

The pilot test of M-LOSS involved 139 respondents (70 parents, 69 teachers). The respondents' mean age was 40.99 ± 13.26 years old. The majority of respondents were female (70.5%), Malay (61.9%), practiced Islam (68.3%), received tertiary education (78.4%), and worked in the government sector (65.5%).

The respondents' overall acceptance was positive. They required approximately 5–10 min to complete the questionnaire. This indicates that the online survey form was well-structured, and the respondents did not experience any technical difficulties. The items were appropriate to the topic, unambiguous, and easily understood. As a result, a 27-item M-LOSS on knowledge of suicide with dichotomous response option was produced.

3.1.5. Rasch Model Analysis ($n = 750$)

Overall, 750 responses were received between March to June 2021. The respondents' mean age and standard deviation were 41.61 (6.97). More than half of the respondents were parents/caretakers (65.5%), whereas the others were teachers (34.5%). Most of the respondents were female (73.6%), Malay (88.0%), and practiced Islam (90.7%). A large portion of the respondents had completed tertiary education (76.7%) and were working in the government sector (68.1%). A majority of the respondents come from an urban school setting (58.3%) compared to a rural school setting (41.7%). There was an almost equal percentage of participants from national secondary schools (39.6%) and full boarding schools (37.2%), while there were fewer participants from religious secondary schools (23.2%).

In terms of item polarity, the first Rasch model analysis of the 27-item M-LOSS found that only one item, Item 18, had a low correlation (PTMEA Corr. = 0.16) (see Supplementary Materials Table S3). The item was removed, and the analysis was repeated with the 26-item M-LOSS. The remaining 26 items were found to have positive correlations ranging between 0.20 and 0.46. As for the INFIT and OUTFIT item fit analyses, the MNSQ values for all 26 items were within the optimal range, varying between 0.90 to 1.11 for the INFIT MNSQ and between 0.83 to 1.17 for the OUTFIT MNSQ. The ZSTD scores were ignored, since all the MNSQ values were acceptable (see Supplementary Materials Table S4). Thus, all the remaining 26 items fit the model and could be used for this study (see Appendix A Table A1).

As for the PCAR, the amount of raw variance explained by the measures is 29.0%, and the amounts of raw variance explained by persons and items were 10.7% and 18.4%, respectively. Although the unexplained variance in the first contrast was 5.3%, which marginally exceeded the cut-off value by 0.3%, the eigenvalue of the first contrast was 2.0, which met the requirement of <3.0. Therefore, the data fit the Rasch model, as proven by the unidimensionality of the instrument (see Supplementary Materials Table S5).

The person separation index for this current sample is 1.52, which suggests that the 26-item M-LOSS can separate the respondents into at least two strata based on their ability. The item separation index was 12.47, indicating an excellent level of separation. The person reliability index achieved the minimum acceptable value of 0.70, whereas the item reliability was near-perfect at 0.99. Overall, the separation and reliability indexes for the 26-item M-LOSS instrument were satisfactory for both person and item measures (see Supplementary Materials Table S6).

The Wright map charts the person and item measures based on their ability and difficulty level. In Figure 2, a bell-shaped distribution can be seen for the person measures (left), with the respondents at the top having higher literacy on suicide and vice versa. The item measures (right) show that Items 27, 5, 3, and 12 are the easiest and do not coincide with any respondents' ability levels. Conversely, Items 24 and 26 are the two most difficult items within the scale. However, these items are still within the ability range of the respondents. There are several items at a similar position, especially within the +1 to +2 logits (e.g., Items 17, 25, 7, and 8), signifying a possible redundancy from the measurement perspective. Finally, a gap of more than a logit was found above Items 24 and 26.

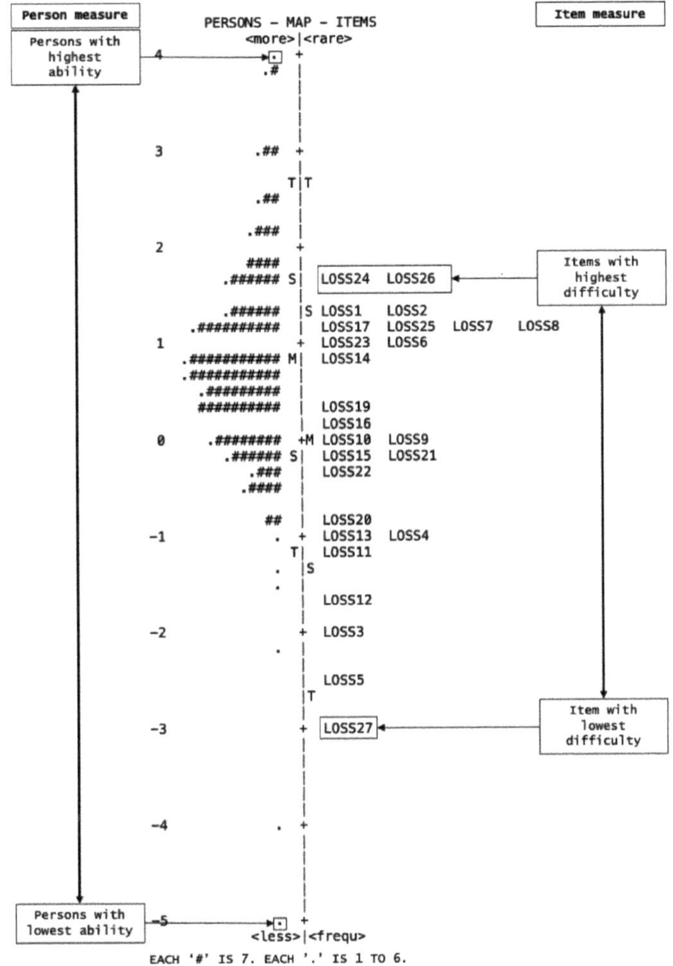

Figure 2. The Wright map for 26-item M-LOSS.

3.2. Suicide Literacy and Mental Health Literacy (n = 867)

A total of 882 responses were collected between July to September 2021. After the data cleanup, 867 responses were retained for further analysis. The respondents' mean age (SD) was 43.80 (8.35) years. The majority of the respondents were parents and caretakers (64.6%), female (71.5%), Malay (87.1%), and practiced the Islamic faith (89.9%). Three quarters of respondents had completed tertiary education (75.2%), were working in the government sector (66.7%), and had a household income within the middle 40% (M40) income bracket (45.4%). The respondents were mostly from SMK (39.3%) in urban areas (58.1%). Only several respondents self-reported a personal history of mental health (2.4%), had had known contact with mental health issues (18.5%), had experience in assisting those with mental health issues (25.0%), or had attended formal mental health first aid training (10.1%) (see Supplementary Materials Table S7).

The mean scores for MAKS-M and M-LOSS for the 867 parents, caregivers, and teachers are tabulated in Tables 1 and 2, respectively. The overall mean score (SD) for MAKS-M is 43.85 (SD = 4.07); the mean score for Part A is 21.36 (SD = 2.55), and that for Part B is 22.46 (SD = 2.66). For M-LOSS, the respondents' overall mean score is 14.05 (SD = 2.61), equivalent to 54.0%. Respondents obtained the lowest score in CN (42.3%), followed by SS (53.4%) and RF (60.4%), and the highest score was in TP (72.5%).

Table 1. Responses frequency and mean scores for 12-item MAKS-M (*n* = 867).

Item	Part	Strongly Disagree n (%)	Disagree n (%)	Don't Know/Neutral n (%)	Agree n (%)	Strongly Agree n (%)	Mean Score Mean (SD)
MAKS-M 1		16 (1.8)	149 (17.2)	294 (33.9)	259 (29.9)	149 (17.2)	
MAKS-M 2		6 (0.7)	23 (2.7)	127 (14.6)	280 (32.3)	431 (49.7)	
MAKS-M 3	A	12 (1.4)	186 (21.5)	207 (23.9)	304 (35.1)	158 (18.2)	
MAKS-M 4		1 (0.1)	18 (2.1)	95 (11.0)	382 (44.1)	371 (42.8)	21.36 (2.55)
MAKS-M 5		13 (1.5)	111 (12.8)	269 (31.0)	338 (39.0)	136 (15.7)	
MAKS-M 6		7 (0.8)	157 (18.1)	206 (23.8)	268 (30.9)	229 (26.4)	
MAKS-M 7		1 (0.1)	9 (1.0)	54 (6.2)	312 (36.0)	491 (56.6)	
MAKS-M 8		7 (0.8)	45 (5.2)	97 (11.2)	299 (34.5)	419 (48.3)	
MAKS-M 9	B	1 (0.1)	15 (1.7)	141 (16.3)	288 (33.2)	422 (48.7)	
MAKS-M 10		2 (0.2)	16 (1.8)	133 (15.3)	295 (34.0)	421 (48.6)	22.46 (2.66)
MAKS-M 11		34 (3.9)	125 (14.4)	143 (16.5)	313 (36.1)	252 (29.1)	
MAKS-M 12		15 (1.7)	75 (8.7)	138 (15.9)	366 (42.2)	273 (31.5)	
Overall mean score of mental health literacy							43.82 (4.07)

Note: SD = Standard deviation. Items are reverse scored.

Table 2. Correct responses and mean scores for 26 item M-LOSS (*n* = 867).

Items	Themes	Correct Answer	Participants with Correct Answers n (%)	Mean Score Mean (SD)
M-LOSS 1		False	526 (60.7)	
M-LOSS 2		False	528 (60.9)	
M-LOSS 3		False	72 (8.3)	
M-LOSS 4		False	163 (18.8)	
M-LOSS 5	Causes and nature (CN)	False	51 (5.9)	4.23 (1.82)
M-LOSS 6		False	477 (55.0)	
M-LOSS 7		False	511 (58.9)	
M-LOSS 8		False	492 (56.7)	
M-LOSS 9		True	558 (64.4)	
M-LOSS 10		False	292 (33.7)	
M-LOSS 11		False	128 (14.8)	
M-LOSS 12		True	782 (90.2)	
M-LOSS 13	Risk factors (RF)	True	700 (80.7)	4.24 (1.24)
M-LOSS 14		True	430 (49.6)	
M-LOSS 15		True	603 (69.6)	

Table 2. *Cont.*

Items	Themes	Correct Answer	Participants with Correct Answers n (%)	Mean Score Mean (SD)
M-LOSS 16		True	528 (60.9)	
M-LOSS 17		False	505 (58.2)	
M-LOSS 18		False	351 (40.5)	
M-LOSS 19	Signs and symptoms (SS)	True	683 (78.8)	
M-LOSS 20		False	284 (32.8)	2.67 (0.97)
M-LOSS 21		True	608 (70.1)	
M-LOSS 22		True	386 (44.5)	
M-LOSS 23		False	584 (67.4)	
M-LOSS 24	Treatment and prevention (TP)	False	520 (60.0)	2.91 (1.07)
M-LOSS 25		False	584 (67.4)	
M-LOSS 26		True	839 (96.8)	
Overall mean score of suicide literacy				14.05 (2.61)

Note: SD = Standard deviation.

The independent t-test and one-way ANOVA results for MAKS-M are presented in Table 3. The analysis revealed significant differences in knowledge on mental health stigma mean scores based upon sex ($p = 0.018$), knowing someone with mental disorders ($p < 0.001$), having experience in assisting psychiatric patients ($p = 0.002$), and having attended formal training in psychological first aid ($p = 0.038$).

Table 3. Comparison of mean between sociodemographic and MAKS-M ($n = 867$).

Variables	Part A Mean (SD)	Part B Mean (SD)	MAKS-M Mean (SD)
Role [a]			
Parents and caregivers	21.31 (2.59)	22.33 (2.62)	43.65 (4.03)
Teachers	21.45 (2.48)	22.70 (2.71)	44.15 (4.13)
Sex [a]			
Male	21.04 (0.16) *	22.30 (2.47)	43.34 (3.85) *
Female	21.49 (2.55)	22.53 (2.73)	44.02 (4.14)
Ethnicity [b]			
Malay	21.39 (2.54)	22.53 (2.64)	43.91 (4.06)
Chinese	21.16 (2.51)	22.12 (2.65)	43.29 (3.65)
Indian	21.00 (2.16)	23.75 (2.06)	44.75 (4.11)
Other Bumiputera	21.50 (3.00)	21.63 (2.75)	43.13 (4.69)
Others	20.09 (1.38)	22.73 (2.83)	42.82 (3.55)
Religion [b]			
Islam	21.40 (2.54)	22.53 (2.64) *	43.94 (4.06)
Christian	20.94 (2.63)	21.49 (2.69)	42.43 (4.18)
Buddhist	21.09 (2.75)	22.03 (2.68)	43.12 (3.91)
Hindu	21.67 (2.08)	24.33 (2.08)	46.00 (4.00)
Others	20.00 (1.41)	23.00 (1.41)	43.00 (2.83)
Education [b]			
No formal education	22.00 (1.41)	26.00 (0.00) **	48.00 (1.41) **
Primary education	19.73 (2.57)	20.64 (2.66)	40.36 (4.32)
Secondary education	21.21 (2.72)	22.18 (2.76)	43.39 (4.21)
Tertiary education	21.43 (2.49)	22.57 (2.60)	44.00 (4.07)
Occupation sector [b]			
Unemployed	21.51 (2.26)	22.14 (2.77)	43.65 (3.97)
Government	21.35 (2.55)	22.52 (2.67)	43.87 (4.07)
Private	21.61 (2.64)	22.39 (2.39)	44.00 (3.81)
Self-employed	20.81 (2.76)	22.52 (2.78)	43.33 (4.61)
Pensioner	21.44 (2.75)	22.50 (2.79)	43.94 (4.52)

Table 3. *Cont.*

Variables	Part A Mean (SD)	Part B Mean (SD)	MAKS-M Mean (SD)
Household income bracket [c,d]			
B40	21.30 (2.56)	21.03 (2.83) ***	43.33 (4.29) **
M40	21.36 (2.53)	22.70 (2.52)	44.06 (3.82)
T20	21.54 (2.62)	22.91 (2.40)	44.45 (4.10)
School locality [a]			
Urban	21.41 (2.51)	22.58 (2.62)	43.99 (4.05)
Rural	21.29 (2.61)	22.29 (2.70)	43.59 (4.10)
School type [b,e]			
SMK	21.37 (2.42)	22.39 (2.72)	43.76 (4.05)
SBP	21.48 (2.57)	22.70 (2.54)	44.18 (4.00)
SMA	21.17 (2.72)	22.20 (2.71)	43.37 (4.23)
Personal history of mental illness [a]			
Yes	22.29 (2.69)	22.67 (2.89)	44.95 (4.51)
No	21.34 (2.55)	22.46 (2.65)	43.80 (4.06)
Had known someone with mental illness [a]			
Yes	22.23 (2.66) ***	22.89 (2.50) *	45.12 (3.89) ***
No	21.17 (2.49)	22.36 (2.68)	43.53 (4.06)
Had assisted those with mental Illness [a]			
Yes	21.83 (2.52) **	22.55 (2.62)	44.38 (3.99) *
No	21.20 (2.54)	22.43 (2.67)	43.64 (4.08)
Had attended formal psychological first aid training [a]			
Yes	21.90 (2.59) *	23.02 (2.34) *	44.92 (3.58) **
No	21.30 (2.54)	22.40 (2.68)	43.70 (4.11)

Note: SD = Standard deviation. [a] independent t-test, [b] one-way ANOVA, [c] Welch's test [d] Malaysian household income stratification. Bottom 40% (B40) (RM4849 and below); middle 40% (M40) (RM4950–RM10959); top 20% (T20) (RM10960 and above). [e] School type: national secondary schools (SMK); fully residential schools (SBP); and religious secondary schools (SMA). * mean difference is significant at $p \leq 0.05$ level, ** mean difference is significant at $p \leq 0.01$ level, *** mean difference is significant at $p \leq 0.001$ level.

Additionally, religion ($p = 0.047$), education level ($p = 0.007$), household income bracket ($p < 0.001$), knowing someone with psychiatric illness ($p = 0.024$), and having attended formal psychological first aid training ($p = 0.021$) were also found to have significantly different mean scores for knowledge on mental health diagnoses.

Finally, factors that have a significant mean difference in the overall mental health literacy include the level of education ($p = 0.003$), household income ($p = 0.012$), close contact with psychiatric patients ($p < 0.001$), experience in assisting psychiatric patients ($p = 0.019$), and formal training in psychological first aid ($p = 0.008$).

Further analysis using the Tukey–Kramer post hoc test for the level of education suggested that only the subgroup of respondents receiving primary education and tertiary education were found to be significantly different ($p = 0.017$). Respondents with tertiary-level education were found to score better in mental health diagnoses, with a mean difference of 3.639. A post hoc test for household income demonstrated a significant difference in the mean score for mental health diagnoses between B40 and M40 households ($p = 0.002$), with mean difference of 0.672, and B40 and T20 households ($p = 0.002$), with a mean difference of 0.887. The mean score differences for overall mental health literacy between respondents from B40 and M40 were significant ($p = 0.040$), with a mean difference of 0.732, and B40 and T20 income brackets were also statistically significant at $p = 0.026$, with a mean difference of 1.122. Post hoc analysis on religion revealed no significant pairwise mean difference between each subgroup despite the significant one-way ANOVA.

For the suicide literacy, we reported a significant mean difference between the school type and the knowledge on SS ($p = 0.003$) and the overall knowledge on suicide ($p = 0.035$); the post hoc test revealed a difference in mean score between respondents from SBP and SMA for mean scores of SS ($p = 0.002$) with a mean difference of 0.296 and overall suicide literacy ($p = 0.037$) with mean difference of 0.574. There is no statistically significant difference for SMK compared to SBP or SMA. The results are tabulated in Table 4.

Table 4. Comparison of mean between sociodemographic and M-LOSS ($n = 867$).

Variables	Causes and Nature Mean (SD)	Risk Factors Mean (SD)	Signs and Symptoms Mean (SD)	Treatment and Prevention Mean (SD)	M-LOSS Mean (SD)
Role [a]					
Parents and caregivers	4.29 (1.84)	4.19 (1.28)	2.68 (0.98)	2.92 (1.06)	14.08 (2.66)
Teachers	4.13 (1.77)	4.33 (1.16)	2.64 (0.96)	2.91 (1.09)	14.00 (2.51)
Sex [a]					
Male	4.24 (1.83)	4.17 (1.25)	2.70 (0.95)	2.98 (1.03)	14.09 (2.58)
Female	4.23 (1.82)	4.27 (1.23)	2.65 (0.98)	2.89 (1.08)	14.04 (2.62)
Ethnicity [b]					
Malay	4.26 (1.79)	4.24 (1.22)	2.67 (1.00)	2.91 (1.07)	14.08 (2.60)
Chinese	3.82 (1.80)	4.39 (1.41)	2.73 (0.91)	2.84 (1.01)	13.78 (2.58)
Indian	4.50 (3.00)	4.75 (1.26)	3.00 (1.16)	2.50 (1.29)	14.75 (3.86)
Other Bumiputera	4.29 (1.08)	4.13 (1.27)	2.54 (0.82)	2.98 (1.11)	13.92 (2.69)
Others	3.64 (2.25)	4.09 (1.70)	2.82 (0.88)	3.36 (0.67)	13.91 (2.70)
Religion [b]					
Islam	4.26 (1.80)	4.23 (1.23)	2.66 (0.99)	2.92 (1.07)	14.08 (2.61)
Christian	3.98 (2.00)	4.18 (1.36)	2.61 (0.81)	2.94 (1.05)	13.71 (2.61)
Buddhist	3.97 (1.73)	4.50 (1.19)	2.76 (0.92)	2.74 (1.05)	13.97 (2.48)
Hindu	5.00 (3.46)	4.33 (1.16)	2.67 (1.16)	2.00 (1.00)	14.00 (4.36)
Others	4.50 (2.12)	4.00 (2.83)	3.00 (1.41)	3.50 (0.71)	15.00 (2.83)
Education [b]					
No formal education	4.00 (0.00)	4.00 (2.83)	2.50 (0.71)	2.50 (0.71)	13.00 (2.83)
Primary education	4.36 (1.69)	4.55 (0.82)	2.55 (0.93)	3.00 (0.89)	14.45 (2.34)
Secondary education	4.24 (1.80)	4.18 (1.31)	2.79 (1.04)	2.93 (1.06)	14.14 (2.67)
Tertiary education	4.23 (1.83)	4.25 (1.22)	2.63 (0.95)	2.91 (1.08)	14.02 (2.60)
Occupation sector [b,c]					
Unemployed	4.38 (1.94)	4.12 (1.25)	2.64 (1.17)	2.97 (1.02)	14.11 (2.86)
Government	4.19 (1.85)	4.27 (1.25)	2.67 (1.95)	2.93 (1.08)	14.07 (2.61)
Private	4.24 (1.68)	4.12 (1.15)	2.65 (0.90)	2.91 (1.05)	13.92 (2.43)
Self-employed	4.48 (1.65)	4.41 (1.17)	2.70 (1.01)	2.83 (1.01)	14.41 (2.45)
Pensioner	3.83 (1.58)	4.11 (1.41)	2.56 (0.92)	2.39 (1.30)	12.89 (2.56)
Household income bracket [b,d]					
B40	4.31 (1.80)	4.24 (1.25)	2.74 (1.01)	2.92 (1.04)	14.21 (2.59)
M40	4.16 (1.83)	4.25 (1.26)	2.63 (0.93)	2.94 (1.07)	13.97 (2.52)
T20	4.27 (1.83)	4.23 (1.13)	2.58 (0.98)	2.82 (1.15)	13.90 (2.93)
School locality [a]					
Urban	4.23 (1.80)	4.26 (1.23)	2.71 (1.00)	2.90 (1.07)	14.11 (2.60)
Rural	4.23 (1.85)	4.21 (1.25)	2.60 (0.94)	2.93 (1.06)	13.98 (2.62)
School type [b,c,e]					
SMK	4.14 (1.83)	4.31 (1.23)	2.67 (0.94) **	3.01 (1.00)	14.13 (2.47) *
SBP	4.35 (1.80)	4.22 (1.21)	2.78 (1.03)	2.88 (1.08)	14.23 (2.71)
SMA	4.20 (1.82)	4.16 (1.28)	2.48 (0.92)	2.81 (1.15)	13.65 (2.64)
Personal history of mental illness [a]					
Yes	4.76 (1.73)	3.95 (1.12)	2.67 (0.91)	2.57 (0.98)	13.95 (2.11)
No	4.22 (1.81)	4.25 (1.24)	2.67 (0.98)	2.92 (1.07)	14.06 (2.62)

Table 4. *Cont.*

Variables	Causes and Nature Mean (SD)	Risk Factors Mean (SD)	Signs and Symptoms Mean (SD)	Treatment and Prevention Mean (SD)	M-LOSS Mean (SD)
Had known someone with mental illness [a]					
Yes	4.44 (1.79)	4.10 (1.22)	2.68 (0.93)	2.97 (1.12)	14.19 (2.66)
No	4.19 (1.82)	4.27 (1.24)	2.66 (0.98)	2.90 (1.06)	14.02 (2.60)
Had assisted those with mental illness [a]					
Yes	4.21 (1.76)	4.19 (1.19)	2.68 (1.00)	2.96 (1.06)	14.04 (2.61)
No	4.24 (1.84)	4.26 (1.25)	2.66 (0.97)	2.90 (1.07)	14.06 (2.61)
Had attended formal psychological first aid training [a]					
Yes	4.36 (1.82)	4.11 (1.22)	2.69 (0.85)	2.92 (1.03)	14.09 (2.51)
No	4.22 (1.82)	4.25 (1.24)	2.66 (0.99)	2.91 (1.07)	14.05 (2.62)

Note: SD = Standard deviation. [a] independent t-test, [b] one-way ANOVA, [c] Welch's test, [d] Malaysian household income stratification. Bottom 40% (B40) (RM4849 and below); middle 40% (M40) (RM4950–RM10959); top 20% (T20) (RM10960 and above). [e] School type: national secondary schools (SMK); fully residential schools (SBP); and religious secondary schools (SMA). * mean difference is significant at $p \leq 0.05$ level, ** mean difference is significant at $p \leq 0.01$ level.

Finally, Pearson's correlation analysis findings are tabulated in Table 5. The correlation between age and M-LOSS is significant and negatively correlated ($r = -0.076$, $p = 0.025$). However, the strength of correlation was deemed weak. The finding also indicates that other variables are not significantly correlated.

Table 5. Correlation between age, MAKS-M, and M-LOSS ($n = 867$).

Variables	Age	MAKS-M	M-LOSS
Age (years)			
MAKS-M	−0.065		
M-LOSS	−0.076 *	0.007	

* Correlation is significant $p \leq 0.05$ level.

4. Discussion

4.1. Rasch Model Analysis of 26-Item M-LOSS

The current study aimed to investigate the psychometric properties of the M-LOSS using Rasch model analysis to develop a reliable and validated instrument to measure the level of suicide knowledge among the Malaysian population. Item polarity and item fit analysis suggested that one item (Item 18, "not all people who attempt suicide plan their attempt in advance") has a relatively low correlation. Although the correlation value of Item 18 was positive, indicating the item was aligned with the overall measure, the relatively low correlation value denotes possible inconsistency of indicator polarity in the scale. This indicates the item is not working well with the others. Consequently, Item 18 was removed [28]. Otherwise, the remaining 26 items showed positive PTMEA Corr. values within the acceptable range with stable INFIT and OUTFIT MNSQ values, signifying that all items fit the Rasch model and are appropriate to the instrument measure.

As for the separation and reliability indexes, the study achieved satisfactory values for both the person and item measures. In this study, based on the 26-item M-LOSS, a minimum of two-person strata can be generated using this instrument. Furthermore, adequate person reliability indicates that the items are consistent in establishing hierarchy based on the person's ability. The excellent item separation and reliability, on the other hand, imply that the items developed are varied in difficulty level while maintaining consistency and replicability of item placement across different samples [28].

For the spread of items in the Wright map, a gap of more than one logit difference was detected above Item 24. This gap may be caused by the borderline person separation index, since no item of appropriate difficulty coincided with the respondents' ability at such level. Thus, the scale was unable to stratify individuals with such level of ability. For improvement, more difficult items can be included to better assess and differentiate respondents with high abilities [28]. On the other hand, the cluster of items between the +1 to +2 logits is deemed psychometrically redundant (e.g., Items 17, 25, 7, and 8), as these items measure a similar ability level. Although these items can be dropped, we argue their importance, as they measure different categories of suicide knowledge. For instance, Items 5 and 3 evaluate the causes and nature of suicide, Item 12 assesses the risk factor, while Item 27 assesses the treatment and prevention of suicide [18]. Bond et al. also added that psychometric and theoretical redundancy are unalike perspectives. Therefore, the initial item development and rationalization should be taken into consideration [28].

4.2. Mental Health Literacy and Suicide Literacy

The current study population scored an average mental health literacy level of 43.82. This is comparable to a separate study among Malaysian teachers by Tay et al., with a mean score of 42.32 [35]. The study sample scored relatively higher than the Lebanese community but lower than studies in Jordan and the United Kingdom [36–38]. It is worth noting that the latter two studies recruited caregivers of patients with psychosis and healthcare workers for their studies, in which they not only worked closely with patients with underlying psychiatric conditions but also received formal training and psychoeducation on various mental health issues. This could explain their population's higher mental health literacy level compared to ours. Although the majority of Malaysian and Lebanese share an identical Islamic theological view on mental health (mental illnesses are associated with sins, divine punishments, and demonic possession), it is argued that Malaysian Muslims are more supportive of psychiatric patients and consider the illness as an opportunity to strengthen their connection with God, as compared to Arabic Muslims, for whom mental illnesses are perceived as shameful and disgraceful to the family [36,39,40].

For the suicide literacy, our sample scored only 54.0% correct responses, which is slightly lower than the mean score of community samples worldwide (58.2%) [41]. Still, this result is deemed comparable with the scores obtained from other populations—South Indian (50.9%), Chinese (53.0%), and Jordanian (55.0%) [20,42,43]. Several other studies reported lower rates of suicide literacy among Turkish (36.9%) and Bangladeshis (43.3%) [44,45] and higher scores among the Australian (>60.0%) and German (58.3%) populations [18,46,47]. This could be due to the lack of emphasis on mental health and suicide prevention programs in emerging and developing countries (EDCs) as opposed to developed countries (i.e., Germany and Australia). Additionally, the EDCs focus on other emerging issues on human development (e.g., poverty, illiteracy, overpopulation), communicable diseases, and maternal and child health, consequently deprioritizing the mental health sector [48]. However, several studies on suicide literacy have reported a similar trend of items under the TP theme being items with the most correct answers, while other subthemes scored lower [41]. This result is also reflected in this study. This may be due to the low item difficulty or a knowledge gap between different domains of suicide. Ludwig et al. further reasoned the items within the TP subtheme were too general and easy to identify [47]. This is thought to suffice, as the general population should be aware of the mental health services available and are not expected to have clinical knowledge such as psychotherapy and pharmacology. Nevertheless, the poor responses in the remaining subthemes are a call to highlight the nature, etiologies, risk factors, and warning signs of suicide in future suicide awareness and prevention programs to enable early recognition and intervention of suicidal individuals.

Individuals familiar with psychiatric illnesses and their interventions (i.e., had close contacts with psychiatric illnesses, had experience in assisting those with psychiatric illnesses, and attended formal psychological first aid training) were found to have significantly higher mean scores of mental health literacy than those who are not. This outcome is aligned with the findings by Doumit et al., wherein personal experiences enabled individuals to familiarize themselves with the symptoms and signs of mental health illnesses and recognize the appropriate diagnoses [36]. The caregiving experience also improved empathy and compassion. In turn, this helps improve knowledge on mental health stigma and reduce prejudice against those suffering from mental disorders [38].

Moreover, we also identified respondents having tertiary education as having scored better in mental health diagnoses and as having better overall mental health literacy than those who completed only primary level education. Similarly, those within the B40 income bracket scored significantly lower than their counterparts. Our findings support a previous study that identified higher education level and socioeconomic status as significant predictors for better mental health literacy [38]. These findings suggest a gap in accessibility for mental health awareness programs between different education levels and socioeconomic statuses. Individuals with higher education levels have more opportunities and exposure to mental health education and professional services, leading to more positive views on mental health issues and familiarity with diagnoses. Hence, more effort should be placed to improve mental health literacy among those with lower education levels from low-income households.

Our study also found that women have higher mental health stigma knowledge than men. Similarly, several other studies found that women show less stigma than men [36,49]. Doumit et al. argued that women are more empathetic and optimistic towards psychiatric patients, which could lead to such an observation [36]. On the other hand, most studies did not find any significant differences in mental health stigma between sexes. While both sexes exhibit mental health stigma, a qualitative study revealed that men have negative beliefs about diagnosis and treatment, whereas women's stigmas heavily focus on society's perceptions [50]. The varying outcomes between sex and mental health stigma knowledge warrant more research to shed light on these mixed findings.

For suicide literacy, the mean scores for the SS subtheme and overall suicide literacy among parents, caregivers, and teachers from SBP are significantly higher than those from SMA. A possible explanation is that boarding school students face additional risks of mental health issues such as homesickness, inadequate social support, and higher academic pressure. This could eventually lead to a higher prevalence of depression, anxiety, and stress among boarding school students [51,52]. Thus, parents and teachers of SBP are more aware of the warning signs of suicide for early detection of suicidal teenagers.

Religion could also be a barrier in discussing suicidality among SMA parents, caregivers, and teachers. Compared to other religions, Islam is firmer regarding the sins of suicide and self-harm, leading to increased social stigma and them being consequently regarded as taboos that should not be discussed openly among the community [53]. However, we have yet to identify any studies investigating how school types could influence the parents', caregivers', and teachers' reception of mental health literacy programs. Thus, this is an interesting proposal to investigate in the future to further clarify our findings.

The study also observed a significant yet negligible negative correlation between age and the level of suicide literacy. The association between age and suicide literacy remained incongruent in other studies. A study by Öztürk reported no significant association between the two variables, while several other studies reported a significant association between age and suicide literacy [44,46,47]. They reasoned that due to the increased exposure to suicide awareness programs and openness in discussing suicide, the younger generation embraces the topic of suicide better than the older generation. However, younger people were thought to have limited understanding and minimal encounters with mental illnesses and suicide, leading to lower knowledge [43]. In this light, further research is required to reconcile the apparent disparity between this information.

Interestingly, no significant correlation was found between mental health literacy and suicide literacy. The current suicide prevention strategy is said to be too focused on treating mental disorders due to the high association of mental disorders and suicide [54]. Suicide prevention strategies should not only concentrate on mental health treatments but also include a diverse range of information to further enhance specific knowledge on suicide. For example, Mishara and Chagnon outlined that suicide is not only the direct consequence of mental illnesses (e.g., cognitive distortion, delusions, and psychotic command hallucinations), but other possible causes include the consequence of living with mental disorders (e.g., social stigma, hopelessness, social and relationship issues, dependency, unemployment, etc.), treatment inadequacy and its possible adverse effects, and additional crisis [54]. This calls for a need for adolescents diagnosed with mental disorders, their parents, caregivers, and teachers to be educated on suicide prevention, since they are at higher risk of suicide. Batterham et al. reported the clinical samples were found to have lower suicide literacy compared to the community sample, indicating that specific psychoeducation programs on suicide prevention for at-risk adolescents are required [55]. The lack of correlation between mental health and suicide literacy necessitates an emphasis on suicidality as separate psychoeducation, not just as an extension of mental health education.

4.3. Strengths and Limitations

This study is the first to translate and validate the LOSS questionnaire to evaluate suicide literacy levels among the Malaysian population and investigate its correlation with mental health literacy. This study's multistage stratified random sampling enabled us to receive responses from a geographically dispersed population while maintaining adequate coverage across West Malaysia. However, our study faced several limitations. Firstly, we were unable to identify the exact response rate for this study due to the limited functionality in tracking the accessibility of the survey, which may put our study at risk of non-response bias. Furthermore, there is a scarceness in Malay-translated instruments to measure general mental health literacy among the Malaysian population. Although MAKS-M was reported to have less satisfactory reliability, it is a promising tool that enables the evaluation of stigma-related mental health knowledge and diagnoses. Therefore, improvement should be made to enhance the psychometric properties of this instrument. Furthermore, the interpretation of the LOSS score is reliant on comparison with other studies, due to the limited studies on suicide literacy in Malaysia, we could not compare these findings with other local populations. We anticipate that the current findings may serve as a useful tool for assessing suicide knowledge among Malaysians and as a foundation for evidence-based development of mental health and suicide awareness programs.

5. Conclusions

The current study has translated and validated the 26-item M-LOSS among Malaysian parents, caregivers, and teachers using Rasch model analysis to evaluate their suicide knowledge. The instrument was unidimensional, with adequate separation and reliability indexes, and all items had an acceptable correlation of 0.20 or above. All 26 items fit the model with INFIT and OUTFIT MNSQ within the optimal range, resulting in the M-LOSS having an overall favorable psychometric property. Furthermore, there may be value in directing mental health and suicide literacy programs towards parents, caregivers, and teachers who are male, who have lower education levels and socioeconomic status, who are unfamiliar with mental disorders, who did not receive formal training on psychological first aid, and who are in religious secondary schools. Lastly, no significant correlation was found between suicide literacy and mental health literacy.

Supplementary Materials: The following supporting information can be downloaded at: https://www.mdpi.com/article/10.3390/healthcare10071304/s1, Table S1: Content validity of 27-item M-LOSS; Table S2: Face validity of 27-item M-LOSS; Table S3: Item fit analysis for 27-item M-LOSS; Table S4: Item fit analysis for 26-item M-LOSS; Table S5: Principal component analysis of residual (PCAR) of the 26-item M-LOSS; Table S6: Separation and reliability indexes of the 26-item M-LOSS; Table S7: Respondents' sociodemographic ($n = 867$).

Author Contributions: Conceptualization, A.A.R. (Asrenee Ab Razak), H.S.K., A.A.R. (Azriani Ab Rahman), M.H., R.S.B. and F.A.G.; methodology, A.A.R. (Asrenee Ab Razak) and H.S.K.; formal analysis, P.K.A.P. and A.K.G.; investigation, P.K.A.P.; data curation, H.S.K. and P.K.A.P.; writing—original draft preparation, P.K.A.P.; writing—review and editing, A.A.R. (Asrenee Ab Razak), H.S.K. and A.K.G.; supervision, A.A.R. (Asrenee Ab Razak) and H.S.K.; project administration, A.A.R. (Asrenee Ab Razak); funding acquisition, A.A.R. (Asrenee Ab Razak), H.S.K., A.A.R. (Azriani Ab Rahman), M.H., R.S.B. and F.A.G. All authors have read and agreed to the published version of the manuscript.

Funding: This research was funded by the Ministry of Higher Education Malaysia under the Transdisciplinary Research Grant Scheme (TRGS), grant number (TRGS/1/2020/USM/02/4/1).

Institutional Review Board Statement: The study was conducted in accordance with the Declaration of Helsinki and approved by the Human Research Ethics Committee of Universiti Sains Malaysia (protocol code: USM/JEPeM/21020179, date of approval: 13 June 2021).

Informed Consent Statement: Informed consent was obtained from all subjects involved in the study.

Data Availability Statement: Not applicable.

Acknowledgments: The authors would like to acknowledge the key informants who volunteered to help making this project a success. We also thank all the participants who agreed to participate in the study voluntarily and contribute to the development of our knowledge in mental health and suicide literacies.

Conflicts of Interest: The authors declare no conflict of interest.

Appendix A

Table A1. The final version of the validated 26-item M-LOSS.

Bil. No.	Item / *Items*	Jawapan / *Answers*	
1.	Jika anda bertanyakan kepada seseorang secara langsung "adakah anda berasa ingin membunuh diri?" ianya mungkin mendorong seseorang itu cuba untuk membunuh diri. *If you asked someone directly 'Do you feel like killing yourself?' it will likely lead that person to make a suicide attempt.*	Betul *True*	Salah *False*
2.	Individu yang cuba membunuh diri hanya bertujuan untuk memanipulasi dan mencari perhatian daripada orang lain. *Those who attempt suicide do so only to manipulate others and attract attention to themselves.*	Betul *True*	Salah *False*
3.	Hanya segelintir individu yang berfikir untuk membunuh diri. *Very few people have thoughts about suicide.*	Betul *True*	Salah *False*
4.	Jika dinilai oleh pakar psikiatri, setiap individu yang mati akibat membunuh diri akan didiagnosis sebagai kemurungan. *If assessed by a psychiatrist, everyone who suicides would be diagnosed as depressed.*	Betul *True*	Salah *False*
5.	Individu yang mempunyai kecenderungan untuk membunuh diri akan sentiasa berkecenderungan untuk membunuh diri dan melayani fikiran mereka. *A suicidal person will always be suicidal and entertain thoughts of suicide.*	Betul *True*	Salah *False*
6.	Bercakap tentang bunuh diri meningkatkan risiko untuk membunuh diri. *Talking about suicide always increases the risk of suicide.*	Betul *True*	Salah *False*
7.	Motif dan sebab bunuh diri mudah untuk dikenal pasti. *Motives and causes of suicide are readily and easily established.*	Betul *True*	Salah *False*
8.	Liputan media berkenaan isu bunuh diri akan mendorong orang ramai untuk mencuba membunuh diri. *Media coverage of suicide will inevitably encourage other people to attempt suicide.*	Betul *True*	Salah *False*
9.	Kebanyakan orang yang cuba membunuh diri gagal dalam cubaan membunuh diri. *Most people who attempt suicide fail to kill themselves.*	**Betul** ***True***	Salah *False*
10.	Individu yang membunuh diri mempunyai penyakit mental. *A person who suicides is mentally ill.*	Betul *True*	Salah *False*
11.	Kebanyakan orang yang membunuh diri mengalami gangguan psikotik (e.g., berhalusinasi, delusi). *Most people who suicide is psychotic.*	Betul *True*	Salah *False*

Table A1. *Cont.*

Bil. No.	Item / Items	Jawapan / Answers	
12.	Individu yang mengalami masalah dalam hubungan atau kewangan mempunyai risiko yang lebih tinggi untuk membunuh diri. *People with relationship problems or financial problems have a higher risk of suicide.*	**Betul** / *True*	Salah / *False*
13.	Individu yang pernah ada cubaan untuk membunuh diri akan lebih cenderung untuk mencuba membunuh diri berbanding individu yang tidak pernah cuba untuk membunuh diri. *A person who has made a past suicide attempt is more likely to attempt suicide again than someone who has never attempted.*	**Betul** / *True*	Salah / *False*
14.	Lelaki lebih cenderung membunuh diri berbanding wanita. *Men are more likely to suicide than women.*	**Betul** / *True*	Salah / *False*
15.	Individu yang bimbang atau gelisah lebih berisiko untuk membunuh diri. *People who are anxious or agitated have a higher risk of suicide.*	**Betul** / *True*	Salah / *False*
16.	Terdapat hubung kait yang kuat antara ketagihan alkohol dan bunuh diri. *There is a strong relationship between alcoholism and suicide.*	**Betul** / *True*	Salah / *False*
17.	Kebanyakan mangsa bunuh diri adalah berumur kurang daripada 30 tahun. *Most people who suicide is younger than 30.*	Betul / True	**Salah** / *False*
18.	Individu yang bercakap tentang membunuh diri jarang berbuat sedemikian. *People who talk about suicide rarely commit suicide.*	Betul / True	**Salah** / *False*
19.	Inidividu yang berkeinginan untuk membunuh diri boleh mengubah keputusan dalam sekelip mata. *People who want to attempt suicide can change their mind quickly.*	**Betul** / *True*	Salah / *False*
20.	Kebanyakan individu yang membunuh diri tidak merancang kehidupan masa hadapan. *Most people who suicide doesn't make future plans.*	Betul / True	**Salah** / *False*
21.	Tindakan membunuh diri jarang berlaku tanpa amaran. *Suicide rarely happens without warning.*	**Betul** / *True*	Salah / *False*
22.	Risiko untuk membunuh diri adalah tinggi terutamanya dalam tempoh kemurungan beransur pulih. *A time of high suicide risk in depression is at the time when the person begins to improve.*	**Betul** / *True*	Salah / *False*
23.	Tiada apa yang boleh dilakukan untuk menghentikan percubaan membunuh diri sekiranya mereka telah nekad mengambil keputusan untuk membunuh diri. *Nothing can be done to stop people from making the attempt once they have made up their minds to kill themselves.*	Betul / True	**Salah** / *False*
24.	Hanya pakar sahaja yang boleh membantu individu yang ingin membunuh diri. *Only experts can help people who want to suicide.*	Betul / True	**Salah** / *False*
25.	Individu yang mempunyai pemikiran untuk membunuh diri tidak seharusnya memberitahu orang lain mengenai perkara tersebut. *People who have thoughts about suicide should not tell others about it.*	Betul / True	**Salah** / *False*
26.	Berjumpa dengan perawat psikiatri atau psikologi dapat membantu untuk mencegah seseorang daripada membunuh diri. *Seeing a psychiatrist or psychologist can help prevent someone from suicide.*	**Betul** / *True*	Salah / *False*

Note: Correct answers for each item are in shown in bold.

References

1. United Nations (UN). *Policy Brief: Education during COVID-19 and Beyond*; United Nations (UN): New York, NY, USA, 2020.
2. Chottera, S.; April, M.; Bright, D.; Sedky, K. Case Series, Social Distancing: Mental Health Implications for Children, Adolescents, and Families—Pediatric and Psychiatric Perspectives. *World Soc. Psychiatry* **2021**, *2*, 159–162. [CrossRef]
3. Hermosillo-De-la-torre, A.E.; Arteaga-De-luna, S.M.; Acevedo-Rojas, D.L.; Juárez-Loya, A.; Jiménez-Tapia, J.A.; Pedroza-Cabrera, F.J.; González-Forteza, C.; Cano, M.; Wagner, F.A. Psychosocial Correlates of Suicidal Behavior among Adolescents under Confinement Due to the COVID-19 Pandemic in Aguascalientes, Mexico: A Cross-Sectional Population Survey. *Int. J. Environ. Res. Public Health* **2021**, *18*, 4977. [CrossRef] [PubMed]
4. Mayne, S.L.; Hannan, C.; Davis, M.; Young, J.F.; Kelly, M.K.; Powell, M.; Dalembert, G.; McPeak, K.E.; Jenssen, B.P.; Fiks, A.G. COVID-19 and Adolescent Depression and Suicide Risk Screening Outcomes. *Pediatrics* **2021**, *148*, e2021051507. [CrossRef] [PubMed]
5. Ministry of Health Malaysia (MOH). Kesihatan Mental dan Sokongan Psikososial Ketika Pandemik COVID-19. Available online: https://covid-19.moh.gov.my/semasa-kkm/2021/06/mhpss-kesihatan-mental-dan-sokongan-psikososial-ketika-pandemik-covid-19 (accessed on 1 December 2021).

6.	Aida, J.; Azimah, M.N.; Mohd Radzniwan, A.R.; Iryani, M.D.T.; Ramli, M.; Khairani, O. Barriers to the Utilization of Primary Care Services for Mental Health Prolems among Adolescents in a Secondary School in Malaysia. *Malays. Fam. Physician* **2010**, *5*, 31–35. [PubMed]

7.	Thai, T.T.; Vu, N.L.L.T.; Bui, H.H.T. Mental Health Literacy and Help-Seeking Preferences in High School Students in Ho Chi Minh City, Vietnam. *Sch. Ment. Health* **2020**, *12*, 378–387. [CrossRef]

8.	Umpierre, M.; Meyers, L.v.; Ortiz, A.; Paulino, A.; Rodriguez, A.R.; Miranda, A.; Rodriguez, R.; Kranes, S.; McKay, M.M. Understanding Latino Parents' Child Mental Health Literacy: Todos a Bordo/All Aboard. *Res. Soc. Work. Pract.* **2015**, *25*, 607–618. [CrossRef]

9.	Sanati, A. Does Suicide Always Indicate a Mental Illness? *Lond. J. Prim. Care* **2009**, *2*, 93–94. [CrossRef]

10.	Milner, A.; Sveticic, J.; de Leo, D. Suicide in the Absence of Mental Disorder? A Review of Psychological Autopsy Studies across Countries. *Int. J. Soc. Psychiatry* **2013**, *59*, 545–554. [CrossRef]

11.	Hawton, K.; van Heeringen, K. Suicide. *Lancet* **2009**, *373*, 1372–1381. [CrossRef]

12.	Perez, N.M.; Jennings, W.G.; Piquero, A.R.; Baglivio, M.T. Adverse Childhood Experiences and Suicide Attempts: The Mediating Influence of Personality Development and Problem Behaviors. *J. Youth Adolesc.* **2016**, *45*, 1527–1545. [CrossRef]

13.	Fuller-Thomson, E.; Baird, S.L.; Dhrodia, R.; Brennenstuhl, S. The Association between Adverse Childhood Experiences (ACEs) and Suicide Attempts in a Population-Based Study. *Child Care Health Dev.* **2016**, *42*, 725–734. [CrossRef]

14.	Yamaguchi, S.; Foo, J.C.; Nishida, A.; Ogawa, S.; Togo, F.; Sasaki, T. Mental Health Literacy Programs for School Teachers: A Systematic Review and Narrative Synthesis. *Early Interv. Psychiatry* **2020**, *14*, 14–25. [CrossRef]

15.	Voracek, M.; Loibl, L.M.; Swami, V.; Vintilă, M.; Kõlves, K.; Sinniah, D.; Pillai, S.K.; Ponnusamy, S.; Sonneck, G.; Furnham, A.; et al. The Beliefs in the Inheritance of Risk Factors for Suicide Scale (BIRFSS): Cross-Cultural Validation in Estonia, Malaysia, Romania, the United Kingdom, and the United States. *Suicide Life-Threat. Behav.* **2008**, *38*, 688–698. [CrossRef] [PubMed]

16.	Siau, C.S.; Wee, L.H.; Yacob, S.; Yeoh, S.H.; Binti Adnan, T.H.; Haniff, J.; Perialathan, K.; Mahdi, A.; Rahman, A.B.; Eu, C.L.; et al. The Attitude of Psychiatric and Non-Psychiatric Health-Care Workers Toward Suicide in Malaysian Hospitals and Its Implications for Training. *Acad. Psychiatry* **2017**, *41*, 503–509. [CrossRef] [PubMed]

17.	Siau, C.S.; Wee, L.H.; Ibrahim, N.; Visvalingam, U.; Yeap, L.L.L.; Wahab, S. Gatekeeper Suicide Training's Effectiveness among Malaysian Hospital Health Professionals: A Control Group Study with a Three-Month Follow-Up. *J. Contin. Educ. Health Prof.* **2018**, *38*, 227–234. [CrossRef]

18.	Chan, W.I.; Batterham, P.; Christensen, H.; Galletly, C. Suicide Literacy, Suicide Stigma and Help-Seeking Intentions in Australian Medical Students. *Australas. Psychiatry* **2014**, *22*, 132–139. [CrossRef]

19.	Linacre, J.M. Sample Size and Item Calibration Stability. *Rasch Meas. Trans.* **1994**, *7*, 328.

20.	Ram, D.; Chandran, S.; Gowdappa, H.B. Suicide and Depression Literacy among Health Professions Students in Tertiary Care Centre in South India. *J. Mood Disord.* **2017**, *7*, 149–155. [CrossRef]

21.	Evans-Lacko, S.; Little, K.; Meltzer, H.; Rose, D.; Rhydderch, D.; Henderson, C.; Thornicroft, G. Development and Psychometric Properties of the Mental Health Knowledge Schedule. *Can. J. Psychiatry* **2010**, *55*, 440–448. [CrossRef]

22.	Pheh, K.S.; Anna Ong, W.H.; Low, S.K.; Tan, C.S.; Kok, J.K. The Malay Version of the Mental Health Knowledge Schedule: A Preliminary Study. *Malays. J. Psychiatry* **2017**, *26*, 10–14.

23.	Polit, D.F.; Beck, C.T. The Content Validity Index: Are You Sure You Know What's Being Reported? Critique and Recommendations. *Res. Nurs. Health* **2006**, *29*, 489–497. [CrossRef]

24.	Polit, D.F.; Beck, C.T.; Owen, S.V. Focus on Research Methods: Is the CVI an Acceptable Indicator of Content Validity? Appraisal and Recommendations. *Res. Nurs. Health* **2007**, *30*, 459–467. [CrossRef] [PubMed]

25.	Yusoff, M.S.B. ABC of Content Validation and Content Validity Index Calculation. *Educ. Med. J.* **2019**, *11*, 49–54. [CrossRef]

26.	Yusoff, M.S.B. ABC of Response Process Validation and Face Validity Index Calculation. *Educ. Med. J.* **2019**, *11*, 55–61. [CrossRef]

27.	Mohamad Marzuki, M.F.; Yaacob, N.A.; Yaacob, N.M. Translation, Cross-Cultural Adaptation, and Validation of the Malay Version of the System Usability Scale Questionnaire for the Assessment of Mobile Apps. *JMIR Hum. Factors* **2018**, *5*, e10308. [CrossRef] [PubMed]

28.	Bond, T.G.; Yan, Z.; Heene, M. *Applying the Rasch Model: Fundamental Measurement in the Human Sciences*, 4th ed.; Routledge: Abingdon, UK, 2020. [CrossRef]

29.	Tennant, A.; Conaghan, P.G. The Rasch Measurement Model in Rheumatology: What Is It and Why Use It? When Should It Be Applied, and What Should One Look for in a Rasch Paper? *Arthritis Care Res.* **2007**, *57*, 1358–1362. [CrossRef]

30.	Boone, W.J.; Staver, J.R.; Yale, M.S. *Rasch Analysis in the Human Sciences*; Springer: Berlin, Germany, 2013.

31.	Reckase, M.D. Unifactor Latent Trait Models Applied to Multifactor Tests: Results and Implications. *J. Educ. Stat.* **1979**, *4*, 207–230. [CrossRef]

32.	Souza, M.A.P.; Coster, W.J.; Mancini, M.C.; Dutra, F.C.M.S.; Kramer, J.; Sampaio, R.F. Rasch Analysis of the Participation Scale (P-Scale): Usefulness of the P-Scale to a Rehabilitation Services Network. *BMC Public Health* **2017**, *17*, 934. [CrossRef]

33.	Bond, T.G.; Fox, C.M. *Applying the Rasch Model: Fundamental Measurement in the Human Sciences*; Lawrence Erlbaum Associates, Inc: Mahwah, NJ, USA, 2001.

34.	Kwak, S.G.; Kim, J.H. Central Limit Theorem: The Cornerstone of Modern Statistics. *Korean J. Anesthesiol.* **2017**, *70*, 144–156. [CrossRef]

35. Tay, K.W.; Ong, A.W.H.; Pheh, K.S.; Low, S.K.; Tan, C.S.; Low, P.K. Assessing the Effectiveness of a Mental Health Literacy Programme for Refugee Teachers in Malaysia. *Malaysian J. Med. Sci.* **2019**, *26*, 120–126. [CrossRef]
36. Doumit, C.A.; Haddad, C.; Sacre, H.; Salameh, P.; Akel, M.; Obeid, S.; Akiki, M.; Mattar, E.; Hilal, N.; Hallit, S.; et al. Knowledge, Attitude and Behaviors towards Patients with Mental Illness: Results from a National Lebanese Study. *PLoS ONE* **2019**, *14*, e0222172. [CrossRef] [PubMed]
37. Dalky, H.F.; Abu-Hassan, H.H.; Dalky, A.F.; Al-Delaimy, W. Assessment of Mental Health Stigma Components of Mental Health Knowledge, Attitudes and Behaviors Among Jordanian Healthcare Providers. *Community Ment. Health J.* **2020**, *56*, 524–531. [CrossRef] [PubMed]
38. Sin, J.; Murrells, T.; Spain, D.; Norman, I.; Henderson, C. Wellbeing, Mental Health Knowledge and Caregiving Experiences of Siblings of People with Psychosis, Compared to Their Peers and Parents: An Exploratory Study. *Soc. Psychiatry Psychiatr. Epidemiol.* **2016**, *51*, 1247–1255. [CrossRef] [PubMed]
39. Subramaniam, M.; Abdin, E.; Picco, L.; Pang, S.; Shafie, S.; Vaingankar, J.A.; Kwok, K.W.; Verma, K.; Chong, S.A. *Stigma towards People with Mental Disorders and Its Components—A Perspective from Multi-Ethnic Singapore. Epidemiology and Psychiatric Sciences*; Cambridge University Press: Cambridge, UK, 2017. [CrossRef]
40. Ng, C.H. The Stigma of Mental Illness in Asian Cultures. *Aust. N. Z. J. Psychiatry* **1997**, *31*, 382–390. [CrossRef] [PubMed]
41. Calear, A.L.; Batterham, P.J.; Trias, A.; Christensen, H. The Literacy of Suicide Scale: Development, Validation, and Application. *Crisis* **2021**. [CrossRef] [PubMed]
42. Han, J.; Batterham, P.J.; Calear, A.L.; Wu, Y.; Shou, Y.; van Spijker, B.A.J. Translation and Validation of the Chinese Versions of the Suicidal Ideation Attributes Scale, Stigma of Suicide Scale, and Literacy of Suicide Scale. *Death Stud.* **2017**, *41*, 173–179. [CrossRef]
43. Al-Shannaq, Y.; Aldalaykeh, M. Suicide Literacy, Suicide Stigma, and Psychological Help Seeking Attitudes among Arab Youth. *Curr. Psychol.* **2021**. *ahead of print*. [CrossRef]
44. Öztürk, A. Evaluation of Suicide Knowledge Level and Stigma Attitudes Towards People Who Committed Suicide in University Students. *J. Psychiatr. Nurs.* **2018**, *9*, 96–104. [CrossRef]
45. Arafat, S.M.Y.; Hussain, F.; Hossain, M.F.; Islam, M.A.; Menon, V. Literacy and Stigma of Suicide in Bangladesh: Scales Validation and Status Assessment among University Students. *Brain Behav.* **2021**, *12*, e2432. [CrossRef]
46. Khan, M.M. Suicide Prevention and Developing Countries. *J. R. Soc. Med.* **2005**, *98*, 459–463. [CrossRef]
47. Ludwig, J.; Dreier, M.; Liebherz, S.; Härter, M.; von dem Knesebeck, O. Suicide Literacy and Suicide Stigma–Results of a Population Survey from Germany. *J. Ment. Health* **2021**. [CrossRef] [PubMed]
48. Li, J.; Li, J.; Thornicroft, G.; Huang, Y. Levels of Stigma among Community Mental Health Staff in Guangzhou, China. *BMC Psychiatry* **2014**, *14*, 231. [CrossRef] [PubMed]
49. Elkington, K.S.; Hackler, D.; McKinnon, K.; Borges, C.; Wright, E.R.; Wainberg, M.L. Perceived Mental Illness Stigma Among Youth in Psychiatric Outpatient Treatment. *J. Adolesc. Res.* **2012**, *27*, 290–317. [CrossRef] [PubMed]
50. Wahab, S.; Rahman, F.N.A.; Wan Hasan, W.M.H.; Zamani, I.Z.; Arbaiei, N.C.; Khor, S.L.; Nawi, A.M. Stressors in Secondary Boarding School Students: Association with Stress, Anxiety and Depressive Symptoms. *Asia-Pac. Psychiatry* **2013**, *5* (Suppl. 1), 82–89. [CrossRef]
51. Mahfar, M.; Noah, S.M.; Senin, A.A. Development of Rational Emotive Education Module for Stress Intervention of Malaysian Boarding School Students. *SAGE Open* **2019**, *9*, 1–16. [CrossRef]
52. Gearing, R.E.; Lizardi, D. Religion and Suicide. *J. Relig. Health* **2009**, *48*, 332–341. [CrossRef]
53. Calear, A.L.; Batterham, P.J.; Christensen, H. Predictors of Help-Seeking for Suicidal Ideation in the Community: Risks and Opportunities for Public Suicide Prevention Campaigns. *Psychiatry Res.* **2014**, *219*, 525–530. [CrossRef]
54. Mishara, B.L.; Chagnon, F. Understanding the relationship between mental illness and suicide and the implications for suicide prevention. In *International Handbook of Suicide Prevention (Research, Policy and Practice)*, 2nd ed.; O'Connor, R.C., Platt, S., Gordon, J., Eds.; John Wiley & Sons, Ltd.: Hoboken, NJ, USA, 2011. [CrossRef]
55. Batterham, P.J.; Han, J.; Calear, A.L.; Anderson, J.; Christensen, H. Suicide Stigma and Suicide Literacy in a Clinical Sample. *Suicide Life-Threat. Behav.* **2019**, *49*, 1136–1147. [CrossRef]

Review

Anticipatory Burden in Adult-Child Caregivers: A Concept Analysis

Hangying She [1,*] and Yuncheng Man [2]

[1] Frances Payne Bolton School of Nursing, Case Western Reserve University, Cleveland, OH 44106, USA
[2] Case School of Engineering, Case Western Reserve University, Cleveland, OH 44106, USA; yxm278@case.edu
* Correspondence: hangying.she@case.edu

Abstract: This study aims to analyze the concept of anticipatory burden in adult-child caregivers. A systematic literature review was performed using four databases, Pubmed, CINAHL, PsycINFO and Medline, with the keywords of "anticipatory burden" and "anticipated burden". Simplified Wilson's classic concept analysis modified by Walker and Avant was employed to identify the attributes, antecedents and consequences of anticipatory burden in the adult-child caregivers. Eighteen articles were analyzed. Attributes of anticipatory burden in adult-child caregivers were found to be: (1) subjective burden, (2) anticipation, (3) overestimation, (4) inability, and (5) family relationship. Antecedents were identified as: (1) potential care recipients, (2) caregiving willingness, and (3) a lack of resources. Consequences included: (1) prediction of caregiving willingness, (2) impacts on caregivers' health, (3) intervention promotion, and (4) behavioral changes. As the adult-child caregiver is one of the main types of family caregivers for the fast-growing aging population, it is important to understand the attributes, antecedents, and consequences of their anticipatory burden. Based on the results of this study, resources such as intervention, policy, and counseling services are recommended to help adult-child caregivers lower their anticipatory burden and get better prepared for providing family care.

Keywords: adult-child caregiver; anticipatory burden; caregiving burden

Citation: She, H.; Man, Y. Anticipatory Burden in Adult-Child Caregivers: A Concept Analysis. *Healthcare* **2022**, *10*, 356. https://doi.org/10.3390/healthcare10020356

Academic Editors: Carlos Laranjeira and Ana Querido

Received: 29 December 2021
Accepted: 8 February 2022
Published: 11 February 2022

Publisher's Note: MDPI stays neutral with regard to jurisdictional claims in published maps and institutional affiliations.

1. Introduction

Around 17% of all adult children end up providing care to their aging parents [1]. The adult-child caregivers can anticipate future caregiving burdens, which would negatively affect their own health and the care recipients. The concept of caregiving burden first appeared as family burden in the literature and was primarily associated with the healthcare cost borne by family members [2]. Nowadays, as the understanding of caregiver burden has evolved, it is described as a multifaceted phenomenon. Its definition now includes the concepts of objective burden (loss of time, energy, and social life), as well as subjective burden (emotional and relational stress) [3,4]. Moreover, financial burden is another important aspect of caregiving burden apart from the objective and subjective burden [5]. Caregiver burden has been extensively studied in the aforementioned three dimensions; however, for those adult children who are not yet caregivers but will become family caregivers for their aging parents in the future, the anticipatory burden of becoming future family caregivers is less understood.

Studying the anticipatory burden of adult-child caregivers is particularly crucial since the population aged over 65 is growing at a fast pace [6], and in many countries and cultures, adult children are obligated to provide care for their dependent parents [5,7–9]. For instance, Chinese parental respect is central in the Confucius culture. More importantly, as a retrospective report found that 37% to 39% of caregivers had considered the possibility of becoming caregivers before they actually carried out the role [10], adult children often plan to care for their aging parents before it happens, which significantly alters their current-state behaviors and even results in deteriorated physical and/or mental health conditions in their current stage [11].

The concept of anticipatory burden in caregiving has been developed in the literature, as mentioned in the concept analysis of anticipatory anxiety [11], anticipatory grief [12], anticipatory decision making [13], and pre-death grief [14]. However, none of these studies have provided a consistent conceptualization of anticipatory burden, specifically in the adult children population. Further, there is a lack of clarity between the existing definitions of what constitutes anticipatory burden in adult-child caregivers. Therefore, this study aims to provide a clear definition of the anticipatory burden in adult-child caregivers and a basis for its assessment and management.

2. Materials and Methods

2.1. Study Design

This systematic literature review uses a concept analysis that derives the attributes, antecedents, and consequences of anticipatory burden in adult-child caregivers, using simplified Wilson's classic concept analysis [15] modified by Walker and Avant [16].

2.2. Literature Inclusion and Exclusion Criteria

This study was conducted in accordance with the guidelines of systematic literature reviews, following the Preferred Reporting Items for Systematic Review and Meta-Analysis (PRISMA) statement [17]. The inclusion criteria were English-language, peer-reviewed studies that appeared from literature search with contents or outcomes related to the keywords. The exclusion criteria were studies that lack appropriate subjects (e.g., the main subject is anticipatory/anticipated grief, anticipatory/anticipated anxiety, caregiver burden, or financial burden, etc.), contents or conceptual definitions, and degree theses were excluded.

2.3. Study Details

We searched "anticipatory burden" or "anticipated burden" in four databases—Pubmed, CINAHL, PsycINFO, and Medline—with keywords limited to the title and the abstract. Publications from 1 January 1982 to 1 December 2021 were searched, limited to those in the English language and peer-reviewed. Further, expert consultation and reference tracing were carried out for seeking potentially relevant publications. Moreover, regular alerts have been established for a few selected databases, such as PubMed, to update the literature review before manuscript submission.

The literature search was solely performed by a single author (H.S.) in this work. A total of 274 articles were extracted as primary sources using the keywords listed above. Following the exclusion of duplication, 221 secondary sources were extracted, and their titles and abstracts were carefully reviewed. Of these, 160 articles without mentioning anticipatory burden as a concept or without describing the phenomenon of anticipatory burden were excluded. Thereafter, following careful full-text screening, 63 articles without a clear definition, attribute, antecedent, consequence, or empirical referent were excluded. Finally, 18 articles were included in this study (Figure 1). The specific process is as follows:

- Select a concept. The concept of anticipatory burden in adult-child caregivers was selected.
- Determine the aims of the concept analysis. The aim of this study is to provide a clear definition of the anticipatory burden in adult-child caregivers and a basis for its assessment and management.
- Identify previous definitions and uses of the concept. Two previous definitions and five uses of the concept in different contexts were identified by reviewing seven of the included articles.
- Determine the defining attributes. Five defining attributes were determined. Subjective burden is based on the nature of anticipation, where the expected burden-generating

events have not happened yet. Anticipation is based on the nature of this concept, as mentioned by Laditka, S.B. and M. Pappas-Rogich [11]. Overestimation is based on the inaccuracy of anticipation, as mentioned by Huang et al. [18]. Inability is based on the nature of burden that will be generated when adult-children are unable to tackle all the caregiving demands. The family relationship was mentioned by Feeney, J.A. and L. Hohaus [19], as a family relationship will influence the willingness (Antecedents) of caregiving.

- Present a model case of the concept.
- Present borderline, related, contrary, invented, and illegitimate cases.
- Identify antecedents and consequences of the concept. Three antecedents were identified: Potential care recipients, caregiving willingness, and a lack of resources. Four consequences were identified: Prediction of caregiving willingness, impacts on caregivers' health, intervention promotion, and behavioral changes.
- Define empirical referents.

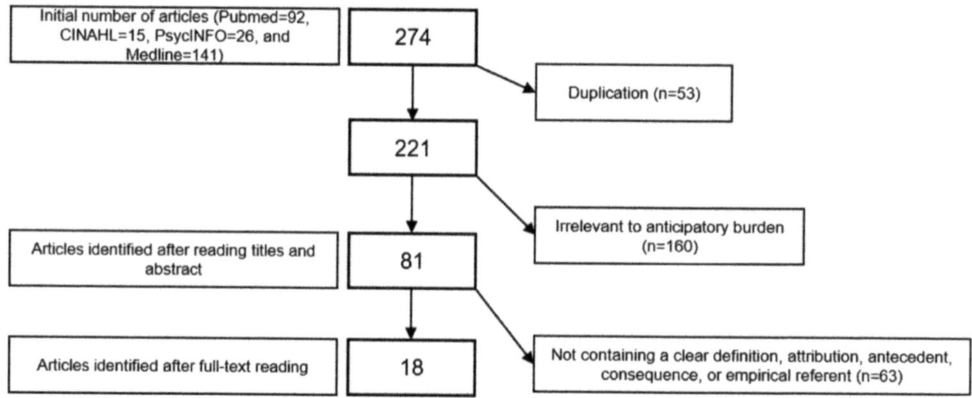

Figure 1. Flow diagram of study selection process.

3. Results

3.1. Previous Definitions and Uses of Anticipatory Burden

Anticipatory burden has been defined and used elsewhere. Feeney et al. described anticipated burden as the burden of spouse caregivers who perceived their partners in severe health conditions in the future [19]. Sekhon et al. studied defined anticipatory burden in a random control trial (RCT) as the accessibility of the intervention and the required amount of effort for participation [20]. Notably, in this study, the concept of anticipatory burden was utilized to determine if low accessibility is a significant factor leading to reduced participation or non-adherence to the intervention among the participants. Another study by Naidoo et al. confirmed that reducing anticipatory burden in RCT could encourage participation [21]. Furthermore, Wilson et al. described the anticipatory burden in the context of anticipatory prescribing, where nurses still experience emotional burden even though the medication was administrated with considerable caution [22]. Additionally, Huang et al. described the impact of anticipatory financial burden on a patient's decision to undergo contralateral prophylactic mastectomy [18]. The recent COVID-19 pandemic has brought up anticipatory burden in various aspects. Kashyap et al. described the anticipatory burden arising in the healthcare system during the COVID-19 pandemic [23]. Kozloff et al. used anticipatory burden to describe the burden of people with schizophrenia and related disorders due to the COVID-19 pandemic [24].

3.2. Attributes of Anticipatory Burden in Adult-Child Caregivers

According to the concept analysis procedural [16], determining the defining attributes of a concept is the core of concept analysis. Defining attributes are the essential characteristics of a concept. It helps differentiate the concept of anticipatory burden in adult-child caregivers from other related concepts. The following attributes of anticipatory burden in adult-child caregivers were identified by reviewing various pieces of literature (Figure 2).

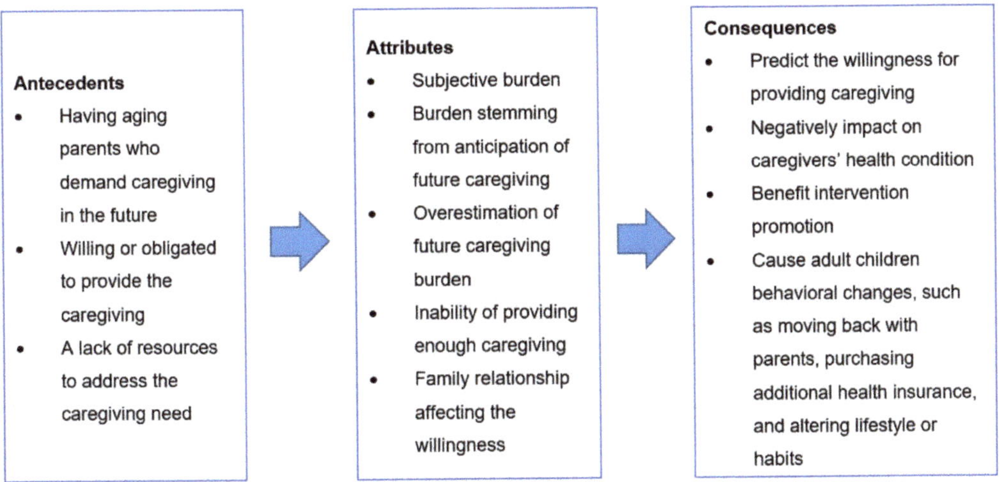

Figure 2. Concept diagram of "anticipatory burden in adult-child caregivers".

(1)   Subjective burden

Adult children who anticipate taking care of their aged parents in the future have not entered the family caregiver role; thus, their anticipatory burden is purely subjective and caused by emotional and relational stress. Since they do not undertake caregiving tasks at the current stage, no objective burden is associated with the anticipatory burden.

(2)   Anticipation

The anticipatory burden in adult-child caregivers is mainly tied to the anticipation of the caregiving burden in the future, whether foreseeable or addressable at the current stage [11]. Adult children often start planning ahead and are stressed by the objective, subjective, and financial burdens of providing caregiving to their aging parents. It is necessary to provide practical social support to adult children undertaking the family caregiver role to help them identify what they need and better prepare them both physically and mentally.

(3)   Overestimation

As described by Huang at al., there is a mismatch between the anticipatory burden and the actual burden when patients decide to have their surgeries [18]. Similarly, adult children sometimes overestimate the caregiving burden for their parents before they actually undertake the family caregiver role. This is primarily due to the fact that they lack certain knowledge on aging population healthcare and also resources.

(4)   Inability

Once adult children enter the role of the family caregiver and provide caregiving to their parents, they have to carry out the duty on a constant basis. As a result, they may anticipate that they are unable to maintain such care on the long run and therefore experience anticipatory burden. On one hand, there is a lack of resourcefulness in the adult children; on the other hand, they may have overestimated the actual caregiving burden.

(5)   Family relationship

A tight family relationship can increase the willingness of adult children to take care of their parents [19]. Anticipatory burden is generated only if the adult children decide to care their aging parents in the future, and thus, the family relationship is an important factor here.

3.3. Cases

3.3.1. Model Case

A model case refers to a case that presents all the key attributes of the concept [16]. This study presents two model cases based on the observation of daily life.

A 45 year-old man lives with his wife and two school-age daughters. Both him and his wife are employed full-time. His father has lived alone for more than 20 years and has recently been diagnosed with early stage dementia. After being aware of the trajectory of dementia (the continuous deteriorating physical and mental health condition of his father) and the future increasing caregiving burden that accompanied it, he was stressed so much that he could not sleep for a week because of the anticipatory burden, such as the caring demands, life–work balance issues, and decision-making conflicts in the future.

On the other hand, a 30-year-old women is employed full-time and unmarried. She lives in town, while her parents live in a rural area. Her mother had cardiovascular disease for more than 5 years, and his father has lived with diabetes for more than 10 years. Both of her parents are still able to take care of themselves. However, she can foresee the deep waters she will be in as her parents are growing older and their health condition continues to drop as the diseases develop. Should she move back with her parents to take care of them and give up her career, or keep her job and leave her parents to the paid care assistant they may barely know? Moreover, leaving her parents to someone unknown leads to more emotional burden due to blame and guilt generated by the filial piety culture. The anticipatory burden and the conflict between the two decisions makes her worried and stressed.

3.3.2. Contrary Case

A contrary case does not display attributes of the concept but instead presents attributes contrary to it. Its presentation allows the attributes of the concept being used to be better understood and clarified.

A 78-year-old woman lives in a nursing home. She feels isolated one day and wants the company of a nurse who used to take good care of the old woman. However, the nurse recently left this job and therefore did not respond to the demand. In this case, the old woman has emotional demand, but the nurse, the previous formal caregiver of her, does not have anticipatory burden, and therefore she simply ignored the demand.

3.3.3. Related Case

A related case contains part of the key attributes [16]. For example, the anticipatory grief. Lindemanm, a psychiatrist, shows in his article [25] that a wife refused her husband returning home following retiring from the military. The reason is that the wife did not anticipate the husband surviving from the war and had gone through the anticipatory grief and psycho-emotionally abandoned the marital relationship with him.

3.3.4. Borderline Case

A borderline case includes some of the attributes of the concept but not all [16]. In this study, the borderline case addresses the attributes of subjective burden, anticipation, and inability.

An employee recently lost his job, and he can no longer afford health insurance. He takes certain medication that was covered by health insurance, and he now has to pay from his own pocket. He is frustrated about not being able to find a new job to cover his bills; he anticipates that if he cannot find a job soon, he will soon go bankrupt.

3.3.5. Invented Case

An invented case is demonstrated by the concept being taken out of the context of our own experience [16]. In the context of parenting, the parents experience anticipatory burden by anticipating issues in educational competition, physical and mental development, and campus bullying as their children grow up.

3.4. Identification of Antecedents and Consequences of Anticipatory Burden in Adult-Children Caregivers

3.4.1. Antecedents

Antecedents refer to additional conditions or events before the occurrence of the concept. Based on the literature review, the following antecedents were identified:

(1) Potential care recipients;
(2) Caregiving willingness;
(3) A lack of resources.

First, the aging parents may have continuously deteriorating health conditions and therefore demand caregiving from their children. Second, adult children are willing or obligated to provide caregiving to their aging parents regardless of whether they have enough resources. Third, adult children experience anticipatory burden in caring for their aging parents due to a lack of resources for addressing their needs.

3.4.2. Consequences

Consequences are those elements or conditions that occur as a result of the concept [16]. Based on the literature review, the following consequences were identified:

(1) Prediction of caregiving willingness;
(2) Impacts on caregivers' health;
(3) Intervention promotion;
(4) Behavioral changes.

Feeney et al. found that anticipatory burden is associated with the willingness to take care of a spouse or partner [19]. Similarly, the anticipatory burden provides a reliable prediction of the willingness of adult children to care for their aging parents. Moreover, anticipatory burden can negatively impact adult-child caregivers' health, as indicated by the research of Kumari et al., as subjective burden often leads to poor physical and mental health [26]. Further, based on evidence by Rizzieri et al. showing that the availability of less-burdensome therapies reduced the following caregiving burden, the reduction in the anticipatory burden can benefit intervention promotion in both the adult-child caregivers and the aging parents [27]. Last but not least, adult children change their behaviors after they experience the anticipatory burden of taking care of their aging parents. For example, they may correct their parents' bad habits or unhealthy lifestyle [28], or they may purchase additional medical insurance for their parents due to the financial pressure of medical care [29].

3.5. Definition of Anticipatory Burden in Adult-Children Caregivers

Based on the previous definitions of anticipatory burden and merging them with the key attributes identified in this study, the anticipatory burden (subjective) was generated by anticipating two things in the future (Figure 3). Firstly, the adult-children anticipate that they cannot change the events/facts of aging and health deterioration of their parents. Both aging and health deterioration of their parents lead to potential caregiver burden in the future. Secondly, the adult-children anticipate that they cannot address the caregiver burden (objective) in the future generated by the aforementioned events/facts.

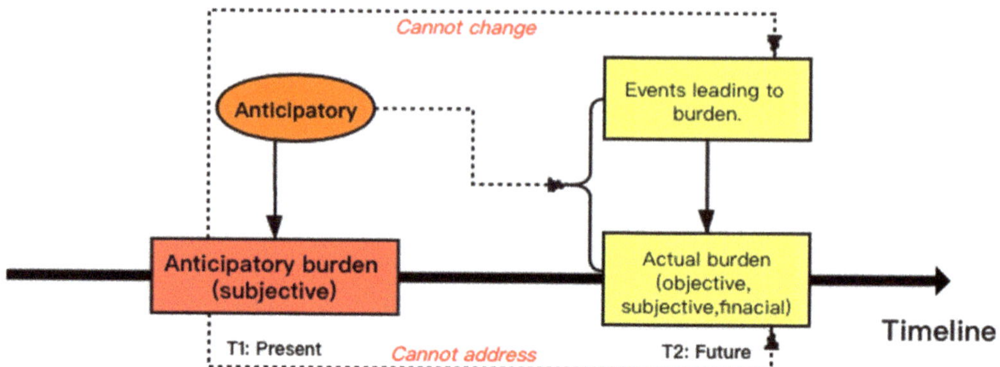

Figure 3. Theoretical model of anticipatory burden in adult-child caregivers.

3.6. Empirical Referents

Empirical referents are used in the final step in the conceptual analysis to recognize the characteristics of the defining attributes and to show that the properties of the concept exist in the actual field [16]. They are particularly useful to address the question: "If we are going to measure this concept or determine its existence in the real world, how do we do so?" Existing instruments for measuring caregiving burden are mostly based on existing objectives [30–34]; they are not designed to determine subjective burden, such as the anticipatory burden in adult-child caregivers. The Zarit Burden Interview is the most popular instrument for determining caregiver burden, especially subjective burden [35]. This self-report instrument contains 22 items assessing the perceived burden in caregivers. However, it has not been applied to determine anticipatory burden in adult-child caregivers.

4. Discussion

This concept analysis concluded that the anticipatory burden in adult-child caregivers is defined as the subjective burden stemming from their planning to take care of their aging parents in the future. Three antecedents, five attributes, and four consequences of the anticipatory burden in adult-child caregivers were generated (Table 1).

Three antecedents of anticipatory burden were generated. Two of them were generated from previous definitions and uses: (1) Specific events will happen in future; (2) These events may not be addressed. In the context of adult-child caregiver, these two antecedents will be (1) Caregiving willingness; (2) Lacking resources. The third antecedent, "Potential care recipients", was based on the specific context—the adult-child caregiver. This finding could help researchers design inclusion criteria when recruiting the target population for a study about anticipatory burden in adult-child caregivers. Additionally, researchers could also develop or promote interventions to help adult children release their anticipatory burden, such as distributing information about how to promote their parent(s)' well-being and about how to obtain caregiving resources.

Five attributes of anticipatory burden were generated. The "subjective burden", "anticipation", and "inability" could help develop items in the measurement instrument. For example, "subjective burden" could be reflected in the language "Do you feel/think/believe ... "; "anticipation" could be reflected in the language "the potential event(s) in the future"; "inability" could be reflected as "something is out of control/lack of resources to something/cannot handle something". The "family relationship" and "overestimation" could be used in the covariates or factors analyses of anticipatory burden in the future study.

Table 1. Previous definitions and uses, attributes, antecedents, attributes, and consequences of anticipatory burden in adult-child caregivers.

Dimension	Sub-Dimensions	Key Findings in the Literature
Previous Definitions and Uses	Definitions	(1) In the context of spousal caregivers; (2) In the context of random control trial participants. [19,20]
	Uses	(1) In the context of random control trial participants; (2) In the context of anticipatory prescribing by nurses; (3) In the context of anticipatory financial burden on a patient's decision; (4) In the context of the healthcare system during the COVID-19 pandemic; (5) In the context of people with schizophrenia and related disorders due to the COVID-19 pandemic. [18,21–24]
Attributes	Subjective burden	N.A.
	Anticipation	The anticipatory burden in adult-child caregivers is mainly tied to the anticipation of the caregiving burden in the future. [11]
	Overestimation	The mismatch between the anticipatory burden and the actual burden. [18]
	Inability	N.A.
	Family relationship	A tight family relationship can increase the willingness of adult children for taking care of their parents. [19]
Antecedents	Potential care recipients	N.A.
	Caregiving willingness	N.A.
	A lack of resources	N.A.
Consequences	Prediction of caregiving willingness	The anticipatory burden is associated with the willingness of taking care of a spouse or partner. [19]
	Impacts on caregivers' health	The anticipatory burden can negatively impact adult-child caregivers' health. [26]
	Intervention promotion	The availability of less-burdensome therapies reduced the following caregiving burden. [27]
	Behavioral changes	(1) correct parents' bad habits or unhealthy lifestyles; (2) purchase additional medical insurance for their parents. [28,29]
Empirical Referents		Existing instruments for measuring caregiving burden are mostly based on existing objectives. The Zarit Burden Interview determines subjective caregiver burden, however, has not been applied in the anticipatory burden in adult-child caregivers. [30–35]

Four consequences of anticipatory burden were generated. Though anticipatory burden is based on the willingness to be an adult-child caregiver, an overly high anticipatory burden may decrease the willingness and make the adult children become hesitant when they consider taking on the caregiver role. It may also impact adult children's health, such as anxiety, worry, and sleep problems, etc. These two negative consequences can be demonstrated as the problems needing to be tacked in policy-making. They can also be used as a reference in mental health consulting if the clients may have caregiving concerns. The anticipatory burden could also help promote interventions. There is no existing intervention target at the potential adult-child caregiver. However, there are lots of interventions about caregivers. The existing intervention could be promoted or revised into a new version to help the potential adult-child caregiver release their anticipatory burden. The last consequence, which is behavior changes, can also help as a reference to develop or promote the interventions.

A limitation of this work is that only a few terms were used when searching the database. Other terms such as "anticipatory grief" may help retrieve more studies for the conceptualization. Another limitation is that some key aspects such as cultural differences and differences in family structure, varying from country to country and family to family, were not analyzed. Future work will preferentially focus on measuring thoughts, feelings, daily experiences, and health of adult-child caregivers from diversified backgrounds to determine their caregiving demands, perceived stress, resourcefulness, and negative physical or physiological symptoms (if any).

5. Conclusions

As taking care of an increasingly aging population presents a global challenge, it is important to study and understand family caregivers who will have a central role in caring for the aging population. This study provides a conceptual analysis that identifies the meaning and properties of anticipatory burden in adult-child caregivers. The definition of anticipatory burden in adult-child caregivers was found as the present subjective burden that adult-children are perceiving, which stems from their plan to take care of their parents and the resulting burden in the future. Based on these findings, resources such as intervention, policy, and counseling services are recommended to help the adult-children to reduce their anticipatory burden and be better prepared to become family caregivers.

Author Contributions: H.S. conceptualized the study, performed the literature review and analysis, prepared the figures and table, and wrote the manuscript. Y.M. edited the manuscript. All authors have read and agreed to the published version of the manuscript.

Funding: This research received no external funding.

Institutional Review Board Statement: Not applicable.

Informed Consent Statement: Not applicable.

Data Availability Statement: Not applicable.

Acknowledgments: The authors acknowledge with gratitude the kind suggestions from Susan R. Mazanec (Frances Payne Bolton School of Nursing, Case Comprehensive Cancer Center, Case Western Reserve University) on this work. The authors acknowledge with gratitude the kind suggestions from three anonymous referees, who have significantly contributed to the improved legibility of this manuscript.

Conflicts of Interest: The authors declare no conflict of interest.

References

1. Wettstein, G.; Zulkarnain, A. *How Much Long-Term Care Do Adult Children Provide*; Issue Brief; Center for Retirement Research at Boston College: Newton, MA, USA, 2017; Number 17-11, pp. 1–4.
2. Grad, J.; Sainsbury, P. Problems of caring for the mentally ill at home. *Proc. R. Soc. Med.* **1966**, *59*, 20–23. [CrossRef] [PubMed]
3. Walker, A.J.; Pratt, C.C.; Eddy, L. Informal caregiving to aging family members: A critical review. *Fam. Relat.* **1995**, *44*, 402–411. [CrossRef]
4. Montgomery, R.J.; Gonyea, J.G.; Hooyman, N.R. Caregiving and the experience of subjective and objective burden. *Fam. Relat.* **1985**, *34*, 19–26. [CrossRef]
5. Zhan, H.J. Chinese caregiving burden and the future burden of elder care in life-course perspective. *Int. J. Aging Hum. Dev.* **2002**, *54*, 267–290. [CrossRef]
6. Weisman, J. *Aging Population Poses Global Challenges*; The Washington Post: Washington, DC, USA, 2005; p. A01. Available online: https://www.washingtonpost.com/wp-dyn/articles/A55582-2005Feb1.html?referrer%3Demail&sub=AR. (accessed on 14 December 2021).
7. Cicirelli, V.G. Attachment and obligation as daughters' motives for caregiving behavior and subsequent effect on subjective burden. *Psychol. Aging* **1993**, *8*, 144. [CrossRef] [PubMed]

8. Lee, Y.-R.; Sung, K.-T. Cultural influences on caregiving burden: Cases of Koreans and Americans. *Int. J. Aging Hum. Dev.* **1998**, *46*, 125–141. [CrossRef]
9. Togonu-Bickersteth, F. Conflicts over caregiving: A discussion of filial obligations among adult Nigerian children. *J. Cross-Cult. Gerontol.* **1989**, *4*, 35–48. [CrossRef]
10. Chappell, N.; Litkenhaus, R. *Informal Caregivers to Adults in British Columbia*; Joint Report by the University of Victoria: Victoria, BC, Canada, 1995.
11. Laditka, S.B.; Pappas-Rogich, M. Anticipatory caregiving anxiety among older women and men. *J. Women Aging* **2001**, *13*, 3–18. [CrossRef]
12. Fulton, R.; Gottesman, D.J. Anticipatory grief: A psychosocial concept reconsidered. *Br. J. Psychiatry* **1980**, *137*, 45–54. [CrossRef]
13. Slyer, J.T.; Archibald, E.; Moyo, F.; Truglio-Londrigan, M. Advance care planning and anticipatory decision making in patients with Alzheimer disease. *Nurse Pract.* **2018**, *43*, 23–31. [CrossRef] [PubMed]
14. Lindauer, A.; Harvath, T.A. Pre-death grief in the context of dementia caregiving: A concept analysis. *J. Adv. Nurs.* **2014**, *70*, 2196–2207. [CrossRef] [PubMed]
15. Wilson, J. *Thinking with Concepts*; Cambridge University Press: Cambridge, UK, 1963; ISBN 978-052-109-601-0.
16. Walker, L.O.; Avant, K.C. *Strategies for Theory Construction in Nursing*; Pearson/Prentice Hall: Upper Saddle River, NJ, USA, 2005.
17. Page, M.J.; McKenzie, J.E.; Bossuyt, P.M.; Boutron, I.; Hoffmann, T.C.; Mulrow, C.D.; Shamseer, L.; Tetzlaff, J.M.; Akl, E.A.; Brennan, S.E.; et al. The PRISMA 2020 statement: An updated guideline for reporting systematic reviews. *BMJ* **2021**, *372*, n71. [CrossRef] [PubMed]
18. Huang, J.; Chagpar, A. Impact of anticipated financial burden on patient decision to undergo contralateral prophylactic mastectomy. *Surgery* **2018**, *164*, 856–865. [CrossRef]
19. Feeney, J.A.; Hohaus, L. Attachment and spousal caregiving. *Pers. Relatsh.* **2001**, *8*, 21–39. [CrossRef]
20. Sekhon, M.; Cartwright, M.; Lawes-Wickwar, S.; McBain, H.; Ezra, D.; Newman, S.; Francis, J.J. Does prospective acceptability of an intervention influence refusal to participate in a randomised controlled trial? An interview study. *Contemp. Clin. Trials Commun.* **2021**, *21*, 100698. [CrossRef]
21. Naidoo, N.; Nguyen, V.T.; Ravaud, P.; Young, B.; Amiel, P.; Schante, D.; Clarke, M.; Boutron, I. The research burden of randomized controlled trial participation: A systematic thematic synthesis of qualitative evidence. *BMC Med.* **2020**, *18*, 6. [CrossRef]
22. Wilson, E.; Morbey, H.; Brown, J.; Payne, S.; Seale, C.; Seymour, J. Administering anticipatory medications in end-of-life care: A qualitative study of nursing practice in the community and in nursing homes. *Palliat. Med.* **2015**, *29*, 60–70. [CrossRef]
23. Kashyap, S.; Gombar, S.; Yadlowsky, S.; Callahan, A.; Fries, J.; Pinsky, B.A.; Shah, N.H. Measure what matters: Counts of hospitalized patients are a better metric for health system capacity planning for a reopening. *J. Am. Med. Inform. Assoc.* **2020**, *27*, 1026–1131. [CrossRef]
24. Kozloff, N.; Mulsant, B.H.; Stergiopoulos, V.; Voineskos, A.N. The COVID-19 Global Pandemic: Implications for People with Schizophrenia and Related Disorders. *Schizophr. Bull.* **2020**, *46*, 752–757. [CrossRef]
25. Lindemann, E. Symptomatology and management of acute grief. *Am. J. Psychiatry* **1944**, *101*, 141–148. [CrossRef]
26. Kumari, R.; Kohli, A.; Malhotra, P.; Grover, S.; Khadwal, A. Burden of caregiving and its impact in the patients of acute lymphoblastic leukemia. *Ind. Psychiatry J.* **2018**, *27*, 249–258. [CrossRef] [PubMed]
27. Rizzieri, A.G.; Verheijde, J.L.; Rady, M.Y.; McGregor, J.L. Ethical challenges with the left ventricular assist device as a destination therapy. *Philos. Ethics Humanit. Med.* **2008**, *3*, 20. [CrossRef] [PubMed]
28. Oliveira, D.; Sousa, L.; Orrell, M. Improving health-promoting self-care in family carers of people with dementia: A review of interventions. *Clin. Interv. Aging* **2019**, *14*, 515–523. [CrossRef]
29. Chandra, A.; Young, J.S.; Dalle Ore, C.; Dayani, F.; Lau, D.; Wadhwa, H.; Rick, J.W.; Nguyen, A.T.; McDermott, M.W.; Berger, M.S.; et al. Insurance type impacts the economic burden and survival of patients with newly diagnosed glioblastoma. *J. Neurosurg.* **2019**, *133*, 89–99. [CrossRef]
30. Appleton, S.; Adams, R.; Porter, S.; Peacock, M.; Ruffin, R. Sustained improvements in dyspnea and pulmonary function 3 to 5 years after lung volume reduction surgery. *Chest* **2003**, *123*, 1838–1846. [CrossRef]
31. Cedano, S.; Bettencourt, A.R.; Traldi, F.; Machado, M.C.; Belasco, A.G. Quality of life and burden in carers for persons with chronic obstructive pulmonary disease receiving oxygen therapy. *Rev. Lat.-Am. Enferm.* **2013**, *21*, 860–867. [CrossRef]
32. Elmståhl, S.; Malmberg, B.; Annerstedt, L. Caregiver's burden of patients 3 years after stroke assessed by a novel caregiver burden scale. *Arch. Phys. Med. Rehabil.* **1996**, *77*, 177–182. [CrossRef]
33. Lee, E.; Lum, C.M.; Xiang, Y.T.; Ungvari, G.S.; Tang, W.K. Psychosocial condition of family caregivers of patients with chronic obstructive pulmonary disease in Hong Kong. *East Asian Arch. Psychiatry* **2010**, *20*, 180–185.
34. Pinto, R.A.; Holanda, M.A.; Medeiros, M.M.; Mota, R.M.; Pereira, E.D. Assessment of the burden of caregiving for patients with chronic obstructive pulmonary disease. *Respir. Med.* **2007**, *101*, 2402–2408. [CrossRef]
35. Zarit, S.H.; Reever, K.E.; Bach-Peterson, J. Relatives of the impaired elderly: Correlates of feelings of burden. *Gerontologist* **1980**, *20*, 649–655. [CrossRef] [PubMed]

MDPI
St. Alban-Anlage 66
4052 Basel
Switzerland
www.mdpi.com

Healthcare Editorial Office
E-mail: healthcare@mdpi.com
www.mdpi.com/journal/healthcare